Experimental Hydraulics

Methods, Instrumentation, Data Processing and Management

Volume II: Instrumentation and Measurement Techniques

IAHR Monograph
Series editors

Peter A. Davies
Department of Civil Engineering,
The University of Dundee,
Dundee,
United Kingdom

Robert Ettema
Department of Civil and Environmental Engineering,
Colorado State University,
Fort Collins,
USA

The **International Association for Hydro-Environment Engineering and Research** (**IAHR**), founded in 1935, is a worldwide independent organisation of engineers and water specialists working in fields related to hydraulics and its practical application. Activities range from river and maritime hydraulics to water resources development and eco-hydraulics, through to ice engineering, hydroinformatics and continuing education and training. IAHR stimulates and promotes both research and its application, and, by doing so, strives to contribute to sustainable development, the optimisation of world water resources management and industrial flow processes. IAHR accomplishes its goals by a wide variety of member activities including: the establishment of working groups, congresses, specialty conferences, workshops, short courses; the commissioning and publication of journals, monographs and edited conference proceedings; involvement in international programmes such as UNESCO, WMO, IDNDR, GWP, ICSU, The World Water Forum; and by co-operation with other water-related (inter) national organisations. www.iahr.org

Supported by
Spain Water
and IWHR, China

Experimental Hydraulics

Methods, Instrumentation, Data Processing and Management

Volume II: Instrumentation and Measurement Techniques

Editors

Jochen Aberle
Leichtweiß-Institute for Hydraulic Engineering and Water Resources, Technische Universität Braunschweig, Braunschweig, Germany
Norwegian University of Science and Technology, Trondheim, Norway

Colin D. Rennie
Department of Civil Engineering, University of Ottawa, Ottawa, Canada

David M. Admiraal
Civil Engineering Department, University of Nebraska, Lincoln, NE, USA

Marian Muste (Editor-in-Chief)
IIHR – Hydroscience & Engineering, University of Iowa, Iowa City, IA, USA

CRC Press
Taylor & Francis Group
Boca Raton   London   New York   Leiden

CRC Press is an imprint of the
Taylor & Francis Group, an **informa** business

A BALKEMA BOOK

Cover Illustration: Graphics assembled by David P. Herwaldt (IIHR – Hydroscience & Engineering) and Marian Muste with photographs courtesy of: Geir Mogen / NTNU Oceans, Colin D. Rennie, Dongsu Kim, John Sloat (WaterCube LLC), and IIHR – Hydroscience & Engineering.

Applied for

Published by: CRC Press/Balkema
　　　　　　Schipholweg 107C, 2316 XC Leiden, The Netherlands
　　　　　　e-mail: Pub.NL@taylorandfrancis.com
　　　　　　www.crcpress.com – www.taylorandfrancis.com

First issued in paperback 2020

© 2017 Taylor & Francis Group, London, UK
CRC Press/Balkema is an imprint of the Taylor & Francis Group, an informa business

No claim to original U.S. Government works

Visit the Taylor & Francis Web site at
http://www.taylorandfrancis.com

and the CRC Press Web site at
http://www.crcpress.com

Library of Congress Cataloging-in-Publication Data

Typeset by Integra Software Services Private Ltd

ISBN 13: 978-1-138-02753-4 (hbk)(set of 2 volumes)
ISBN 13: 978-1-138-03216-5 (hbk)(vol 1)
ISBN 13: 978-0-367-57326-3 (pbk)(vol 2)
ISBN 13: 978-1-138-03815-8 (hbk)(vol 2)

About the IAHR Book Series

An important function of any large international organisation representing the research, educational and practical components of its wide and varied membership is to disseminate the best elements of its discipline through learned works, specialized research publications and timely reviews. IAHR is particularly well-served in this regard by its flagship journals and by the extensive and wide body of substantive historical and reflective books that have been published through its auspices over the years. The IAHR Book Series is an initiative of IAHR, in partnership with CRC Press/ Balkema – Taylor & Francis Group, aimed at presenting the state-of-the-art in themes relating to all areas of hydro-environment engineering and research.

The Book Series will assist researchers and professionals by advancing and transferring contemporary knowledge needed for research, education and engineering practice. The series includes Design Manuals and Monographs. The Design Manuals, usually prepared by multiple authors, guide the application of theory and research findings to engineering practice; and, the Monographs give state-of-the-art coverage of various significant topics in water engineering.

The first and highly successful IAHR book was *"Turbulence Models and their Application in Hydraulics"* by W. Rodi, published in 1984 by Balkema. *"Turbulence in Open Channel Flows"* by I. Nezu and H. Nakagawa, also published by Balkema (in 1993), had an important impact on the field and, during the period 2000–2010, further authoritative texts (published directly by IAHR) included *Fluvial Hydraulics* by S. Yalin and A. Da Silva and *Hydraulicians in Europe* by W. Hager. All of these publications continue to strengthen the reach of IAHR and to serve as important intellectual reference points for the Association.

Since 2011, the Book Series is once again a partnership between CRC Press/Balkema – Taylor & Francis Group and the Technical Committees of IAHR. The present book is an exciting further contribution to IAHR's Book Series, substantially aiding water engineering research, education and practice, and showcasing the expertise IAHR fosters.

Series editors
Peter A. Davies
Robert Ettema

Editors Biography Volume II

Jochen Aberle
Lead Editor

Jochen Aberle is Professor at the Technische Universität Braunschweig, Germany, and at the Norwegian University of Science and Technology. His research expertise includes river morphology, sediment transport, rough bed flows, flow-vegetation-sediment interaction, fluid-structure interaction, and experimental methods in the field and laboratory. He has published more than 100 articles in peer-reviewed journals and conference proceedings. He is an Associate Editor of the Journal of Hydraulic Research, and was chair of the IAHR Experimental Methods and Instrumentation Technical Committee.

Colin D. Rennie

Colin D. Rennie is a Professor of Civil Engineering, at the University of Ottawa, Canada. He has expertise in areas of river morphology and engineering, environmental hydraulics, sediment transport, turbulence, and aquatic habitat. His research focuses on channel morphodynamics and mixing processes, utilizing high-resolution field measurements with acoustic instruments, laboratory physical models, and three-dimensional numerical modelling. He has published 150+ refereed articles, including papers in premiere journals such as Nature. He is an Associate Editor of the Journal of Hydraulic Engineering, and a past-Chair of the IAHR Experimental Methods and Instrumentation Technical Committee.

David M. Admiraal

David M. Admiraal is an Associate Professor in the Civil Engineering Department at the University of Nebraska–Lincoln. His research expertise includes laboratory and field investigations of sediment transport, river hydraulics, and hydraulic structures. Most of his work incorporates experimental modeling. Dr. Admiraal has over 50 journal and conference publications in a wide variety of hydraulic engineering topics. He is actively involved in American Society of Civil Engineers' Technical Committee on Hydraulic Measurements and Experimentation and has served as the chair of the committee. He has also acted as co-chair for two conferences on Hydraulic Measurements and Experimental Methods.

Marian Muste
Editor-in-Chief

Marian Muste is Research Engineer with the IIHR-Hydroscience & Engineering, and Adjunct Professor in the Department of Civil & Environmental Engineering at the University of Iowa. Dr. Muste's areas of research include experimental river mechanics (laboratory and field investigations), instrumentation development and implementation (especially image-, acoustic-, and laser-based), uncertainty analysis, and hydroinformatics. He contributed to 185 peer-reviewed papers and 75 technical reports. He was chair for the International Association of Hydro-Environmental Engineering & Research's (IAHR) Experimental Methods and Instrumentation Technical Committee from 2000 to 2007.

Galileo Galilei (1564 – 1642)

"Measure what is measurable, and make measurable what is not so."

Contents Volume I and II

Volume I

Preface and Contributions   xxix
List of Contributing Authors Vol. I   xxxv

1   **Introduction**   1
   1.1   Book Overview   1
   1.2   The Role of Experiments in Hydraulics   2
   1.3   Approach   4
   1.4   Structure of Volume I   6
   References   9

2   **Hydraulic Flows: Overview**   11
   2.1   Introduction   11
   2.2   Turbulent Flows in Hydraulic Engineering 12
       2.2.1   Wall-bounded shear flows   12
       2.2.2   Free-shear flows   15
       2.2.3   Flow near interfaces   19
   2.3   Turbulence Mechanics: Concepts and Descriptive Frameworks   20
       2.3.1   Reynolds-averaging Navier-Stokes (RANS) framework   21
       2.3.2   Double-averaging Navier-Stokes (DANS) framework   26
       2.3.3   Statistical characterization of turbulence   30
             2.3.3.1   Definitions and basic concepts   30
             2.3.3.2   Statistical moments   31
             2.3.3.3   Correlation functions   33
             2.3.3.4   Structure functions   34
             2.3.3.5   Spectra of hydrodynamic variables 35
             2.3.3.6   Turbulence scales   37
             2.3.3.7   Statistical theories   39
       2.3.4   Structural characterization of turbulence   43
       2.3.5   Mixing, diffusion, and dispersion   46
       2.3.6   Key working hypotheses pertinent for experimentation 49

2.4   Open-Channel Flows   51
  2.4.1   Classification   52
  2.4.2   Boundary layer concepts applied to open-channel flows   55
    2.4.2.1   General concepts   55
    2.4.2.2   Bed shear stress   56
    2.4.2.3   Velocity distribution   56
      2.4.2.3.1   Velocity distributions in channels
            with smooth beds   57
      2.4.2.3.2   Velocity distributions in channels
            with rough beds   58
      2.4.2.3.3   Velocity defect law and log-wake law   58
    2.4.2.4   Mean velocity and flow resistance coefficients   60
    2.4.2.5   Uniform open channel flow equations   61
  2.4.3   Relevant 1D, 2D, and 3D equations for open-channel flows   63
    2.4.3.1   RANS equations with heat transfer   64
    2.4.3.2   Closure relationships for turbulence models   65
      2.4.3.2.1   Algebraic models   66
      2.4.3.2.2   One-equation models   67
      2.4.3.2.3   Two-equation turbulence models   69
        2.4.3.2.3.1 $k-\varepsilon$ Turbulence model   70
        2.4.3.2.3.2   Values of constants
               in the k-ε model   71
      2.4.3.2.4   Higher-order models   72
    2.4.3.3   Determination of the boundary conditions   73
    2.4.3.4   One-dimensional shallow-water flow equations:
          the St. Venant equations   74
  2.4.4   Secondary flows in open-channel flows   76
    2.4.4.1   Secondary flows classifications   76
    2.4.4.2   Origin of secondary flows in channels: vorticity
          conservation   79
    2.4.4.3   Analytical models for secondary currents   80
2.5   Complex Flows   81
  2.5.1   Mobile boundary channels and sediment transport   81
    2.5.1.1   Introduction to sediment transport and related
          phenomena   81
    2.5.1.2   Sediment and fluid properties relevant for sediment
          transport   83
    2.5.1.3   Initiation of motion, bedload rates, and bedforms   89
      2.5.1.3.1   Modified Shields diagram   90
      2.5.1.3.2   Shields-Vanoni-Parker (SVP) river
            regime diagram   93
      2.5.1.3.3   Movable bed laboratory models   94
      2.5.1.3.4   Bedload transport   95

| | | 2.5.1.3.5 | Two-dimensional (2D) transport of bedload | 99 |

2.5.1.3.5 Two-dimensional (2D) transport of bedload ... 99
2.5.1.3.6 Conservation of sediment mass: the Exner equation ... 101
2.5.1.3.7 Bedforms ... 101
2.5.1.4 Suspended sediment transport ... 106
2.5.1.4.1 Rousean profile ... 108
2.5.1.4.2 Rousean profile in stratified flows ... 109
2.5.1.4.3 Non-equilibrium suspensions ... 112
2.5.1.4.4 Transport of sediment mixtures in suspension ... 114
2.5.1.4.5 Sediment resuspension by unsteady flows ... 115
2.5.2 Flows in vegetated channels ... 116
2.5.3 Aerated flows ... 120
2.5.4 Ice-laden flows ... 121
2.A Appendix ... 122
2.A.1 Ratio of Nikuradse equivalent roughness size and sediment size for rivers ... 122
2.A.2 Equivalent roughness of bed forms ... 122
Notation ... 124
References ... 129

3 Similitude ... 143
3.1 Introduction ... 143
3.2 Basics ... 145
3.3 Dynamic Similitude from Flow Equations ... 147
3.4 Water Flow ... 154
3.4.1 Flow processes ... 154
3.4.2 Dynamic similitude ... 155
3.5 Multi-Phase Flow and Transport Processes ... 158
3.5.1 Other processes ... 158
3.5.2 Dynamic similitude for sediment transport ... 158
3.6 Addressing Similitude Shortcomings ... 160
3.6.1 Avoidance ... 160
3.6.2 Hybrid approaches ... 161
3.6.3 Similitude compromise ... 165
3.6.4 Geometric distortion ... 165
3.A Appendix: Dimensional Analysis ... 167
Notation ... 172
References ... 173

4   **Selection and Design of the Experiment**   177
    4.1   The Experimental Process   177
            4.1.1   Introduction   177
            4.1.2   Experiment phases   178
            4.1.3   Safety considerations   181
    4.2   Experimental Setup Components   182
            4.2.1   Introduction   182
            4.2.2   Facilities   182
            4.2.3   Measurement systems   184
            4.2.4   Measurement system response and characteristics   190
            4.2.5   Measurement environment   193
    4.3   Laboratory Facilities   194
            4.3.1   Introduction   194
            4.3.2   Flumes and basins   195
                    4.3.2.1   Dimensions   195
                    4.3.2.2   Layout   196
                    4.3.2.3   Flow boundary conditions   197
                    4.3.2.4   Special facilities   199
            4.3.3   Flume assembly   200
                    4.3.3.1   Flow circulation system   201
                    4.3.3.2   Flow conditioning   202
    4.4   Instrument Selection   202
            4.4.1   Preliminary considerations   202
            4.4.2   Instrument spatial and temporal resolutions   205
            4.4.3   Practical relationships for estimating turbulent flow
                     scales   207
            4.4.4   Selection of the instruments and settings for measurements
                     in turbulent flows   213
                    4.4.4.1   Attaining the required spatio-temporal resolution 214
                    4.4.4.2   Attaining the required frequency content for
                               the data   216
                    4.4.4.3   Summary for instrument selection in turbulent
                               flows   217
    4.5   From Signals to Data   218
            4.5.1   Signal classification   218
            4.5.2   Signal digitization   222
            4.5.3   Representation of continuous and discrete data   224
            4.5.4   Selection of an optimum data-acquisition system   225
    References   234

5   **Experiment Execution**   237
    5.1   Instrument-Flow and Facility-Flow Interactions   237

5.1.1	Introduction	237
5.1.2	Types of interactions	237
5.1.3	Measurements in flows with velocity gradients	239
5.1.4	Influence of turbulence levels	240
5.2	Conducting the Experiment	244
5.2.1	Activities	244
5.2.2	Instrument calibration	245
5.2.3	Establishing the measurement procedure	246
5.2.4	Data acquisition for turbulent flows	247
5.2.4.1	Sampling rate	247
5.2.4.2	Sampling duration	248
5.2.5	Preliminary runs	252
5.2.6	Flow control	254
5.2.7	Quality control	255
5.2.8	Record keeping	257
5.3	Field Experiments	261
5.3.1	Introduction	261
5.3.2	Planning	263
5.3.3	Execution	264
5.3.4	Health and safety aspects	264
5.3.4.1	Health and safety of participants	265
5.3.4.2	Health and safety of the public	266
5.3.4.3	Protection of the instrumentation	266
5.3.4.4	Consideration of the environment	267
5.4	Complex Experiments	267
5.4.1	Sediment transport	267
5.4.1.1	Introduction	267
5.4.1.2	Experimental flow complexity	268
5.4.2	Gravity currents	270
5.4.2.1	Introduction	270
5.4.2.2	Experimental flow complexity	271
5.4.3	Flow through vegetation	272
5.4.3.1	Introduction	272
5.4.3.2	Experimental complexity	274
5.4.4	Aerated flows	275
5.4.4.1	Introduction	275
5.4.4.2	Experimental flow complexity	276
5.4.4.3	Measurement approaches	278
5.4.5	Ice-covered flows	280
5.4.5.1	Introduction	280
5.4.5.2	Laboratory Investigation	281
5.4.5.3	Field investigation	281
5.5	Interaction of Experiments with Numerical Modeling	283

	5.5.1	Introduction	283
	5.5.2	Composite modeling	284
References			286

6   Data Analysis — 295

6.1		Introduction	295
6.2		Basic Concepts, Terminology, and Notation in Probability and Statistics	297
	6.2.1	Randomness, sampling, population, and homogeneity	297
	6.2.2	Probability and conditional probability; distributions, quantiles, moments, and expectations	299
	6.2.3	Sampling statistics and distributions, statistical independence, and conditional sampling	302
	6.2.4	Bias, variance, and formulating estimators	304
6.3		Descriptive Statistics and Exploratory Data Analysis	305
	6.3.1	Histograms, kernel density estimates, sample moment statistics, and quantiles	306
	6.3.2	Unsupervised learning: principal component analysis and clustering	309
	6.3.3	Data conditioning (or validation): outlier detection and data replacement	313
6.4		Hypotheses, Statistical Significance, and Interval Estimates	318
6.5		Bootstrapping	321
6.6		Regression	322
	6.6.1	The linear model and interval estimates	322
	6.6.2	The coefficient of determination and correlation	325
	6.6.3	Miscellaneous topics in linear regression	326
	6.6.4	Model transformations and nonlinear regression	326
	6.6.5	Residuals and bootstrapping regression	328
	6.6.6	Spurious correlations	329
6.7		Bayesian Inference	329
	6.7.1	Bayesian linear regression	330
	6.7.2	Comments on Bayesian applications in hydraulics	332
6.8		Extended Examples in Regression	333
	6.8.1	The velocity profile in a uniform open-channel suspension flow	333
	6.8.2	Hydraulic geometry of alluvial channels	336
	6.8.3	Pressure-flow scour	338
	6.8.4	Comments on Bayesian and bootstrapped estimates	341
6.9		Classification Analysis: Logistic Regression, Linear Discrimination Analysis, and Tree Classification	341
	6.9.1	Extended example: Classification of bedform channel regimes	344

6.10		Machine (or Statistical) Learning Approaches	346
	6.10.1	Artificial neural networks	347
	6.10.2	Relevance vector machines	349
	6.10.3	Other general issues in machine learning techniques	351
	6.10.4	Cross-validation	351
	6.10.5	Extended example: prediction of sand transport in channels using machine learning	352
6.11		Data Conditioning: Time Series and Filtering	355
	6.11.1	Linear time-invariant filters	357
	6.11.2	Filter performance: power-gain and phase frequency response	358
	6.11.3	Ringing and end effects, and other aspects	361
6.12		Time Series and Spectral Analysis	362
	6.12.1	Tests of stationarity	364
	6.12.2	Autocorrelations and integral time scales	365
	6.12.3	Spectral analysis and the periodogram	368
	6.12.4	Smoothing periodogram estimates, confidence intervals and spectral resolution	369
	6.12.5	Treatment of irregularly sampled data, normalization, and zero padding	371
	6.12.6	Other techniques in time series analysis	372
	6.12.7	Extended example: Autocorrelation and power spectral estimates	372
6.13		Spatial Interpolation, Kriging, and Spatial Derivatives	377
	6.13.1	The ordinary kriging model	378
	6.13.2	The variogram and its modeling	379
	6.13.3	Kriging predictions and standard prediction error	381
	6.13.4	Evaluation of spatial derivatives and uncertainties	382
6.14		Identification of Coherent Structures	383
	6.14.1	Introduction	383
	6.14.2	Common educing methods	385
	6.14.3	Educing methods based on invariants of the velocity gradient tensor	386
	6.14.4	The proper orthogonal decomposition (POD) technique	388
6.15		Final Comments	396
6.A		Appendix A	396
	6.A.1	Some density distributions used in statistical inference	396
	6.A.2	Confidence and prediction intervals in simple linear regression	397
	6.A.3	Confidence interval of ratios	398
References			399
7		Uncertainty Analysis for Hydraulic Measurements	407
7.1		Introduction	407

		7.1.1	Standardized methods for uncertainty analysis	408
	7.2	Concepts and Terminology		409
		7.2.1	Measurement and uncertainty analysis	409
		7.2.2	Errors and uncertainties	410
		7.2.3	Propagation of uncertainties	413
	7.3	Uncertainty Analysis Implementation		413
		7.3.1	Implementation steps	414
		7.3.2	Additional considerations	418
	7.4	Uncertainty Inferences Using Intercomparison Experiments		424
		7.4.1	Overview of intercomparison experiments	424
		7.4.2	Method implementation	425
		7.4.2.1	Error model	425
		7.4.2.2	Uncertainty estimation using individual repeated measurements	426
		7.4.2.3	Uncertainty estimates for sets of repeated measurements	427
	7.5	Practical Issues		430
	References			432
8	**Hydroinformatics Applied to Hydraulic Experiments**			437
	8.1	Introduction		437
	8.2	Hydroinformatics		438
	8.3	Digital Environmental Observatories		439
		8.3.1	Concepts and terminology	439
		8.3.2	Integrated datasets and models delivered as computer services	443
	8.4	Outlook		445
	References			446
	Subject Index			449

Volume II

	Preface and Contributions		xxix
	List of Contributing Authors Vol. II		xxxv
1	**Introduction**		1
	1.1	Book Overview	1
	1.2	The Role of Instrumentation and Measurement Techniques	2
	1.3	Approach	3
	1.4	Structure of Volume II	5
	References		6

2   **Flow Visualization**   9
      2.1   Introduction   9
      2.2   Fundamentals   10
              2.2.1   The physics of flow tracking with tracers   10
              2.2.2   Flow lines   12
              2.2.3   Optical considerations and imaging hardware   13
      2.3   Flow Visualization Techniques   15
              2.3.1   Flow visualization with aqueous tracers   15
                        2.3.1.1   Dyes and other aqueous flow tracers   16
                        2.3.1.2   Aqueous tracer injection and imaging   17
                        2.3.1.3   Quantitative concentration measurements
                                   using laser-induced fluorescence   20
              2.3.2   Flow visualization with discrete particles   21
                        2.3.2.1   Visualization of particle trajectories   21
                        2.3.2.2   Hydrogen bubble visualization   21
      2.4   Visualization of Flows Near Solid Surfaces   23
      2.5   Understanding Flow Topology from Flow Visualization Data   25
              2.5.1   Vortex topology   25
              2.5.2   Critical point theory   27
      2.A   Appendix   30
              2.A.1   Summary and specifications for pulsed and continuous lasers   30
      References   31

3   **Velocity**   35
      3.1   Introduction   35
      3.2   Acoustic Backscattering Instruments (ABIs) for Fine-Scale Flow
             Measurements   37
              3.2.1   Introduction   37
              3.2.2   Principles of Acoustic Backscattering Instruments   38
                        3.2.2.1   Backscattering   38
                        3.2.2.2   Doppler shift frequency   39
              3.2.3   Instrument types   41
                        3.2.3.1   Monostatic instruments   41
                                   3.2.3.1.1   Monostatic profiler   42
                        3.2.3.2   Multistatic instruments   43
                                   3.2.3.2.1   Operating principle of multistatic
                                                instruments   44
                        3.2.3.3   Point measurement instruments (ADVs)   46
                                   3.2.3.3.1   Three-receiver configuration   46
                                   3.2.3.3.2   Four-receiver configuration   48
                        3.2.3.4   Acoustic Doppler Velocity Profiler (ADVP)   50
              3.2.4   Error sources   52
                        3.2.4.1   Doppler noise   52

| | | | 3.2.4.1.1 | Identification and quantification of the different Doppler noise sources | 53 |

3.2.4.1.1   Identification and quantification of the different Doppler noise sources   53

3.2.4.1.2   Doppler noise correction based on two-point cross correlation   54

3.2.4.2   Aliasing   56

3.2.4.3   Spikes   57

3.2.4.4   Spatial averaging and other noises   58

3.3   Acoustic Instruments for Mean Flow Characterization in Field Conditions: Acoustic Doppler Current Profilers (ADCP)   59

3.3.1   Introduction   59

3.3.2   ADCP systems and measurement outcomes   61

3.3.2.1   Configurations   61

3.3.2.2   Processing techniques used for ADCP to estimate Doppler shift (Velocity measurement)   63

3.3.2.2.1   Narrowband   63

3.3.2.2.2   Broadband   64

3.3.2.3   Measuring velocities   66

3.3.2.3.1   Measuring velocity profiles   66

3.3.2.3.2   Unmeasured zones   67

3.3.2.3.3   Beam configurations and velocity profile estimation   68

3.3.3   Deployment methods   70

3.3.3.1   Manned boat   71

3.3.3.2   Tethered boat   71

3.3.3.3   Remote controlled boat   73

3.3.3.4   Boat velocity   74

3.3.3.5   Compass consideration   75

3.3.4   Error sources   75

3.3.5   Horizontal ADCP   77

3.4   Acoustic Travel-Time Tomography   79

3.A   Appendix   81

3.A.1   ABIs: Backscattering   82

3.A.2   ABIs: Extraction of the Doppler shift frequency   83

3.A.3   ABIs: Sampling volumes ("gates")   84

3.A.4   ABIs: Transducer characteristics   84

3.A.5   ABIs: Range velocity ambiguity   85

3.A.5.1   Monostatic profilers (UVPs)   85

3.A.5.2   Multistatic profilers   85

3.5   Point Velocimeters for Field Applications   86

3.5.1   Introduction   86

3.5.2   Available instrumentation   86

3.5.2.1   Mechanical current meters   87

3.5.2.2   Electromagnetic current meters   89

		3.5.2.3	Acoustic Doppler velocimeters	89
	3.5.3	Measurement considerations		90
		3.5.3.1	Sampling duration	90
			3.5.3.1.1 Sampling duration for stream gauging	90
			3.5.3.1.2 Sampling duration for the characterization of turbulence	91
		3.5.3.2	Deployment methods	92
		3.5.3.3	Calibration and testing	92
3.6	Laser-Doppler Velocimetry/Anemometry			94
	3.6.1	Introduction		94
	3.6.2	Fundamental concepts of Doppler velocimetry		95
		3.6.2.1	Doppler shift and Doppler frequency	95
		3.6.2.2	Multiple velocity components	99
		3.6.2.3	Summary of practical issues	99
	3.6.3	Basic instrumentation configurations: lasers and optics		101
		3.6.3.1	Typical components	101
		3.6.3.2	Summary of practical issues	104
	3.6.4	Tracer particles		108
		3.6.4.1	Types of tracer particles	108
		3.6.4.2	Summary of practical issues	110
	3.6.5	Signal processing and calculation of flow velocity statistics		111
		3.6.5.1	Determination of the Doppler frequency	111
		3.6.5.2	Spatial and temporal resolution	114
		3.6.5.3	Calculation of statistics in turbulent flows	117
		3.6.5.4	Summary of practical issues	121
	3.6.6	Final remarks		123
3.B	Appendix			124
	3.B.1	Fringe model		124
	3.B.2	Directional ambiguity		125
	3.B.3	Tracer particles – light scattering		127
	3.B.4	Tracer particles – quantification of tracking ability		128
3.7	Image-Based Velocimetry Methods			131
	3.7.1	Introduction		131
	3.7.2	Particle Image Velocimetry (PIV)		132
		3.7.2.1	Introduction	132
		3.7.2.2	PIV measurement process details	134
			3.7.2.2.1 Selection of tracer particles	134
			3.7.2.2.2 Light sheet formation and image capture	134
			3.7.2.2.3 Image processing to measure the velocity field	136
		3.7.2.3	Spatio-temporal resolution	140
		3.7.2.4	Measurement noise	144

	3.7.2.5	Sources of error	145
	3.7.2.6	Practical PIV configurations	148
		3.7.2.6.1 Single camera PIV (2C-2D)	149
		3.7.2.6.2 Stereoscopic PIV (3C-2D)	151
		3.7.2.6.3 Tomographic PIV (3C-3D)	152
		3.7.2.6.4 Special application – field PIV	153
3.7.3	Particle Tracking Velocimetry (PTV)		156
	3.7.3.1	Introduction	156
	3.7.3.2	Methodology	157
	3.7.3.3	Sources of error	162
	3.7.3.4	Averaging of velocity data	163
	3.7.3.5	Multi-frame tracking	164
	3.7.3.6	3-D Particle tracking (for complex flows)	166
3.7.4	Image velocimetry applied to large scales		167
	3.7.4.1	Introduction	167
	3.7.4.2	Basics of Large-Scale PIV (LSPIV)	168
		3.7.4.2.1 Mapping relation	168
		3.7.4.2.2 Image processing	170
		3.7.4.2.3 Sample measurements	171
		3.7.4.2.4 Error sources in LSPIV	174
	3.7.4.3	Basics of Space-Time Image Velocimetry (STIV)	177
		3.7.4.3.1 Generation of space-time image (STI)	177
		3.7.4.3.2 STIV algorithm	179
	3.7.4.4	Traceability of surface features	180
	3.7.4.5	Performance of LSPIV and STIV	181
		3.7.4.5.1 Comparison between LSPIV and STIV	181
		3.7.4.5.2 Comparison with other velocity measurement methods	181
		3.7.4.5.3 Comparison with discharge measurements	181
	3.7.4.6	Application of PTV to large-scale measurements (LPTV)	183
3.C	Appendix		185
	3.C.1	Light scattering in PIV	185
	3.C.2	Velocimeter comparison	186
3.8	High-Frequency Radar		187
3.9	Drifters and Drogues		189
3.10	Dilution Method		191
	References		193
4	**Topography and Bathymetry**		**211**
4.1	Introduction		211

4.2   Terrestrial Laser Scanning: Topographic Measurement
      and Modelling   211
      4.2.1   Introduction   211
      4.2.2   Principles of measurement   213
              4.2.2.1   Laser ranging   213
              4.2.2.2   Beam divergence and reflectivity   215
              4.2.2.3   Laser scanning   216
      4.2.3   Available instrumentation   217
      4.2.4   Deployment methods   219
      4.2.5   Spatial and temporal resolution   220
      4.2.6   Post-processing and data modelling   223
4.3   Ultrasonic Sensing   224
      4.3.1   Introduction   224
      4.3.2   Principles of sonar measurements   225
      4.3.3   Types of sonar devices   226
      4.3.4   Spatial and temporal resolution   227
      4.3.5   Error sources   229
4.4   Photogrammetry   229
      4.4.1   Introduction   229
      4.4.2   Principles of measurement   230
      4.4.3   Available instrumentation   234
      4.4.4   Spatial and temporal resolution   235
      4.4.5   Error sources   235
4.5   Other Surface Profiling Methods   236
      4.5.1   Introduction   236
      4.5.2   Optical displacement meters   236
      4.5.3   Mechanical profilers   238
      4.5.4   Surveying   239
4.6   Grain Size Distribution   240
      4.6.1   Introduction   240
      4.6.2   Principles of sampling   240
      4.6.3   Volumetric sampling and sieving   241
              4.6.3.1   Sampling   241
              4.6.3.2   Sieving   241
      4.6.4   Surface sampling and sieving   242
              4.6.4.1   Line-by-number sampling   242
              4.6.4.2   Photo sieving   244
4.A   Appendix   247
      4.A.1   Structure from Motion and MultiView Stereo   247
      4.A.2   Sediment size classification   248
      4.A.3   Spatial variability   251
References   253

5   **Sediment Transport**   261
    5.1   Introduction   261
    5.2   Bedload   261
          5.2.1   Physical traps and samplers   261
                  5.2.1.1   Principles of measurement   261
                  5.2.1.2   Samplers and deployment methods   262
                            5.2.1.2.1   Sediment retention basins   262
                            5.2.1.2.2   Trough or pit samplers   263
                            5.2.1.2.3   Bedload samplers   264
                  5.2.1.3   Spatial and temporal resolution   265
                  5.2.1.4   Error sources   265
          5.2.2   Passive acoustic measurements   266
                  5.2.2.1   Principles of measurement   266
                  5.2.2.2   Available samplers and deployment methods   267
                  5.2.2.3   Calibrations and error sources   268
          5.2.3   Active acoustic measurements   269
                  5.2.3.1   Principles of measurement   269
                  5.2.3.2   Available samplers and deployment methods   269
                  5.2.3.3   Spatial and temporal resolution   270
                  5.2.3.4   Calibration and error sources   270
          5.2.4   Monitoring of bed form movement   271
                  5.2.4.1   Principles of measurement   271
                  5.2.4.2   Available technologies   272
                  5.2.4.3   Spatial and temporal resolution   273
                  5.2.4.4   Error sources   273
          5.2.5   Bedload particle tracers   274
                  5.2.5.1   Principles of measurement   274
                  5.2.5.2   Available particle tracing methods   274
                  5.2.5.3   Spatial and temporal resolution   275
                  5.2.5.4   Error sources   275
    5.3   Suspended Load   276
          5.3.1   Physical sampling for suspended sediment   276
                  5.3.1.1   Collecting a representative sample   276
                  5.3.1.2   Depth integrating vs. point integrating samplers 277
                  5.3.1.3   Choosing a sampler   278
                  5.3.1.4   Sampling throughout a channel cross-section   278
                  5.3.1.5   Error sources   279
          5.3.2   Optical measurements   280
                  5.3.2.1   Optical backscatter   280
                  5.3.2.2   Laser diffraction   282
                  5.3.2.3   Focused beam reflectance   283
          5.3.3   Acoustic methods   283

		5.3.3.1	Acoustic backscattering	283
		5.3.3.2	Principles of measurement	284
		5.3.3.3	Instruments	284
			5.3.3.3.1 Acoustic Doppler Current Profiler (ADCP)	284
			5.3.3.3.2 Acoustic Doppler Velocity Profiler (ADVP)	285
		5.3.3.4	Error sources	286
5.A	Appendix			286
	5.A.1	Concentration measurements with Acoustic Backscattering Instruments (ABIs)		286
		5.A.1.1	Principles of concentration measurements with Acoustic Backscattering Instruments (ABIs)	286
		5.A.1.2	Acoustic inversion methods	288
			5.A.1.2.1 Iterative inversion method	288
			5.A.1.2.2 Explicit inversion method	288
		5.A.1.3	Attenuation compensation using hardware	289
			5.A.1.3.1 Backward and forward scattering	289
			5.A.1.3.2 Two-frequency method	289
	5.A.2	Concentration measurements with ADCPs		290
		5.A.2.1	Scattering formulations	290
		5.A.2.2	Obtaining concentration from backscattered intensity	294
		5.A.2.3	Obtaining concentration from attenuation	297
	References			298
6	**Auxiliary Hydraulic Variables**			**309**
6.1	Introduction			309
6.2	Water Depth			309
	6.2.1	Principles of measurement		309
	6.2.2	Available instrumentation		311
		6.2.2.1	Mechanical instruments and methods	311
		6.2.2.2	Acoustic and optical instruments and methods	313
		6.2.2.3	Spatial and temporal resolution	313
		6.2.2.4	Error sources	314
6.3	Water Surface and Bed Slope			315
	6.3.1	Principles of measurement		315
6.4	Pressure			315
	6.4.1	Principles of measurement		315
	6.4.2	Available instrumentation		316
		6.4.2.1	Steady pressure measurements	316
		6.4.2.2	Fluctuating pressure measurements	318
			6.4.2.2.1 Strain gauge transducers	318

		6.4.2.2.2	Piezoelectric transducers	319
		6.4.2.2.3	Piezo-resistive transducers	319
	6.4.2.3	Pressure sensitive paint		321
	6.4.3	Spatial and temporal resolution		321
	6.4.4	Error sources		322
6.5	Bed Shear Stress			322
	6.5.1	Principles of measurement		322
	6.5.2	Available instrumentation		324
		6.5.2.1	Direct techniques	324
			6.5.2.1.1 Shear plates	324
			6.5.2.1.2 Surface coating methods	325
		6.5.2.2	Indirect techniques	326
			6.5.2.2.1 Obstacle flow techniques	326
			6.5.2.2.2 Heat and mass transfer techniques	328
			6.5.2.2.3 Pulsed wall probes	329
			6.5.2.2.4 Micro-optical shear stress sensor	329
	6.5.3	Spatial and temporal resolution		330
		6.5.3.1	Spatial resolution	330
		6.5.3.2	Temporal resolution	330
	6.5.4	Limitations and error sources		331
6.6	Drag Forces			332
	6.6.1	Principles of measurement		332
	6.6.2	Available instrumentation		335
	6.6.3	Spatial and temporal resolution		335
	6.6.4	Error sources		335
6.7	Conductivity-Temperature-Depth Probes for Fluid Density			336
References				338

7	**Discharge**		**343**
7.1	Introduction		343
7.2	Intrusive Flowmeters		348
	7.2.1	Weirs	348
	7.2.2	Sluice gates	349
	7.2.3	Differential pressure flow meters: orifice-plate and Venturi meters	349
7.3	Non-Intrusive Flowmeters		350
	7.3.1	Electromagnetic discharge meters	350
	7.3.2	Acoustic travel-time discharge meters	350
7.4	Discrete Streamflow Measurements Based on Velocity Integration		351
	7.4.1	Mid-section and mean-section methods	351
	7.4.2	ADCP transects	353
7.5	Continuous Streamflow Monitoring Using Stage Measurements		356
	7.5.1	Simple stage-discharge ratings	356

7.5.2   Complex stage-discharge ratings   360
7.6   Continuous Streamflow Monitoring Using Velocity Measurements   364
7.6.1   Index-velocity method   364
7.6.2   Hydraulics-based velocity models   367
7.6.3   Hybrid approaches for velocity models   369
7.7   Practical Issues   371
References   373

8   Autonomous Underwater Vehicles as Platforms for Experimental
Hydraulics   377
8.1   Introduction   377
8.2   Autonomous Underwater Vehicles   377
8.3   Horizontal and Vertical Positioning   380
8.3.1   Operational modes   381
8.3.2   Horizontal positional aiding   382
8.3.3   Kalman filters   383
8.3.3.1   Defining the Kalman filter   384
8.3.3.2   Kalman filter equations   384
8.4   Vehicle and Mission Design Considerations for Data Collection   385
8.4.1   Vehicle induced flow field   385
8.4.1.1   Boundary layer   386
8.4.1.2   Pressure field   387
8.4.2   Mission design   388
8.5   Mapping Under Ice   389
8.5.1   Ice environments   389
8.5.2   Under-ice AUV operations   390
8.5.3   Case study: Erebus Glacial Tongue   391
8.5.3.1   Correction for sound velocity   391
8.5.3.2   Applied filters   393
References   395

Subject Index   399

Preface and Contributions

Hydraulics is the branch of engineering and science associated with all forms of water movement in natural and constructed channels and conduits. Physical interactions between many types of gaseous, liquid, and solid materials with water flows are also relevant to the field of hydraulics. Notable examples include sediment transport, ice flows, entrained air flows, and flows of water of varying salinity or turbidity. In addition, conventional hydraulics also addresses the transport of energy, heat, and dissolved chemicals by water. Related topics include water waves, thermal plumes in lakes and streams, and transport of contaminants and nutrients by rivers. Most problems in hydraulics are much too complex to be solved by direct mathematical formulation, making laboratory and field experiments crucial for understanding and advancing our knowledge about various flows.

This two-volume book is a comprehensive guide to designing, conducting, and interpreting laboratory and field experiments on a broad range of topics associated with modern hydraulic studies. The book provides guidance on the selection of appropriate laboratory facilities, instrumentation and experimental techniques. Advice is also given on performing experiments. Furthermore, methods are described for analysing data, assessing uncertainties in measurements, and presenting results derived from observations and experiments. The book considers experiments performed under a range of conditions, from well-equipped, well-staffed experimental facilities to environments where accessibility to advanced instrumentation and expertise is limited. Although the book focuses primarily on laboratory experiments, including hydraulic physical modelling, it also applies to field experiments of varying complexity and accessibility.

A comprehensive guide to experimental hydraulics is much needed. Most readily available guides are already several decades old, and do not reflect the remarkable developments that have occurred in contemporary hydraulics, especially regarding digital instrumentation and handling of large, high-density experimental datasets. The experts comprising the book's editorial team, aware of this major gap in literature on hydraulics, decided to embark on a concerted effort to write an authoritative and up-to-date reference text on hydraulic experiments. The book is the result of the first substantial effort in the community of hydraulic engineering to assemble in one place descriptions of all the components of hydraulic experiments along with a concise outline of essential hydraulic theory.

Aiming to ensure readability and accessibility to information, this book is organized into two volumes with content that follows a uniform, multi-layered approach (i.e., body text, summary tables, examples, appendices) across the chapters. The first volume focuses on fundamental hydraulics and on the phases of conducting and analysing the results of hydraulic experiments. The second volume concentrates on flow visualization and measurement methods and comprehensively describes the measurement instrumentation available for hydraulic experiments. The coverage of established experimental hydraulics subjects (such as conventional measurement instruments) is briefer, while descriptions of contemporary topics in both experimental methods and instrumentation are more extensive.

Target readers are college faculty, researchers, practitioners, and students involved in various aspects of experimental hydraulics. The book's primary purpose is to provide readers a clear and readily accessible guide to hydraulic experiments, the necessary steps when conducting them, and the extensive information about the principles, capabilities, and limitations of available instrumentation (particularly contemporary technologies). It is assumed that readers understand elementary fluid mechanics and statistics and have a general technical background. Ample references are given to complementary books and articles on engineering hydraulics and experimentation, enabling the interested reader to delve further into the topics covered.

This book was written by a team of more than 45 experts who have been at the forefront of research, teaching or practice in experimental hydraulics. Most of the writers have also been active members of the International Association for Hydro-Environment Engineering and Research's (IAHR) Technical Committee on Experimental Methods and Instrumentation in the last three decades. The need for this book was conceived, discussed, and developed during multiple meetings held at various conferences and through extensive correspondence. During these discussions, an editorial group was established to handle critical editorial duties. Most of the credit in creating the book, though, must be given to the dedicated team of authors. Their desire to advance experimental hydraulics, and their consistently good-natured collaboration, led to the book's realization. The authors contributing to chapters in this volume are named in the table below. Additionally, the affiliations for the contributing authors for this volume are subsequently listed.

The editorial group also thanks Professor Peter Davies, the former IAHR Book Editor, and Janjaap Blom, Senior Publisher at Taylor & Francis, for their guidance and support, as well as patience. Many other people also should be acknowledged and thanked for contributing figures and background material used in preparing the book and for reviewing portions of the book. The summary table below recognizes the volunteers who provided assistance to the authors.

Everybody involved in preparing this book trusts that it achieves the vision of an authoritative and accessible reference resource for experimentation in hydraulics. Readers are invited to contact the editorial team regarding various omissions, misprints, and suggestions for improvements to this edition of the book. Future editions will address such items and include on-going developments in instrumentation and experimental techniques.

Marian Muste
Editor-in-Chief

List of authors contributing to the book's chapters and sections

Chapter/ Section	Title	Contributing Author(s)
	VOLUME II	
Chapter 1. Introduction		**Coordinators: Aberle J., Muste M., Rennie C.D.**
1.1	Book overview	Muste M., Aberle J., Rennie C.D.
1.2	The role of instrumentation and measurement techniques	Ettema R., Aberle J., Muste M.
1.3	Approach	Aberle J., Muste M., Rennie C.D.
1.4	Structure of Volume II	Aberle J., Rennie C.D., Muste M.
Chapter 2. Flow visualisation		**Coordinator: Buchholz J.**
2.1	Introduction	Buchholz J.
2.2	Fundamentals	Buchholz J.
2.3	Flow visualization techniques	Buchholz J.
2.4	Visualization of flows near solid surfaces	Buchholz J.
2.5	Understanding flow topology from flow visualization data	Buchholz J.
2.A.1	Summary and specifications for pulsed and continuous lasers	Ferreira R., Aleixo R.
Chapter 3. Velocity		**Coordinators: Aberle J., Rennie C.D., Admiraal D., Muste M.**
3.1	Introduction	Aberle J., Rennie C.D., Lemmin U.
3.2	Acoustic Backscattering Instruments (ABIs) for fine-scale flow measurements	Lemmin U.
3.3	Acoustic instruments for mean flow characterization in field conditions: Acoustic Doppler Current Profilers (ADCP)	Szupiany R.N., Garcia C.M., Oberg K.
3.3.5	Horizontal ADCP	Hoitink A.J.F.
3.4	Acoustic travel time tomography	Hoitink A.J.F.
3.A.1	ABIs: Backscattering	Lemmin U.
3.A.2	ABIs: Extraction of Doppler shift frequency	Lemmin U.
3.A.3	ABIs: Sampling volumes ("gates")	Lemmin U.
3.A.4	ABIs: Transducer characteristics	Lemmin U.
3.A.5	ABIs: Range velocity ambiguity	Lemmin U.
3.5	Point velocimeters for field applications	Jamieson E.
3.6	Laser-Doppler velocimetry/ anemometry	Ferreira R., Aleixo R.
3.B.1	Fringe model	Ferreira R., Aleixo R.
3.B.2	Directional ambiguity	Ferreira R., Aleixo R.
3.B.3	Tracer particles – light scattering	Ferreira R., Aleixo R.
3.B.4	Tracer particles – quantification of tracking ability	Ferreira R., Aleixo R.
3.7	Image based velocimetry methods	Cameron S., Fujita I., Muste M., Admiraal D.M.

3.7.1	Introduction	Admiraal D.M.
3.7.2	Particle Image Velocimetry (PIV)	Cameron S., Admiraal D.M.
3.7.3	Particle Tracking Velocimetry (PTV)	Admiraal D.M.
3.7.4	Image velocimetry applied to large scales	Fujita I., Muste M.
3.C.1	Light scattering in PIV	Cameron S.
3.C.2	Velocimeter comparison	Muste M.
3.8.	High-frequency radar	Hoitink A.J.F.
3.9	Drifters and drogues	Rennie C.D.
3.10	Dilution method	Rennie C.D.

Chapter 4. Topography and bathymetry   **Coordinator: Aberle J.**

4.1	Introduction	Aberle J.
4.2	Terrestrial laser scanning: Topographic measurement & modelling	Brasington J.
4.3	Ultrasonic sensing	Rüther N.
4.4	Photogrammetry	Henning M., Detert M., Aberle J.
4.5	Other surface profiling methods	Aberle J.
4.6	Grain-size distribution	Detert M., Bezzola G.R., Weitbrecht V.
4.A.1	Structure from Motion and MultiView Stereo	Detert M, Henning M.
4.A.2	Sediment size classification	Detert M., Bezzola G.R., Weitbrecht V.
4.A.3	Spatial variability	Detert M., Bezzola G.R., Weitbrecht V.

Chapter 5. Sediment transport   **Coordinator: Rennie C.D.**

5.1	Introduction	Rennie C.D., Aberle J.
5.2	Bedload	Rickenmann D., Rennie C.D., Aberle J., Muste M.
5.2.1	Physical traps and samplers	Rickenmann D.
5.2.2	Passive acoustic measurements	Rickenmann D.
5.2.3	Active acoustic measurements	Rennie C.D.
5.2.4	Monitoring of bed form movement	Aberle J., Muste M.
5.2.5	Bedload particle tracers	Rickenmann D.
5.3	Suspended load	Wren D.G., Moore S.A., Lemmin U.
5.3.1	Physical sampling for suspended load	Wren D.G.
5.3.2	Optical measurements	Wren D.G.
5.3.3	Acoustic methods	Moore S.A., Lemmin U.
5.3.3.1	Acoustic backscattering	Moore S.A., Lemmin U.
5.3.3.2	Principles of measurement	Moore S.A.
5.3.3.3	Instruments	Moore S.A., Lemmin U.
5.3.3.3.1	Acoustic Doppler Current Profiler (ADCP)	Moore S.A.
5.3.3.3.2	Acoustic Doppler Velocity Profiler (ADVP)	Lemmin U.
5.3.3.4	Error sources	Moore S.A.
5.A.1	Concentration measurements with Acoustic Backscattering Instruments (ABIs)	Lemmin U.
5.A.2	Concentration measurement with ADCPs	Moore S.A.

Chapter 6. Auxiliary hydraulic variables Coordinator: Aberle J.

6.1	Introduction	Aberle J.
6.2	Water depth	Aberle J.
6.3	Water surface and bed slope	Aberle J.
6.4	Pressure	Detert M., Weitbrecht V.
6.5	Bed shear stress	Aberle J., Rowinski P., Henry P-Y., Detert M.
6.6	Drag forces	Henry P-Y., Aberle J.
6.7	Conductivity-temperature-depth probes for fluid density	Rennie C.D.

Chapter 7. Discharge Coordinators: Rennie C., Hoitink A.J.F., Muste M.

7.1	Introduction	Hoitink A.J.F., Muste M., Rennie C.D.
7.2	Intrusive flowmeters	Rennie C.D.
7.3	Non-intrusive flowmeters	Rennie C.D., Muste M.
7.4	Discrete streamflow measurements based on velocity integration	Muste M., Rennie C.D.
7.5	Continuous streamflow monitoring using stage measurements	Muste M., Hoitink A.J.F.
7.6	Continuous streamflow monitoring using velocity measurements	Hoitink A.J.F., Muste M.
7.7	Practical issues	Hoitink A.J.F.

Chapter 8. Autonomous underwater vehicles as platforms for experimental hydraulics Coordinators: Forrest A., Rennie C.D., Aberle J.

8.1	Introduction	Forrest A., Leong Z., Hamilton A.K., Laval B.E.
8.2	Autonomous underwater vehicles	Forrest A., Leong Z., Hamilton A.K., Laval B.E.
8.3	Horizontal and vertical positioning	Forrest A., Leong Z., Hamilton A.K., Laval B.E.
8.4	Vehicle and mission design considerations for data collection	Forrest A., Leong Z., Hamilton A.K., Laval B.E.
8.5	Mapping under ice	Forrest A., Leong Z., Hamilton A.K., Laval B.E.

ACKNOWLEDGEMENTS

Some of the chapters and sections benefitted from contributions made by colleagues who are not included in the authorship. The acknowledged colleagues are:

Front Matter: Rennie photo credit davidtaylorphotostudio.com

Chapters 1, 3, 7: Robert Ettema (editing tasks)

Chapter 3: Dr. Ana Margarida Ricardo for the support in preparing the elements of Annex 3.B.4 and for the valuable discussions about the contents of the LDV chapter. Dr. Elsa Carvalho for sharing some of her LDV results that were used to illustrate the LDV data processing.

Chapter 7: LeCoz, J. (editing tasks)

All chapters: Heng-Wei Tsai, Carl Smith, and Mary Windsor – students affiliated with IIHR- Hydroscience & Engineering, The University of Iowa (editing and compilation tasks)

SAMPLE CITATIONS FOR BOOK CHAPTERS AND SECTIONS:

Chang K.A. & Socolofsky S.A. (2017). Aerated flows, Section 2.5.3 in Experimental Hydraulics, Volume I; M. Muste, D.A. Lyn, D.M. Admiraal, R. Ettema, V. Nikora, and M.H. Garcia (Editors); Taylor & Francis, New York, NY.

Moore, S.A. & Lemmin, U. (2017). Acoustic backscattering, Section 5.3.3.1 in Experimental Hydraulics, Volume II; J. Aberle, C.D. Rennie, D.M. Admiraal, M. Muste (Editors); Taylor & Francis, New York, NY.

List of Contributing Authors Vol. II

Jochen Aberle   Leichtweiß-Institute for Hydraulic Engineering and Water Resources
Technische Universität Braunschweig
Braunschweig, Germany
Email: j.aberle@tu-braunschweig.de

Department of Civil and Environmental Engineering
Norwegian University of Science and Technology
Trondheim, Norway
Email: jochen.aberle@ntnu.no

David Admiraal   Department of Civil Engineering
University of Nebraska-Lincoln
Lincoln, U.S.A.
Email: dadmiraal@unl.edu

Rui Aleixo   Interdepartmental Centre for Industrial Research in Building and
Construction
University of Bologna
Bologna, Italy
Email: rui.aleixo@unibo.it

GHT Photonics
Padua, Italy
Email: ferreira_aleixo@yahoo.co.uk

Gian Reto Bezzola   Hazard Prevention Division
Federal Office for the Environment (FOEN)
Berne, Switzerland
Email: gianreto.bezzola@bafu.admin.ch

James Brasington   School of Geography
Queen Mary, University of London
London, E1 4NS, UK
Email: j.brasington@qmul.ac.uk

James Buchholz   Department of Mechanical and Industrial Engineering IIHR – Hydroscience
& Engineering
The University of Iowa
Iowa City, U.S.A.
Email: james-h-buchholz@uiowa.edu

Stuart Cameron — School of Engineering
University of Aberdeen
Aberdeen, Scotland
Email: s.cameron@abdn.ac.uk

Martin Detert — Laboratory of Hydraulics, Hydrology and Glaciology (VAW)
ETH Zurich
Zurich, Switzerland
Email: detert@vaw.baug.ethz.ch

Robert Ettema — Department of Civil and Environmental Engineering
Colorado State University
Fort Collins, U.S.A.
Email: rettema@engr.colostate.edu

Rui M.L. Ferreira — CERIS – Instituto Superior Técnico
Universidade Lisboa
Lisbon, Portugal
Email: ruimferreira@tecnico.ulisboa.pt

Alexander Forrest — Environmental Dynamics Laboratory, Department of Civil &
Environmental Engineering
University of California
Davis, U.S.A.
Email: alforrest@ucdavis.edu

Tahoe Environmental Research Center, John Muir Institute of the
Environment
University of California
Davis, Incline Village, U.S.A.

Australian Maritime College
University of Tasmania
Launceston, Tasmania, Australia

Ichiro Fujita — Department of Civil Engineering
Kobe University
Kobe, Japan
Email: ifujita@kobe-u.ac.jp

Carlos M. Garcia — School of Exact, Physical and Natural Sciences
Córdoba National University
Córdoba, Argentina
Email: carlos.marcelo.garcia@unc.edu.ar

Institute for Advanced Studies for Engineering and Technology (IDIT –
CONICET – UNC)
Córdoba, Argentina

Andrew K.
Hamilton — Department of Civil Engineering
University of British Columbia
Vancouver, Canada
Email: andrew@madzu.com

Department: Department of Geography and Environmental Studies
Carleton University
Ottawa, Canada

Martin Henning
Hydraulic Engineering in Inland Areas
Federal Waterways Engineering and Research Institute
Karlsruhe, Germany
Email: martin.henning@baw.de

Pierre-Yves Henry
Department of Civil and Environmental Engineering
Norwegian University of Science and Technology
Trondheim, Norway
Email: pierre-yves.henry@ntnu.no

A.J.F. (Ton) Hoitink
Hydrology and Quantitative Water Management Group, Department of Environmental Sciences
Wageningen University and Research
Wageningen, Netherlands
Email: ton.hoitink@wur.nl

Elizabeth C. Jamieson
National Hydrological Services (Water Survey of Canada)
Environment and Climate Change Canada, Government of Canada
Ottawa, Canada
Email: elizabeth.jamieson@canada.ca

Bernard E. Laval
Department of Civil Engineering
University of British Columbia
Vancouver, Canada
Email: blaval@civil.ubc.ca

Ulrich Lemmin
ECOL-ENAC
École Polytechnique Fédérale de Lausanne (EPFL)
Lausanne, Switzerland
Email: ulrich.lemmin@epfl.ch

Zhi Leong
Australian Maritime College
University of Tasmania
Launceston, Tasmania, Australia
Email: zhi.leong@utas.edu.au

Stephanie A. Moore
National Hydrological Services (Water Survey of Canada)
Environment and Climate Change Canada, Government of Canada
Ottawa, Canada
Email: stephanie.moore2@canada.ca

Marian Muste
IIHR Hydroscience & Engineering, Civil & Environmental Engineering Department
University of Iowa
Iowa City, Iowa 52242, U.S.A.
Email: marian-muste@uiowa.edu

Kevin Oberg
Office of Surface Water
U.S. Geological Survey
Urbana, U.S.A.
Email: kaoberg@usgs.gov

Colin D. Rennie
Civil Engineering
University of Ottawa
Ottawa, Canada
Email: Colin.Rennie@uottawa.ca

Dieter
Rickenmann

Mountain Hydrology and Mass Movements
Swiss Federal Research Institute WSL
Birmensdorf, Switzerland
Email: dieter.rickenmann@wsl.ch

Pawel M.
Rowinski

Hydrology and Hydrodynamics
Institute of Geophysics, Polish Academy of Sciences
Warsaw, Poland
Email: p.rowinski@igf.edu.pl

Nils Rüther

Department of Civil and Environmental Engineering
Norwegian University of Science and Technology
Trondheim, Norway
Email: nils.ruther@ntnu.no

Ricardo N.
Szupiany

College of Engineering and Water Sciences
Littoral National University
Santa Fe, Argentina
Email: rszupian@fich.unl.edu.ar

National Scientific and Technical Research Council (CONICET)
Buenos Aires, Argentina

Volker Weitbrecht   Laboratory of Hydraulics, Hydrology and Glaciology (VAW)
ETH Zurich
Zurich, Switzerland
Email: weitbrecht@vaw.baug.ethz.ch

Daniel G. Wren

National Sedimentation Laboratory
U.S. Department of Agriculture
Oxford, USA
Email: Daniel.Wren@ars.usda.gov

Chapter 1

Introduction

1.1 BOOK OVERVIEW

The purpose of this book is to aid contemporary laboratory and field experimentation in civil engineering hydraulics. In so doing, it presents current and foreseeable practice in the design and conduct of experiments in hydraulics, and includes guidelines for selecting and operating instrumentation, for analyzing measurements, and for the use of consistent methods for quantifying and documenting the quality of measurements.

Experiment design and conduct are commonly influenced by the availability of instrumentation and facilities, and by the expertise of the experimenter. Accordingly, as instrumentation and facilities have evolved, the level of sophistication of hydraulic experiments has increased, as has the requisite level of expertise. The range of experiment capabilities has expanded correspondingly – experimenters today can measure, record and observe far more than ever before. Additionally, improvements in methods for quantifying measurement reliability provide a more solid technical foundation for experimental findings.

This book is published in two volumes to make it easily accessible without eliminating important information related to hydraulics experimentation. The two volumes have the following emphases:

Volume I: **Fundamentals and Methods** – guidelines for the design and conduct of hydraulic experiments and the analysis and presentation of measured data

Volume II: **Instrumentation and Measurement Techniques** – information about the principles, capabilities, and limitations of hydraulics instruments as well as guidance regarding selection of instrumentation and techniques for measurement and flow visualization

The ensuing sections of this chapter introduce the structure of Volume II and present the topics associated with this volume, including the role of instruments, their capabilities and limitations, and their proper use in experimental studies. The focus is on instrumentation and experimental methods applied to unidirectional flows, although many of the instruments and experimental methods covered can also be used in experiments on multi-directional and complex flows such as oscillatory flows, flow through vegetation, or ice-covered flows.

Contemporary hydraulic research is embedded in a cross-disciplinary framework focusing on fundamental hydraulic processes and their interactions with many different environmental processes. Much present hydraulic research involves investigation of water flow associated with physical, chemical, and biological processes at multiple scales, typically in the context of protecting, restoring, and managing environmental quality (*e.g.*, Singh & Hager, 1996; Rubin & Atkinson, 2001; Rowinski, 2007). Detailed descriptions of chemical and biological instruments and methods are beyond the scope of this book. However, hydraulics instruments are often used by researchers in disciplines besides hydraulics (*e.g.*, ecologists and biologists) to determine various physical aspects of flowing water; notably, flow velocity. This book is designed to be of use to researchers in many fields involving water flow.

This book is a concise, comprehensive guide to instruments or measurement methods for specific experimental situations. In this regard, an important purpose of this book is to provide incisive guidance on capabilities, limitation, and implementation of hydraulics instruments and measurement techniques; this guidance aids the selection and use of instruments and measurement techniques. In accordance with this purpose, the chapters and sub-sections of Volume II have been prepared by experienced hydraulic experimenters.

1.2 THE ROLE OF INSTRUMENTATION AND MEASUREMENT TECHNIQUES

Hydraulics instrumentation provides the means to observe, sense, and control flow processes in the laboratory and the field. Instrumentation and measurement techniques are indispensable when quantification of the observed processes is required. Experiments continue to play a crucial role in contemporary hydraulics, because many aspects of water flow remain to be elucidated or documented in research or practical application (*e.g.*, Ettema *et al.*, 2000; Novak *et al.*, 2010; Frostick *et al.*, 2011, 2014; Sukhodolov, 2015).

The development of hydraulics instruments is tightly connected with the evolving needs of experimentation related to fundamental science and practical applications. Hydraulic experiments have often posed significant challenges in terms of instrumentation capability for measuring flow (*e.g.*, non-uniform, unsteady, turbulent) and transport processes (*i.e.*, the movement of sediment, air, heat and chemicals in water). Another continuous motivation for instrumentation development is the measurement and monitoring of liquid-related transport processes for agricultural, industrial and other socio-economic purposes. Instruments for measuring water flow have existed for almost three centuries, and have advanced substantially in capabilities ever since. Indeed the history of flow instruments pre-dates the equations of fluid motion (*e.g.*, see Rouse & Ince, 1957). Notably, Henri de Pitot in 1732 introduced the stagnation tube and the manometer, and the notion of velocity head. Shortly thereafter, in 1738, Daniel Bernoulli presented the concept of pressure head and the piezometer.

Since the early work of Charles-Augustin de Coulomb during the 1780s, and the subsequent work of numerous physicists and electrical engineers, the principles of electricity have been used for instruments to measure flow properties. In this context, other scientific areas stimulated the development of new instruments, along with advanced methods for data acquisition and data processing.

In line with substantial advances in the sophistication of instrumentation and measurement techniques, it is nowadays possible to measure more flow and environmentally related properties, and to do so with remarkably high spatial and temporal resolution. Experiments involving complex flows, and requiring high degrees of temporal and spatial resolution, are more and more common, both in the laboratory and in the field (see Section 5.4, Volume I). Due to the considerable progress in technologies underpinning hydraulics instruments it is increasingly possible to conduct *in situ* field experiments that were inconceivable two to three decades ago, including turbulence measurements (*e.g.*, Sukhodolov, 2012), secondary flows (*e.g.*, Venditti *et al.*, 2014) and sediment transport (*e.g.*, Williams *et al.*, 2015). Such developments have been possible only with the assimilation of advanced methods for evaluating measurement reliability and for handling and visualizing the massive amounts of data that experiments may produce, as discussed in Volume I.

There is continued need for the advances in instrumentation and methods to be matched with practical guidance on conducting hydraulics experiments. Researchers and practitioners require adequate knowledge of how selected characteristics of flows can be measured, appreciating the limitations of instruments, and the ability to judge quality or reliability of data and observations obtained from experiments. Addressing this need can be challenging, as the working principles of sophisticated instruments and aspects of experimental design are not commonly part of undergraduate or graduate education. Effective application of instruments and methods often requires learning principles from other fields (*e.g.*, electronics, acoustics and optics). Instruments should never be used by simply applying a *plug-and-play* approach, as science and engineering practice, after all, relies on sound data. Laboratory and field experiments and monitoring essentially seek to enlighten cause-and-effect relationships and, thereby, to strengthen analytical thinking. This intent, which also entails accurate post-processing of measured data, drives successful, pertinent, and accurate experimentation.

1.3 APPROACH

Volume II of this book presents contemporary knowledge and methods regarding instrumentation and measurement techniques for hydraulic experiments. It assembles, in one place, essential information on available instruments, theoretical fundamentals and measurement techniques. The intent is to provide necessary information for the selection and use of instruments required for various aspects of a hydraulic experiment. The volume's chapters are structured to be convenient for experimenters wishing to readily access this information. Presently, much practical information on many contemporary instruments for experimental studies in environmental hydraulics resides in scattered scientific papers, and is consequently far less accessible.

A large number of commercial instruments are available for hydraulic experiments. Technological advancements continue to evolve instruments and experimental methods. Additionally, many instruments are used beyond their specified limits in order to gain additional information from measurements; *e.g.*, estimating suspended-sediment concentration from measurements with acoustic Doppler current profilers. Consequently, researchers often modify existing instrumentation or develop custom-made instruments. The present volume reviews instruments commonly used in hydraulic experiments, describing measurement principles, available spatial and temporal resolution, and potential error sources. Use herein of manufacturers' names does not endorse any specific product or company, rather it helps experimenters connect general principles to instrument configurations and performance characteristics of specific, commercially available instruments.

Inevitably over time some instruments have been replaced with improved instruments. For example, about sixty years ago the hot-film anemometer was introduced as the first instrument that could be used for measuring velocity fluctuations associated with turbulence in water flow. It since has been replaced by laser and acoustic-based instruments. Over sixty years ago, the electric analog, or electric resistance analog, was used as an experimental means for studying water flow through various forms of channel and permeable media. It too is no longer used in hydraulic experiments. On the other hand, the Pitot tube and piezometer remain as reliable, often-used instruments, and point gauges can still be considered as state of the art (improved reading of scales).

Although this book could serve as a textbook for a course on instruments and methods in hydraulic experiments, it was designed to be a convenient handbook. Therefore, its style is concise and makes frequent use of tables and figures to present information. The book's contributors have gone to great lengths to provide specific guidance on instruments and methods currently used in hydraulic experimentation. However, given that the subject comprises a wide variety of topics and applications, the depth of the description varies from section to section, commensurate with the complexity, usage and novelty of the technique or procedure under discussion.

This volume groups instruments and experimental methods in separate chapters addressing the following facets of hydraulic experiments: i) flow velocities; ii) topography and bathymetry; iii) sediment transport; iv) additional hydraulics variables, such as, temperature, pressure, and bed shear stress; and, v) water discharge. In addition, Chapters 2 and 8 describe general visualization methods and the use of autonomous, underwater vehicles, respectively. The instruments and measurement techniques are described irrespective of the measurement situation (*i.e.*, laboratory or field), because many instruments (or families of instruments) are used in field and laboratory situations. Typically, only slight, if any, modification is needed to apply an instrument or technique to each situation.

The consistency of terminology and notation is an important feature of this book. Common notations include symbols for variable types (vectors, tensors, or complex quantities), dimensionless numbers, operators, Greek symbols, subscripts and superscripts for variables. However, due to the wide range of instruments and multitude of parameters, it was not possible to use identical notation throughout all chapters. The book uses Standard International (SI) units. Numbering of the figures, tables, equations, and examples includes the number of the section in which they are found. For example, the first figure of Section 3.2 of Chapter 3 is identified as Figure 3.2.1. To improve accessibility, references and appendices immediately follow the chapters in which they are mentioned.

1.4 STRUCTURE OF VOLUME II

The two volumes of this book comprise a total of sixteen chapters. The first eight chapters are in Volume I and deal with the theoretical foundation of hydraulics and experimental methods, including similitude analysis, an important tool for the design of experiments. The eight chapters forming Volume II describe flow visualization, measurement instrumentation, and experimental techniques that lead to measurements. The chapters in Volume II group instruments based on the type of variable measured. While structured in volumes, experimental methods and instrumentation are integral parts of hydraulic experiments, and so there are many cross-references between the volumes.

As apparent from the present volume, hydraulic experiments involve many instruments and make use of many measurement principles. Therefore, many experienced experimenters have contributed to this volume. The contributions of each author are listed in a table at the volume's outset, wherein the authorship for individual volumes, chapters, and sections is specified.

Chapter 2 *Flow Visualization* describes contemporary flow-visualization principles and methods used to illuminate flow patterns in transparent fluids. Most advanced methods use sophisticated imaging devices and different materials and techniques for illumination. The resulting visualization information is frequently interpreted in terms of fundamental concepts of fluid flow. The topics covered in this chapter range from principles of visualization to descriptions of common visualization methods. The chapter includes imaging hardware, and discusses interpretation of flow visualization images and data.

Chapter 3 *Velocimetry* is an essential, core chapter. Flow velocity measurement usually is of major importance in all studies of flow processes. This chapter describes different instruments and approaches used in velocimetry. The instruments utilize sophisticated measurement principles, which are amply explained.

Sections 3.2 to 3.4 present velocity instruments based on acoustic principles. Acoustic Backscattering Instruments (ABIs) include, for example, acoustic Doppler velocimeters (ADV) and acoustic Doppler current profilers (ADCP), which are widely used today in both the laboratory and in the field. Section 3.4 presents principles of acoustic travel-time tomography; although at present this method is rarely used in freshwater environments, it can be advantageous for measuring discharge in shallow flows. Section 3.5 reviews point velocimeters, with an emphasis on field applications. Although point velocimeters can also be used in laboratory environments, this emphasis reflects the common use of point velocimeters for fieldwork.

Sections 3.6 and 3.7 present laser- and image-based velocimetry methods. Section 3.6 explains the principles of Laser Doppler Velocimetry (LDV) and covers many practical issues. Image-based velocimetry methods, such as Particle Image Velocimetry (PIV) and Particle Tracking Velocimetry (PTV), are described in Section 3.7, which includes detailed information on the application of image-based velocimetry to large-scale particle image velocimetry (LSPIV); drones make this latter technique increasingly useful. Sections 3.8 to 3.10 respectively discuss velocity measurements with high frequency radars, drifters and drogues, and dilution methods.

Chapter 4 *Topography and Bathymetry* summarizes instruments and methods used to obtain information on topography, bathymetry, and grain-sizes across various spatial and temporal scales. Such information is required for generating digital elevation models (DEM) and for determining roughness coefficients and morphology associated with river reaches. The methods described in this section are terrestrial laser scanning, ultrasonic sensing, photogrammetry, additional profiling methods, and methods for determining grain-size distributions of sediment.

Chapter 5 *Sediment Transport* covers instruments and methods for measuring sediment transport. Sediment erosion and deposition may lead to channel changes which in turn may modify the flow field. Moreover, flows carrying sediments are often dynamic; *i.e.*, unsteady, and flow acceleration and deceleration during the unsteady phase may initiate or terminate sediment suspension, generate or modify bed forms, and change bed and channel topography. This chapter describes long-standing and novel instruments and techniques for measuring bedload and suspended load transport rates. These include physical samplers as well as instruments based on acoustics (*e.g.*, ADCPs) or optics (*e.g.*, optical backscatter turbidity meters). Acoustic methods in particular have evolved for use in dynamic conditions. They can be used as a measurement triad, measuring the following quantities at the same location and at the same time with a resolution of turbulence scales: flow velocity, shear, turbulence; sediment transport (sediment concentration, erosion and deposition fluxes, sediment size); and, bed parameters (bed constitution, position, shape).

Chapter 6 *Auxiliary Hydraulic Variables* describes instruments and methods for determining additional hydraulic variables. Auxiliary variables considered are water depth, water surface elevation, pressure, bed shear stress, drag forces, and the physical properties of water. Many instruments and techniques exist for measurement of these variables. Therefore, detailed descriptions of all of them are beyond the scope of this chapter, but the reader is guided to pertinent information in other sources.

Chapter 7 *Discharge* addresses water-discharge estimation, and describes measurement methods rather than instruments. Additionally, the chapter only briefly reviews the well-documented methods for discrete and continuous discharge measurement in laboratory and field situations. Instead, this chapter focuses on contemporary methods for discharge monitoring in field conditions using new sensing technology (acoustic- or image-based) and combinations of measurement and numerical approaches.

Chapter 8 *Autonomous Underwater Vehicles* outlines the promising use of autonomous underwater vehicles (AUV). These devices are basically underwater robots that serve as platforms for experimental hydraulics in sophisticated or extreme conditions (*e.g.*, under ice in the Arctic). This chapter gives glimpses into the future and shows prospective innovative approaches for hydraulic experiments.

REFERENCES

Ettema, R., Arndt, R., Roberts, P., & Wahl, T. (2000) Hydraulic modeling: Concepts and practice. *ASCE manuals and reports on engineering practice*. 97. Reston, ASCE.

Frostick, L.E., McLelland, S.J., & Mercer, T.G. (eds.) (2011) *Users Guide to Physical Modelling and Experimentation: Experience of the HYDRALAB Network*. IAHR Design Manual. Boca Raton, Taylor & Francis.

Frostick, L.E., Thomas, R.E., Johnson, M.F., Rice, S.P., & McLelland, S.J. (eds.) (2014) *Users Guide to Ecohydraulic Modelling and Experimentation: Experience of the Ecohydraulic*

Research Team (PISCES) of the HYDRALAB Network. IAHR Design Manual. Leiden, CRC Press/Balkema.

Novak, P., Guinot, V., Jeffrey, A., & Reeve, D.E. (2010) *Hydraulic Modelling - An Introduction*. London, Spon Press.

Rouse, H. & Ince, S. (1957) *History of Hydraulics*. Iowa City, The University of Iowa.

Rowinski, P. (2007) Environmental Hydraulics. *Acta Geophys*, 55(1), 1–2.

Rubin, H. & Atkinson, J. (2001) *Environmental Fluid Mechanics*. New York, Marcel Dekker Inc.

Singh, V.P. & Hager, W.H. (eds.) (1996) *Environmental Hydraulics*. Dordrecht, Kluwer Academic Publishers.

Sukhodolov, A.N. (2012) Structure of turbulent flow in a meander bend of a lowland river. *Water Resour Res*, 48, W01516, doi:10.1029/2011WR010765.

Sukhodolov, A.N. (2015) Field-based research in fluvial hydraulics: Potential, paradigms and challenges. *J Hydraul Res*, 53(1), 1–19.

Venditti, J.G., Rennie, C.D., Bomhof, J., Bradley, R.W., Little, M., & Church, M. (2014) Flow in a bedrock canyon. *Nature*, 513, 534–537. doi:10.1038/nature13779.

Williams, R.D., Rennie, C.D., Brasington, J., Hicks, M., & Vericat, D. (2015) Linking the spatial distribution of bedload transport to morphological change during high-flow events in a shallow braided river, *J Geophys Res Earth Surf*, 120, 604–622. doi:10.1002/2014JF003346.

Chapter 2

Flow Visualization

2.1 INTRODUCTION

The study of fluid flow is inherently visual. Its component processes of flow conveyance, mass transport, and momentum transport leading to bed shear stresses and hydrodynamic loads on structures in complex flow fields are best understood within the context provided by a depiction of the organized features in the flow. This description can serve as a basis for planning and interpreting the physics underlying more detailed, quantitative measurements and can constitute a substantial contribution in itself. Direct experimental visualization of flow, which is the focus of this chapter, can be an effective means for generating this picture.

This chapter is an in-depth review of contemporary flow-visualization principles and methods and time-tested methods that continue to be very useful. Contemporary visualization practice makes use of sophisticated imaging devices and a variety of techniques and materials for illuminating flow features. The resulting visualization information is frequently interpreted in terms of fundamental concepts of fluid flow. Topics covered include fundamental principles governing the behavior of flow visualization tracers and imaging hardware, common visualization methods, and the interpretation of flow visualization data.

In the early 16th century, Leonardo da Vinci made his famous sketch of a water jet issuing into a pool (Figure 2.1.1a), likely the first documented application of flow visualization to the study of turbulence. From observation of the flow, he was able to deduce the roles of large and small eddying motions on the transport of suspended debris, and he began to establish qualitative rules for the motion of water. Similarly, in his famous experiment, Osborne Reynolds (Reynolds, 1883) used dye visualization to reveal the motion of fluid flow in a pipe, and demonstrated transition to turbulence, as shown in Figure 2.1.1b. These examples suggest that the power of direct visualization of the flow to elucidate a basic understanding of the fluid motion rivals even modern quantitative flow measurement technologies. In fact, the coherent nature of turbulent shear flows had gone unnoticed for decades while high-frequency-bandwidth point velocity measurements were the primary means of flow interrogation, until the revealing visualizations of organized mixing layer structure were presented by Brown and Roshko (1974). While it is often desirable to use flow visualization simply to acquire a preliminary understanding of flow patterns, deeper insights about a flow can be obtained if one has sufficient creativity and patience (Sigurdson, 2003), and many results have been archived for their didactic and pleasing artistic qualities (Van Dyke, 1982; Samimy *et al.*, 2004).

(a)   (b)

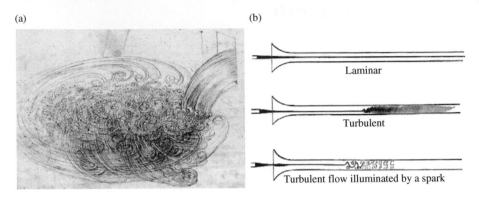

Figure 2.1.1 (a) Leonardo da Vinci's sketch of "Water falling upon water", c. 1508–09; (b) Osborne Reynolds' illustrations of dye visualization in a pipe (adapted from Reynolds, 1883).

Given the importance of visualizing the spatial structure of fluid flow, fluid mechanicians and hydraulicians have developed and refined many approaches to flow visualization and several reviews and texts have described these methods (Clayton & Massey, 1967; Werlé, 1973; Merzkirch, 1987; Gad-el-Hak, 1988; Yang, 1989; Freymuth, 1993; Crimaldi, 2008; Smits & Lim, 2012). This chapter presents fundamental principles behind the application and interpretation of flow visualization, and introduces several advanced techniques used in water, while providing citations for more in-depth study.

2.2 FUNDAMENTALS

Flow visualization by the methods presented in this chapter usually involves selectively marking water with a passive tracer, illuminating the tracer, imaging the light scattered by the tracer, and interpreting the resulting patterns. Fundamental concepts associated with the behavior of tracers, descriptions of their patterns, and the application of imaging hardware are discussed in this section.

2.2.1 The physics of flow tracking with tracers

Tracers added to the flow for water flow visualization consist of either suspended particles or bubbles, or aqueous dye solutions. Ideally, the tracer will convect at the local flow velocity without influencing it. Such a *passive* tracer should be neutrally buoyant such that buoyancy and inertial effects are minimized and, in the case of dyes, ideally should diffuse in such a way as to track the quantity of interest (*e.g.*, momentum, vorticity, or a scalar quantity for which the tracer is a surrogate).

In order for discrete suspended tracers to move with the local water velocity, the response of the particle to gravitational or flow acceleration must be such that the particle velocity does not deviate significantly from that of the surrounding flow. For particles of microscopic dimensions, Stokes' drag law typically applies, yielding the steady-state relative velocity U_r of the particle with respect to the local fluid velocity under the action of a constant imposed acceleration $\vec{a}$ (Raffel *et al.*, 2007):

$$\vec{U}_r = \frac{d_p^2\left(\rho_p - \rho_f\right)}{18\mu_f}\vec{a} \tag{2.2.1}$$

where d_p is the particle diameter, and ρ and μ represent density and dynamic viscosity, respectively, with subscripts p and f indicating particle and fluid properties, respectively. The relaxation time for the first-order response of a particle subjected to a step change in local flow velocity is $\tau = d_p^2\rho_p/(18\mu)$. This equation assumes the particle density to be much larger than the fluid density, and is generally conservative in water. The ratio of this time scale to a representative time scale of the flow (*e.g.*, associated with a relevant flow velocity spectrum) is known as the Stokes number $S_k = \tau/\tau_f$. A practical recommendation is that particles be selected such that $S_k \leq 0.1$ for good tracking of the flow (McKeon *et al.*, 2007).

The use of dye as a tracer requires careful consideration of its transport in water, in relation to flow quantities that are tracked, particularly if the intention is to characterize the vorticity field. As discussed by Lim (2012), dyes (passive scalars) and vorticity obey different transport equations with important distinctions. The vorticity transport equation for a fluid with constant properties is given by:

$$\frac{\partial\vec{\omega}}{\partial t} + (\vec{u}\cdot\nabla)\vec{\omega} = (\omega\cdot\nabla)\vec{u} + v\nabla^2\vec{\omega} \tag{2.2.2}$$

whereas the transport of a passive scalar is given by:

$$\frac{\partial C}{\partial t} + (\vec{u}\cdot\nabla)C = D\nabla^2 C \tag{2.2.3}$$

The left sides of Equations (2.2.2) and (2.2.3) contain the material derivatives of vorticity and scalar concentration, respectively. Two important differences are apparent.

First, the transport of vorticity (momentum) and the dye are governed by different transport properties: kinematic viscosity (v) and mass diffusivity (D), respectively. Their ratio is given by the Schmidt number $Sc = v/D$. $Sc = O(10^3)$ for dye in water, so that the vorticity diffuses much more rapidly than the dye. Thus, when dye is initially injected into rotational regions of a laminar flow field (such as a boundary layer), the observation of high concentrations of dye further downstream (often in a separated shear layer or wake) does not necessarily imply large vorticity values. It is common in free shear flows to observe intricate dye patterns with sharp interfaces, which may elicit inferences of a particular type of motion (*e.g.*, a spiral pattern may suggest rotation). However, in reality, the vorticity may have diffused away and that particular region of the flow may be relatively inactive. Thus, caution must be exercised when interpreting dye patterns in still images. However, as pointed out by Sigurdson (1997), if the dye is introduced into regions of concentrated vorticity, it will continue to track the regions of highest vorticity, even if the vorticity magnitude has significantly diminished at some later time. In turbulent flows, where the macroscopic motion of eddies becomes more important to large-scale transport than the molecular diffusion mechanisms parameterized by v and D, the *turbulent* Schmidt number (based on corresponding turbulent diffusive mechanisms) is usually close to one so that distinctions between the diffusion of dye and vorticity are much smaller and often unimportant.

On the other hand, stretching within the flow field can skew the relationship between vorticity and dye concentration in the opposite direction, and this phenomenon governs the second difference between Equations (2.2.2) and (2.2.3). The first term on the right side of Equation (2.2.2) describes vorticity tilting and stretching, which provides a mechanism for altering the vorticity of a fluid element that is not present in the dye transport equation (Equation 2.2.3). Consequently, stretching of vortex lines intensifies the vorticity field but not the dye concentration. Similarly, given the vector nature of the vorticity field, it is possible for regions of vorticity with opposite rotation to merge and become mutually annihilated, whereas the tracer does not exhibit this behavior (Sigurdson, 1997).

2.2.2 Flow lines

Closely tied to the visual nature of our understanding of water flows are various lines defined in the flows by the motion of the water or flow tracers. Four such lines are defined as follows (*e.g.*, White, 2011):

- *streamline*: a line that is tangent everywhere to the local velocity vector.
- *streakline*: a line that is defined by the locus of all fluid *particles* passing through a given point in the flow field.
- *pathline*: the trajectory traversed by an individual fluid particle over a finite time interval.
- *timeline*: a curve formed by the locus of fluid particles at an instant of time that originally defined a straight line (usually perpendicular to the primary direction of flow).

If computed or quantitatively-measured flow fields are available, calculation of streamlines can be achieved by integration of the velocity field. However, a single instantaneous image of a flow tracer pattern does not directly provide velocity data, and therefore is not useful for determining streamlines unless the flow field is steady, in which case streamlines, streaklines, and pathlines are coincident. If a passive continuous tracer, such as dye, is released from a localized dye injection port (a common means of introducing dye into the flow) so as to form a thin filament of dye within the flow, then, by definition, the filament approximates a streakline since all of the dyed fluid elements have passed through the same point in the flow. An example is shown in Figure 2.2.1a, where dye is injected from small tubes placed on a dune crest (Kadota & Nezu, 1999). Alternatively, if a discrete tracer particle is exposed on a single image for a sufficiently long duration in order to trace out its trajectory in the image, that trajectory is, by definition, a pathline. The arc length of the pathline equals the product of its average velocity and the exposure time. Figure 2.2.1b contains short pathlines created by particles suspended in an unsteady flow of water over a circular cylinder.

Timelines are created by releasing a line of continuous or discrete tracers, usually oriented perpendicular to the primary flow direction. For example, timelines, such as those shown in Figure 2.2.1c are generated using the hydrogen bubble method described in Section 2.3.2.2. If the lines are created at known, constant time intervals, then the initial distance between them is proportional to the local flow velocity if the flow is steady and uniform. Variations in their separation further downstream can be used to interpret the velocity field. Thus, streaklines, pathlines, and timelines are intrinsic outcomes of many flow visualization techniques.

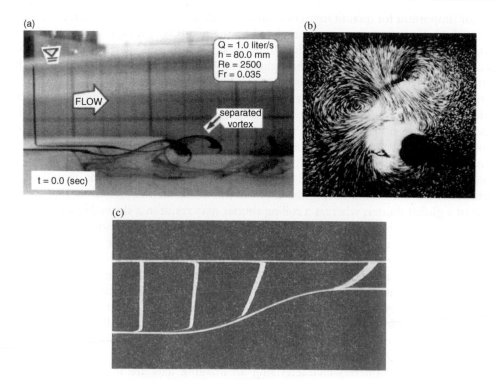

Figure 2.2.1 Observational lines delineating flow features: a) streaklines formed by injection of a dye filament into the flow over a dune crest (Kadota & Nezu, 1999 copyright International Association for Hydro-Environment Engineering and Research (IAHR), reprinted by permission of Taylor & Francis Ltd, http://www.tandfonline.com on behalf of IAHR.); b) pathlines generated from a time-exposure of particle motion in an oscillatory flow around a cylinder (Williamson, 1985); and c) timelines created using hydrogen bubble visualization in a contraction republished with permission of American Society of Mechanical Engineers ASME from Schraub et al., 1995; permission conveyed through Copyright Clearance Center, Inc.

2.2.3 Optical considerations and imaging hardware

Selection of appropriate imaging and illumination hardware and its proper use can have a profound impact on the quality of flow visualization data, its usefulness for interpreting the flow field, and its esthetic qualities. In cases where quantitative measurements are needed (such as in concentration measurements by laser-induced fluorescence), appropriate hardware selection is even more important. The two most important and expensive components of the camera are the imaging sensor (including any onboard image processing hardware) and the lens. Electronic imaging sensor technology has advanced significantly in recent decades and will continue to grow, driven by multiple areas of application. For many flow visualization applications, relatively inexpensive consumer-grade cameras can now be used.

The two types of image sensors typically found in scientific or consumer-grade cameras used for flow visualization are charge coupled devices (CCD) and complementary metal oxide semiconductors (CMOS). Recently, a hybrid technology, called scientific CMOS (sCMOS) has emerged that combines advantages of both technologies. Historically, the strengths of CCD cameras have been their high and uniform sensitivity across the entire

sensor (important for quantitative concentration measurements), since signal conditioning and conversion of charges accumulated on the pixels (resulting from photon interactions) into voltages is accomplished by a single unit. In general, pixel-to-pixel
uniformity is therefore high, but throughput is low due to the serial processing of the
data. Alternatively, on CMOS sensors, signal processing and conversion hardware is
integrated into each pixel, resulting in a massively parallel system yielding high frame
rates, at the expense of reduced pixel area – and thus reduced light sensitivity – due to the
added hardware on each pixel. It should also be noted that, while all CCD sensors
employ a *global shutter*, in which all pixels are exposed simultaneously, some CMOS
sensors use a *rolling shutter*, in which individual pixels are sampled sequentially, and
thus, different pixels are sampled at different times during the frame exposure. If
detectable motion occurs during the exposure, it would be manifested as blur in the
case of a global shutter; whereas a rolling shutter may result in a skewed image.

Other key attributes to consider when selecting an imaging sensor are the image
resolution, frame rate, sensitivity, dynamic range, and whether the sensor is monochromatic or color. Sensitivity is related to the ability of the pixel to absorb incident
photons. It is dependent on the size of the pixel as well as the spectrum of the incident
light. Sensor dynamic range (dB) is a measure of the ratio of the maximum pixel signal
(intensity) to the level of the noise floor. The maximum signal is governed by the pixel's
capacity to store electrons excited by incident photons, and the noise floor is affected by
the stochastic nature of photon incidence, thermal effects, amplification of the analog
signal, and analog to digital conversion.

The primary parameters characterizing an imaging lens are its focal length and
maximum aperture. Whereas typical imaging lenses are complex devices constructed
from multiple glass elements, to first order, their function can be elucidated using thin
lens theory. For an ideal, thin, parabolic lens, the thin lens formula illustrates the
relationship between lens focal length f, object distance o (the distance between the
lens and the object being imaged) and the image distance i (the distance between the lens
and the image sensor for a focused image):

$$\frac{1}{o} + \frac{1}{i} = \frac{1}{f} \tag{2.2.4}$$

The linear magnification of the object (the ratio of the object size to that of the image) is
$M = i/o$. Lens aperture governs the amount of light reaching the sensor. Lenses are
characterized by the f-number (N, often written as $f/\#$, where # is the f-number), which
is the ratio $N=f/D$, where D is the aperture diameter. Lenses capable of small f-numbers
therefore typically[1] transmit a large amount of light to the image sensor, permitting
lower illumination intensities, shorter exposure times, and lower image noise; however,
they tend to be more expensive due to the larger lens diameter required and, when
operated at low f-number, have limited depth of field (the distance between the nearest
and furthest objects or features that appear in focus in the image). Just as in conventional photography, the f-number is an important parameter used to control what will
and what will not be resolved within the field of view. Whereas it may be important to
resolve the full three-dimensional structure of a flow feature (thus requiring a large
depth of field, corresponding to a large f-number), it can also be helpful to deliberately
defocus background artifacts which are not relevant to the flow field by using a small f-
number and thus also a small depth of field.

If consumer-grade video or still cameras are used for flow visualization, it is important to use a camera that allows manual control of focus, lens aperture, and internal image processing (*e.g.*, exposure, white balance, etc.). Flow visualization images generally exhibit very different qualities from the typical types of photography subjects, which involve solid objects. The automatic features on consumer cameras tend to become confused, perpetually adjusting the optics and image processing, and resulting in out-of-focus images, incorrectly-exposed features of interest, or time variations in focus and image intensity. Moreover, digital consumer cameras employ various algorithms for image compression and other proprietary image processing algorithms that might alter image features during the recording process. Careful evaluation of a consumer imaging device's characteristics should be made before using it for scientific studies.

Illumination is typically achieved using either a diffuse, broadband source such as incandescent, fluorescent or LED lighting, or continuous-wave or pulsed lasers. The former are generally much less expensive and are suitable for volumetric illumination; whereas the latter can support calibrated intensity measurements, and can be focused into thin sheets or swept through an imaging plane to extract planar slices of the flow (see Section 3.7.2.2.2, Volume II). The simplest way to generate a laser sheet is to pass it through a cylindrical lens, with beam thickness often controlled using a long-focal-length spherical lens. For some applications – especially for preliminary exploration of a flow field, a glass chemistry stirring rod can provide a convenient and very inexpensive means to generate a light sheet for lower-powered lasers. Alternatively, a Powell lens may be used as it produces a more spatially-uniform light sheet, or the beam may be swept through the field of view using a rotating mirror, or multiple rotating mirrors such as the device developed by Delo *et al.* (2004).

Lasers employed in flow visualization are either pulsed (*e.g.*, Nd:YAG or Nd:YLF, with pulse durations usually on the order of nanoseconds) or continuous wave (*e.g.*, argon ion or diode-pumped solid-state lasers), in which the camera shutter regulates image exposure time. Willert *et al.* (2010) have also proposed a circuit in which a pulsed light sheet can be inexpensively generated using high powered LEDs with light sheet optics. Lasers frequently used in flow visualization and other flow measurements are summarized in Appendix 2.A.

2.3 FLOW VISUALIZATION TECHNIQUES

In this section, we discuss techniques employed to visualize water flows. The techniques are summarized in Table 2.3.1, with references. Additional flow visualization techniques – some for use in air – are included in Table 2.3.1, for which the interested reader is advised to consult the listed references.

2.3.1 Flow visualization with aqueous tracers

The use of dyes to identify flow features is, in many cases, one of the least expensive and easiest visualization methods to implement. Three-dimensional flow patterns can often be revealed with relatively little effort, where quantitative point measurement techniques are generally unable to elucidate the three-dimensional structure, and quantitative planar or volumetric velocimetry methods are much more costly and time-intensive to

Table. 2.3.1 Summary of flow visualization techniques.

Technique	Fluid/Domain	Reference	Notes
Aqueous tracer injection	Liquids Lab/Field	Section 2.3.1.2	
Smoke injection	Gases Lab/Field	Merzkirch (1987)	
Laser-induced fluorescence	Liquids/Gases Lab	Section 2.3.1.3	Applications for quantitative concentration measurements in water are discussed. Also useful for qualitative identification of flow structure.
Particle tracing	Liquids/Gases Lab/Field	Section 2.3.2.1	
Hydrogen bubble visualization	Water Lab	Section 2.3.2.2	Low-speed flows
Tufts	Liquids/Gases Lab/Field	Section 2.4	
Oil film	Liquids/Gases Lab	Section 2.4	Useful for identifying time-averaged flow separation and attachment, and the topological structure of the flow.

implement. Dye visualization can be used in preliminary experiments to identify target areas for measurements with more laborious quantitative measurement techniques or, with great care, can be used to elucidate subtle details of the flow field.

2.3.1.1 Dyes and other aqueous flow tracers

For qualitative characterization of the dynamics and structure of a hydraulics flow field, the two requirements for a flow tracer are that a) it is easily visible, and b) it reasonably faithfully tracks the motion of water, notwithstanding the caveats discussed in Section 2.2.1. Many options can be considered, including laboratory chemicals and household liquids. Lim (2012) provides a practical review of the options.

Fluorescent dyes are frequently used to characterize either the motion of a flow or as a surrogate for concentration measurements, as described in Section 2.3.1.3. The most common of these dyes are fluorescein, rhodamine B, and rhodamine 6G. As summarized by Crimaldi (2008), fluorescein absorbs incident radiation in a band centered on 490 nm (blue-green; similar to the 488 nm emission line of an argon ion laser), and emits in a band around 515 nm (green). Rhodamine B has a peak absorption wavelength of 555 nm (yellow green) and peak emission about 580 nm (orange). However, as noted by Lim (2012), Rhodamine B exhibits a dark red color when illuminated with an argon ion laser. If a long-pass optical filter, with a cutoff wavelength between the excitation wavelength and the peak of the fluorescence band, is applied to the camera when imaging fluorescent dyes with narrow-bandwidth laser excitation, the excitation wavelength can be removed to enhance contrast. This is necessary for making quantitative concentration measurements as is discussed in Section 2.3.1.3, but the enhanced image quality is also useful in qualitative applications, especially when the water contains significant suspended material. Broadband illumination using white light sources can also be quite effective with fluorescent dyes as long as illumination is appropriately introduced such as to maintain adequate contrast between the fluorescing dye and the image background, as discussed below. The most frequently-used non-fluorescent tracer is food coloring (which can be obtained in aqueous or powdered form).

It is desirable that the tracer, to the extent possible, match the density of the water into which it is introduced so that it illuminates or tags the flow feature of interest. This is especially true at low flow velocities, since the inertial forces in the flow become very small, and the buoyancy forces become more important, causing the dye to rise or sink more rapidly with respect to the fluid motion. If the dye is denser than water (usually the case for aqueous dyes) its density can be reduced by adding alcohol to the dye. Temperature-related density differences can be minimized by submerging sealed canisters containing the dye within the working fluid until thermal equilibrium is achieved.

2.3.1.2 Aqueous tracer injection and imaging

Since flow visualization data are acquired through line of sight, it is necessary to introduce flow tracers judiciously in order to maintain visual contrast, and to avoid obscuring tagged flow structures in complex, three-dimensional flows.

Dye is usually introduced into the flow either by direct injection through a probe or rake placed in the flow as shown in Figure 2.3.1a, through the slow release of dye painted onto the surface of a model as in Figure 2.3.1b, or through dye ports built into the surface of a model as in Figure 2.3.1c. The use of one or more probes for dye injection can provide great versatility since it allows the probe to be easily moved within the flow to examine flow patterns in different regions, or to optimize the position of the probe in order to examine a particular flow feature. However, direct injection inevitably results in a non-zero momentum flux from the probe. As demonstrated by Lim (2012), when dye is injected parallel to the flow, it is necessary to match the dye velocity to the flow velocity to minimize instabilities that result from shear flow, which may appear as small vortices in the streaklines.

Contrast between the dye and background can be achieved through careful arrangement of lighting and preparation of background surfaces, as illustrated in Figure 2.3.2. In laboratory flumes, solid-color black or white, adhesive-backed shelf liner vinyl provides an ideal background for high contrast in flow visualization images. Black should be used for fluorescent dyes or highly reflective tracers (such as milk), and white provides the best contrast for darker colored dyes. Shelf liner vinyl should be placed on the inside surface of glass-walled flumes to avoid compromising the background intensity due to reflections from the front side of the glass. When using diffuse broadband illumination, care must be exercised to avoid illuminating the background (for dark backgrounds) or the window through which the flow is being imaged, to reduce glare and maximize image contrast. This can be accomplished using shades to control the illuminated region. Ideally the camera should be shrouded since its reflection in the viewing window may be visible in the image. Figure 2.3.2 shows an example configuration for good image quality.

Figure 2.3.3 illustrates the use of dye visualization in a laboratory model with thermally-stratified flow (Ettema et al., 2005). Red dye is injected into the warmer upper layers, and blue dye is injected into the cooler lower layers. Without protection, the intake withdraws predominantly warm (red) surface water (Figure 2.3.3a);

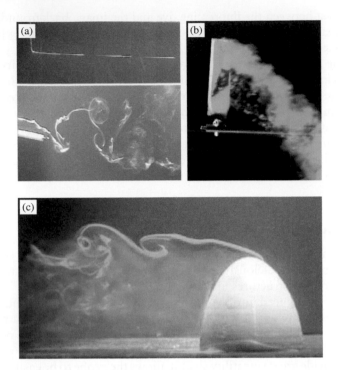

Figure 2.3.1 Examples of flow visualization using dye: a) (top) a dye probe fabricated from a bent hypodermic tube periodically injecting fluorescein to produce streaklines in water, (bottom) a dye probe injecting milk near the trailing edge of an oscillating fin to visualize the wake; b) Visualization using a mixture of fluorescein dye and water-soluble white glue painted onto the leading edge (left edge) of a rotating blade (image credit: Kevin Wabick), and c) a fluorescein dye filament introduced from a port built into a model, illustrating growth of the shear layer instability and entrainment into the recirculation region (image credit: Craig Wojcik, Seyed Hajimirzaie). All images are visualized with diffuse, broadband (white) illumination.

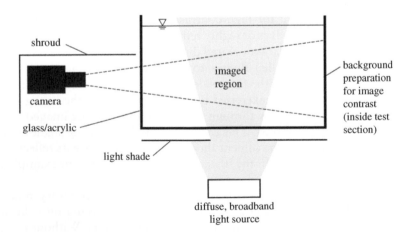

Figure 2.3.2 An example configuration for generating quality dye visualization images with white light. The cross section of the flow facility is depicted.

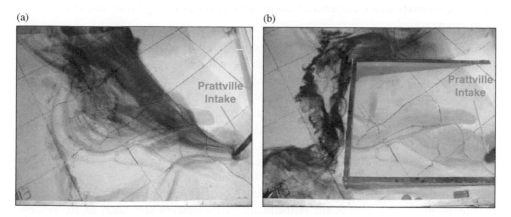

Figure 2.3.3 Visualization of the flow toward an intake located in a thermally stratified flow: a) the intake without protection (in its natural state); The intake withdraws predominantly warm (red) surface water. b) the intake conditioned with a bottom curtain. Blue dye is released into the colder bottom layer and red dye into the warmer surface layer. (Ettema *et al.*, 2004)

however, the installation of a bottom curtain results in the withdrawal of the cooler bottom layer (Figure 2.3.3b).

In addition to resolving details of the flow structure, dye visualization can be used to obtain bulk flow measurements. For example, Ghilardi *et al.* (2014) imaged potassium permanganate (dark violet in color) injected uniformly across an open channel flume to measure mean flow velocity through an array of boulders which inhibited access by other velocity measurement instrumentation. Aqueous tracers also have often been applied to understand flow structure in natural fluvial environments. For example, Roy *et al.* (1999) employed flow visualization to correlate flow structures with velocity time series measured using an electromagnetic current meter (ECM). Differences in turbidity were used to characterize water motion at a river confluence, and milk was used to visualize flow structures generated by a pebble cluster as shown in Figure 2.3.4. Similarly, Constantinescu *et al.* (2009) used localized dye injection in field experiments within a groyne field to validate large eddy simulations of mixing layer structure.

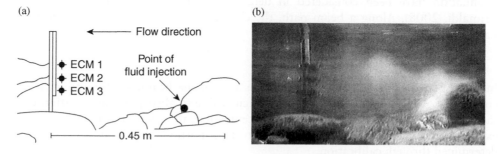

Figure 2.3.4 Flow visualization in the wake of a cluster (adapted from Figures 4 and 5 of Roy *et al.*, 1999). a) schematic showing the bed topography in the imaging region and the point of injection of milk (tracer); b) an eddy shedding from the top of the obstacle.

2.3.1.3 Quantitative concentration measurements using laser-induced fluorescence

With careful calibration and experimental control, image intensity can be used to infer concentrations of fluorescent dyes introduced into the flow. A detailed account of the underlying theory, methods, and hardware, along with a summary of applications can be found in Crimaldi (2008), and the reader is referred to that work and the included references for in-depth discussions of the technique and associated hardware. In this section, a brief overview of the central concepts is presented, along with examples of its application.

Laser induced fluorescence measurements may be performed for points, lines, planes, and volumes within the flow; however, *planar* laser induced fluorescence (PLIF) is most frequently used. The imaging configuration for quantitative PLIF is the same as that for qualitative flow visualization using laser sheets for illumination; however, accurate measurements require significant refinements. A fluorescent dye such as fluorescein, rhodamine B or rhodamine 6G is used as the tracer. Most PLIF experiments use argon ion lasers as the illumination source due to their long-term stability and superior beam characteristics (Crimaldi, 2008). The 488 nm argon ion emission is also very close to the peak in the absorption band of fluorescein. Pulsed lasers (*e.g.*, Nd:YAG) are also often used but usually must be operated at high intensities in order to produce sufficient energy, due to their very short pulse durations. This can cause nonlinear saturation effects in the dye, as described below, and pulse-to-pulse intensity variations may also be important, especially when the lasers are not operated near their design repetition rate.

The method of laser induced fluorescence is based on the principle that fluorescence intensity is dependent on excitation intensity and dye concentration (Crimaldi, 2008):

$$F \propto \frac{I_e}{1 + I_e/I_{sat}} C \approx I_e C \tag{2.3.1}$$

where F is the emission intensity, I_e is the excitation intensity, I_{sat} is the saturation intensity of the dye, and C is the dye concentration. The final, approximate, expression in Eq. 2.3.1 is valid when $I_e \ll I_{sat}$, and laser intensity is usually controlled such that the linear relationship is valid. However, use of Eq. 2.3.1 requires knowledge of the spatial distribution of the excitation intensity, which can vary due to several factors investigated in detail by Ferrier *et al.* (1993). These include light sheet expansion, defects or dust on the light sheet optics, and light attenuation through the fluid. The effects of light attenuation have been considered in detail (Walker, 1987; Ferrier *et al.*, 1993; Crimaldi, 2008). Along a beam path, the change in intensity is given by the Beer-Lambert law (Walker, 1987):

$$dI_e(z) = -\varepsilon C(z) I_e(z) dz \tag{2.3.2}$$

In Eq. 2.3.2, z is the coordinate along the axis of the beam, $C(z)$ is the local dye concentration, and ε is the extinction coefficient due to absorption and scattering of the incident beam. For a uniform dye concentration, the excitation intensity decays exponentially in z. In general, the constant of proportionality in Eq. 2.3.1 must be determined through calibration.

Laser-induced fluorescence measurements may be combined with simultaneous and co-located measurements of the flow velocity in order to compute fluxes of the dissolved scalar. For example, Crimaldi *et al.* (2007) measured the velocity within a small volume (point) using a laser Doppler velocimetry probe while obtaining dye-concentration measurements in the same volume. Since the fluorescence emission was at a different wavelength than the laser excitation, the light scattered by suspended particles could be isolated with a short-pass optical filter and used for velocity measurements while the fluorescence emission was long-pass filtered and converted to concentration using a photomultiplier tube. Likewise, laser-induced fluorescence may be combined with simultaneous particle image velocimetry measurements to obtain spatial distributions of scalar fluxes.

It is also possible to obtain quantitative measures of concentration without the use of fluorescence. For example, Uijttewaal *et al.* (2001) studied mass exchange between a model groyne field and river by imaging the time evolving, depth-averaged intensity of an initially uniform distribution of potassium permanganate.

2.3.2 Flow visualization with discrete particles

Imaging of discrete particles suspended in the flow can reveal fluid motion either through particle trajectories in a time-exposure image, the evolution of lines of particles, or their displacements between images (the basis of particle image velocimetry and particle tracking velocimetry). Near neutrally-buoyant (in water) particles can be made of glass (*e.g.*, hollow spheres), polymers, and other materials. Bubbles with diameters on the order of microns typically have small enough buoyancy-driven terminal velocities to make them useful as flow tracers. In natural environments, existing suspended material may also suffice. While the relaxation time described in Section 2.2.1 is of relevance to the shortest time scales of the flow, for uniformly-dispersed particles, the settling velocity (Eq. 2.2.1 under gravitational acceleration) introduces another time scale, based on the settling time of the particles in the experimental facility, which may limit the duration of the experiment.

2.3.2.1 Visualization of particle trajectories

The imaging and quantitative analysis of suspended tracer particle displacements is the basis of particle image velocimetry (PIV), which has been in widespread use for the past two decades (see also Section 3.7.2, Volume II). However, well-seeded individual PIV images generally provide no information about flow structure. Prior to the advent of PIV, qualitative (or even quantitative) visualization of the flow field was often achieved by long-exposure images of suspended particles to form short pathlines such as those imaged by Williamson (1985) in an oscillatory flow over a cylinder, shown in Figure 2.2.1b. A striking recent application of the technique can be found in the sneeze visualizations of Bourouiba *et al.* (2014) shown in Figure 2.3.5.

2.3.2.2 Hydrogen bubble visualization

Bubbles can be used as flow tracers as long as they are sufficiently small such that buoyancy effects do not result in bubble velocities deviating significantly from the local flow velocity. One of the most frequently-applied methods is the hydrogen bubble method, in which small (O(10) micron diameter) bubbles are generated on a thin wire using electrolysis. The underlying principle is relatively simple and the method is inexpensive to implement; however, properly tuning the system typically requires great care, as described in detail by Sabatino *et al.* (2012), who provide a detailed

Figure 2.3.5 Pathlines formed by time exposure of droplets ejected during a sneeze (Bourouiba *et al.*, 2014).

description of the application of the technique. Schraub *et al.* (1965) also describe the technique and provide numerous examples of its application.

To apply the hydrogen bubble method, water must have some dissolved electrolyte such that electrical conduction can occur at a rate needed to generate bubbles in sufficient concentration. A thin wire is stretched perpendicular to the primary flow direction, and a voltage is intermittently applied to the wire. The wire serves as the negative electrode in an electrolysis reaction in which hydrogen is formed on the wire, while oxygen is formed at another electrode placed away from the domain of interest; note that two molecules of hydrogen gas are produced for every one of oxygen, which is one of the primary motivations for making the wire the hydrogen-producing electrode. Short bursts of current create lines of bubbles along the wire that are convected away from the wire, forming a timeline in the flow. Sustained application of voltage forms a sheet of bubbles that deforms with the fluid motion. A thin layer of insulating material applied to portions of the wire results in bubble formation only on the exposed portions of the wire. This can be used to generate streaklines in the flow, analogous to the smoke wire technique used in wind tunnels. Figure 2.3.6 illustrates the application of a hydrogen bubble wire parallel to the surface of a flat plate to reveal the structure of a turbulent boundary layer (Schraub *et al.*, 1965). The figure shows that by controlling the spatial and temporal formation of bubbles, as described above, squares of bubbles are released from the wire. The squares of bubbles are distorted as they are advected through the flow, revealing shear and rotation in the flow.

To minimize the effects of the wire wake deficit, inhibit vortex shedding from the wire, and to produce bubbles sufficiently small such that their buoyancy-driven ascent velocity is small, the wire must be very thin (typically 50 micron diameter or less). This limits the volumetric rate at which bubbles can be generated, and thus the method is effective only in relatively low-speed flows. The thin wires tend to be very fragile; therefore, it can be challenging to apply sufficient tension to keep the wire taut without breaking it. Sabatino *et al.* (2012) recommend platinum wires for a good balance of bubble quality, wire strength, conductivity, solderability, and cost. Although other materials, such as stainless steel and tungsten, have been used, they tend to produce large bubbles, yielding poor image quality. Aluminum and steel can also be used and are less expensive, but they rapidly corrode in the presence of an electrolyte.

Be aware that the hydrogen bubble method is inherently dangerous since the flowing water forms part of the circuit, and several tens of DC volts are typically applied in

Figure 2.3.6 Eddy structure within a turbulent boundary layer revealed using hydrogen bubble visualization (republished with permission of American Society of Mechanical Engineers ASME from Schraub et al., 1995; permission conveyed through Copyright Clearance Center, Inc.).

order to obtain a sufficient bubble formation rate. Budwig and Peattie (1989) reduced the risk by setting the anode to ground potential. The signal from a function generator or digital pulse generator is usually used to control switching of the bubble wire circuit, and a high-power transistor circuit provides the necessary voltage and current amplification to drive the electrolysis reaction. A simple MOSFET-based amplifier circuit is given in Budwig and Peattie (1989).

2.4 VISUALIZATION OF FLOWS NEAR SOLID SURFACES

Flow close to solid surfaces, especially three-dimensional surfaces, can pose visualization challenges. Three-dimensional flows often are distinguished by lines of flow separation and attachment that delineate distinct regions of the flow and, along with critical points, are important for characterizing the flow structure, as elaborated in this section and further discussed in Section 2.5.2. Relatedly, three-dimensional flows typically involve turbulence structures that periodically develop, disperse and dissipate, thereby complicating flow visualization; *e.g.*, in the case of the highly turbulent flow field at a pier with a scour hole (Ettema *et al.*, 2006).

Since flow separation lines are associated with shear stress minima, mobile materials tend to accumulate in them, and therefore surface shear stress is a useful complement to flow measurement. In wind tunnel experiments, a very common method for visualizing this surface stress distribution is the oil film method in which a film of oil mixed with fine powder (to make the film opaque) is applied to the surface. The shear stress distribution on the surface redistributes the oil, creating visible patterns on the surface (see, for example, Merzkirch, 1987, Section. 2.4 or Section 6.5.2, Volume II). Although less common in water, similar approaches have also been taken by applying paint or oil

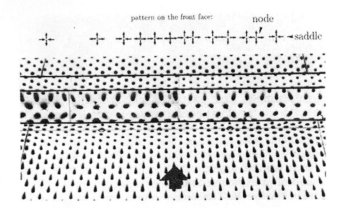

Figure 2.4.1 Crystal violet dye visualization of the flow approaching a two-dimensional ridge. The orientation of the surface shear stress is indicated by the elongation of the dye spots on the surface, which reveal a sequence of alternating nodes and saddles (See Section 2.5.2) on the upstream face of the ridge (republished with permission of American Society of Mechanical Engineers ASME from Martinuzzi & Tropea, 1993; permission conveyed through Copyright Clearance Center, Inc.).

to the surface (*e.g.*, investigation of surface flow patterns on a marine propeller by Bhattacharyya *et al.*, 2015). Martinuzzi and Tropea (1993) gained similar information on submerged surfaces in water by applying drops of crystal violet dye and recording their shear-stress-driven deformation, as shown in Figure 2.4.1.

An alternative for obtaining a qualitative assessment of the surface flow patterns is the use of tufts (sometimes called *tell-tales*) – highly flexible filaments mounted on the surface of the body which bend in the direction of the local flow. Simple tuft visualizations may be achieved by taping or gluing short pieces of thread or yarn to the surface (as is done in wind tunnels); however, Schostek (2012) has recently developed a method for fabricating and mounting polymer tufts for use in water, as shown in Figure 2.4.2.

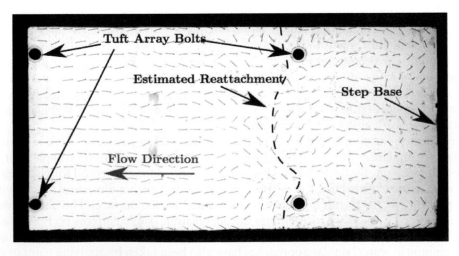

Figure 2.4.2 Surface flow patterns revealed by polymer tufts mounted downstream of a backward-facing step (Schostek, 2012).

Tufts may also be placed on wires or grids suspended in the bulk flow in order to visualize flow patterns away from surfaces.

2.5 UNDERSTANDING FLOW TOPOLOGY FROM FLOW VISUALIZATION DATA

Flow visualization provides an intuitive means to understand the spatial patterns and topology of flow motion and transport. While observation of flow patterns is often qualitative, it is possible for the qualitative picture to be augmented by theoretical considerations that can lead to a more fully three-dimensional description of the flow. This section discusses interpretation methods based on the solenoidal property of vorticity and on topological constraints or requirements for critical points in the velocity field at three-dimensional surfaces.

2.5.1 Vortex topology

Swirling flows identified in image sequences (or single images, subject to the caveats discussed in Section 2.2.1) signal the presence of vorticity. Since the velocity field can be reconstructed from the full three-dimensional vorticity field using the Biot-Savart law (*e.g.*, Saffman, 1992), vortex-dominated flows (that is, flows in which the vorticity tends to be concentrated into discrete vortex structures that govern its evolution) can usually be described effectively by a vortex *skeleton* model. Construction of such a model can be an invaluable tool for developing insight into transport and flow/structure interactions within a flow. Central to the vortex description of the flow are the concepts of a *vortex line*: a curve everywhere tangent to the local vorticity vector at an instant in time; and a *vortex tube*: a tube-like surface formed from the set of all vortex lines passing through a closed circuit at an instant in time (Marshall, 2001). If the primary vortical structures in the flow can be idealized as vortex tubes while still representing the salient features of the flow, a significant reduction of the complexity of the flow description is achieved (in comparison to the continuous velocity or vorticity field), which is often sufficient to identify physical mechanisms governing observed phenomena or to establish the basis for a quantitative model to describe transport within the flow.

The solenoidal property of the vorticity field ($\nabla \cdot \vec{\omega} \equiv 0$, where $\vec{\omega}$ is the vorticity vector), constrains a vortex tube to have constant strength along its length (where *strength* is quantified as the vortex circulation, defined as the closed-circuit integral of the tangential velocity component around the perimeter of the vortex tube: $\Gamma = \oint \vec{u} \cdot \vec{dl}$). This conclusion requires that, in bounded flow domains, vortex tubes either form closed loops in the flow, or begin and end on boundaries (Saffman, 1992), which provides a powerful constraint on the possibilities for the vortex topology of a flow, given a partial knowledge of the locations and orientations of vortices acquired through flow visualization.

Often, quantitative measurements of the vorticity distribution can significantly augment the understanding that a qualitative visualization of the flow can provide, and so

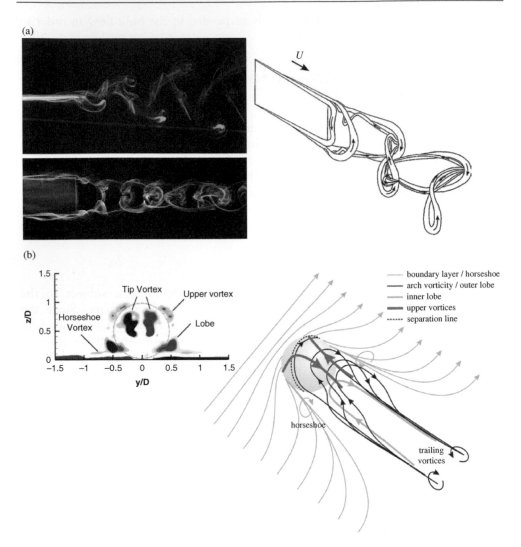

Figure 2.5.1 Vortex skeleton models derived from flow visualization data. a) Fluorescein and Rhodamine B dye in the wake of a low-aspect-ratio pitching panel, illuminated by a halogen light (reproduced from Buchholz & Smits, 2005, with the permission of AIP Publishing), and the corresponding vortex model (Buchholz & Smits, 2006); b) Sample distribution of the streamwise vorticity component, measured using PIV, in the near wake of a bed-mounted spherical obstacle, and corresponding vortex model. (Hajimirzaie et al., 2014, with permission of Springer).

this section adopts a broad definition of flow visualization to include any means of visualizing the flow field: qualitative or quantitative; experimental or numerical. Figure 2.5.1 illustrates two examples of vortex skeletons derived from experimental qualitative and quantitative flow visualizations. Figure 2.5.1a contains dye visualization of a low-Reynolds-number wake generated by a low-aspect-ratio panel pitching in a uniform flow (Buchholz & Smits, 2006). The wake is comprised of a *chain* of horse-shoe-shaped vortex loops that each ultimately connect back to the panel, on which they originate. This model illuminates the mechanisms by which thrust is produced, and establishes objectives for enhancing the propulsive efficiency.

It should be noted that the constituent elements of the vortex skeleton shown in Figure 2.5.1a are, strictly speaking, not vortex tubes. Their strengths are indicated, in a qualitative sense, by their drawn thicknesses, which vary along their lengths. This variation in strength occurs because the streamwise legs of the horseshoe structures are connected by a weak shear layer (the boundary layer shed from the trailing edge of the panel in between discrete vortex shedding events), which contains vortex lines that cross the boundaries of the drawn vortex "tubes," thus changing their strengths.

Figure 2.5.1b shows the flow structures created by water flowing over a deeply-submerged, spherical, bed-mounted obstacle (Hajimirzaie *et al.*, 2014). In this case, particle image velocimetry measurements were obtained in several planes adjacent to and downstream of the obstacle to construct the model. In this case, the model contains representative vortex lines, which are more fundamental flow constituents than vortex tubes, that enable more subtle details of the flow field to be described. For example, in the present case, they are used to represent transverse boundary layer vorticity on the bed which accumulates upstream of the obstacle and turns into the streamwise direction to form the horseshoe vortex structure that wraps around the base of the sphere. Since the vortices produce large velocity gradients, they can impose large localized bed shear stresses, and therefore knowledge of the vortex structure facilitates the interpretation of sediment deposition patterns around the obstacle, and informs models of bedload sediment transport in fluvial environments with boulders. The concept of a "vortex" is not well-defined due to the continuous variation of rotation and shear throughout the regions of the flow containing vorticity, and these kinds of vortex skeleton models are not rigorous, but rather *representative*, and provide a means for understanding the mechanisms governing flow evolution and the underlying transport processes. They provide a conceptual, tractable framework to guide further measurement or analysis of a flow.

2.5.2 Critical point theory

Another powerful tool for the characterization of flow structure is the identification of critical points on real or imaginary surfaces in the flow (Perry & Chong, 1987). While in fluid flows, the vector fields of interest are typically the velocity vector (with corresponding field lines given by streamlines) or surface shear stress, the principles apply to any vector field. A critical point, or singular point, is defined as a point at which the velocity magnitude is zero and the slope of the streamlines is indeterminate. Critical points can be classified according to the behavior of the streamlines as they approach the critical point, which is shown by Chong *et al.* (1990) to be governed by the discriminant of the velocity gradient tensor at the critical point. Critical points can be characterized as either nodes (with streamlines representative of a source, sink, vortex, or combination thereof), or saddles – points at which streamlines converge and diverge along different directions (Hunt *et al.*, 1978; Foss, 2004).

Understanding the locations of critical points provides a foundation for understanding the structure of the velocity field since there are a limited number of ways in which the streamlines can be connected. Whereas direct measurement of a flow domain may be incomplete, topological considerations may be employed in order to help determine the number and types of critical points existing in the flow field, as described by Hunt *et al.* (1978) and Foss (2004), and a more complete picture can thus be obtained. Any surface may be shown to be topologically equivalent to the inside or outside surface of a spherical shell in which some number of *holes* are punched out, and some number of *handles* are added (a handle is analogous to the surface of a closed-loop handle of a coffee mug). That is, the surface of a sphere can be continuously deformed to conform to the surface of interest. If holes are added to the sphere (*e.g.*, a single hole transforms the surface of the sphere into a sheet, and two holes form a tubular surface), it is necessary that the flow direction is uniformly inward or outward on the edges of each of the holes. Then an Euler characteristic (topological shape), X, of the resulting surface is defined (Hunt *et al.*, 1978) as

$$X = 2 - 2 \, \Sigma(\text{handles}) - \Sigma(\text{holes}) \tag{2.5.1}$$

The Euler characteristic provides a constraint on the number and types of critical points on the surface. The simplest vector field defined on an unmodified sphere ($X=2$), or its topological equivalent, must have a point at which field lines emerge (a node) and a point at which they terminate (another node), forming the so-called "hairy sphere" as shown in Figure 2.5.2a. The Euler characteristic is related to the number of nodes and saddles by the Poincaré-Bendixon theorem:

$$X = \Sigma(\text{nodes}) - \Sigma(\text{saddles}) \tag{2.5.2}$$

Clearly, the basic case of the hairy sphere, with two nodes and no saddles, satisfies Equation 2.5.2, where $X=2$. If a saddle were added, it would be necessary to add one more node such as to close all of the field lines, as illustrated in Figure 2.5.2b.

Equations (2.5.1) and (2.5.2), combined with a qualitative or quantitative visualization of some portion of the velocity field or surface shear stress, provide a framework for predicting the unobserved critical points in the flow, and therefore also the topology of the streamlines. Effective application of critical points to gain an understanding of the flow requires judicious selection of the surface on which the analysis will be conducted. Foss (2004, 2007) outlines a particularly useful approach which is briefly summarized here in the following example, and the reader is encouraged to consult the examples provided in those publications.

A frequently-cited example of significant importance to experimental hydraulics is flow over a fully-submerged, bed-mounted obstacle within an open channel or wind or water tunnel (*e.g.*, the spherical obstacle in Figure 2.5.1b). If the surface is defined to consist of the side walls, top, and bed of the channel or tunnel, conforming also to the shape of the protruding obstacle as shown by the shaded surface in Figure 2.5.2c for a prismatic obstacle, the surface may be constructed from a surface with two holes cut for the inflow and outflow. Therefore, according to Eq. 2.5.1, the Euler characteristic is $X=0$, and the number of nodes must equal the number of saddles. The

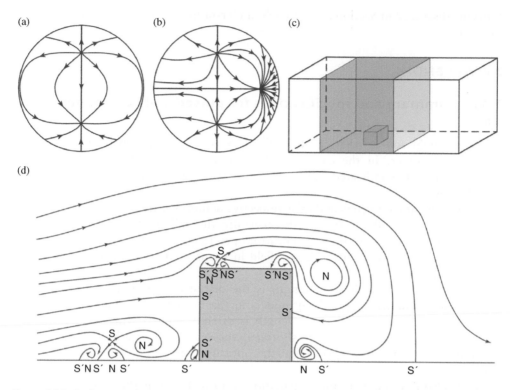

Figure 2.5.2 Surface topologies and critical point patterns. a) a spherical surface with two nodes (a source and a sink); b) the same spherical surface with the addition of a saddle, and an additional node – still in compliance with the Poincaré-Bendixon theorem; c) a tubular surface (shaded, X=0) deformed to be compliant with the surface of a duct with prismatic protrusion on one of the walls; d) The pattern of critical points on the planar surface formed by collapsing the surface in c (following Hunt et al., 1978).

surface visualization of Figure 2.4.1 reveals the pattern of nodes and saddles on the upstream surface of a two-dimensional ridge in channel flow, clarified by the interpretation of critical points at the top of the figure. As illustrated by Foss (2004), the spherical surface may also be evacuated such that it collapses into a plane parallel to the flow. In this case, Equation (2.5.2) is modified such that critical points within the plane are counted twice since they intercept both parallel faces of the flattened sphere, whereas critical points on the edge of the plane (the seam) are counted only once[2]. Again, applying two holes to permit inflow at the upstream edge of the plane and outflow at the downstream edge, the resulting Euler characteristic is still X=0. Figure 2.5.2d shows an interpretation, based on the analysis by Hunt *et al.* (1978), of the flow patterns in this plane for a surface-mounted prismatic obstacle. The inferred critical points are labeled. Assigning contributions of +2 for nodes (N), +2 for saddles in the interior of the flow (S), and +1 for saddles on the boundary (S'), it is evident that Equations (2.5.1) and (2.5.2) are satisfied (X = 2 – 2 holes = 0; X = 2 x nodes – 2 x interior saddles – boundary saddles = 2x9 – 2x2 – 14 = 0). In this case, the nodes represent vortices, and this topological picture of the flow can be reconciled with the concepts discussed in Section 2.5.1 to help understand the three-dimensional vortex structure.

2.A APPENDIX

2.A.1 Summary and specifications for pulsed and continuous lasers

Lasers are light sources that can be defined by two main parameters: wavelength (color) and power. In the case of pulsed lasers, the energy per pulse and its duration are often indicated, allowing the power associated with each pulse to be calculated.

Other relevant parameters and their respective definitions are identified below.

Frequency linewidth: is the width (normally the full width at half-maximum, FWHM) of the laser optical spectrum. More precisely, it is the width of the power spectral density of the emitted electric field in terms of frequency, wavenumber or wavelength, λ.

Coherence length: Coherence length is the distance over which the phases of multiple wavelengths (longitudinal modes) in the output beam stay reasonably in phase with each other. Coherence length is directly related with spectral width. The wider the spectral width, the shorter is the coherence length. The longer the coherence length the easier to design the optical configuration of the laser-based system.

M^2: is also referred as beam propagation ratio or beam quality is defined as the beam parameter product divided by λ/π (ISO Standard 11146 (2005)). The beam parameter product, the product between the beam radius (measured at the waist) and the beam divergence half-angle (measured at the far-field) is obtained considering a diffraction-limited Gaussian beam with the same wavelength. For the case of a diffraction-limited beam, if $M^2 = 1$, the beam is Gaussian. The Gaussian mode corresponds to a particular solution of the Helmholtz electromagnetic wave equation within the paraxial approximation. The more general solution is described by the Hermite-Gaussian modes or Transverse Electromagnetic Modes (TEM). These modes are often referred as $TEM_{m,n}$ where m and n and indices of the mode and are related with the zeros of the Hermite polynomials, m refers to the number of intensity minima in the direction of the electric field oscillation, and n refers to the number of minima in the direction of the magnetic field oscillation (Someda, 2006). A Gaussian beam is also denoted as TEM_{00}. Hermite-Gaussian modes consist in different configurations of light spots and shadows, and were first visualized by Boyd and Gordon (1961). More information regarding this topic can be found in different sources e.g., Csele (2004), Someda (2006), Laufer (2008) and Silvfast (2008). For Hermite-Gaussian beams, related to a $TEM_{m,n}$ resonator mode, one has an $M^2 = (2m + 1)$ in the x direction, and $M^2 = (2n + 1)$ in the y direction.

Tables 2.A.1.1 and 2.A.1.2 show some typical values of the different parameters for pulsed and continuous wave lasers. These values are merely guiding values as the laser manufacturers are continuously increasing the range of various lasers sources (e.g., wavelengths, power output, etc.).

Table 2.A.1.1 Sample of selected pulsed lasers for PIV systems

Laser Type	Wavelength[1] (nm)	Frequency (MHz)	Power (W)	Energy per pulse (mJ)	Repetition rate (Hz)
Pulsed diode laser	808	3.71×10^8	300	0.2–30	0.1–50
Diode pumped Nd: YLF	527	5.69×10^8	15–30	> 60	0.1–1000
Nd:Yag	532	5.64×10^8	15	> 30	0–100

1 It is possible to achieve many different wavelengths with the different lasers types. Here only some are presented as examples.

Table 2.A.1.2 Sample of continuous wave lasers for LDV systems

Laser Type	Wavelength[1] (nm)	Frequency (MHz)	Frequency Linewidth[2] (MHz)	Coherence length[3] (m)	Power (mW)
He-Ne	632.8	4.74×10^8	~1500	~ 0.3	1–50
	1150	2.61×10^8			
	3391	8.84×10^7			
He-Cd	441.6	6.79×10^8	~5000	~0.1	10–50
	457.9	6.55×10^8			70–280
Ar-Ion	465.8	6.44×10^8	~5000	~0.1	40–200
	488	6.14×10^8			1000–4000
	514.5	5.83×10^8			1000–4000
Diode Pumped Solid State (DPSSL)	488	6.14×10^8	<1[4]	> 100	50–1500
	515	5.82×10^8			

1 Some lasers produce multiple wavelengths. Here only selected wavelengths are presented for illustration purposes.
2 Order of magnitude. It depends on the laser beam conditioning and from manufacturer to manufacturer.
3 Order of magnitude.
4 Some manufacturers have reported values of about 1 kHz.

NOTES

1   While this is generally the case, the amount of light transmitted by the lens is also dependent on the number of glass elements that comprise the lens, and their composition. Each lens element will absorb and reflect some of the light incident upon it.
2   Hunt *et al.* (1978) invoke a different interpretation of such a surface, in which nodes and saddles in the interior of the surface increment X by one; whereas critical points on the boundary, called "half"-saddles or –nodes contribute ½.

REFERENCES

Bhattacharyya, A., Neitzel, J.C., Steen, S., Abdel-Maksoud, M., & Krasilmikov, V. (2015) Influence of flow transition on open and ducted propeller characteristics. In: *Proceedings of the 4th International Symposium Marine Propellers, Austin, Texas, USA, June 2015.*

Bourouiba, L., Dehandschoewercker, E., & Bush, J.W.M. (2014) Violent expiratory events: On coughing and sneezing. *J Fluid Mech*, 745, 537–563.

Boyd, G.D. & Gordon, J.P. (1961) Confocal multimode resonator for millimeter through optical wavelength masers. *Bell Syst Tech J*, 40, 489–508

Brown, G.L. & Roshko, A. (1974) Density effects and large structure in turbulent mixing layers. *J Fluid Mech*, 64, 775–816.

Budwig, R. & Peattie, R. (1989) Two new circuits for hydrogen bubble visualization. *J Phys E*, 22, 250–254.

Buchholz, J.H.J. & Smits, A.J. (2005) Gallery of Fluid Motion: Wake of a low aspect ratio pitching plate. *Phys Fluids*, 17, 091102 (2005), doi:10.1063/1.1942512.

Buchholz, J.H.J. & Smits, A.J. (2006) On the evolution of the wake structure produced by a low aspect ratio pitching panel. *J Fluid Mech*, 546, 433–443.

Buchholz, J.H.J. & Smits, A.J. (2008) The wake structure and thrust performance of a rigid low-aspect-ratio pitching panel. *J Fluid Mech*, 603, 331–365.

Chong, M.S., Perry, A.E., & Cantwell, B.J. (1990) A general classification of three-dimensional flow fields. *Phys Fluids A*, 2(5), 765–777.

Clayton, B.R. & Massey, B.S. (1967) Flow visualization in water: A review of techniques, *J Sci Instrum*, 44, 2–11.

Constantinescu, G., Sukhodolov, A., & McCoy, A. (2009) Mass exchange in a shallow channel flow with a series of groynes: LES study and comparison with laboratory and field experiments. *Env Fluid Mech*, 9, 587–615.

Crimaldi, J.P. (2008) Planar laser induced fluorescence in aqueous flows. *Exp Fluids*, 44, 851–863.

Crimaldi, J.P., Koseff, J.R., & Monismith, S.G. (2007) Structure of mass and momentum fields over a model aggregation of benthic filter feeders. *Biogeosciences*, 4, 269–282.

Csele, M. (2004) *Fundamentals of Light Sources and Lasers*. 1st edition. Hoboken, John Wiley & Sons, Inc.

Delo, C.J., Kelso, R.M., & Smits, A.J. (2004) Three-dimensional structure of a low-Reynolds-number turbulent boundary layer. *J Fluid Mech*, 512, 47–83.

Ettema, R., Kirkil, G., & Muste, M. (2006) Similitude of large-scale turbulence in scour around cylinders. *J Hydraul Eng*, 132(1), 33–40.

Ettema, R., Muste, M., Odgaard, J., and Lai, Y. (2004). "Lake-Almanor cold-water feasibility study: hydraulic model", IIHR – Hydroscience & Engineering Limited Distribution Report, 438, IIHR – Hydroscience & Engineering, The University of Iowa, Iowa City, IA.

Ferrier, A.J., Funk, D.R., & Roberts, P.J.W. (1993) Application of optical techniques to the study of plumes in stratified fluids. *Dynam Atmos Oceans*, 20, 155–183.

Foss, J.F. (2004) Surface selections and topological constraint evaluations for flow field analyses. *Exp Fluids*, 37(6), 883–898.

Foss, J.F. (2007) Topological considerations in fluid mechanics measurements. In: Tropea, C., Yarin, A.L. & Foss, J.F. (eds.) *Springer Handbook of Experimental Fluid Mechanics*. Berlin, Springer, pp. 909–918.

Freymuth, P. (1993) Flow visualization in fluid mechanics. *Rev Sci Instrum*, 64(1), 1–18.

Gad-el-Hak, M. (1988) Visualization techniques for unsteady flows: An overview. *ASME J Fluids Eng*, 110, 231–243.

Ghilardi, T., Franca, M.J., & Schleiss, A.J. (2014). Bulk velocity measurements by video analysis of dye tracer in a macro-rough channel. *Meas Sci Tech*, 25, 1–11.

Hajimirzaie, S.M., Tsakiris, A.G., Buchholz, J.H.J., & Papanicolaou, A.N. (2014) Flow characteristics around a wall-mounted spherical obstacle in a thin boundary layer. *Exp Fluids*, 55,1762. doi:10.1007/s00348-014-1762-0

Hunt, J.C.R., Abell, C.J., Peterka, J.A., & Woo, H. (1978) Kinematical studies of the flows around free or surface-mounted obstacles; applying topology to flow visualization. *J Fluid Mech*, 86(1), 179–200.

International Organization for Standardization (2005) ISO Standard 11146. *Lasers and laser-related equipment—Test methods for laser beam widths, divergence angles and beam propagation ratios*. Geneva, ISO.

Kadota, A. & Nezu, I. (1999) Three-dimensional structure of space-time correlation on coherent vortices generated behind dune crest. *J Hydraul Res*, 37(1), 59–80.

Laufer, G. (2008) *Introduction to Optics and Lasers in Engineering*. Revised edition. Cambridge, Cambridge University Press.

Lim, T.T. (2012) Dye and smoke visualization. In: Smits, A.J. & Lim, T.T. (eds.) *Flow Visualization: Techniques and Examples*. Second Edition. London, Imperial College Press.

Marshall, J.S. (2001) *Inviscid, Incompressible Flow*. New York, John Wiley & Sons, Inc.

Martinuzzi, R. & Tropea, C. (1993) The flow around surface-mounted, prismatic obstacles placed in a fully-developed channel flow. *ASME J Fluids Eng*, 115, 85–92.

McKeon, B.J., Comte-Bellot, G., Foss, J.F., *et al.* (2007) Velocity, vorticity, and Mach number. In: Tropea, C., Yarin, A.L. & Foss, J.F. (eds.) *Springer Handbook of Experimental Fluid Mechanics*, Berlin, Springer, pp. 215–472.

Merzkirch, W. (1987) *Flow Visualization*, 2nd edition. New York, Academic Press.

Perry, A.E. & Chong, M.S. (1987) A description of eddying motions and flow patterns using critical-point concepts. *Annu Rev Fluid Mech*, 19, 125–155.

Raffel, M., Willert, C.E., Wereley, S.T., & Kompenhans, J. (2012) *Particle Image Velocimetry, a Practical Guide*. 2nd Edition. Berlin, Springer.

Reynolds, O. (1883). An experimental investigation of the circumstances which determine whether the motion of water in parallel channels shall be direct or sinuous and of the law of resistance in parallel channels. *Phil Trans Royal Soc*, 174, 935–982.

Roy, A.G., Biron, P.M., Buffin-Bélanger, T., & Levasseur, M. (1999) Combined visual and quantitative techniques in the study of natural turbulent flows. *Water Resour Res*, 35(3), 871–877.

Sabatino, D.R., Praisner, T.J., Smith, C.R., & Seal, C.V. (2012) Hydrogen bubble visualization. In: Smits, A.J. & Lim, T.T. (eds.) *Flow Visualization, Techniques and Examples – Chapter 2*. London, Imperial College Press.

Saffman, P.G. (1992) *Vortex Dynamics*. Cambridge, Cambridge University Press.

Samimy, M., Breuer, K.S., Leal, L.G., & Steen, P.H. (eds.) (2003) *A Gallery of Fluid Motion*. Cambridge, Cambridge University Press.

Schostek, M.A. (2012). Design and Use of Servo-Driven Actuators for Spanwise-Varying Control of a Backward-Facing Step Flow. M.Sc. Thesis, Edmonton, Canada, The University of Alberta.

Schraub, F.A., Kline, S.J., Henry, J., Runstadler, P.W., & Littell, A. (1965) Use of hydrogen bubbles for quantitative determination of time-dependent velocity fields in low-speed water flows. *Trans ASME J Basic Eng*, 87(2), 429–444.

Sigurdson, L.W. (1997) Flow Visualization in Turbulent Large-Scale Structure Research, In: Flow Visualization Society of Japan (eds.) *Atlas of Visualization, Vol. III – Chapter 6*. CRC Press, Boca Raton, FL.

Sigurdson, L.W. (2003) Flow visualization in turbulent large-scale structure and flow control research. In: *Proceedings of the 7th Asian Symposium on Visualization*, 3–7 November, 2003, Singapore. CD-ROM: ISBN: 981-04-9686-9, (9 pages).

Silfvast, W.T. (2008) *Laser Fundamentals*. 2nd edition. Cambridge, Cambridge University Press.

Smits, A.J. & Lim, T.T. (eds.) (2012) *Flow Visualization, Techniques and Examples*. 2nd edition. London, Imperial College Press.

Someda, C.G. (2006) *Electromagnetic Waves*. 2nd edition. Boca Raton, CRC Press.

Uijttewaal, W.S.J., Lehmann, D., & van Mazijk, A. (2001) Exchange processes between a river and its groyne fields: Model experiments. *J Hydraul Eng*, 127(11), 928–936.

Van Dyke, M. (1982) *An Album of Fluid Motion*. Stanford, Parabolic Press.

Walker, D.A. (1987) A fluorescence technique for measurement of concentration in mixing liquids. *J Phys E*, 20, 217–224.

Werlé, H. (1973) Hydrodynamic flow visualization. *Annu Rev Fluid Mech*, 5, 361–382.

Willert, C., Stasicki, B., Klinner, J., & Moessner, S. (2010). Pulsed operation of high-power light emitting diodes for imaging flow velocimetry. *Meas Sci Technol*, 21, 075402.

Williamson, C.H.K. (1985) Sinusoidal flow relative to cylinders. *J Fluid Mech*, 155, 141–174.

White, F.M. (2011) *Fluid Mechanics*. New York, McGraw Hill.

Yang, W.-J. (1989) *Handbook of Flow Visualization*. New York, Hemisphere.

Chapter 3

Velocity

3.1 INTRODUCTION

Flow velocity measurements are an essential component of experimental hydraulics and hydraulic model validation. Therefore it is important to measure flow velocities with sufficient accuracy and spatio-temporal resolution (see Chapter 2, Volume I). However, since most flows in hydraulic studies are turbulent and such measurements are carried out under a wide range of flow cases (e.g. in rivers, lakes and oceans, over and through hydraulic structures, open-channel flows in laboratory flumes), this can prove, at times, to be a complex operation. Many different velocimeters are available for this purpose, enabling the measurement of point-based velocities, instantaneous velocity profiles, as well as planar and volume based measurements of flow velocities.

Tables 3.1.1 and 3.1.2 provide an overview of instruments commonly used for flow velocity measurements. The tables indicate the features of each instrument: i) measurement principle (*e.g.*, mechanical, pressure, thermal, electromagnetic, acoustic and optical); ii) main application domain (laboratory or field); iii) interaction with flow (*i.e.*, intrusive or non-intrusive instruments, see Section 5.1.2, Volume I); iv) dimension resolution (*i.e.*, 1D-3D; see Table 4.4.2, Volume I); and, v) achievable sampling rate. Most of the instruments are described in this chapter, with the exception of instruments based on the quantification of other variables, such as pressure (*e.g.*, Section 6.4, Volume II). For completeness, information about velocity measuring instruments not presented here can be found in the references given in Tables 3.1.1 and 3.1.2 or in the ensuing text.

Additional methods and instrumentation exist for velocity measurements. However, not all of them are directly applicable in water flows. For example, LIDAR (*e.g.*, Sathe & Mann, 2013) and SODAR (*e.g.*, Coulter & Kallistratova, 2004) techniques listed in Table 4.4.2 in Volume I are mainly used in atmospheric studies to measure wind speed (*e.g.*, Fernando *et al.*, 2007; Lang & McKeogh, 2011). Hot-wire anemometry, also referred to as Constant Temperature Anemometry, CTA (see Comte-Bellot, 1976; Goldstein, 1996; McKeon *et al.*, 2007 for an overview) finds it main application in wind-tunnel experiments, because CTA wire can easily break in flowing water. Nonetheless, CTA has been used in water flows (*e.g.*, Nezu & Nakagawa, 1993) but has been steadily replaced with more modern techniques during the past decades. Other measurement techniques and instruments have been developed using medical imaging techniques. However, they are rarely used in hydraulic studies and therefore are not in the scope of this chapter; examples are ultrasound imaging velocimetry or echo-PIV (Poelma, 2016), magnetic resonance velocimetry (*e.g.*, Elkins & Alley, 2007; Oojj

Table 3.1.1 Instruments for velocity measurements – point-based measurements

Instrument	Sensor type/ Domain	Flow interaction/ Sampling[1]	Sampling rate[2]	Reference
Point measurements				
Multiple instruments	Various principles, laboratory and field	Intrusive, non-intrusive, 1D-3D	low to high	Section 3.5
Acoustic Backscattering Instruments (ABIs) Acoustic Doppler Velocimeter (ADV)	Acoustic, laboratory and field	Non-intrusive, 2D, 3D	high	Sections 3.2, 3.5
Laser Doppler Velocimetry (LDV); Laser Doppler Anemometry (LDA)	Optical, laboratory	Non-intrusive, 1D-3D	high	Section 3.6
Pitot tube	Pressure, laboratory	Intrusive, 1D	low	Section 6.4.2
Thermal anemometry (Hot wire-HW)	Thermal, laboratory	Intrusive, 3D	high	Comte-Bellot (1976), Goldstein (1996), McKeon et al. (2007)

1 1D, 2D, and 3D define the dimension of the velocity vector(s).
2 "high" means that turbulence characteristics can be sampled, while "low" means that only time-averaged values can be recorded.

Table 3.1.2 Instruments for velocity measurements – profilers, planar and volume measurements

Instrument	Sensor type/ Domain	Flow interaction/ Sampling[1]	Sampling rate[2]	Reference
Profilers				
Acoustic Doppler Current Profiler (ADCP)	Acoustic, field	Non-intrusive, 2D -3D	low	Section 3.3
Acoustic Doppler Velocity Profiler (ADVP)	Acoustic, laboratory	Non-intrusive, 1D-3D	high	Section 3.2
Ultrasonic Velocity Profiler (UVP)	Acoustic, field	Non-intrusive, 1D	high	Section 3.2
Planar and volume measurements				
Particle Image Velocimetry (PIV) Particle Tracking Velocimetry (PTV)	Optical, laboratory	Non-intrusive, area to volume, 2D-3D	high	Section 3.7.2 Section 3.7.3
Large Scale Particle Image Velocimetry (LSPIV)	Optical, field	Non-intrusive, area, 2D	low	Section 3.7.4
Space-Time Image Velocimetry (STIV)	Optical, field	Non-intrusive, area, 2D	low	Section 3.7.4

1 1D, 2D, and 3D define the dimension of the velocity vector(s).
2 "high" means that turbulence characteristics can be measured, while "low" means that only time-averaged values can be recorded.

et al., 2011) or X-ray particle image velocimetry (*e.g.*, Fouras *et al.*, 2007; Heindel, 2011).

Due to the importance of flow velocity measurements in hydraulic experiments, this chapter is significantly larger than the other chapters in this volume. The following nine sections describe the working principles of the instruments as well as relevant background on how to obtain velocities from the measured signals; the appendices give further information. Improved versions or new generations of instruments are continually being developed. Therefore, as instrument performance evolves, the reader is encouraged to monitor the scientific literature for the latest developments; *e.g.*, Koca *et al.* (2017) with respect to the Vectrino Profiler.

Sections 3.2 to 3.3 provide an overview of Acoustic Backscattering Instruments (ABIs) used for velocity measurement. The background of Acoustic Doppler Velocimeters (ADVs), Ultrasonic Velocity Profilers (UVPs) and Acoustic Doppler Velocity Profilers (ADVPs) is presented in Section 3.2. Section 3.3 discusses Acoustic Doppler Current Profilers (ADCPs), and Section 3.4 deals with Acoustic Travel Time Tomography. Section 3.5 presents point velocimeters for field applications. Although this section focuses on field applications, it should be noted that essentially the same velocity instruments are used in laboratory studies, such as mechanical and electromagnetic current meters.

Sections 3.6 and 3.7 describe Laser Doppler Velocimetry (LDV) and image-based velocimetry methods, respectively. Phase Doppler Anemometry (PDA) is similar to LDA and therefore is not extensively covered in this book. In fact, the only difference between PDA and LDV is that PDA allows simultaneous measurement of particle (or air-bubble) size using phase Doppler principles (Arndt *et al.*, 2007). The image-based methods explained in Section 3.7 include Particle Image Velocimetry (PIV), Particle Tracking Velocimetry (PTV) and specific considerations with respect to the application of image velocimetry to large scales (*i.e.*, field applications). Laser Speckle Velocimetry (LSV) and Planar Doppler Velocimetry, PDV, also referred to as Doppler Global Velocimetry, DGV, (see Samimy & Wernet, 2000; McKeon *et al.*, 2007 for a description) are not detailed here due to their similarity to PIV. Note that many image-based methods employ flow-visualization techniques (see Chapter 2, Volume II) and photogrammetric methods (see Section 4.4, Volume II). For example, the flow visualization technique Planar Laser-Induced Fluorescence (PLIF) may also be used for flow velocity measurements and is described in Chapter 2, Volume II.

Sections 3.8 - 3.10 briefly review high-frequency radar, drifters and drogues, as well as dilution methods. The latter method also can be applied for discharge measurements (see Chapter 7, Volume II).

3.2 ACOUSTIC BACKSCATTERING INSTRUMENTS (ABIs) FOR FINE-SCALE FLOW MEASUREMENTS

3.2.1 Introduction

Acoustic Backscattering Instruments (ABIs) have considerably evolved over the past decades and are today one of the leading tools used to study water flows under laboratory and field conditions. Although these instruments are relatively simple to operate, their measurement principle is quite complex, since it is related to the physics of the interaction of sound waves with small particles suspended in the flow. Besides being non-intrusive, ABI-based instruments present a number of advantages over traditional techniques (Thorne & Hanes, 2002; Hurther *et al.*, 2011). For example, they can measure 3D velocities resolving turbulence scales, they only require a unique initial calibration, and they are robust.

ABIs measure velocities in a point, along a line, or within volumes using acoustic backscattering as a basic principle. Therefore, they are first introduced in this section. ABIs not only perform well for the characterization of turbulence and mean flow in laboratory studies, but also in the field where the presence of particulate matter often makes measurements by other techniques difficult (Pinkel, 1979; Sheng & Hay, 1988; Hay & Sheng, 1992; Lhermitte & Lemmin, 1994; López & Garcia, 1999; Thorne *et al.*, 1991; Thorne & Hardcastle, 1997; Nikora & Goring, 2000; Hurther & Lemmin, 2003; Zappa *et al.*, 2003; Davies & Thorne 2005).

Velocity information obtained with ABIs is based on the analysis of the frequency of the signal returned to the sensor by particles interacting with sound waves (see Section 3.2.2). Analysis of other aspects of the returned signal allows to infer characteristics of the fine particles suspended in the flow, as explained in Section 5.3.3 (Volume II). This multi-tasking capability of ABIs makes them attractive for co-located flow and sediment measurements. Due to their relatively high spatio-temporal resolution, these instruments are well suited for simultaneous measurements of turbulent flow, sediment in suspension, dynamics of particles moving along the bed and the location of the non-moving bed. Thus, ABIs are unique in that only one instrument is needed to investigate all aspects of sediment transport processes over a wide range of scales.

The focus of this section is on so-called monostatic and multistatic instruments used under laboratory conditions. In a monostatic instrument, the emitter and the receiver transducer are located at the same position (*e.g.*, Ultrasonic Velocity Profiler (UVP); see Section 3.2.3.1), whereas in multistatic instruments, separate transducers are used for the emission of the sound pulse and its reception (*e.g.*, Acoustic Doppler Velocimeters (ADVs); Section 3.2.3.3 and Acoustic Doppler Velocity Profilers (ADVPs); Section 3.2.3.4). Error sources are discussed in Section 3.2.4. A Cartesian reference system *(x, y, z)* is employed in this section, with *x* being the longitudinal (positive in the direction of the mean flow), *y* the transversal and *z* the vertical pointing axis. The total flow velocity vector $\vec{V}$ is composed of three components (u, v, w) in the three axis directions respectively, as defined above. The Reynolds decomposition $\vec{V} = \overline{V} + V'$ will be applied, where the overbar denotes the time-averaged component of the velocity, and the prime symbol indicates deviation of the instantaneous velocity from the time-averaged value.

3.2.2 Principles of Acoustic Backscattering Instruments

3.2.2.1 Backscattering

The interaction of sound waves with suspended particles is at the basis of ABIs. In water, sound waves are generally scattered from omnipresent density interfaces, such as solid particles, air bubbles, plankton, density differences created locally by dissolved salts, or thermal gradients. When a plane sound wave strikes a suspended particle, a scattered wave is produced only if the size of the particle is much smaller than the acoustic wavelength. The amount of the scattered acoustic energy is very small

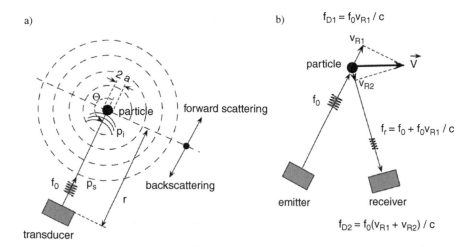

Figure 3.2.1 Acoustic backscattering a) from a suspended particle at rest and b) from a moving particle. In the figure, f_0 denotes the carrier frequency $f_0 = c/\lambda_0$ (λ_0 = wavelength and c = speed of sound); p_i = pressure at the sphere; p_s = backscatter pressure; Θ = backscattering angle (which is equal to π when the emitter and receiver are located in the same position); r = distance of the sphere from the transducer; a = particle diameter; f_r = backscatter frequency; v_{R1} = radial particle speed relative to the sound source; v_{R2} = radial target speed relative to the receiver, f_D = Doppler frequency.

compared to the emitted energy. It is scattered in all directions, and the intensity and the direction of propagation of the emitted wave are hardly affected by the scattering.

Backscattering is defined as the scattering into the space between the sound source (the transmitter emitting a sound pulse of duration τ of typically four to eight sinusoidal waves at a carrier frequency $f_0 = c/\lambda_0$; c = speed of sound; λ_0 = wavelength) and a plane at the level of the particle, perpendicular to the axis between the source and the particle (Figure 3.2.1a). The signal due to backscattering captured by the ABI transducer is a voltage V proportional to the instantaneous backscattered pressure. The output voltage can be used to extract the velocity vector of the moving particles, the concentration of the suspended particles and the distance to the bed. Different types of ABIs have been developed to provide this information, as will be discussed below.

The ABIs principle of operation requires determining the pressure produced by the sound wave on the particles encountered along its propagation path. For example, the pressure p_i generated by the incident wave on the surface of a rigid, immobile sphere of radius a suspended in a fluid, produces a backscatter pressure, p_s, and a forward pressure that continues to propagate in the medium, as depicted in Figure 3.2.1a. Additional information about the signal transformation from pressure to instrument signal is given in Appendix 3.A.1.

3.2.2.2 Doppler shift frequency

Irrespective of their configuration and processing technique, the detection of the Doppler shift is the key feature of ABI-based instruments. To demonstrate the principle of operation, one can assume an ABI configuration as illustrated in

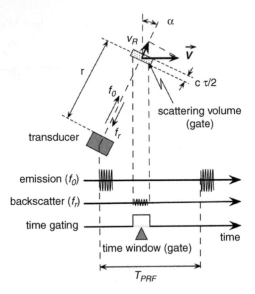

Figure 3.2.2 Principle of operation of a single gate pulse-to-pulse coherent monostatic acoustic Doppler velocimeter.

Figure 3.2.2 where a pulse emitted from a transducer into the water is repeated at a lower frequency, called the "pulse repetition frequency" (f_{PRF}). In the time window between two pulse emissions of duration $T_{PRF} = 1/f_{PRF}$, the ABI receives waves of frequency f_r, backscattered from particles in the water (Figure 3.2.2). If the frequency of a sound wave emitted from a fixed source is f_0, then a particle moving with a radial speed v_{R1} relative to the sound source will experience additional $v_{R1}T/\lambda_0$ waves during time T due to the Doppler effect. For a stationary sound receiver, the sound waves that will be scattered by the target moving with a radial speed v_{R2} can be considered as being emitted by a moving source (Figure 3.2.1b). This causes a second Doppler effect and the total Doppler shift frequency f_D becomes

$$f_D = f_0 \left(\frac{v_{R1} + v_{R2}}{c - v_{R2}} \right) \tag{3.2.1}$$

In the application of this equation to an ABI, it can be assumed that v_{R2} is small compared to c, so that it can be dropped from the denominator. Information about water velocity can be obtained from the radial velocities in the above equation by tracking the particles in the water. Note that ABIs measure the velocity of these particles moving with the fluid, rather than the fluid velocity. Assuming that these particles truly follow the water movement with negligible inertial lag, their velocity is considered to be identical to the fluid velocity. The determination of the Doppler shift frequency from the echos' signal is explained in more detail in Appendix 3.A.2. Appendix 3.A.3 defines the sampling or scattering volume (also called a gate; see Figure 3.2.2), in which the Doppler shift frequency is determined. Appendix 3.A.4 gives the characteristics of ABI transducers.

3.2.3 Instrument types

This section will present the system configurations that are most often used in geophysical and laboratory flow studies.

3.2.3.1 Monostatic instruments

In a monostatic instrument, the emitter and the receiver transducer are located in the same position. In this configuration, $v_{R1} = v_{R2}$ in eq. (3.2.1) and the Doppler shift frequency can be given as

$$f_D = f_0\left(\frac{2v_R}{c}\right) \tag{3.2.2}$$

Only the component of the target velocity directed toward and away from the sound source, called v_R, contributes to the Doppler shift frequency (Figure 3.2.2). This yields for the radial velocity of interest

$$v_R = \frac{cf_D}{2f_0} \tag{3.2.3}$$

In hydrodynamic applications, a vertical profile of the horizontal flow velocity u is often sought. However, a vertically oriented transducer cannot provide a horizontal velocity component, since no radial velocity component is produced. Measurements with monostatic ABIs, therefore, always have to be taken along a line inclined at an angle α to the vertical.

To investigate the properties of a 2D flow, three transducers with different inclinations must be used, since the measured radial velocities v_R contain information about two velocity vectors, u in the direction of the mean flow and w in the vertical direction, as expressed by

$$
\begin{aligned}
v_{R1} &= u\sin\alpha - w\cos\alpha \\
v_{R2} &= w \\
v_{R3} &= -u\sin\alpha - w\cos\alpha
\end{aligned}
\tag{3.2.4}
$$

where the subscripts 1 to 3 correspond to the different transducers (Figure 3.2.3b) and v_{R1} and v_{R2} are not to be confused with v_{R1} and v_{R2} in Equation (3.2.1). For steady uniform flow, the mean (time-averaged) velocities, variances and co-variances are deduced by

$$\overline{u} = \frac{\overline{v_{R1}} - \overline{v_{R3}}}{2\sin\alpha}; \quad \overline{w} = \frac{\overline{v_{R1}} - \overline{v_{R3}}}{2\cos\alpha} = \overline{v_{R2}}$$

$$\overline{u'w'} = \frac{\overline{v_{R3}'^2} - \overline{v_{R1}'^2}}{2\sin 2\alpha}$$

$$\sqrt{\overline{u'^2}} = \sqrt{\frac{\overline{v_{R1}'^2} + \overline{v_{R3}'^2} - 2\overline{v_{R2}'^2}\cos^2\alpha}{2\sin^2\alpha}}$$

$$\sqrt{\overline{w'^2}} = \sqrt{\overline{v_{R2}'^2}}$$

$$\tag{3.2.5}$$

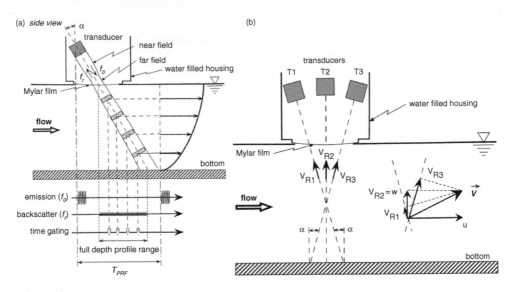

Figure 3.2.3 a) Principle of operation of a monostatic acoustic Doppler velocity profiler. For clarity, only a limited number of gates is shown. b) Configuration of three monostatic profilers for the determination of two mean velocity components. Transducer 1 is tilted along the flow, transducer 2 is vertical, and transducer 3 has the same tilt as transducer 1, but is directed against the flow. Transducers have been placed in a water-filled housing above the water surface in order to avoid flow disturbance by the instrument hardware and velocity measurements in the near field of the transducer beam.

where the overbar denotes time-averaging and the prime denotes the fluctuating velocity. This procedure is strictly applicable to uniform flow only. However, it has also been found to work correctly in gradually varying non-uniform flow.

3.2.3.1.1 Monostatic profiler

The monostatic configuration has mainly been used in profilers. A profiler is an ABI that provides quasi-instantaneous backscattered information at a number of consecutive gates ranging from 1 to j aligned along the acoustic beam of the transducer (Figure 3.2.3a). This concept was initially developed in the medical field for blood flow studies (Baker & Watkins, 1967). A profile of the velocity distribution that may cover the whole water column can be obtained if, after each emitted pulse, a number of gates ranging from 1 to j are sampled in a closely spaced sequence (Lhermitte & Lemmin, 1990, 1994; Zedel et al., 1996; Lemmin & Rolland, 1997). This profile is almost instantaneous (Figure 3.2.3a). Using again three different transducer orientations, the means of two flow velocity components can be extracted for all gates in the profile by applying the above Equations (3.2.5) to each gate ranging from 1 to j (Figure 3.2.3a). Lhermitte and Lemmin (1990) suggested that an inclination of 30° gives good results for both components. For a typical mounting of the three transducers, there is only one gate in the whole profile where the two velocity components are determined in the same volume (Figure 3.2.3b). Thus, only mean parameters under uniform flow conditions can be obtained with this type of profiler. A number of different monostatic profilers

Figure 3.2.4 Ultrasonic Velocity Profiler (UVP): a) several transducers are mounted on a bar to measure mean longitudinal velocity profiles simultaneously, and b) the transducers are connected to the processing unit and the data are visualized and recorded on the computer. Photo: Olivier Mariette, met-flow.com

are on the market today typically under the name of "Ultrasonic Velocity Profiler" (UVP; see Figure 3.2.4). It should be noted that the time interval between the emitted pulse and the sampling volume increases in proportion to the depth range over which velocity information is sought in monostatic profilers. The maximum depth of a profile and the maximum velocity that can be measured are linked by the range velocity ambiguity, as discussed in Appendix 3.A.5.1).

3.2.3.2 Multistatic instruments

The limitations of monostatic ABIs in studying details of turbulent flows are overcome by extending the methodology of the monostatic systems described above to pulse-to-pulse coherent multibeam systems, also called multistatic systems. In multistatic systems, separate transducers are used for the emission of the sound pulse and its reception (Figure 3.2.5). The most common instrument based on this principle is the Acoustic Doppler Velocimeter (ADV). Unlike the three monostatic profiler configuration (Figure 3.2.3b), sound pulses are only emitted from a central transducer for such a configuration. This transducer can work as a receiver for the backscattered sound waves between

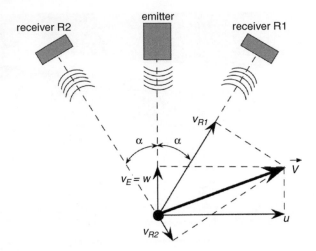

Figure 3.2.5 Configuration of a single gate pulse-to-pulse coherent bistatic Acoustic Doppler Velocimeter (ADV) using one emitter and two receivers.

two emitted sound pulses, as it does in the monostatic mode. The additional transducers placed around this emitter serve as receivers for the backscattered waves coming from the same pulse of sound waves. All receivers simultaneously receive the backscattered waves from the same gate volume. Separate hardware for each receiver within the instrument system allows the application of the pulse-pair algorithm given above for each receiver, thus providing a Doppler shift frequency for each receiver. The hardware of the instrument system synchronizes the information obtained from the different receiver circuits. Different geometries for the arrangement of these transducers can be envisioned.

For convenience and ease of signal extraction, these receivers have to be inclined toward the central emitter and are normally placed symmetrically around the emitter. In this case, sound waves that are backscattered from the gate volume in the central beam have the same time of flight to all the surrounding receivers. It is obvious from Figure 3.2.5 that a bistatic configuration with one pair of opposingly-inclined receivers, inclined at angle α, is sufficient for the determination of two instantaneous velocity components. In order to obtain a third velocity component, at least one additional inclined receiver is needed.

3.2.3.2.1 Operating principle of multistatic instruments

The operating principle of multistatic instruments (*e.g.*, ADVs) will first be explained here for a simple bistatic transducer geometry combining one emitter and two symmetrically placed receivers (Figure 3.2.5). In practical applications, the emitter and thus the axis of the instrument can be oriented vertically with respect to the mean flow, allowing a vertical profile of the two velocity components by changing the position of the gate volume in the vertical. This is an advantage over a monostatic system where profiles are obtained along beams that have to be inclined.

The vertically pointing emitter is only sensitive to the backscattered waves related to the vertical velocity. The two inclined receivers receive signals that contain contributions from the horizontal and the vertical velocity components. These have to be separated by proper signal treatment algorithms. The transducer that is positively inclined with respect to a given flow direction receives from a given scattering volume, a signal containing a Doppler shift frequency that can be expressed as

$$f_{D1} = f_{r1} - f_0 = \frac{f_0}{c}(v_E + v_{R1})$$
(3.2.6)

With $v_{R1} = u_1 \sin \alpha_1 + w_1 \cos a_1$; $v_E = w_1$, one can write

$$f_{D1} = \frac{f_0}{c}\left(u_1 \sin \alpha_1 - w_1(1 + \cos \alpha_1)\right)$$
(3.2.7)

where u_1 and w_1 are the instantaneous horizontal and vertical velocities, α_1 is the Doppler angle of the positively inclined subsystem and f_{r1} is the received sound frequency. It can be shown that $f_{D1}(c/f_0)$ corresponds to the projection of the velocity along the bisector of the emitted and scattered sound wave paths. At the same time, the upstream, negatively inclined transducer receives from the same scattering volume a signal that is shifted by f_{D-} with respect to the emitted frequency. This can be given as

$$f_{D2} = \frac{f_0}{c}\left(-u_2 \sin \alpha_2 - w_2(1 + \cos \alpha_2)\right)$$
(3.2.8)

If the upstream and the downstream subsystems are symmetrical with respect to the central emitter, then $\alpha_1 = \alpha_2 = \alpha$. Since the signal in both receivers is received from the same gate volume

$$u_1 = u_2 = u \text{ and } w_1 = w_2 = w$$
(3.2.9)

Under these conditions, the unknown instantaneous velocity components u and w can be calculated by

$$u = \frac{c\,(f_{D1} - f_{D2})}{2\,f_0 \sin \alpha} \text{ and } w = \frac{c\,(f_{D1} + f_{D2})}{2\,f_0\,(\cos \alpha + 1)}$$
(3.2.10)

The two Doppler shift frequencies f_{D1} and f_{D2} are extracted from the time-series of the backscattered signals by the pulse-pair algorithm, as indicated above. As in the case of monostatic instruments, an average over N_{pp} pulse returns has to be made for the statistic stability of the velocity estimates. A number of $N_{pp} = 32$ was suggested as the lower limit (Rolland & Lemmin, 1997).

For an ABI configuration with one vertically pointing emitter and receivers inclined at an angle α with respect to the vertical and positioned in a vertical plane that is placed at an angle β with respect to the (x-z) plane, the measured Doppler shift frequency relates to the particle velocities as

$$f_D = \frac{f_0}{c}[u \cos \beta \sin \alpha + v \sin \beta \sin \alpha + w\,(\cos \alpha + 1)]$$
(3.2.11)

The above equation is generally valid for all bistatic ABIs (Figure 3.2.6). As will be seen below, the algorithms used to estimate the Doppler shift frequency are instrument dependent. Multistatic ABIs that have been developed and are commercialized as point

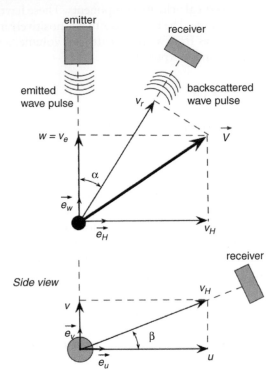

Figure 3.2.6 General principle of an Acoustic Doppler Velocimeter (ADV) when the flow vector $\vec{V}$ is not aligned with the transducer plane. In the side view, the projection of the flow vector $\vec{V}$ in this plane is shown.

measurement instruments measuring in a single gate volume, are called Acoustic Doppler Velocimeters (ADVs). Multistatic profilers have also been developed (Rolland & Lemmin, 1997).

3.2.3.3 *Point measurement instruments (ADVs)*

There are several Acoustic Doppler Velocimeters (ADVs) on the market with different configurations. In the following, details will be provided for the most commonly used configurations. Information about ADVs specifically developed for field measurements can be found in Section 3.5.

3.2.3.3.1 *Three-receiver configuration*

With an instrument based on the three-transducer configuration with one emitter and two receivers in one plane, as described above, two velocity components can be determined. If all three velocity components are sought, an additional receiver transducer has to be placed outside this plane. In the standard configuration, the central emitter is surrounded by three receivers, symmetrically spaced at 120° angles, as indicated in Figure 3.2.7. This configuration is typically used in ADVs (Lohrmann

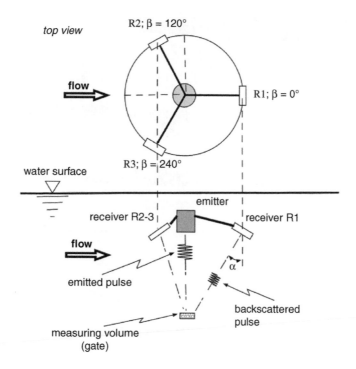

Figure 3.2.7 Configuration of a typical three-receiver ADV instrument for single gate 3D velocity measurements.

et al., 1994; see example in Figure 3.2.8). For this ADV configuration, the above equations can be extended in the following way (Blanckaert & Lemmin, 2006):

$$\begin{bmatrix} u \\ v \\ w \end{bmatrix} = \begin{bmatrix} \sin\alpha & 0 & \cos\alpha + 1 \\ -1/2\sin\alpha & \sqrt{3}/2\sin\alpha & \cos\alpha + 1 \\ -1/2\sin\alpha & -\sqrt{3}/2\sin\alpha & \cos\alpha + 1 \end{bmatrix}^{-1} \frac{c}{f_0} \begin{bmatrix} f_{D1} \\ f_{D2} \\ f_{D3} \end{bmatrix} = G_3^{-1} \frac{c}{f_0} \begin{bmatrix} f_{D1} \\ f_{D2} \\ f_{D3} \end{bmatrix}$$

(3.2.12)

or

$$\begin{bmatrix} u \\ v \\ w \end{bmatrix} = G_3^{-1} \frac{c}{f_0} \begin{bmatrix} f_{D1} \\ f_{D2} \\ f_{D3} \end{bmatrix}$$

(3.2.13)

with G_3 being the geometrical transformation matrix which relates to the arrangement of the three receivers with respect to the central emitter. From these equations, the time-averaged velocity components, the fluctuating components, Reynolds stresses (Lohrmann

Figure 3.2.8 Three three-receiver ADVs mounted to a rod for deployment in the field for synchronous measurements. Photo: Vladimir Nikora.

et al., 1995) and higher order correlations can be determined. The geometrical transformation matrix G_3 is supplied for each instrument by its manufacturer.

3.2.3.3.2 Four-receiver configuration

In the four-receiver configuration, two sets of opposing receivers form two independent, perpendicular planes (Figure 3.2.9). If the two planes are oriented along the longitudinal and the transverse flow directions, respectively, the longitudinal system measures the streamwise u and vertical w_l velocities, as discussed above, based on data obtained from receivers R1 and R2. The transverse system provides the transverse v

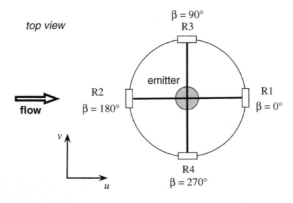

Figure 3.2.9 Top view of a typical four-receiver ADV configuration.

and vertical w_t velocities based on data obtained from receivers R3 and R4. This configuration therefore provides a redundancy for the simultaneously measured vertical velocity components w_l and w_t that can be exploited for noise removal, as will be dealt with later in Section 3.2.4.1.

A more recent generation of ADVs groups four receivers around the central emitter, as discussed above (trade name Vectrino from Nortek; Figure 3.2.10) instead of the three-receiver configuration of earlier systems. For the given geometry of the sensor head, specified by the geometrical transformation matrix G_4 that is determined by the manufacturer, the flow velocity components, u, v and w can be calculated from the beam velocities according to:

$$\begin{bmatrix} u \\ v \\ w_l \\ w_t \end{bmatrix} = \begin{bmatrix} \sin \alpha & 0 & \cos \alpha + 1 \\ -0 & \sin \alpha & \cos \alpha + 1 \\ -\sin \alpha & 0 & \cos \alpha + 1 \\ 0 & -\sin \alpha & \cos \alpha + 1 \end{bmatrix}^{-1} \frac{c}{f_0} \begin{bmatrix} f_{D1} \\ f_{D2} \\ f_{D3} \\ f_{D4} \end{bmatrix} = G_4^{-1} \frac{c}{f_0} \begin{bmatrix} f_{D1} \\ f_{D2} \\ f_{D3} \\ f_{D4} \end{bmatrix} \qquad (3.2.14)$$

or

$$\begin{bmatrix} u \\ v \\ w_l \\ w_t \end{bmatrix} = G_4^{-1} \frac{c}{f_0} \begin{bmatrix} f_{D1} \\ f_{D2} \\ f_{D3} \\ f_{D4} \end{bmatrix} \qquad (3.2.15)$$

where G_4 is the geometrical transformation matrix that relates to the arrangement of the four receivers with respect to the central emitter.

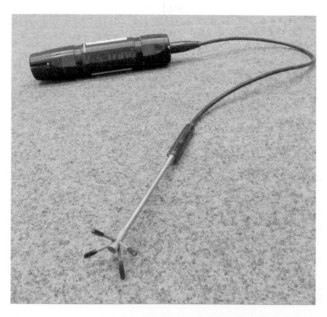

Figure 3.2.10 Vectrino four-receiver ADV. In this version, the four-receiver head is connected to the electronic housing by a 1 m long cable. This allows for a more flexible mounting of the transducer head without submerging the electronic housing.

3.2.3.4 Acoustic Doppler Velocity Profiler (ADVP)

ABIs also exist as Acoustic Doppler Velocity Profilers (ADVPs) that can measure instantaneous 2D to 3D velocity profiles, using a central emitter and several receivers symmetrically placed around the emitter (Rolland & Lemmin, 1997; Hurther & Lemmin, 2001). The profiles are obtained by gating the backscattered signal according to the time of flight. The angle α between the axes of the emitter and a receiver (Figure 3.2.11) changes along the profile. As with the ADV, the vertical flow velocity, w, is simultaneously measured in the two receiver planes, thus allowing the same data treatment for both instrument types. An example for such an instrument is shown in Figure 3.2.12.

For profilers, the emitter has to be a narrow beam transducer with the characteristics discussed above. However, the receiver transducers have to be wide-angle transducers that allow receiving signals from the whole profiling range of the instrument (Figure 3.2.11). The combination of the distance of the receiver transducers to the

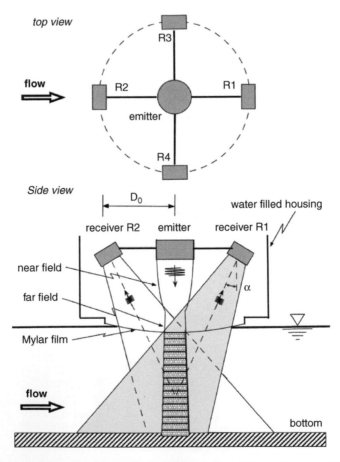

Figure 3.2.11 Configuration of a four-receiver pulse-to-pulse coherent Acoustic Doppler Velocity Profiler (ADVP). Note that the receiver transducers have to have a wide opening angle in order to receive backscattering information over the full water depth.

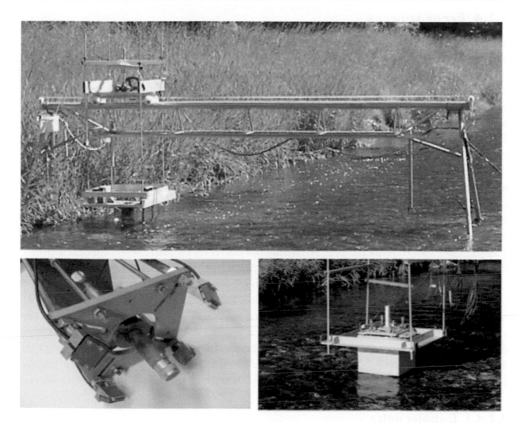

Figure 3.2.12 The four-receiver ADVP in a field measurement campaign on the Venoge river, Vaud, Switzerland. Bottom left: The plate with the emitter in the center and the four receivers. The transducer geometry is adjustable to match the flow field (length of profile). Bottom right: The transducer plate is mounted inside a waterfilled housing with a Mylar window on the bottom. Top: The housing is fixed on a carriage which can slide along a bridge in order to take profiles across a river. The transducers are connected by cable to the electronics on shore. Data are visualized and recorded in real time on a computer. Photo: Mario Franca

emitter, the opening angle of the receiver transducers and the inclination of the receiver transducers determines the profiling range of the profiler. The angle α changes along the profile with distance from the transducer hardware. In order to maximize the profiling range and to minimize the flow disturbance by the hardware, it was suggested that the transducer head be mounted in a separate chamber positioned above the free surface. Contact with the free surface is made via a Mylar sheet at the bottom of the chamber that barely touches the flow surface (Figure 3.2.3 and Figure 3.2.11; Rolland & Lemmin, 1997; Hurther & Lemmin, 2001). In this way, only a thin, near surface layer of the flow, typically 3 to 5 mm thick, is perturbed by the instrument. Furthermore, this configuration avoids measuring in the near field region where the phase fronts may be severely disturbed (see Appendix 3.A.4). Appendix 3.A.5.2 explains how the range velocity ambiguity for an ADVP is determined.

3.2.4 Error sources

Velocity from the Doppler frequency does not require calibration since ABIs remain calibrated throughout their lifetime. The application of this principle to hydraulic flow measurements, however, is sensitive to certain system errors. Unfortunately, the precision of ABI measurements in turbulent flows is affected by random spikes, Doppler noise, spatial averaging of mean gradients and unresolved fluctuating scales smaller than the sampling volume. Random spikes that are not inherent to the measuring principle of ABIs mainly occur in poor measurement environments due to lack of acoustic targets in the measuring volume. In contrast, Doppler noise is inherent to the measuring principle (Garbini *et al.*, 1982). It occurs even if a sufficiently large number of acoustic targets are homogeneously distributed in the sampling volume. If measurements are made sufficiently far from the bed, acoustic Doppler instruments perform well for direct shear Reynolds stress measurements in bed-shear dominated turbulence, as found in rivers or macro-tidal estuaries (Lane *et al.*, 1998; Kim *et al.*, 2000; Nikora & Goring, 2000; Zappa *et al.*, 2003). Near the bed, ABIs may not correctly measure when there is a strong velocity gradient within the measuring volume (the gate; see Appendix 3.A.3) (Dombroski & Crimaldi, 2007), since the backscattered signal is spatially averaged over the measuring volume. Acoustic reflection from the bed may also generate noise and spikes. Due to the inherent contribution of random Doppler phase noise within the sampling gate volume, ABIs can only resolve a limited fraction of the inertial subrange depending on the dimensions of the gate volume. Furthermore, the spatial averaging that always occurs within the gate volume can lead to an underestimate of the kinetic energy contained in those turbulence scales that are smaller than the gate volume dimensions.

3.2.4.1 Doppler noise

In addition to the true Doppler frequency $\widetilde{f}_D$, the measured Doppler frequencies f_D contain a noise contribution σ. One can therefore write $f_D = \widetilde{f}_D + \sigma$. In turbulent flows, the measured Doppler frequency can be decomposed in the classical manner,

$$f_D = \overline{f}_D + f_D' = \overline{f}_D + \left(\widetilde{f}_D' + \sigma \right) \tag{3.2.16}$$

With a good approximation, the noise contribution σ has the following characteristics, as experimentally verified by Hurther and Lemmin (2001):

- Its energy content is uniformly distributed over the investigated frequency domain (white noise). It has a flat power spectrum over the investigated frequency range.
- It is unbiased ($\overline{\sigma} = 0$). Therefore, it does not affect the mean velocity estimates.
- It is statistically independent of the corresponding true Doppler frequency ($\overline{\sigma \, \widetilde{f}_D} = 0$).
- The energy content of the noise is $\overline{\sigma}^2 > 0$, and affects turbulence measurements ($\overline{f_D'^2} = \overline{\widetilde{f}_D'^2} + \overline{\sigma}^2$)
- If the different receivers i, j are identical and ideal, their noise contribution has the same energy content ($\sigma_i^2 = \sigma_j^2 = \sigma^2$).
- The noise contribution of the different receivers is statistically independent ($\overline{\sigma_i \sigma_j} = 0$, if $i \neq j$).

Noise effects are of particular importance if turbulence measurements are evaluated, since they directly affect the fluctuating quasi-instantaneous flow field. In that case, certain mechanisms inherent to the measurement technique may reduce the measurement precision. The following noise sources affecting turbulence measurements taken with acoustic Doppler systems have been identified theoretically and experimentally:

1. The Doppler ambiguity process that is characterized by the amplitude modulation of the backscattered signal related to the transit time of the acoustical targets through the measuring volume.
2. The spatial averaging of the instantaneous velocity field (a larger number of targets are present instantaneously) that is taken over the sampling volume weighted by the directivity function of the emitter.
3. The effect of the mean flow shear stress present within the sampling volume.
4. The phase distortion effect of the emitted wave front that occurs mainly in the near field.
5. The effect of turbulent scales that are of the same order of magnitude or smaller than the sampling volume's transverse size.
6. The electronic circuitry's sampling errors linked to the A/D conversion.

Except for spatial averaging (source 2, as given above), all other noise sources enter as additional variance terms in the measured fluctuating velocity variances and therefore are statistically independent. It has to be realized that corrections can only be made for mean turbulence parameters, not for individual measured values. Correction methods for mean turbulence parameters expressed as additional variances broadly fall into the categories outlined below.

3.2.4.1.1 Identification and quantification of the different Doppler noise sources

The identification and quantification of the different Doppler noise sources was carried out by Garbini et al. (1982), Cabrera et al. (1987), Lhermitte and Lemmin (1990, 1994), Zedel et al. (1996), Voulgaris and Trowbridge (1998), McLelland and Nicholas (2000). The different noise terms are estimated as functions of the sampling volume's distance from the transducer (sources 1, 3, 4 and 5, as given above). They are then added to provide the total noise variance σ.

In monostatic ABIs and in multistatic ABIs with three receivers (ADVs), this approach is the only way to determine noise effects on the measurements. These estimates depend on the characteristics of the flow and the instrument and are not very accurate. Since the noise contribution cannot correctly be determined, the accuracy of the turbulence measurements cannot be estimated either. It has to be considered that the noise variance affects the radial velocities along the axes of the transducers. The noise correction of the orthogonal flow velocity components depends on the noise variance of the radial velocities and the transformation matrix G (see Equation (3.2.13)) that is used to convert the radial velocities into orthogonal components. Therefore, the noise level of the two horizontal components is equal and it is typically at least an order of magnitude larger than that of the vertical component. Voulgaris and Trowbridge (1998) indicate that for a SonTek ADV, the variance of the vertical component is $O(10^{-1})$ x σ^2_{rms} and it is $O(10^1)$ x σ^2_{rms} for the horizontal components.

No direct correction of the turbulence power spectral density is possible, even if the noise variances are evaluated correctly.

3.2.4.1.2 Doppler noise correction based on two-point cross correlation

The Doppler noise correction based on a two-point cross-correlation was proposed by Garbini *et al.* (1982) who used a UVP. It assumes that the noise sources (sources 1, 3, 4 and 5, as given above) between two points in the velocity profile are uncorrelated. This method can be applied directly to the data and does not require any predictions or assumptions, as do the methods in Section 3.2.4.1.1. Its application to four-receiver bistatic ABIs was developed in Hurther and Lemmin (2001). The measured turbulent velocity variances are affected by an additional noise variance σ

$$\overline{\langle u'^2 \rangle} = \overline{\langle \widetilde{u}'^2 \rangle} + a\langle \sigma^2 \rangle, \quad \overline{\langle v'^2 \rangle} = \overline{\langle \widetilde{v}'^2 \rangle} + a\langle \sigma^2 \rangle \text{ and } \overline{\langle w'^2 \rangle} = \overline{\langle \widetilde{w}'^2 \rangle} + b\langle \sigma^2 \rangle$$

(3.2.17)

where the tilde denotes the true flow quantity. In these equations it is assumed that the Doppler angles of the four receivers are identical and that the noise variance is the same for all receivers. This implies that the receiver transducers are identical and ideal, as was verified by Hurther and Lemmin (2001). The coefficients a and b are related to the geometrical configuration of the instrument. By the same approach, it is found that the mean Reynolds stresses are noise free: $\overline{\langle u'^2 w'^2 \rangle} = \overline{\langle \widetilde{u}'^2 \widetilde{w}'^2 \rangle}$ and $\overline{\langle v'^2 w'^2 \rangle} = \overline{\langle \widetilde{v}'^2 \widetilde{w}'^2 \rangle}$. This also holds true for all tri-covariance terms.

The aim of the correction is to eliminate the noise terms from the above equations (3.2.17). For the correction, use will be made of the fact that a four-receiver ABI provides a redundant and independent measure of the quasi-instantaneous vertical velocity in the longitudinal (subscript l) and in the transversal (subscript t) planes

$$\langle w'_l \rangle = \langle \widetilde{w}'_l \rangle + b\langle n_l \rangle$$

(3.2.18)

$$\langle w'_t \rangle = \langle \widetilde{w}'_t \rangle + b\langle n_t \rangle$$

(3.2.19)

where n_l and n_t represent the noise signals due to the Doppler phase noise contribution. The cross correlation, $\langle R_{lt} \rangle$, of these two signals is

$$\langle R_{lt} \rangle(\tau) = \overline{\langle \widetilde{w}'_l \rangle(t) \langle \widetilde{w}'_t \rangle(t + \tau)}$$

(3.2.20)

if the noise signals of the two velocity estimates are uncorrelated (τ denotes the lag). The magnitude of the cross spectrum $\langle S_{lt} \rangle$ can then be obtained as

$$\langle S_{lt} \rangle = \left| \int_{-\infty}^{\infty} \langle R_{lt} \rangle(\tau) \exp(-2\pi f \tau) d\tau \right|$$

(3.2.21)

and the geometrically weighted noise spectrum $b\langle N \rangle$ can be expressed as

$$b\langle N \rangle(f) = \langle S_l \rangle(f) - \langle S_{lt} \rangle(f) = \langle S_t \rangle(f) - \langle S_{lt} \rangle(f)$$

(3.2.22)

where $\langle S_l \rangle (f)$ and $\langle S_t \rangle (f)$ are the power spectral densities of the fluctuating vertical components measured in the longitudinal and the transversal direction, respectively. The noise variance common to all measured velocity components is then

$$\langle \sigma \rangle^2 = \frac{1}{b} \int_{-\infty}^{\infty} [\langle S_l \rangle (f) - \langle S_{lt} \rangle (f)] df = \frac{1}{b} \int_{-\infty}^{\infty} [\langle S_t \rangle (f) - \langle S_{lt} \rangle (f)] df \qquad (3.2.23)$$

It was shown in Hurther and Lemmin (2001) that

$$\overline{\langle w_l'^2 \rangle} = \overline{\langle w_t'^2 \rangle} \qquad (3.2.24)$$

This indicates that the receivers can be considered identical due to the same noise contribution in terms of their energy. Furthermore, for all f

$$\langle S_{lt} \rangle (f) < \langle S_l \rangle (f) \qquad (3.2.25)$$

$$\langle S_l \rangle (f) = \langle S_t \rangle (f) \qquad (3.2.26)$$

These two relations show that the magnitude of the cross spectrum is lower than that of the power spectral densities of the vertical velocity component calculated from the longitudinal and the transversal planes. Since the "true" velocity spectra of the two velocity measurements have to be identical, any difference can only originate from the uncorrelated noise signals between the two independent measurements of $\langle w' \rangle$. Then, multiplying the noise spectrum with the weighting factors of the instrument transformation matrix given in G_4, and supplied by the instrument manufacturer, the geometrically weighted noise spectra for the vertical, as well as for the horizontal velocity components are determined and the velocities spectra are corrected by subtracting the corresponding noise spectra. The corrected turbulence intensities, $\overline{(u'^2)}^{1/2}$, $\overline{(v'^2)}^{1/2}$ and $\overline{(w'^2)}^{1/2}$ are obtained from the integration of the corresponding corrected spectra. Due to the geometrical configuration of the sensor head (Figure 3.2.11), the horizontal velocity components are more sensitive to noise and uncertainties than the vertical components. The noise level of the horizontal velocity components is amplified with respect to the noise level of the vertical component by $1/\tan^2(\alpha/2)$ (Blanckaert & Lemmin, 2006) which is equal to 13.9 for the Vectrino and may be as high as 20 in profiling ADVPs. Therefore this method is sensitive to errors in the angle estimate.

This Doppler noise correction method is uniquely based on the measurements and does not depend on the flow or the instrument characteristics. The quality of the measurements and the system can be assessed by comparing the two redundant velocities and the corresponding noise levels.

Blanckaert and Lemmin (2006) suggested an optimized noise correction method by turning the sensor head shown in Figure 3.2.11 by 45° with respect to the mean flow direction. In this case, in addition to the Reynolds stresses, all velocity variances can be determined noise-free without using the geometrical correlation. Furthermore, this approach allows multiple noise estimates, thereby enabling to check the system and the data quality. However, it requires a G_4 matrix that is different from the one given above in Equation (3.2.15) and that is not supplied by commercial instrument manufacturers.

Another cross-correlation technique was proposed in Hurther and Lemmin (2008) using the quasi-instantaneous and co-located bi-frequency capability of the ADVP presented above. In this case, the velocity redundancy is obtained from the measurement of the same radial velocity components at two slightly different carrier frequencies. The advantage of this method over the one of Hurther and Lemmin (2001) is the use of a single transducer at two different frequencies, thus avoiding the use of an additional receiver and hardware components in the system. Typically, these noise reduction methods reduce the relative errors in the second and third order turbulent velocity moments to values between 10% and 30%, respectively.

3.2.4.2 Aliasing

The range-velocity ambiguity discussed in Appendix 3.A.5 is one of the most restricting aspects of pulsed coherent Doppler velocity techniques. It leads to a compromise that limits the maximum measurable velocity $v_{R\ max}$ determined by the pulse repetition frequency f_{PRF}. If the true velocity v_R is higher than this maximum (e.g., the bold x in Figure 3.2.13), it is aliased into the range $(-v_{R\ max}, v_{R\ max})$ and results in incorrect velocity estimates (e.g., the circled x in Figure 3.2.13). The true velocity v_R is related to the measured velocity v_{Rm} by $v_R = v_{Rm} + 2n\ v_{R\ max}$, where n, the Nyquist number, specifies the aliasing interval. The aliased fraction of the velocity data can be detected from an abrupt velocity change of amplitude equal to $\pm\ v_{R\ max}$. For this type of aliasing, the analysis of the Doppler frequency or phase tracking methods in the interval $(-\pi, \pi)$ allow removing the discontinuity and recovering the aliased velocity data (Rolland, 1994; Franca & Lemmin, 2006). Combining the expression that relates the Doppler frequency to the phase angle, (Equation (3.A.2.1) in Appendix 3.A.2) and the Nyquist frequency limitation, one obtains

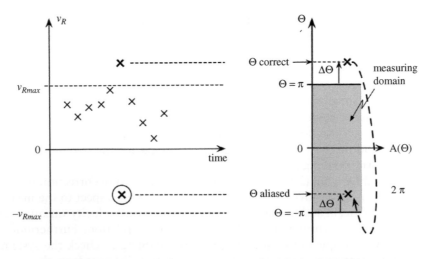

Figure 3.2.13 Sketch of the principle of aliasing in ABI measurements. Note that the scales on the left diagram and on the right diagram are made equal only for the demonstration of the principle.

$$\left| \overline{f_D} \right| = \frac{1}{2\,\pi\,T_{PRF}} \arctan\left(\left\langle \frac{I_{n+1}\,Q_n - I_n Q_{n+1}}{I_n I_{n+1} + Q_n Q_{n+1}} \right\rangle_{Npp}\right) \le \frac{f_{prf}}{2} \qquad (3.2.27)$$

By simplifying this expression, the aliasing restriction can be linked to the arctan function limit

$$\left| \arctan(\varTheta) \right| \le \pi \qquad (3.2.28)$$

With respect to its correct value, an aliased frequency has a lag of 2π. As can be seen in Figure 3.2.13, the correct phase angle which is larger than π falls outside the measuring range, but will be found inside the measuring range with the wrong phase angle value due to aliasing. Therefore, the dealiasing technique consists of adding or subtracting the value of 2π to the value of the arctan function in Equation (3.A.2.1) in Appendix 3.A.2.

This procedure works well in high signal-to-noise ratio (SNR) cases. It is a relative correction method, i.e., after the dealiasing operation, either all velocities are correct or all are incorrect by the same amount. Therefore, dealiasing must be combined with a procedure to determine the absolute velocities. Hurther et al. (2011) proposed a dealiasing method based on the pulse repetition frequency f_{PRF}. As indicated in Appendix 3.A.5, according to Shannon's sampling theorem, the maximum unaliased measurable Doppler shift frequency is limited by $f_{PRF}/2$. Since a phase angle of π corresponds to $f_{PRF}/2$, both methods lead to the same result. Dealiasing is particularly important when a time derivative analysis is carried out on the flow velocity data. In this case, aliased data may produce incorrect results.

3.2.4.3 Spikes

Spikes in time-series may occur as the result of aliasing, as was discussed above. Most of the data records shown in the literature as being polluted by spikes, actually have the characteristics of aliased data. Data quality parameters, such as correlation and SNR are often high in these data sets (Doroudian et al., 2010).

Spikes may also be introduced into the data when obstacles, such as solid particles or air bubbles block the sound path between the emitter and the receiver. If a particular spike is found in all four-receiver beam velocities, the obstacle is located in the incident path of the sound pulse, i.e., between the emitter and the sampling volume. If a spike is only seen in a single beam velocity, then the obstacle is on the beam-dependent back-scattering path. For spherical particles of proper size with respect to the acoustic wavelength, backscattering is spatially homogeneous (Lhermitte & Lemmin, 1993). However, for non-spherical backscattering particles, such as sand, backscattering intensity can vary in the direction of the different receivers.

Furthermore, spikes occur in the data records when there are not enough tracking particles present in the gate volume to produce a measureable Doppler shift frequency. This results in a signal drop-out for all receivers, and an abnormally low SNR. In addition to velocity information, commercial ABIs can also record a correlation coefficient and the SNR, thus allowing for data quality control.

It was shown in the literature (Goring & Nikora, 2002; Cea et al., 2007; Doroudian et al., 2010) that most of the observed spike data fall into an area that is clearly separate from the main data cluster. In those studies, a near Gaussian distribution of the good data was found. Goring and Nikora (2002) developed a spike removal (despiking) technique based on iterative phase-space thresholding as the most suitable solution for spike detection and this technique is now frequently used (Cea *et al.*, 2007). Cea *et al.* (2007) applied several spike removal algorithms. For not too complex flows, they showed that the difference between the results of a minimum/maximum threshold filter (Doroudian *et al.*, 2010) and those of the phase-space algorithm was less than 5%.

Spike data can simply be removed from the time-series. However, this approach limits the data analysis to mean parameters and no time-series analysis can be performed. Spike replacement can be carried out instead by using data splicing techniques such as spline procedures over adjacent points (Goring & Nikora, 2002). If spikes extend over several consecutive data points, a splicing algorithm will probably introduce an artificial distribution. All replacement procedures may locally produce gradients that introduce artificial frequencies in the spectral analysis, since they introduce white noise caused by the steps in the reconstruction technique using the last valid data point. It can therefore be expected that the high frequency end of the spectra is most strongly affected by this operation.

According to Elgar *et al.* (2005) incorrect velocity data can be detected by looking at simultaneously recorded correlation. By setting a threshold level for the correlation, velocity data are eliminated and replaced when the corresponding correlation values fall below the threshold. However, Wahl (2003) and Cea *et al.* (2007) pointed out that in turbulent flows, good data may still exist for low correlation values.

It is important to note that each spike in the streamwise velocity is correlated with an increased amplitude spike in the vertical velocity component. Considering that spike removal procedures (Goring & Nikora, 2002; Cea *et al.*, 2007) are often only applied to the streamwise component of the flow vector, instantaneous Reynolds stress calculations and TKE estimates might be strongly affected, if spikes are not removed from both components. To avoid this problem, spike removal should be carried out on beam radial velocities before flow velocities are calculated (Doroudian *et al.*, 2010).

It has been reported by most authors that spike detection and spike replacement techniques in three-receiver ADV data still remain arbitrary procedures and that none of the currently available methods gives totally satisfactory results. Garcia *et al.* (2007b) developed guidelines for performing turbulence measurements using ADVs. Spike removal, however, depends on the flow conditions and caution should be taken when proposing general guidelines (Doroudian *et al.*, 2010). Doroudian *et al.* (2010) found that both Doppler noise removal and spike removal are necessary to obtain results for turbulence intensities such as $\overline{(u'^2)}^{1/2}$ which correspond to those obtained with well-documented formulae.

3.2.4.4 Spatial averaging and other noises

Garbini *et al.* (1982) have shown that the ambiguity induced by the spatial averaging process can be neglected as long as the sampling volume is sufficiently small to avoid large spatial averaging in the spectrum. Thus, variations in spatial averaging contributions result from changes in the flow characteristics along a vertical profile. Effects of this process are more likely to be found in the near wall region of the flow where strong vertical gradients occur (Dombroski & Crimaldi, 2007; Hurther & Lemmin, 2001). The magnitude of these effects also depends on the roughness of the wall. In ADV Profilers, spatial averaging effects may change along the beam due to the emitter transducer opening angle. The problem of the changing opening angle can be avoided by using a focalized emitter transducer. Hurther and Lemmin (1998) demonstrated that a constant beam width over a range of several decimeters can be obtained with a focalized transducer. This improves the transducer performance, since spatial averaging is constant along the beam due to a constant gate volume. In this case, variations in spatial averaging contributions will only result from changes in flow characteristics.

Doroudian *et al.* (2010) suggested a sampling frequency limit of close to 50 Hz for the Vectrino ADV. This was determined by the deviation of the spectra from the −5/3 slope that occurs due to spatial averaging effects resulting from fluctuating scales that are smaller than the sampling volume (gate) dimensions. According to McLelland and Nicholas (2000), noise in the velocity variance increases with higher sampling frequencies, even though higher sampling frequencies may resolve a larger portion of the turbulence range. For streamwise velocity Vectrino data, noise removal resulted in scatter in the medium and high frequency ranges (Doroudian *et al.*, 2010), indicating that for the Vectrino instrument, this part of the spectrum is also affected by noise from other sources. This additional noise was not seen in the ADVP spectra obtained in the same study. It is therefore instrument dependent.

In this context, it should be noted that the capacity of particles to be true flow tracers for high frequency fluctuations in turbulent flow may be limited. Mei (1996) investigated the effect of the particle's size and inertia on its capacity to track the flow. He found that the best results are obtained for particles with a density ratio particle/water near one and a small particle response time that is related to the square of the particle radius. Increasing the density ratio by using sediment tracer particles, leads to low pass filtering and thus a cut-off for the high frequency fluctuation tracking. Particle size increase has the same effect. For sediment particles of about 100 μm diameter, the cut-off frequency may be around 20 Hz (Pedocchi & Garcia, 2012). Thus, the imperfect flow tracing capacity of heavy sediment particles will induce noise at the high frequency end of energy spectra. Slip velocity between particles and the fluid may be another problem to be considered in turbulent flows (Cisse *et al.*, 2013). Blanckaert and Lemmin (2006) proposed to use small hydrogen bubbles as flow tracers in clear water flows. They demonstrated that signal quality, particularly at the high frequency end of the spectra, greatly improved when these tracers were added.

3.3 ACOUSTIC INSTRUMENTS FOR MEAN FLOW CHARACTERIZATION IN FIELD CONDITIONS: ACOUSTIC DOPPLER CURRENT PROFILERS (ADCP)

3.3.1 Introduction

The acoustic Doppler current profiler (ADCP) was designed and is typically used for measurement of discharge in field conditions. However, the raw data acquired by the instrument are multi-dimensional velocity components. Therefore, ADCP is presented in this section with reference to both velocity and discharge measurements (see also

Chapter 7, Volume II). The principle of operation for each of the ADCP beams is basically similar to a monostatic ABI as described in Section 3.2.2.

Recently, the use of acoustic measurement techniques based on the Doppler effect (see Section 3.2) has become universal practice in hydrology (flow discharge measurements) and fluvial hydraulics (measurement of three-dimensional flow fields). ADCPs are used worldwide to characterize and quantify flow in water bodies such as rivers, waterways and estuaries (Oberg *et al.*, 2005; Oberg & Mueller 2007). Figure 3.3.1 shows different ADCPs presently (2014) used for streamflow measurements made by Teledyne RD Instruments, OTT and SonTek/YSI.

ADCPs are able to perform accurate measurement of both flow discharge and mean velocity flow field (Simpson, 2001; Mueller, 2002; Dinehart & Burau, 2005; Oberg *et al.*, 2005; Garcia *et al.*, 2007a; Jackson *et al.*, 2008). Oberg and Mueller (2007) validated about 100 flow discharge measurements made using the Rio Grande ADCPs, comparing ADCP-measured discharge to discharges measured using conventional discharge-measurement techniques (Turnipseed & Sauer, 2010). ADCPs can measure flow fields with high temporal (1 sec or less) and spatial resolution (depth cells as small as 2 cm), providing additional information in three dimensions that cannot be obtained by conventional current meter methods. Thus, ADCPs can be used to characterize and identify three-dimensional (3-D) turbulent structures in natural flows from a stationary or moving boat (Mueller *et al.*, 2013; Muste *et al.*, 2012). Another important advantage of using ADCPs is that the moving-boat ADCP method facilitates measuring flow discharge in relatively short sampling time during unsteady flow conditions, which is a

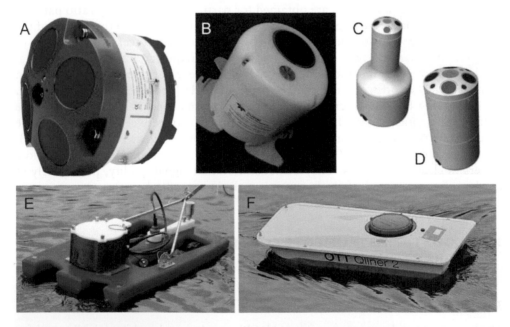

Figure 3.3.1 Commonly-used Acoustic Doppler Current Profilers: A) Teledyne RD Instrument ADCP Río Grande (http://www.teledynemarine.com), B) Teledyne RD Instrument RiverRay (credit Justin A. Boldt), C) Sontek S5 (http://www.sontek.com), D) Sontek M9 (http://www.sontek.com), E) Teledyne RD Instrument StreamPro ADCP and F) OTT QLiner 2 (http://www.ott.com).

major constraint for conventional methods that require substantially longer sampling time.

In addition to quantifying flow discharge, there are many previous studies related to the use of ADCPs for the characterization of the three-dimensional flow velocity components and flow turbulence (Stacey *et al.*, 1999; Schemper & Admiraal, 2002; Nystrom *et al.*, 2002; Howarth, 2002; Kawanisi, 2004). ADCPs have been used to estimate shear stress (Kostaschuk *et al.*, 2004; Szupiany *et al.*, 2012), to characterize reach-scale flow velocity fields (Muste *et al.*, 2004a; Dinehart & Burau 2005; Parsons *et al.*, 2007; Szupiany *et al.*, 2007, 2009, 2012; Rennie & Church, 2010), to perform morphological surveys (Guerrero & Lamberti, 2011; Jamieson *et al.*, 2011a), to estimate suspended sediment transport (Guerrero *et al.*, 2011; Moore *et al.*, 2012; Sassi *et al.*, 2012; Latosinski *et al.*, 2014), to estimate bed load sediment transport (Rennie *et al.*, 2002; Ramooz & Rennie, 2010; Jamieson *et al.*, 2011b; Latosinski *et al.*, 2017), and to estimate longitudinal dispersion coefficient in natural rivers (Carr *et al.*, 2005; Shen *et al.*, 2010).

The U.S. Geological Survey (USGS) has been using ADCPs extensively since the 90's. In 2013, about 90% of all USGS flow discharge measurements were performed by using acoustic instruments. Ninety-eight percent of all discharge measurements of non-wadable streams (those performed from a boat, cableway, or a bridge) were made with an ADCP. The USGS has developed measurement techniques, policies and guidelines for using ADCP for flow discharge measurements, the most important of which is Mueller *et al.* (2013). The results of ADCP field measurements and laboratory tests, along with the techniques and guidelines have been reported by the USGS in technical reports and journal articles (see, for example, USGS, 2002a,b, 2006; Oberg & Mueller, 2007; USGS, 2011; Mueller *et al.*, 2013).

3.3.2 ADCP systems and measurement outcomes

Detailed descriptions of how ADCPs operate and measure flow and boat velocities may be found in manufacturers' user manuals (Sontek/YSI, 2000; Teledyne RD Instruments, 1996), U.S. Geological Survey reports (Simpson, 2001; Mueller & Wagner, 2009), and in various journal papers. This section is not intended to be a comprehensive coverage of information that can be found in other documents, but rather a concise overview of key operational concepts that will help the user understand the capabilities and limitations of ADCPs.

3.3.2.1 Configurations

ADCPs use the Doppler shift principle to quantify the relative velocity between the instrument and the particles moving with the water flow (suspended matter as sediments, bubbles or organic matter; see also Section 3.2). Here it is assumed that: i) the concentration and size of particles suspended over the water column are sufficient to reflect the necessary acoustic energy and allow for the calculation of Doppler shift and, ii) that the particles move at the same speed of the flow. Note that in natural channels, there may be relatively large objects (such as fish, pieces of wood, among others) that do not travel at the flow velocity. In these cases the Doppler shift of the reflected acoustic signal does not correspond with flow velocity, resulting in incorrect velocity measurement and possibly invalid or missing data below the object.

An ADCP typically has 3, 4, or 8 acoustic transducers mounted on the head of the system, located at an angle to the vertical, ranging from 20° to 30° depending on the manufacturer (see Figure 3.3.1). The Teledyne RD Instrument RiverRay (Figure 3.3.1B) includes a phased array (flat surface), Janus four beams at 30° beam angle. Most of the new ADCPs include a vertical beam (not used for flow velocity measurement), which provides more precise bathymetric data and improved characterization of slopes and irregular river beds. Each transducer used for flow velocity measurements generates a pulse of sound with a known acoustic frequency that spreads through the water. This sound pulse is reflected in all directions by the particles. A portion of the reflected energy is backscattered toward the transducer. If the particles are moving with respect to the ADCP along the beam axis, the reflected sound energy will have a different acoustic frequency. This frequency change in pulse, measured by each transducer, can be used to estimate the flow velocity in the radial direction, through the following expression (see also Section 3.2.3.1):

$$V_B = \frac{c\,f_D}{2\,f_0} \tag{3.3.1}$$

where V_B is the flow velocity in the radial direction; c is the speed of sound in the water; f_D is the difference in the acoustic frequency between the emitted (f_0) and return (f_B) pulses due to Doppler shift (f_0-f_B).

Actually, an ADCP measures the flow velocity in the radial direction (along the beam) using data of the phase change ($d\varphi$) between two backscattered acoustic pulses (or samples of the same pulse) using

$$V_B = \frac{\lambda}{4\pi}\frac{d\varphi}{dt} \tag{3.3.2}$$

which is stated as a function of the wavelength of the emitted acoustic signal λ, and the interval time (dt) between the two pulses (or samples of the same pulse; Lhermitte & Serafin, 1984).

Because the ADCP measures radial velocities, at least 3 acoustic beams are required to resolve velocity in 3 dimensions. However, most ADCPs are equipped with four acoustic beams. Additional information concerning the number of beams and their orientation is provided in a subsequent section.

In addition, ADCPs can measure boat speed by analyzing sound pulses reflected from the river bed. The ADCP transmits separate acoustic pulses, with pulse characteristics optimized for this measurement, to determine the velocity of the ADCP (and boat) relative to the river bed. This technique is known as bottom tracking. Unlike water velocity measurements, bottom tracking measurements can be very precise due to the strong acoustic return from the bed (typical accuracy is one order of magnitude better than water velocity accuracy). Water depth also can be measured by additional analysis of the bottom track pulse. Thus ADCPs collect information regarding instrument speed (and position), distance traveled, and water depth in each location where a water velocity measurement is made.

3.3.2.2 Processing techniques used for ADCP to estimate Doppler shift (velocity measurement)

The main processing techniques used by the ADCPs to compute the Doppler shift (and the flow velocity) can be classified as narrowband and broadband. In these two processing techniques, the frequency change (Doppler shift) between the emitted and backscattered signal (Equation (3.3.1)) is calculated indirectly from the determination of the phase change from two consecutive acoustic pulses or samples within individual pulses (see Equation (3.3.2)). Different measurement modes, or processing technique combinations, are implemented by different ADCPs (*e.g.*, Teledyne RDI ADCP Rio Grande and RiverRay or ADCPs Sontek S5 and M9 models) including the use of multi-frequency acoustic pulses (ADCPs Sontek S5 and M9 models) in order to better adapt to the measurement site conditions such as stream depth, water velocity, flow turbulence and characteristics of bed material. Some newer ADCPs (*e.g.*, Sontek S5 and M9 and Teledyne RD Instrument RiverRay) have automatic adaptive configuration. The different available configuration options involve use of different structures of the pulses transmitted and the combination of different number of pulses in the water column at a given time. The main processing techniques are discussed in the following sections.

3.3.2.2.1 Narrowband

A particular feature of this technique is that the echo of each acoustic pulse sent from the transducer is allowed to return to it before the following acoustic pulse is transmitted. Since there is only one acoustic pulse in the water column at any moment, the structure of acoustic pulses with narrowband technology is simple and they can be processed quickly. While the sampling frequency with this technique is high, the accuracy of the measurement is relatively low, and in order to increase accuracy, the ADCP averages many samples before reporting flow velocity.

The narrowband technique can be used as either coherent or incoherent mode (Figure 3.3.2). Using the incoherent mode, a unique acoustic pulse (used for determination of a value of flow velocity) is transmitted with a length equal to the selected cell size. The transmitted pulse can generate an echo in every cell and depth depending on the density of particles. Using the incoherent mode, each pulse (echoes) is processed

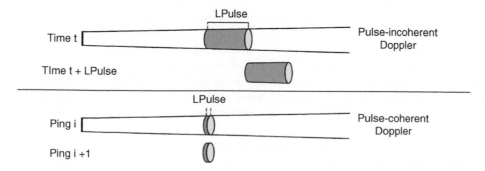

Figure 3.3.2 Narrowband Doppler measurement techniques (modified from Brumley *et al.*, 1996).

independently, measuring phase changes over a fraction of the pulse duration to determine the Doppler frequency shift, *i.e.*, $f_D = d\varphi/T$, where $d\varphi$ is a phase change calculated from performing an autocorrelation on a received waveform and T is a measurement period (Brumley *et al.*, 1996). The incoherent mode with narrowband technology has been used in Sontek/YSI ADCPs (Sontek/YSI, 2000).

The main parameters that users can specify when implementing the narrowband incoherent technique are blanking distance, cell size, and the sampling interval. The acoustic frequency determines the maximum profiling depth, maximum relative velocity, minimum recommended cell size, and specific random noise related to the selected narrowband frequency (SonTek/YSI, 2000). Therefore, users have to determine the acoustic frequency that is the most appropriate for the flow conditions (Mueller *et al.*, 2013).

In narrowband coherent mode, the flow velocity is computed using pulses of shorter duration (allowing for greater depth resolution, see Section 3.3.2.3.2) than pulses generated with incoherent mode. However, unlike the incoherent technique, the phase change is computed between two successive echoes (reflected pulses) at the same depth. Higher accuracy is achieved in coherent than incoherent mode, because the "signal dwell time" (*i.e.*, the observation time to detect a Doppler phase change, or, equivalently, the spectral resolution) in the measurement bin now depends on the time between pulses, not the pulse length. Like the pulse-incoherent system, in the narrowband coherent mode, the echo from each single pulse is allowed to return to the transducer before the next pulse is transmitted. Coherence refers to the fact that coherence must be maintained between the two pulses. The application of this technique is limited and is recommended for low flow velocity due to the presence of ambiguity errors specific to the coherent mode. Because phase change between two backscattered acoustic pulses is periodic, the solution used to determine flow velocity is multi-valued which generates an ambiguity in the velocity determination. Thus, ambiguity errors occur when large flow velocity exceeds the ambiguity velocity, defined as the maximum allowable radial motion for phase measurements to be unambiguous. Using longer interval time (*dt*) between the two pulses (coherent mode), more accurate velocity measurements are possible, but lower ambiguity velocity is required. On the other hand, using short lags usually provides noisy data but no ambiguity errors.

3.3.2.2.2 Broadband

Broadband ADCP technology is different from the narrowband technique (pulse incoherent or coherent) since broadband technology uses two (or more) short pulses in the water column at the same time (Figure 3.3.3) forming a pulse of longer duration (also called a pulse train) with complex structure. Broadband technology was developed and patented by Teledyne RD Instruments (Brumley *et al.*, 1996) to overcome some of the disadvantages of the implementation of the narrowband technology (high noise level avoiding velocity ambiguity). Since the Broadband patent has expired, other manufacturers (*e.g.*, SonTek) have incorporated some features of the Broadband approach into their instruments.

Acoustic pulses of broadband technique are complex. A pulse train typically consists of many short pulses put together and encoded so that many independent velocity samples (obtained from each short pulse) can be obtained from a single pair of pulse trains (Figure 3.3.3 and Figure 3.3.4). In Figure 3.3.3 (Brumley *et al.*, 1996), a

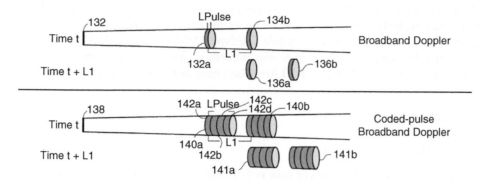

Figure 3.3.3 Broadband Doppler measurement techniques (modified from Brumley et al., 1996).

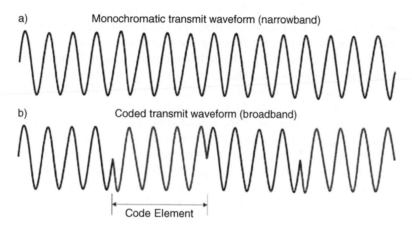

Figure 3.3.4 Code transmitted waveform: a) narrowband and b) broadband.

transducer (138) generates a pulse pair 140a,b that propagates through the water as shown, for example, by the later pulses 141a,b. Each pulse 140 includes four equal-sized code elements 142a,b,c,d that each comprise one or more cycles (or portions thereof) of the transmitted acoustic waveform. The code elements 142 represent phase coding such that each element is either at 0 or 180 degrees of phase. While only two coded-pulses are shown in Figure 3.3.3, the method can be generalized to include more than two pulses.

Due to the complexity of the pulse, the calculation process is slower than narrowband techniques, however, multiple independent samples can be obtained from each of the pairs of coded pulses. Unlike the pulse-coherent method (narrowband), the maximum profiling range of the broadband current profiler is not limited to the pulse repetition interval. The pulse length, or width, is typically much shorter than the depth cell size which results in a large time-bandwidth product (hence the term broadband). Hereinafter, both the broadband ADCP and coded-pulse broadband ADCP systems and methods will generally be referred to as the broadband ADCP unless otherwise indicated.

ADCPs including Broadband technique offer the users multiple water modes (WM) and bottom modes (BM). This allows the user to optimize the measurements for specific flow conditions, *i.e.*, depth and velocity. While Teledyne RDI Rio Grande ADCPs user should select manually water and bottom modes according flow conditions, new instrument (Sontek S5 and M9 and Teledyne RDI RiverRay) have automatic adaptive configuration.

3.3.2.3 Measuring velocities

3.3.2.3.1 Measuring velocity profiles

ADCPs measure vertical profiles of flow velocity by analysis of acoustic signals received at different time intervals, due to the fact that the acoustic pulse travels radially in the water column. The transmitted acoustic pulse propagates at the speed of sound in the water and some portion of the sound is reflected from particles present in suspension back to the transducer. Thus, the reflected acoustic energy from the farthest particles takes longer to return to the transducer than the signal reflected near the transducer. Then, by measuring the round-trip time for the transmitted sound and knowing the speed of sound, the distance along the beam from which the signal was reflected can be obtained. Therefore, whenever the ADCP transmits a ping, it records return signals that can be associated with different distances along the beam, and therefore water depths.

The return signals recorded by the ADCP are analyzed and discretized into ranges that correspond to varying distance from the transducers. These ranges are referred to as depth cells or bins. Figure 3.3.5 explains how observations are compartmented into bins and information for a single transducer is organized to create the bins. The transducer transmits a signal to the water, traveling diagonally in this graph. Immediately after the transmit pulse is complete, the ADCP turns off the transducer and waits for a short time called the blanking period. The blanking period is the time required for transducer

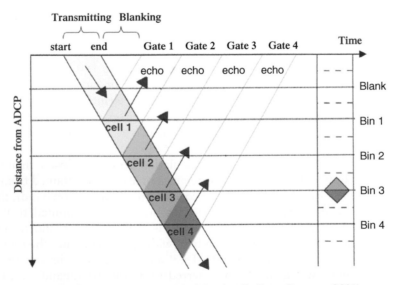

Figure 3.3.5 Relationship between range gates and depth cells (from Simpson, 2001).

vibrations (called ringing) and electronic feedback to die down, and can be computed as a distance which is known as the blanking distance. The ADCP now starts processing the echo corresponding to Range Gate 1. When Gate 1 is complete, the ADCP immediately begins processing Gate 2, and the process continues for the following Gates. The center of Cell 1 contributes the largest fraction of the echo signal to Range Gate 1. The echo from the farthest part of Cell 1 contributes signal only from the leading edge of the transmit pulse. The echo from the closest part of Cell 1 contributes signal only from the trailing edge of the transmit pulse. The velocity in each depth cell is a weighted average using a triangular weight functions (Teledyne RD Instrument, 1996).

3.3.2.3.2 Unmeasured zones

Unfortunately, ADCPs cannot measure in the entire water column (Figure 3.3.6). The unmeasured zones in the water column are divided into the following regions:

a)   A region close to the water surface associated with the immersion depth of the ADCP;
b)   A region immediately below the ADCP transducers, known as the blanking distance, where velocities are not recorded due to the likelihood of errors introduced by ringing and electronic feedback. This distance depends on the water measurement mode and the depth cell size. More recently, it has been shown that

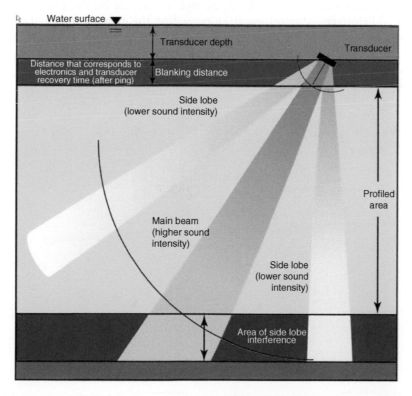

Figure 3.3.6 Acoustic Doppler current profiler beam pattern and location of unmeasured zones (from Mueller *et al.*, 2013).

the ADCP can disturb the flows measured near to the surface and the blanking distance is sometimes increased to minimize this effect (Mueller *et al.*, 2007; Muste *et al.*, 2010).

c)   A region near the streambed (ranging from 6–10 % of the water depth) that cannot be measured because of the interference that is produced by lateral lobes ("side lobes") formed around the beam. This interference occurs when acoustic energy from a side lobe reflects off the streambed and returns to the transducer prior to or at the same time as reflections from the main beam.

3.3.2.3.3 Beam configurations and velocity profile estimation

For most engineering applications, the magnitude of flow velocity in orthogonal components is required. As previously mentioned, the Doppler shift is directional; that is, only the movement of the particles parallel to the acoustic beam (the radial velocity) can be measured. Therefore, at least 3 beams having known orientations are required to calculate the 3-D flow velocity components. Usually, the beams are equally spaced horizontally (typically 90 degrees for a four-beam configuration) and are inclined an angle α with the vertical (20 to 30 degrees, see Figure 3.3.7).

At present (2016), most ADCPs have 4 beams in a "Janus" (looking in opposite directions) configuration (see Figure 3.3.8a). A Janus configuration has the advantage of minimizing errors caused by pitch and roll acting on the probe while also providing additional quality information regarding the measurements. The following equations to obtain flow velocities in the three directions x, y and z are for a typical ADCP configuration with 4 beams where the beam number 3 is pointing upstream:

$$V_x = \frac{V_{B4} - V_{B3}}{2 \sin \alpha} \tag{3.3.3}$$

$$V_y = \frac{V_{B1} - V_{B2}}{2 \sin \alpha} \tag{3.3.4}$$

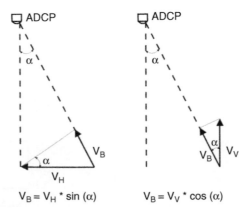

$$V_B = V_H * \sin(\alpha) \qquad V_B = V_V * \cos(\alpha)$$

Figure 3.3.7 Measured radial velocity (V_B) for a) purely horizontal velocity (V_H) or b) purely vertical velocity (V_V).

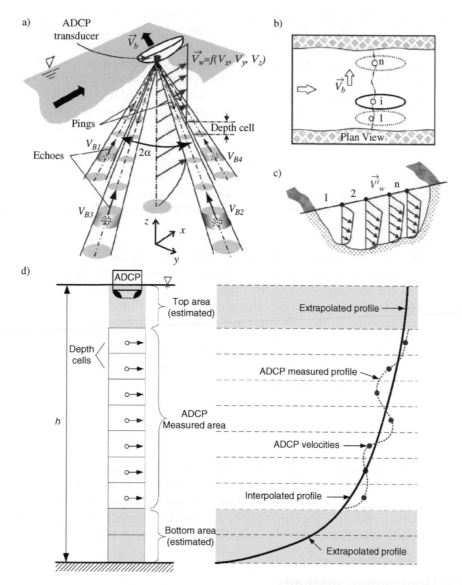

Figure 3.3.8 Measurement of velocities with ADCPs: a) schematic of the 4-beam "Janus" configuration and the end vertical velocity profile outputted by the ADCP (adapted from Muste *et al.*, 2012, permission conveyed through Copyright Clearance Center, Inc.); b) typical ADCP deployment for the acquisition of velocities; c) resultant velocity distribution across the channel cross section; and d) details of the construction of the vertical velocity profile (adapted from Teledyne RD, 1992).

$$V_z = \frac{V_{B1} + V_{B2} + V_{B3} + V_{B4}}{4 \cos \alpha} \tag{3.3.5}$$

where V_{B1}, V_{B2}, V_{B3} and V_{B4} are the radial velocities measured in beam *1, 2, 3* and *4*, respectively (see Figure 3.3.8a) and V_y, V_x, V_z are the streamwise, cross stream and vertical velocity components with respect to the instrument assuming beam 3 is pointed upstream (see Figure 3.3.8a). The angle α is the tilt angle of the beams referenced to vertical (see Figure 3.3.7) and V_b is the boat velocity.

Equations (3.3.3) to (3.3.5) are evaluated for each depth cell (bin) located at the same elevation over of the vertical. Following the velocity determination for all the measured cells on the vertical, a unique vertical velocity profile is reported for the ADCP vertical axis as illustrated in Figure 3.3.8c. As can be observed from the last figure, the ADCP is equivalent to a string of conventional current meters placed over the vertical measuring continuously the water velocity. The depth cell can be varied to accommodate various flow situations, within a wide range, *i.e.*, from a few centimeters to several meters. The measured velocity distribution completed with the extrapolated portions at the top and bottom of the vertical along with precisely known distances along the water column depth, allow to efficiently calculate the discharges between consecutive vertical profiles as described in Chapter 7.4.2, Volume II.

Regardless of the number of beams used, the primary assumption of Equations (3.3.3) to (3.3.5) is that the flow sampled by each beam is the same, *i.e.*, flow through the sampling volume is homogeneous. So if the flow measured by one beam is radically different from the flow measured by the other beams (*e.g.*, strong turbulence effects) this assumption breaks down and the calculated velocity may not be correct. Therefore the velocities measured by the ADCP present greater uncertainty for highly turbulent flows with non-uniform spatial distribution (*e.g.*, wake of a bridge pier).

The redundant beam present in four-beam ADCPs allows for the computation of a new variable called error velocity or difference velocity. Error velocity is defined as the difference in the vertical velocity estimated using two orthogonal pairs of beams, *i.e.*, beams 3 and 4 versus 1 and 2, as it is expressed in the following equation:

$$V_e^u = \frac{V_{B1} + V_{B2}}{2\cos\alpha} - \frac{V_{B3} + V_{B4}}{2\cos\alpha} = \frac{V_{B1} + V_{B2} - V_{B3} - V_{B4}}{2\cos\alpha} \tag{3.3.6}$$

For TRDI ADCPs, the error velocity is then scaled based on the variance of the horizontal velocity measurement (Teledyne RD Instruments, 2010). For SonTek ADCPs, the error velocity is calculated as one-half the value calculated using Equation (3.3.6) and is referred to as a "difference velocity". The error velocity accounts for errors due to both noise in the estimates of individual beam velocities and velocity heterogeneity between beams.

3.3.3 Deployment methods

The methods for deploying an ADCP for measurements will depend on (1) the measurement objectives, (2) site-specific conditions such as hydraulic characteristics (*e.g.*, water velocities or turbulence), and (3) access considerations (boat ramps, bridges, or cableways). ADCPs are most commonly used from a moving boat but are sometimes mounted in a fixed location (downlooking/uplooking/sidelooking) to record velocity profiles over time. The most common types of moving boat ADCP deployments are: manned boats, tethered boats, and remote-controlled boats. Figure 3.3.9 shows some example measurements of cross-section velocities obtained in a large river and a small artificial channel with different deployment methods (manned and tethered boats).

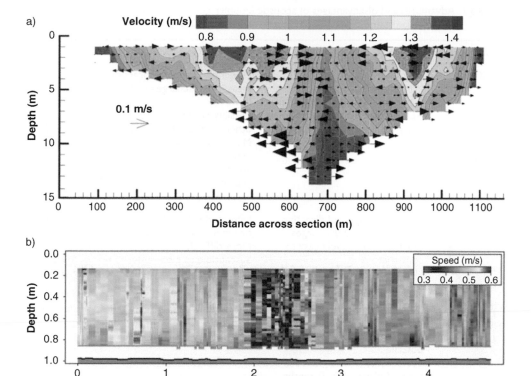

Figure 3.3.9 Examples of ADCP measurements at different scales and deployment methods. a) Parana River (Argentina) downstream of a confluence (cross section velocity component are included) – Q= 11250 m³/s. ADCP 1200 kHz Teledyne RD Instrument – manned boat deployment, and b) income flow to the clarification process of the water treatment plant "Los Molinos", Córdoba, Argentina – Q=2.1 m³/s. ADCP River-Surveyour S5 YSI/Sontek using a "SmartPulse" algorithm with cell sizes as small as 2 cm–tethered boat deployment.

3.3.3.1 Manned boat

Manned boats offer versatility in the selection of measuring sections but may be limiting due to the draft of the boat. ADCPs can be mounted on either side, off the bow (front), or in a well through the hull of a manned boat. Mueller *et al.* (2013), see Table 3.3.1, present advantages and disadvantages for mounting locations on manned boats. A boat equipped with a side-mount for an ADCP is shown in Figure 3.3.10. Additional photographs of a variety of ADCP mounts are available in USGS Open-File Report 01-01 (Simpson, 2001, pp. 58–69), Mueller and Wagner (2009) or at the USGS Hydroacoustics Web page (http://hydroacoustics.usgs.gov/ [accessed 20.02.2017]).

3.3.3.2 Tethered boat

A tethered boat is small unmanned boat equipped with an ADCP, radios or Bluetooth systems for ADCP communication, and a tether for pulling the boat back and forth

Table 3.3.1 Advantages and disadvantages of acoustic Doppler current profiler (ADCP) mounting locations on manned boats (from Mueller & Wagner, 2009).

Mounting location	Advantages	Disadvantages
Side of boat	Easy to deploy Mounts are easy to construct and are adaptable to a variety of boats ADCP draft measurement can be easily obtained	Moderate chance of directional bias in measured discharges with some boats and flows Possibly closer to ferrous metal (engines) or other sources of electromagnetic fields (EMF) Moderate-low risk of damage to ADCP from debris or obstructions in the water Susceptible to roll-induced bias in ADCP depths
Bow (front) of boat	Minimizes chance of directional bias in measured discharges Mounts are relatively easy to construct Usually far from ferrous metal or electromagnetic fields	Increased risk of damage to ADCP from debris or obstructions in the water More difficult to measure ADCP depth Susceptible to pitch-induced bias in ADCP depths, particularly at high speeds or during rough conditions (waves)
Well in center of boat	Protected from debris and obstructions Accurate depth measurements possible Least susceptible to pitch-and-roll-induced bias in ADCP depths	Often requires special modifications to boat

Figure 3.3.10 Boat equipped with side-mounted ADCP for flow measurements.

Figure 3.3.11 Tethered boat being deployed from the down-stream side of a bridge and a temporary, bank-operated cableway.

across the stream. A typical application involves the tethered boat being deployed from the downstream side of a bridge but it can also be deployed from a permanent cableway or a temporary, bank-operated cableway (Figure 3.3.11). The user communicates with the ADCP by means of wireless radio modems (900 Mhz or Bluetooth radios) without requiring a direct cable connection. Perhaps the most common deployment method for ADCP measurements is the tethered boat deployment. It is used in a variety of settings but most often they are used from downstream side of bridges for convenience. In that case special consideration should be taken (see Mueller *et al.*, 2013 for details) to avoid effects of flow non-homogeneity. Tethered boats are relatively easy to deploy but may be limiting in that it may not be possible to access the best measurement section with a tethered boat.

3.3.3.3 Remote controlled boat

Unmanned, remote-controlled boats allow the deployment of ADCPs where deployment with a manned boat or tethered boat may not be feasible or ideal. Similar to (but smaller than) manned boats, a remote-controlled boat has self-contained motors and a remote-controlled system for maneuvering the boat across the river (Mueller *et al.*, 2013). Although their use is not widespread, there are situations where they are routinely used. They offer the possibility of making measurements in hazardous situations where a manned boat measurement cannot be made safely. One of the important considerations for remote-controlled boat measurements is the reliability of the boat, controls, and the skill of the operator.

3.3.3.4 Boat velocity

ADCPs always measure velocity relative to the instrument itself. Therefore for moving boat measurements described above, the ADCP will actually measure the water speed in the river plus the boat speed. In order to compute the true water velocity with respect to a fixed reference system, the boat velocity must be known. Two methods for measuring the boat speed are available and commonly used: bottom tracking and Global Positioning Systems (GPS). In bottom tracking, the ADCP transmits sound pulses separately from the water velocity pulse. These pulses are typically much longer than water velocity pings (on the order of 20-30% of the water depth) and are reflected from the streambed and analyzed using the Doppler shift principle. This velocity is used to determine the true water velocity from the measured velocity by removing the velocity of the ADCP (boat). This method assumes that the streambed is stationary. If the streambed is not moving, this measurement is typically a very accurate measurement; however, sediment transport can cause a bias in bottom tracking that results in a false upstream velocity of the ADCP causing a low bias in the measured water velocity and corresponding discharge.

The alternative approach to measuring boat speed involves the use of GPS. Two options are available for determining boat velocity using GPS: (1) differentiated position using the GGA NMEA-0183 sentence and (2) GPS receiver computed velocity reported in the VTG NMEA-0183 sentence. The GGA data sentence broadcast by the GPS includes time, position data such as latitude, longitude, and elevation, and information about the satellite constellation in use. When using the GGA sentence to measure boat (and therefore ADCP velocity), the velocity is determined by computing the distance traveled between successive GPS positions and dividing that distance by the time between the GPS position fixes. For this reason, positional accuracy is important to achieving an accurate boat velocity measurement using GGA, and therefore, a differential correction signal is required. To use the GGA sentence, DGPS receivers are required, and receivers should have a 95-percent accuracy of about 1 m or less in the horizontal location (Rennie & Rainville 2006). Many GPS receivers can be used to measure velocity relative to ground with an assessment of the Doppler shift in the satellite carrier phase frequencies, which is typically reported in the VTG sentence. The method uses the actual signal frequency, and not a phase angle, to determine the Doppler shift. No ambiguity can be present in the computed velocities. The use of VTG has the advantage of being minimally affected by multipath and satellite changes because of the short sampling time required. In addition, multipath and ionospheric/atmospheric distortions do not affect the precision of the measurement. Therefore, the Doppler measurement of boat velocity does not require any differential correction.

Specific studies require fixed instrument measurements. In these cases, the ADCPs can operate as vertical upward-looking or horizontal side-looking deployment. The first one is used in rivers to measure flow velocities fluctuation and turbulence parameters (see for example Jackson *et al.*, 2008 and McKay *et al.*, 2013) and, primarily, in marine environment monitoring of waves and currents in the nearshore region (*e.g.*, Essen, 1994; Visbeck & Fischer, 1995; Seim & Edwards, 2007).

The same standard down or up-looking system could be deployed also as a side-looking mode to collect long term monitoring data in rivers and marine environments or to apply index-velocity methods to obtain discharge rating. Two beam specific acoustics instrument (HADCP) are also available for this type of measurement (Ganju *et al.*, 2011; see Section 3.3.5).

3.3.3.5 Compass consideration

The ADCP has an internal fluxgate compass to measure the orientation of the instrument relative to the local ambient magnetic field (magnetic north). When bottom tracking is used, the direction of the boat-velocity vector as measured by bottom tracking and water-velocity vector are referenced to the ADCP. When GPS is used to determine the boat-velocity vector, this vector is referenced to true north as determined from the GPS data. The orientation of the instrument relative to true north must be determined to put the boat-velocity vector and the relative water-velocity vector in the same coordinate system and allow for the computation of the water-velocity vector. This requires that the compass be accurately calibrated for local magnetic conditions. Proper setup and calibration of the ADCP's internal compass, determination of the local magnetic variation, and a slow boat speed are critical to quality velocity and discharge measurements made using GPS data as the boat-velocity reference. The accuracy of the compass in commercially-available ADCPs is approximately ± 1 to 2 degrees. Errors are caused by different factors such as magnetic effects of objects in the vicinity, sudden movements of the compass (due to strong accelerations/decelerations or abrupt pitch/roll) or incorrect local magnetic variation registered by the user. The effect of these errors at flow velocity and discharge is directly proportional to the boat velocity and its variations. For this reason it is recommended to maintain the boat speed slow, steady and practical for the field condition. Unless the measurement conditions do not permit it, bottom track as a reference system is recommended, otherwise, the technological advances of the GPS working in differential form, provide reliable results if the relevant precautions are taken. For more details about limitations, methodologies and errors involved by both references systems see Mueller *et al.* (2013).

3.3.4 Error Sources

Many factors affect the uncertainty of ADCP flow velocity measurements, such as measurement location, velocity fluctuations due to flow turbulence, flow unsteadiness, instrument configuration, and instrument signal processing techniques. Thus, random errors of ADCP flow velocity measurements from moving boats include error contributions from all the processes present during the flow-field sampling (for example, instrument noise, environmental noise, and flow turbulence). The term "environmental noise" is used to refer to the effect measuring conditions have on instrument performance, *e.g.*, wake turbulence, surface waves, and Doppler noise, and the term "flow turbulence" is used to refer to the turbulence generated by flow/bed interaction. Even measurements obtained using an ideal instrument (one that has no electronic noise and that is not adversely affected by measurement conditions) will have random errors because of turbulent fluctuations in the flow. Table 3.3.2 describes the main sources of errors. This table also includes the most common user errors.

Table 3.3.2 shows the significant diversity of sources of error that can affect the flow velocity measurements using ADCP. Recently, different research has focused on quantifying the expected errors for different ADCP configurations and operating conditions in an attempt to characterize the ADCP accuracy (Muste *et al.*, 2004a; Oberg & Mueller, 2007; Gonzalez-Castro & Muste, 2007; Mueller *et al.*, 2007; Mueller & Oberg, 2011; Tarrab *et al.*, 2012; Garcia *et al.*, 2012).

Table 3.3.2 Main sources of errors in flow velocities measurements using ADCP (adapted from Muste *et al.*, 2004a).

Error source	Description	Error Type
Instrument Error	Instrument Noise = Function (frequency, length and power of the transmitted acoustic pulse, processing techniques (narrowband or broadband) and modes (coherent or incoherent), type of phase-encoded broadband pulses, etc.)	Random
	Beam patterns = Function (type and size of the transducer, beam angle, acoustic frequency, etc.).	Bias
	Beam configuration = Function (number of beams, orientation)	Not known
	Sound speed = Function (water temperature and salinity)	Bias
	Transducer noise = Function (electronic transducer, etc.)	Random
	Side-lobe interference = Function (beam geometry and water depth)	Bias
	Heading/pitch/roll = Function (sensor, tracking system, boat speed)	Random
	Boat tracking system = Function (uncertainty of Bottom Track and GPS reference system).	Random
Environmental error	Moving bed = Function (water mode, suspended and bedload sediment transport).	Bias
	Environmental noise = Function (large –*i.e.*, wakes – and small scale turbulence intensity, shear velocity, water depth, etc.)	Random
	Heading/pitch/roll = Function (waves, etc.)	Bias and random
	Flow conditions = Function (geometry, non-uniform flow, unsteadiness, etc.)	Bias and random
User	ADCP mount = Function (material, location in the boat)	Bias
	Software configuration = Function (Communication set up, water and bottom modes, mounting depth, magnetic declination, salinity, shore distance, extrapolation discharge methods)	Bias and random
	Sampling interval = Function (turbulence intensity, flow regime, etc.)	Bias and random
	Boat movement = Function (boat speed, acceleration and course changes)	Bias and random

Reducing ADCP errors (except those related to ADCP sensor design) can be achieved by strictly implementing the recommendations of the manufacturer and USGS in the measurements, which include calibration, post-processing and checkups, among others. Reducing errors generated by external factors (*e.g.*, operator, site selection and inadequate operation configuration) requires complete understanding of ADCP principles and operation and proper selection of parameters and operating modes for each measurement site (Lipscomb, 1995; Muste *et al.*, 2004a).

3.3.5 Horizontal ADCP

The Janus configuration in Figure 3.3.8a has been developed for vertical deployments, with the ADCP transducer directed upward or downward. Customized configurations allow for horizontal deployment of ADCPs, by orienting the beams in a single plane to resolve the streamwise (V_s) and stream-normal (V_n) velocity components along the central axis as illustrated in Figure 3.3.12a. ADCPs configured to be deployed horizontally are referred to as HADCPs. Similar to the Janus configuration for vertical deployment, an additional beam is often included for error estimation. Figure 3.3.12a exemplifies the configuration of a 3-beam HADCP deployed in a channel bend. The horizontal velocity components and error velocity are calculated as:

$$V_s = \frac{V_{B2} - V_{B1}}{2\sin\beta}$$

$$V_n = \frac{-\cos\beta(V_{B1} + V_{B2}) - V_{B3}}{1 + 2\cos^2\beta} \qquad (3.3.7)$$

$$V_e = \frac{V_{B1} + V_{B2}}{2\sin\beta\sqrt{1 + 2\cos^2\beta}} - \frac{V_{B3}}{\tan\beta\sqrt{1 + 2\cos^2\beta}}$$

where β is the angle between the beams.

HADCPs are increasingly used for continuous discharge monitoring in rivers and streams, especially at locations where the relation between water level and discharges is ambiguous (see Chapter 7.5, Volume II). Such conditions may occur in lowland areas

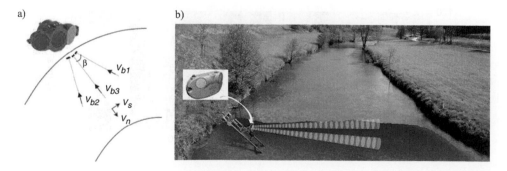

Figure 3.3.12 H-ADCP arrangement for velocity acquisition: a) plan view of a 3-beam HADCP configuration (http://www.teledynemarine.com) in a curved channel (positive beam velocities are toward the transducer); b) actual deployment of a two-beam H-ADCP (http://www.sontek.com).

subject to backwater or drawdown effects, and in harbors where the tidal motion and wind-induced seiches cause oscillations in the discharge. Whereas the continuous increase of the acoustic beam footprint with the distance from the sensor is usually not a concern for vertical ADCP deployments, the horizontal deployment of the HADCP may cause interference of the acoustic beams with the bottom or surface with distance from the sensor location. For large distance from the probe the signal might become nearly extinct as a result of backscattering attenuation. For example, consider an HADCP with a 1.2 degree acoustic beam divergence deployed 1 m from the surface with the probe axis set horizontally. For such setting the acoustic beams will intersect the water surface at a distance 95.5 m distance from the probe. At this distance, the footprint for the beams is 2 m which might be problematic for a shallow stream.

HADCPs are generally considered attractive alternatives for acoustic travel time meters as the latter require installation of equipment at least at two locations (see Section 3.4). Figure 3.3.12b shows a 2-beam HADCP deployed in a stream. These probes are maintained under water for long time intervals favoring the growth of microbial cultures on the surface of the instrument case including the sensors. The probes continue to operate even after the fouling of the transducers occur as long as the backscattering signal is not completely extinct. The mounting of the probe illustrated in the figure designed for permanent deployments with possibilities to allow for adjustments with the stage in the stream.

Figure 3.3.13 illustrates sample time-series of cross-channel flow velocity acquired with an HADCP deployed on the bank of the tidal Berau River (Indonesia) (Hoitink

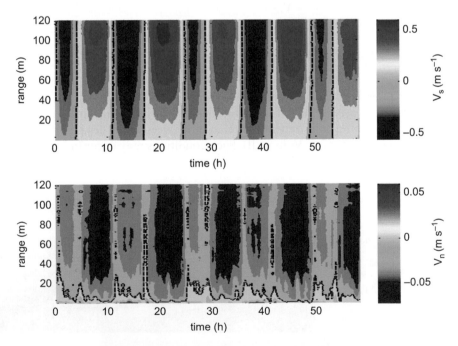

Figure 3.3.13 Sample time-series of streamwise (top) and stream-normal (bottom) velocities acquired with an HADCP deployed at the bank of the Berau River (Indonesia). The river is affected by semidiurnal tides (cf. Hoitink *et al.*, 2009). Solid lines in the bottom figure indicate flow reversal.

et al., 2009). The HADCP are typically deployed near the banks of a channel, because it is logistically easier than a central channel deployment. Moreover, this positioning is also better with respect to the interference with possible vessel traffic in the stream. This in turn results in the HADCP crossing the bank-attached boundary layer with the flow strength increasing with distance from the bank up to the central region of the channel. The bank-attached boundary layer has lower and more irregular flow distribution compared to the mid-channel region, a characteristic that is exacerbated for irregular bank geometry. It is therefore important that the HADCP range extends beyond the bank boundary layer, and reaches the mid channel section where the flow is more spatially coherent. The stronger and more regular flow distribution in the mid-channel area produce a stronger relation between the in-beam measured velocity and the cross-section average velocity than the bank-attached area.

3.4 ACOUSTIC TRAVEL-TIME TOMOGRAPHY

Acoustic tomography is a powerful tool in physical oceanography that has the potential to be applied for estimation of the velocity and, hence, to monitor discharge in shallow flows. It relates the changes of scattered sound waves emitted at one place and received at another to the environmental conditions of the acoustic medium, including temperature, salinity and current velocity. The word tomography refers to section by section imaging of a volume. In the present context, the "images" are created by sound waves travelling back and forth through the stream water body between a pair of acoustic sensors. One of the sensors (acoustic source) is located on stream bank and the other one (the receiver) is located on the opposite bank, along a line crossing the stream at oblique angle, as illustrated in Figure 3.4.1. The smaller the angle between the flow and the central acoustic axis, θ, the more accurate is the estimated bulk flow velocity in theory. In practice, the distance between the sensors is limited by attenuation of the acoustic signal strength.

Figure 3.4.2 exemplifies multiple sound rays as they propagate from source to receiver. The changes in the speed of the sound pulse are related to the velocity of the water body. Temperature and concentration gradients in the water body are also

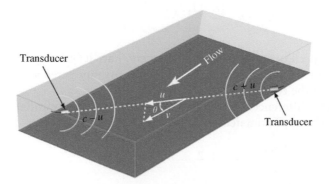

Figure 3.4.1 The angular difference θ between the main flow direction and a linear ray between the transducers. Adopted from Kawanisi *et al.* (2010), with permission from Elsevier.

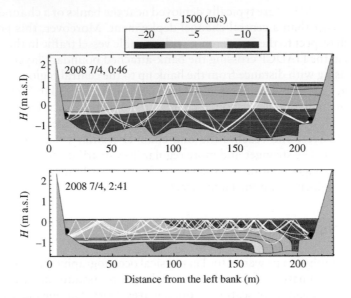

Figure 3.4.2 Examples of acoustic tomography images. Rays refract due to density gradients and reflect at the surface, at the river bed (top panel) and at a pycnocline formed by a salt wedge in a channel (bottom panel). Adopted from Kawanisi et al. (2010), with permission from Elsevier.

changing the speed of sound between the two sensors. The water surface and the bed act as reflectors, as illustrated in the top panel. For strongly stratified flow (bottom panel), the rays may reflect at the pycnocline rather than at the bed. The density change at the interface results in refraction according to the Snell's law.

In a symmetrical deployment, the two transducers are oriented such that the central beam axes of the two sensors coincide. The ray paths Γ_i in both ways are then identical. Indicating the two opposite directions along a path by + and -, the travel time in the two directions reads (Kawanisi et al., 2009):

$$t_i^{\pm} = \int_{\Gamma_i^{\pm}} \frac{ds}{c(x, y) \pm \vec{u}(x, y) \cdot \vec{n}} \quad (i = 1, 2, ..., M) \tag{3.4.1}$$

Herein, c is the speed of sound, ds is an incremental arc length along a sound ray, $\vec{u}$ is the velocity vector and $\vec{n}$ is the unit vector along the ray numbered as i. The line integral starts and ends at the two transducers.

The travel-time difference between two directions along a ray and the average of the travel times can be approximated as (Kawanisi et al., 2009):

$$\Delta t_i = t_i^- - t_i^+ \approx \frac{2L_i u_{m_i}}{c_{m_i}^2} \quad (i = 1, 2, ..., M)$$

$$t_{m_i} = \frac{t_i^- + t_i^+}{2} \approx \frac{L_i}{c_{m_i}} \quad (i = 1, 2, ..., M) \tag{3.4.2}$$

where the path-averaged velocity and sound speed along a ray read:

$$u_{m_i} = \frac{1}{L_i} \int\limits_{\Gamma_i} \vec{u} \cdot \vec{n} ds$$

$$c_{m_i} = \frac{1}{L_i} \int\limits_{\Gamma_i} c ds \qquad (3.4.3)$$

Herein, L is the length of two opposite ray paths. For shallow flows, the vertical flow component can be neglected, and $\vec{u}$ can be considered a horizontal velocity vector. When the rays systematically cover the vertical plane between the two transducers, or, more precisely, if the ray density is uniform across the oblique slice across the shallow flow, then the value of u_m averaged over all rays differs a factor $1/\cos\theta$ from the cross-section averaged velocity $\langle v_m \rangle$. Under those circumstances:

$$\langle v_m \rangle = \frac{\langle u_m \rangle}{\cos\theta} = \frac{1}{\cos\theta}\frac{1}{M}\sum_{i=1}^{M} u_{m_i} = \frac{1}{\cos\theta}\frac{1}{2M}\sum_{i=1}^{M}\frac{c_{m_i}^2}{L_i}\Delta t_i \qquad (3.4.4)$$

The angular difference θ between $\vec{u}$ and $\vec{v}$ is illustrated in Figure 3.4.1. The discharge Q is the product of $\langle v_m \rangle$ and the cross-sectional area. Values of Q are better resolved for smaller θ, yet when θ decreases the travel distance increases, causing additional attenuation of the acoustic signal between the tomographic pair as mentioned previously. To monitor discharge in a 120 m wide channel with 30 kHz transducers, Kawanisi *et al.* (2010) obtained promising results using $\theta = 30°$.

Equation (3.4.4) shows the variables required for estimation of $\langle v_m \rangle$, which include Δt_i, L_i and c_{m_i} for each ray. Measurements of Δt_i are obtained from the transducers utilizing a GPS clock. Estimates of L_i are obtained by constructing the ray paths from the following set of differential equations (Dushaw & Colosi, 1998):

$$\frac{d\varphi}{dr} = \frac{\partial c}{\partial r}\frac{1}{c}\tan\varphi - \frac{\partial c}{\partial z}\frac{1}{c}$$

$$\frac{dz}{dr} = \tan\varphi \qquad (3.4.5)$$

$$\frac{dt}{dr} = \frac{\sec\varphi}{c}$$

Herein, φ is the angle between a ray such as illustrated in Figure 3.4.2 and the central horizontal axis between the transducers. The speed of sound has to be estimated throughout the domain. In well-mixed conditions, rays are straight and the speed of sound can be assumed constant. In stratified conditions, a vertical array of temperature and salinity sensors may be used to estimate the speed of sound field (*e.g.*, Medwin, 1975), assuming horizontal homogeneity.

3.A APPENDIX

This appendix gives more information related to the Acoustic Backscattering Instruments (ABIs) presented in the preceding sections. The notation used in the appendix is based on the corresponding sections and is therefore not always defined here again.

3.A.1 ABIs: Backscattering

Appendix 3.A.1 provides details concerning the determination of the wave pressure in Section 3.2.2. Considering a sphere as a first order approximation for suspended particles (see Figure 3.2.1a), the pressure generated by the incident wave on the surface of a rigid, immobile sphere of radius a suspended in a fluid, is given by

$$p_i = p_0 \exp(ik(r \cos \Theta - ct)) \tag{3.A.1.1}$$

and in the far field ($kr \gg 1$, where k = wavenumber and r = distance of the sphere from the transducer), the pressure of the scattered wave from the sphere can be expressed as

$$p_s(r, \Theta, t) = \frac{p_0 a}{2r} \exp(ik(r \cos \Theta - ct)) f_\infty(\Theta, a) \tag{3.A.1.2}$$

where p_0 = reference pressure, i = imaginary unit, Θ = backscattering angle, c = sound speed, t = time, and $f_\infty(\Theta, a)$ is the form function that describes the distribution of the scattered intensity. For the domain of ABIs, sound speed is mainly affected by temperature, since its dependence on pressure and particle concentration is negligible. However, in the range from 10 to 20°C, the relative variation of sound speed c is only about 3%. Slightly different expressions for p_s are found if the particle is not a rigid sphere. In geophysical and laboratory flows, a large number of particles are suspended in the fluid. It is assumed that the particles are randomly distributed and that they have the same compressibility and density. If the particle concentration is low enough (e.g., 1% by volume), particles are sufficiently far apart to avoid multiple scattering.

In ABIs, a sound pulse of duration τ of several plane waves at a carrier frequency $f_0 = c/\lambda_0$, where λ_0 = wavelength, is emitted from a transducer into the water column (Figure 3.2.1a). One may assume that the particle concentration is uniform, the size distribution is steady, and the relative movement of the particles can be ignored. By also ignoring attenuation, the backscattered sound pressure for a single particle is

$$p_{bs}(r) = \left(\frac{p_0 r_0}{2r^2}\right) D^2 a f_\infty(\Theta, a) \tag{3.A.1.3}$$

where D = transducer directivity function. The backscattering angle Θ is equal to π when the emitter and receiver are located in the same position. For a cloud of uniformly sized scatterers randomly and homogeneously distributed in an inviscid fluid, the wave pressure $p_{bs}(r)$ backscattered from individual particles can be treated independently and the scattered pressure from a control volume may be taken on the average as the incoherent sum of the contribution from each particle. This indicates that the ensemble-averaged mean square of the pressure $<p_{bs}^2(r)>$ coming from a region at distance r obeys

$$\langle p_{bs}^2(r) \rangle = N p_{bs}^2(r) \tag{3.A.1.4}$$

where N is the number of suspended particles insonified by an incident pulse.

Tatarski (1961) suggested that in clear water, sound backscattering is produced by temperature structures occurring at a $\lambda_0/2$ scale. However, backscattering intensity produced by these temperature structures is smaller than the observed backscattering (Shen & Lemmin, 1997a). On the other hand, they found that backscattering by microbubble clusters corresponded better to the observed backscattering in clear water.

The signal due to backscattering captured by the ABI transducer is a voltage V proportional to the instantaneous backscattered pressure. Due to the 'speckle' character of this signal, a single emitted wave pulse does not provide useful information for V. In most cases, the mean square value of the output voltage is used. If S_M is the sensitivity constant of the instrument, then

$$\langle V^2(r) \rangle = S_M \langle p_{bs}^2(r) \rangle \tag{3.A.1.5}$$

The voltage received by the transducer is a function of distance r from the transducer. The system sensitivity S_M relates to the properties of the hardware and the transducer. For most instruments, it can be calculated or calibrated and it can be considered constant.

3.A.2 ABIs: Extraction of the Doppler shift frequency

In an ABI, the Doppler shift frequency can be extracted by mixing the received voltage signal at frequency f_r with the emitted signal at frequency f_0 (see Figures 3.2.1b and 3.2.2), producing the "in-phase" $I(t)$ and the "quadrature" $Q(t)$ outputs that are the real and imaginary parts of the complex echo signal. From this signal, the Doppler shift frequency f_D and its sign and thus the magnitude and the direction of the radial velocity vector can be determined. The Doppler shift frequency is most often obtained by the "pulse-pair method," an auto-covariance algorithm (Zrnic, 1977; Lhermitte & Serafin, 1984) due to its computational simplicity. The pulse-pair estimator calculates the first moment of the Doppler spectrum from estimates of the complex auto-correlation function at lag $T_{PRF} = 1/f_{PRF}$. For a complex time-series signal $z(t) = I(t) + iQ(t)$ with Gaussian in-phase and quadrature components, an estimate of the mean Doppler frequency is obtained by (Lhermitte & Serafin, 1984)

$$\overline{f_D} = \frac{1}{2\pi T_{PRF}} \arctan\left(\left\langle \frac{I_{n+1}Q_n - I_n Q_{n+1}}{I_n I_{n+1} + Q_n Q_{n+1}} \right\rangle_{Npp} \right) \tag{3.A.2.1}$$

where the expression in the brackets on the RHS has been averaged over N_{pp} consecutive pulse-pairs in order to augment the statistical stability of the resultant velocity estimates and limit the noise contribution. This averaging is necessary, since the number of scattering targets in the water is not always large enough to return an incoherent pressure field from a single pulse (Zrnic, 1977). From the initial time-series of N_{pp} consecutive pulses of $I(t)$ and $Q(t)$, a new series of $N_v = N_{tot}/N_{pp}$ radial velocity estimates v_R determined by Equations (3.2.1) and (3.A.2.1) is obtained whose variance can be computed by

$$\sigma_{v_R}^2 = \frac{1}{N_v} \sum_{k=1}^{N_v} (v_{R_k} - \overline{v_R})^2 \tag{3.A.2.2}$$

One can also write a variance

$$\sigma_{N_{pp}}^2 = \frac{c^2 \sigma_f^2}{4 f_0^2} \tag{3.A.2.3}$$

where σ_f is the Doppler frequency estimator variance. Statistically the two variances are linked by

$$\sigma_{v_R}^2 + \langle \sigma_{N_{PP}}^2 \rangle = \text{const} = \sigma_{N_{PP}=N_{tot}}^2 \tag{3.A.2.4}$$

where the angled brackets represent an average over N_v values. $\sigma_{N_{PP}=N_{tot}}^2$ is the variance given by Equation (3.A.2.3) that is calculated with a number of pulse-pairs N_{PP} equal to the total number of couples in the time-series. For studies of turbulent flow, a high sampling rate is desirable. It is therefore of interest to find the lower limit of N_{pp} in order to determine the maximum of the unambiguous velocity data rate. This establishes the final velocity time resolution of f_{PRF}/N_{pp}. For $N_{pp} < 16$, uncertainties cause strong deviations (Garbini et al., 1982). Lemmin and Rolland (1997) showed that the sum of the two terms on the LHS of Equation (3.A.2.4) remains reasonably constant for $N_{pp} > 32$.

From the Doppler frequency, the radial velocity of a particle cloud can be calculated. In ABIs, different techniques and transducer configurations have been developed to determine the actual flow velocity related to this radial velocity, as detailed in Section 3.2.3.

3.A.3 ABIs: Sampling volumes ("gates")

The signal received by an ABI contains contributions from all density interfaces in the water along the acoustic beam path. By gating the received signals to correspond to the pulse's time of flight from a certain distance r, a small sampling or scattering volume, called a "gate" can be interrogated (Figure 3.2.2).

The height of the gate is given by $c\tau/2$ where τ is the pulse duration. Since a piezo-electrical transducer cannot reach full power output instantaneously, a pulse with a duration of several wave cycles has to be emitted. Typically, in pulse-to-pulse coherent ABIs, the frequency f_0 is in the MHz range and thus the acoustic pulse of four to eight sinusoidal waves has a duration of several μs. This results in a spatial resolution for a gate height in the radial direction in the mm range. The gate diameter is defined by a cylindrical volume whose dimension is related to the beam-opening angle of the transducer (see Appendix 3.A.4 below). The Doppler frequency is determined for each gate separately.

3.A.4 ABIs: Transducer characteristics

To measure the speed of water in a certain direction, an omni-directional sound source is replaced by a narrowly-focused, cylindrical, piezo-electrical transducer that produces a conically- shaped sound beam with a small aperture angle. The radiation pattern of an ultrasonic transducer can be divided into a near field and a far field (see Figures 3.2.3a and 3.2.11). The limit between the near field (that is close to the transducer surface) and the far field is given by the distance $r = A/\lambda$ from the transducer face, with A = surface area of the transducer and λ = acoustic wavelength (Hay, 1991; Lhermitte & Lemmin, 1994). For typical transducers used in pulse-to-pulse coherent ABIs, this distance is > 10 cm. In the near field, the acoustic pressure is complicated, and forms irregular phase fronts that do not allow for clean measurements (Ma *et al.*, 1987; Lhermitte & Lemmin, 1994). In contrast, the far field acoustic pattern is simpler and most acoustic backscattering instruments operate in the far field. The sound wave in this region may be considered as a directional narrow beam with spherical divergence. The beam width is defined as the width between the half power (or – 3dB) points on both sides of the main lobe. The main lobe is the strongest part of the transmitted pulse in the cone centered in the axis of the transducer. For typical transducers used in pulse-to-pulse coherent ABIs, the opening angle of the transducer is about 3°. In general, for piezo-electric transducers, the larger the transducer surface area, the smaller the opening angle. A small beam-opening angle is desirable.

3.A.5 ABIs: Range velocity ambiguity

3.A.5.1 Monostatic profilers (UVPs)

The time interval between the emitted pulse and the sample volume in monostatic ABI profilers increases in proportion to the depth range in which velocity information is sought (Figure 3.2.3a). In order to prevent ambiguous range information, the echo from the maximum depth of interest $D_{R\max}$ must be received by the transducer before the next pulse is emitted. The pulsed transmitting mode leads to an inherent time discretization at $T_{PRF} = 1/f_{PRF}$ (Figure 3.2.3a). Therefore, the pulse repetition frequency f_{PRF} determines the maximum Doppler shift frequency possible and consequently the maximum radial velocity that can be detected before aliasing occurs. Applying Shannon's sampling theorem, the maximum unaliased measurable Doppler shift frequency is limited by $f_{PRF}/2$. Aliasing will lead to false velocity estimates (see Section 3.2.4.2). As a result, in velocity profiling, a trade-off between the maximum radial sampling depth $D_{R\max}$ and the maximum non-aliased radial velocity $v_{R\max}$ exists, known as the "range-velocity ambiguity" (Lemmin & Rolland, 1997)

$$v_{R\max}D_{R\max} = \frac{c^2}{Af_0} \tag{3.A.5.1}$$

with $A = 8$. If the sign of the velocity is known in advance, the maximum velocity can be doubled by implementing an aliasing corrector algorithm, thus extending the Nyquist frequency. In this case $A = 4$ (Lemmin & Rolland, 1997; Rolland, 1994). One way to overcome the range-velocity ambiguity limit is to use a system with a dual f_{PRF}, as discussed by Lhermitte (1999). The range-velocity ambiguity is one of the most limiting aspects of pulsed coherent Doppler velocity techniques.

3.A.5.2 Multistatic profilers

As with the monostatic profiler discussed above, the maximum Doppler shift frequency is limited by the pulse repetition frequency f_{PRF} for a given carrier frequency f_0 in multistatic profilers. Consequently, the maximum profiling depth $D_{R\ max}$ is again limited by $1/f_{PRF}$ and one obtains

$$v_{Rmax} = \frac{c f_{PRF}}{2 f_0}$$

$$D_{Rmax} = \frac{D_0 - (c/f_{PRF})^2}{2(D_0 \cos \alpha - c/f_{PRF})}$$

(3.A.5.2)

where D_0 represents the distance between the transmitter and the receiver (Figure 3.2.11). For the monostatic configuration, $D_0 = 0$ and the combination of the two equations leads to Equation (3.A.5.1). It should be noted that the maximum velocity in the bistatic configuration is calculated independent of the maximum range. Thus, the maximum velocity equation also applies to the fixed geometry of Acoustic Doppler Velocimeters (ADVs).

It can be seen that the range velocity ambiguity is also a function of the Doppler angle α that changes strongly along the profiling depth for Acoustic Doppler Velocity Profilers (ADVPs) (Rolland & Lemmin, 1997). This means that the maximum measurable flow velocity changes along the length of the profile. When an ADVP is mounted downwards looking in an open-channel flow, the upper fraction of the profile is more likely to be affected by aliasing, since the Doppler angle α is smaller and the velocity is usually higher compared to the one in the near-bed region. Larger u_{max} can be measured at small α which means at greater distances. Rolland and Lemmin (1997) found that a minimum angle of $\alpha = 15°$ at the gate furthest away from the transducer has to be respected for turbulence measurements. Due to this angle constraint, the actual maximum profiling range may already be reached before $D_{R\ max}$. In most cases, a bistatic ADVP allows measuring higher velocities than a monostatic Ultrasonic Velocity Profiler (UVP).

3.5 POINT VELOCIMETERS FOR FIELD APPLICATIONS

3.5.1 Introduction

This sub-chapter describes the various field instrumentation used for measuring water velocity at a point in the field. In general, these instruments are of three different types: mechanical current meters (for example, Price AA current meters), electromagnetic current meters (ECMs) and acoustic Doppler velocimeters (ADVs). ADV's were discussed in Section 3.2 and they are mentioned here just for completeness. In most open channel applications, the primary purpose of velocity measurements is to compute streamflow or river discharge using the mid-section method (Terzi, 1981; Rantz et al., 1982). However, the technological advances of ECMs and ADVs have allowed researchers to expand the traditional use of velocity meters to quantify and investigate additional parameters in the field such as turbulence and sediment transport.

3.5.2 Available instrumentation

Mechanical current meters have been used to measure water velocity in the United States since 1896 (Smoot & Novak, 1977) and continue to be a reliable instrument in the field for many professional hydrographers measuring streamflow. ECMs and ADVs were first developed for laboratory applications in the 1980's and 90's, respectively,

Table 3.5.1 Summary of various point velocity meters for field measurements.

Meter Type[1]	Description	Velocity Range (m/s)	Accuracy
Price Type AA	Mechanical current meter with vertical-axis cups	+ 0.03 to + 4.88	2 %
Price Pygmy	Mechanical current meter with vertical axis cups	+ 0.02 to + 0.9	2 %
OTT C31	Mechanical current meter with horizontal axis propeller	+ 0.025 to + 10	± 0.01 s/ ±0.5 pulses
OTT C2	Mechanical current meter with horizontal axis propeller	+ 0.025 to + 5	± 1 %
OTT ADC	Acoustic Doppler velocimeter	–0.2 to + 2.5	1 % ± 0.25 cm/s
OTT MF	Electromagnetic current meter	+ 0 to + 6	2 % ± 1.5 cm/s (at <3m/s)
Marsh McBirney 2000	Electromagnetic current meter	– 0.15 to +6.1	2 % ± 1.5 cm/s
Sontek FlowTracker	Acoustic Doppler velocimeter	± 0.001 to 4	1 % ± 0.25 cm/s
Nortek Vector	Acoustic Doppler velocimeter	± 0.01 to 7	0.5 % ± 0.1 cm/s
Nortek Vectrino	Acoustic Doppler velocimeter	± 0.01 to 4	0.5 % ± 0.1 cm/s

[1] The use of brand names is for identification and example only and does not constitute endorsement by the authors.
Note: Technical specifications are specified according each manufacturer's literature. Price AA and Pygmy specifications as recommended by Gurley Precision Instrument Company.

and shortly after also became field tools for use in oceans, rivers and lakes. ADVs are typically considered an improvement to ECMs for both laboratory and field investigations because ADVs typically have higher accuracy specifications (Table 3.5.1), they can measure three-dimensional velocity components, are non-intrusive, and their calibration does not change (see Section 3.2). Table 3.5.1 summarizes a variety of available technology for point velocity measurements. Details, including the advantages and disadvantages of the three general types of instruments are discussed below.

3.5.2.1 Mechanical current meters

Mechanical current meters measure the velocity of flowing water by means of a rotating element that, when placed in water, will rotate or revolve at a frequency that is directly related to the velocity of the water. Mechanical current meters require calibration. The device is towed through still water at a known speed and the number of rotations is counted or automatically recorded. In some cases, individual meter rating equations are computed for each individual meter through calibration and used to ensure meter accuracy (Engel, 1999). Alternatively, a standard rating table may be used that applies to a whole class of instruments (Hubbard *et al.*, 2001).

There are two general types of mechanical current meters, the cup type with a vertical axis of rotation (Figure 3.5.1A) and the screw or propeller type which rotates about the horizontal axis (Figure 3.5.1B). Vertical-axis current meters are most commonly used in North America, while horizontal-axis current meters are most typically used in Europe, in countries such as Germany and the United Kingdom, and in China.

Vertical-axis meters have cups or vanes, and have the following benefits: (a) they operate at lower velocities than horizontal-axis meters; (b) bearings are well protected

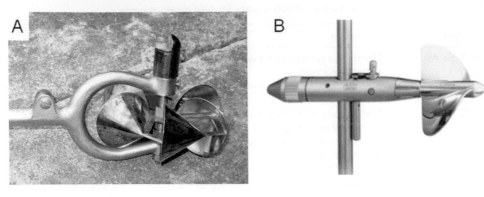

Figure 3.5.1 (A) Price AA vertical-axis current meter with yoke and bucket wheel; (B) OTT C31 horizontal axis-current meter (www.ott.com).

from silt or fine sediments; and (c) the meter is reparable in the field without compromising the rating. On the other hand, horizontal-axis meters: (a) disturb the flow less because the axis is symmetrical with the flow direction; (b) the meter is less likely to be entangled by debris; and (c) bearing friction is less because bending moments are eliminated (Rantz *et al.*, 1982).

Different methods are used to manually count or record the number of revolutions. Typically an electrical contact circuit is used with contact points designed to complete an electrical circuit at selected frequencies of revolution. A contact chamber contains the contact points, and, based on its design, will complete the circuit once or twice per revolution, or once per five revolutions. When the circuit is complete an electrical impulse is produced, which produces an audible click for the operator to count manually or which gets registered on a counting device. The revolutions are then counted over a specified time interval (for example 40s) which can be captured with a stopwatch or other time measurement device.

A major limitation of mechanical current meters is that they do not detect (and therefore cannot account for) velocity (or flow) direction. This can be particularly problematic for measurement sections where flow is not uniform. Fulford *et al.* (1994) summarize laboratory comparison test results for current meters and found that the repeatability and linearity of all the meters tested were good. However, all tested current meters were found to under and over-register velocities. In particular, the errors usually increased as the velocity and angle of the flow increased. At oblique flow angles between $\pm10°$, the vertical axis meter errors ranged from -3.30 % to -0.17 % for the Price type-AA meter and from 7.87 % to 8.92 % for the Price pygmy meter. The horizontal axis meters performed slightly better: at angles between $\pm10°$, errors ranged from 0.58% to 0.91% for the Ott C-31 meter with plastic impeller and from 2.02% to 3.77% for the Chinese LS25-3A meter with plastic impeller.

Another limitation of the traditional rotating-element current meter is that given their mechanical nature, cleaning, maintenance and regular calibration (or quality assurance testing) are essential to ensuring an accurate and reliable velocity measurement. As well, mechanical current meters typically require greater minimum water depths for operation. For example, the Price AA and Pygmy each require a minimum depth of 0.15 m and 0.076 m, respectively; while the acoustic FlowTracker ADV requires only 0.02 m.

3.5.2.2 Electromagnetic current meters

Electromagnetic current meters (ECMs) operate using the principle of electromagnetic induction (or Faraday's law of induction), which states that the voltage induced across any conductor as it moves at right angles through a magnetic field is proportional to the velocity of that conductor (Shercliff, 1962). The ECM generates an electromagnetic field and a voltage is induced by an electrical conductor (in this case the flowing water) moving through the magnetic field. For a given magnetic field strength, the magnitude of the induced voltage is proportional to the velocity of the conductor (or water velocity).

The ECM generates an electromagnetic field using a magnet energized periodically (for example, at 50 Hz), while the voltage generated by the passing fluid is measured using electrodes embedded in the sensor. It is therefore necessary for the fluid to be electrically conductive for the Faraday principle to apply. The electrical current is transported by ions in the fluid. Sea water (or salt water), has high conductivity (~5 S/m), while deionized water has low conductivity.

The highest resolution ECMs measure two velocity components in a small sphere (about 2 cm diameter) at 20 Hz. ECMs are reasonably robust for field applications; however, the presence of the ECM probe disturbs the flow field being measured. As well, the zero reading should be checked periodically in still water, as voltage measurements are prone to zero drift and are sensitive to electronic interference (ECMs should not be used near metallic objects) and improper grounding. One advantage of the ECMs (like ADVs) is that they are easy to clean by simply using clean water and mild soap to remove dirt and nonconductive grease and oil from the probe's electrodes and surface (Fulford *et al.*, 1994).

3.5.2.3 Acoustic Doppler velocimeters

The advantage of both ADVs and ECMs is that they have no moving parts and thereby eliminate uncertainty due to friction and resistance. A detailed description of the working principles of ADVs can be found in Section 3.2. As indicated in that section, ADVs determine water velocity by measuring the change in acoustic frequency (or Doppler shift) from reflections from moving particles or scatters in the flow, which are assumed to be moving at the same velocity as the water. The measured shift in frequency between the transmitted and received signal is proportional to the velocity of the moving particles along the path of the received signal. Figure 3.5.2 presents common field-ADV setups measuring the two-dimensional (2D) (Figure 3.5.2A) or three-dimensional (3D) (Figure 3.5.2B) water velocity at the location where the transmit and receive signals converge (sampling volume), which is a small distance (typically ~ 10cm for field ADVs) away from the transducers. Alternatively, the ADV probe geometry can be set to have diverging beams (Figure 3.5.2C). In this orientation, the flow between the beams is assumed to be homogenous and the water velocity is taken to be the average of the measured velocity along each beam. This assumption would not be appropriate for highly turbulent or multi-dimensional flow.

Most ADVs can be configured for different sampling rates or frequencies. For typical field applications, output sample rates may range from 1 Hz (*i.e.*, SonTek's FlowTracker) to 64 Hz (Nortek's Vector).

Sufficient backscatter is an essential requirement for acoustic measurement devices. In low backscatter conditions, where signal-to-noise ratios (SNR) are low (typically <

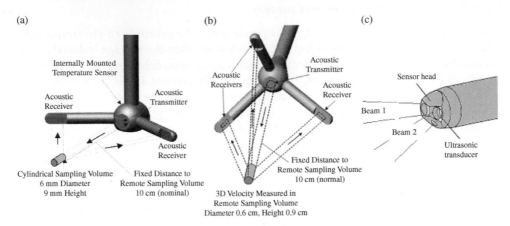

Figure 3.5.2 ADV probe orientation and sampling volume: (A) 2D side looking SonTek FlowTracker (FlowTracker Technical Manual 2009); (B) 3D down looking 10-MHz SonTek ADV (www. sontek.com). (C) 2D side looking OTT transducer head (modified from OTT Operating Instructions)

10 dB), there is the potential for data quality problems. Some ADV manufactures (SonTek/YSI) indicate that the SNR should be greater than 10 dB for optimal data quality, but field and laboratory testing by the United States Geological Survey (USGS) have indicated that the FlowTracker ADV (a SonTek device) may collect reliable data when the SNR is lower: velocity data are unreliable when SNR < 4 dB (Rehmel, 2007).

Boundary effects are another consideration for ADVs which may result in poor or incorrect velocity data to be measured. Boundary effects occur when the sampling volume for an ADV includes a portion of a solid, stationary boundary such as a boulder in the stream. If the sampling volume includes an object then the velocity data will be incorrect and biased low, since a portion of the sampling volume is incorrectly measuring zero velocity at or near the solid boundary or object. Boundary problems are generally more common at shallow depths and in gravel or cobble-bed streams where the bottom is composed of large sediment sizes.

3.5.3 Measurement considerations

3.5.3.1 Sampling duration

Depending on the instrument used and the objectives or purpose of the measurement, different sampling strategies may be employed for discrete (or point) velocity measurements. As the duration of velocity measurements increases, the uncertainty of the velocity data decreases (Carter & Anderson, 1963).

3.5.3.1.1 Sampling duration for stream gauging

Velocity measurements in streams or rivers are often used to gauge the discharge or flow in a water course. Only mean streamwise velocity at a point is required for discharge measurements, and a minimum of 40 seconds of sampling is typically recommended (Terzi, 1981; Rantz *et al.*, 1982). Sampling time uncertainty is also site-dependent, as temporal scales of the flow characteristics are defined by the flow regime (*i.e.*, Reynolds number), geometry of the channel, water depth and bed roughness. Typically, sample duration should be increased if flow is less uniform or at low velocities, where turbulent flow and velocity fluctuations may have more of an effect on mean velocity. Any sampling strategy will have to balance the requirements of the measurement (*e.g.*, mean velocity vs. turbulence characteristics), the spatial resolution required (*e.g.*, more data points = more time) and consideration of the site location and/or flow conditions (*e.g.*, steady vs. unsteady flow conditions).

To evaluate the performance of ADVs for discharge measurements in the field, the USGS carried out validation measurements comparing ADV (FlowTracker) and mechanical current meter (Price AA and Pygmy) measurements at 43 different USGS streamflow-gaging stations (Rehmel, 2007). Results indicated that ADV discharges were not statistically different (within 5% at a 95% confidence interval) from the mechanical current meter discharges. The largest differences (> 8%) were at site locations with higher turbulence (as measured by FlowTracker standard error velocity), where measurement uncertainty is higher.

3.5.3.1.2 Sampling duration for the characterization of turbulence

In all natural water courses and open channels, turbulent flow exists. Turbulence can be characterized by local eddying which results in pulsations or fluctuations in the velocity at any point. Laboratory flume experiments by Pierce (1941) showed that at high velocities (0.3 m/s) pulsations have only a minor effect on the mechanical current meter measurement, while at low velocities (0.03 m/s), the greater magnitude of the pulsations relative to the mean required longer durations for suitable measurements. Fulford (1995) also discusses the effects of pulsating flow on current meter performance.

The measurement of turbulence is also important for understanding the dynamics of flowing water. For quantifying turbulence in open channel flows, time series measurements of velocity are required and should be longer than typically required for discharge measurements. Buffin-Bélanger and Roy (2005) recommend that, for most turbulence statistics, the optimal record length (minimum sampling effort to achieve low standard errors) ranges between 60 and 90 s. The authors collected velocity time series data in three gravel-bed rivers using both ECMs and ADVs, with sample times ranging between 20 and 60 min. and sampling frequencies of 20 and 25 Hz for the ECMs and ADVs, respectively. Ultimately, the sampling frequency of the velocity measurement should be high enough to capture the turbulent fluctuations that are of interest. However, the required measurement durations may vary, depending on the turbulent time scales which in turn depend on the length scales of the channel dimensions. Interestingly, MacVicar *et al.* (2007), in their comparison between ADVs and ECMs for field measurements of velocity and turbulence, found that the older ECM technology provided more reliable estimates of flow parameters in high turbulence. The authors suggest that spectral anomalies in the ADV data, measured increases in noise from the power spectra and decreasing correlations, indicate that ADV error increases as a result of turbulence, particularly in the vertical velocity component. Nikora and Goring (1998) introduce a simple technique for reducing the influence of Doppler noise on estimating turbulence characteristics using ADV velocity data.

3.5.3.2 Deployment methods

A variety of deployment methods exist for point velocity meters, which include the following:

- Wading measurement: The instrument is attached to a wading rod and held stationary by someone wading in the water. Wading methods are typically the preferred choice when stream depth and water velocity allow for safe entry. Figure 3.5.3 shows example results (in graphical form) of a wading measurement with a FlowTracker ADV.
- Cableway, Tagline or Tethered measurement: In deeper channels or flows where wading is not possible or safe, the instrument can be suspended from a cable, line or rod into the flow. A sounding weight is suspended below the instrument to keep it stationary in the flowing water. In this setup, the instrument may be suspended from a cableway or bridge.
- Manned boat: For measurements in streams and rivers that are too deep to wade, and where no cableways or bridges are available, a boat attached to a tag line may be deployed. Similar to a cableway or tethered measurement, the instrument (with sounding weight) is suspended from a cable from the boat. In this set up, it is difficult to know the location of each velocity measurement, which is why a tagline is useful. The advancement of acoustic technology and Acoustic Doppler Current Profilers (ADCPs) in particular, which typically have a built in compass and bottom tracking capabilities (and in some cases incorporate GPS data for positioning), negate the need for a tagline for manned boat deployments and allow the operator to maneuver the boat across the river while accounting for boat position and boat speed (see Section 3.3).
- Under ice: For winter measurements, holes may be drilled or cut into ice to deploy instruments. Care should be taken to ensure that ice thickness is sufficient for working safely. A different yoke design and winter rod set for the Price Type AA current meter may be needed for under ice measurements. Low backscatter conditions are typical of winter conditions; therefore, care should be taken when using acoustic instruments to ensure reliable data quality.

3.5.3.3 Calibration and testing

All meters (mechanical, acoustic and electromagnetic) should be calibrated throughout the range of velocity for which it is to be used, and should not be used outside the range of calibration. Most devices are calibrated by the manufacture within their technical specifications. However, deviations from these specifications, including accuracy limits, have been found before (Fulford, 2001).

Typically, the mechanical nature of traditional mechanical current meters requires that they be calibrated more often than ADVs and ECMs as they are more prone to physical damage and alteration. The calibration strategy may be to individually calibrate each current meter in a tow tank; thereby providing each meter with its own

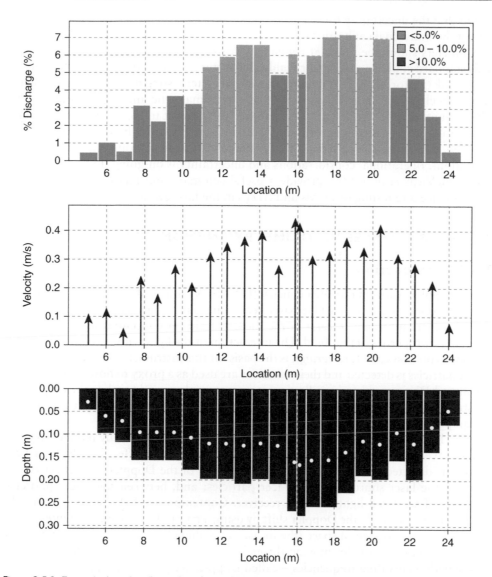

Figure 3.5.3 Example (graphical) results of a wading discharge measurement with a FlowTracker ADV, using the 0.6 method, where one velocity measurement is obtained at each vertical section or panel. The location of each velocity measurement (at 0.6 depth) is indicated by the white circle in the bottom graph of depth.

specific rating table (or meter equation) to convert the number of revolutions to a velocity (Engel, 1999), which is the practice in Canada. On the other hand, the USGS uses a standard rating table for each type of mechanical current meter used (*i.e.*, Price AA vs. Pygmy) (Hubbard *et al.*, 2001). Using linear calibration equations, Smoot and Carter (1968) found no significant difference between individual and group calibrations. However, using a non-linear calibration equation, which provides a more accurate representation of the rotor response at low velocities (< 0.3 m/s), Engel (1999) found that the uncertainty for the group calibration is greater than the uncertainty for any of the individual calibrations.

All meters should be tested at regular intervals against a known velocity standard. This standard could be a laboratory calibrated meter or a laboratory test facility such as a tow tank (Fulford, 2001). A literature review of meter testing for point velocity meters (which include mechanical, acoustic and electromagnetic meters) is provided by Thibodeaux (1994). While little has changed with respect to mechanical meters since this publication, electromagnetic and acoustic meters have become more widely adopted, and at such a rapid pace, that research into understanding their capabilities, limitations, operational considerations and uncertainty is on-going. In their introduction, MacVicar *et al.* (2007) provide a useful summary and literature review of the performance and testing of ECMs and ADVs in the last few decades.

3.6 LASER-DOPPLER VELOCIMETRY/ANEMOMETRY

3.6.1 Introduction

Laser Doppler Anemometry/Velocimetry (LDA/LDV) is an indirect technique to measure flow velocities. It is a particle-based technique, as it requires the presence of small particles (herein termed *tracers*) in the moving fluid, in concentrations that do not change the properties of that fluid. The interaction of the light with individual particles (an optics process called *scattering*), is the basis for this instrument. The motion of such tracer particles is detected and their velocities are used as a proxy to flow velocities. The acronym LDV is used herein, reflecting correct etymology and for the sake of generality-strickly speaking, the prefix "anemo" refers to wind. Exceptions concern references to equipment specifically developed under the label LDA.

Laser light, *i.e.*, coherent (same phase for all waves) and essentially monochromatic (same wavelength) light, is used to generate the Doppler shift effect by which particle motion is detected and quantified. LDV makes use of the Doppler shift effect twice: between the laser source and the tracer particles; and, between the tracer particles (emitters of scattered light) and the receiver. Contemporary LDV systems normally avoid detecting directly the Doppler shift by using a pair of laser beams that intersect in a small but finite region of space (the measuring volume) on which a fringe pattern is originated. Tracer particles moving through the measuring volume scatter the fringe wavelength, generating frequencies – called Doppler frequencies proportional to particle speed.

LDV is one of the least intrusive techniques to measure flow velocities. Its intrusiveness is mostly the result of seeding the flow with particles that may affect flow or fluid properties. If good optical access is guaranteed, no material parts of the instrumentation need be near the region of velocity measurement.

The fundamentals of LDV were established in the 1960's and 1970's, following the founding work of Yeh and Cummings (1964). The currently used dual-beam mode was patented by von Stein and Pfeifer (1969). Contemporary instrumentation has been advanced by improved hardware design and further theoretical investigations, notably those by Foreman *et al.* (1965), Rudd (1969), Lading (1971), Adrian and Goldstein (1971), George and Lumley (1973), Grant and Orloff (1973), Durst and Zaré (1974), Farmer (1976), Durst *et al.* (1976), Neti and Clark (1979) or Yanta and Ausherman

(1981). A fairly sophisticated technique for investigating open-channel flows has been used since the 1980's (Steffler *et al.*, 1985; Nezu & Rodi, 1985; Tominaga & Nezu, 1991; Nezu *et al.*, 1997; Ishigaki *et al.*, 2002; Aberle *et al.*, 2008; Ferreira *et al.*, 2012; Ricardo *et al.*, 2014).

LDV is essentially a hardware-based technique, as it requires complex electronic units and sensitive optical systems, with little software complexity. The technique requires controlled, clean research environments, such as laboratories where the flows are confined in flumes, generally glass-walled, with optimal visualization conditions. In the context of open-channel flows, LDV can, in principle, be deployed in field conditions (Nezu & Nakagawa, 1993). However, the need for optical access, the relative fragility of the optical components, the eventual need of cooling systems and demanding power sources makes this technique less attractive for field studies. Other techniques, notably Acoustic Doppler Velocimetry (Section 3.2), are relatively more intrusive (despite often being defined as non-intrusive) but have more robust hardware.

The purpose of this section is to discuss key aspects of LDV, including the necessary instrumentation, the behavior of admissible tracer particles, and the main signal-processing issues. This section also briefly outlines known data-processing difficulties, notably calculation of power density distribution of turbulence frequencies. Also addressed are certain practical aspects such as the choice of tracer particles and issues concerning positioning of probes. Note that instrument brand names are used herein only for ease of identification and information purposes.

3.6.2 Fundamental concepts of Doppler velocimetry

3.6.2.1 Doppler shift and Doppler frequency

Consider a moving body emitting laser light (Figure 3.6.1). It moves with velocity $u \cdot n_{\mathrm{pr}}$ along the path that connects the body and a receiver; u is the velocity of the body in the referential of the receiver, and n_{pr} is the unit vector that sets the direction of that path.

The light emitted when the emitter body is at a given point A is characterized (and registered at the receiver) by the wavelength λ_1 and by the frequency $f_1 = c/\lambda_1$; where c is the speed of light in the medium (note that c was used to denote speed of sound in the chapters dealing with acoustic instruments). When the emitter body is at A, its distance to the receiver is ds and the travel time of the light from the emitter body to the receiver

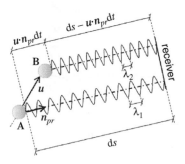

Figure 3.6.1 The Doppler-shift effect exemplified for a moving body emitting laser light to a receiver.

is $dt = ds/c$, because the speed of light is independent of the motion of the emitter. The number of wavelengths between point A and the receiver is $N = ds/\lambda_1 = cdt/\lambda_1$. When the body is at B, dt has elapsed and the distance to the receiver is $ds - \boldsymbol{u}\cdot\boldsymbol{n}_{pr}dt$. At this new position, the same waves are now compressed or expanded (depending on the signal of $\boldsymbol{u}\cdot\boldsymbol{n}_{pr}$) on the path whose length is $ds - \boldsymbol{u}\cdot\boldsymbol{n}_{pr}dt$. The wavelength, as perceived by the receiver, becomes $\lambda_2 = (ds - \boldsymbol{u}\cdot\boldsymbol{n}_{pr}\,dt)/N = \lambda_1(ds - \boldsymbol{u}\cdot\boldsymbol{n}_{pr}dt)/(cdt) = \lambda_1(1 - \boldsymbol{u}\cdot\boldsymbol{n}_{pr}/c)$. Hence, when the particle is at B, the frequency is

$$f_2 = \frac{c}{\lambda_2} = \frac{c}{\lambda_1\left(1 - \frac{\boldsymbol{u}\cdot\boldsymbol{n}_{pr}}{c}\right)} = f_1 \frac{1}{1 - \frac{\boldsymbol{u}\cdot\boldsymbol{n}_{pr}}{c}} \tag{3.6.1}$$

This configures the Doppler-shift effect: the change in frequency perceived by the receiver depends on the velocity of the emitting body along the path that connects the body and the receiver.

Now consider that the emitting body is a tracer particle actually scattering laser light from a source, which is immobile in the reference frame of the receiver (Figure 3.6.2). Employing the same reasoning that led to Equation (3.6.1), the frequency scattered by the particle when at A is related to the frequency of the light source, f_0,

$$f_1 = f_0 \frac{1}{1 + \frac{\boldsymbol{u}\cdot\boldsymbol{n}_{sp}}{c}} \tag{3.6.2}$$

where $\boldsymbol{n}_{sp}$ is the unit vector along the direction that connects the source and the moving body.

A Taylor series expansion leads to $(1 - \boldsymbol{u}\cdot\boldsymbol{n}_{pr}/c)^{-1} = 1 + \boldsymbol{u}\cdot\boldsymbol{n}_{pr}/c + O(\boldsymbol{u}\cdot\boldsymbol{n}_{pr}/c)^2$. Given that $\boldsymbol{u}\cdot\boldsymbol{n}_{pr} \ll c$, the result $(1 - \boldsymbol{u}\cdot\boldsymbol{n}_{pr}/c)^{-1} = 1 + \boldsymbol{u}\cdot\boldsymbol{n}_{pr}/c$ is nearly exact. The same is true for $(1 + \boldsymbol{u}\cdot\boldsymbol{n}_{pr}/c)^{-1} = 1 - \boldsymbol{u}\cdot\boldsymbol{n}_{pr}/c$. Performing these approximations and introducing Equation (3.6.2) into Equation (3.6.1) yields

$$\begin{aligned} f_2 &= f_0\left(1 - \frac{\boldsymbol{u}\cdot\boldsymbol{n}_{sp}}{c}\right)\left(1 + \frac{\boldsymbol{u}\cdot\boldsymbol{n}_{pr}}{c}\right) \\ &= f_0\left(1 + \frac{\boldsymbol{u}\cdot\boldsymbol{n}_{pr}}{c} - \frac{\boldsymbol{u}\cdot\boldsymbol{n}_{sp}}{c} + O\left(\frac{\boldsymbol{u}\cdot\boldsymbol{n}_{pr}\boldsymbol{u}\cdot\boldsymbol{n}_{sp}}{c^2}\right)\right) \end{aligned} \tag{3.6.3}$$

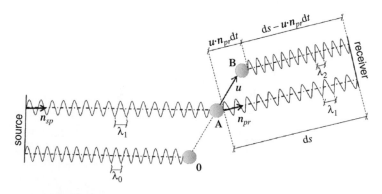

Figure 3.6.2 The Doppler shift exemplified for a tracer particle scattering light to a receiver.

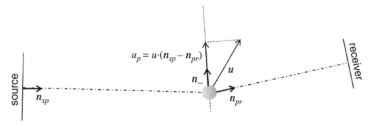

Figure 3.6.3 Geometry of the single beam LDV mode.

Since $\boldsymbol{u}\cdot\boldsymbol{n}_{pr}\, \boldsymbol{u}\cdot\boldsymbol{n}_{sp} \ll c^2$, and as $f_0 = c/\lambda_0$, the difference between the frequency emitted by the source of laser light and the frequency detected at the receiver is

$$f_0 - f_2 = \frac{\boldsymbol{u}\cdot\left(\boldsymbol{n}_{sp} - \boldsymbol{n}_{pr}\right)}{\lambda_0} \tag{3.6.4}$$

Equation (3.6.4) is the mathematical expression of the Doppler shift. If the wavelength of the emitted light is known, the Doppler shift allows for the back-calculation of the absolute value of the velocity of the particle along the direction $\boldsymbol{n}_- = (\boldsymbol{n}_{sp} - \boldsymbol{n}_{pr})/|\boldsymbol{n}_{sp} - \boldsymbol{n}_{pr}|$, identified as $u_p = \boldsymbol{u}\cdot(\boldsymbol{n}_{sp} - \boldsymbol{n}_{pr})$ in Figure 3.6.3. This is the principle behind the first LDV (Yeh & Cummins, 1964), which was used to solve the problem of measuring low velocities in water. In this initial setup they used a 632.8 nm (red) He-Ne laser with 5 mW power output. The emitted frequency was thus 3.5×10^{14} Hz. For a tracer particle velocity of 1.0 m/s and for a receiver aligned with the laser source, the Doppler shift is 1.6×10^6 Hz (1.6 MHz), therefore much smaller than the emitted frequency.

These values are common in water experiments and represent a major problem: the received frequency is $(10^8 \pm 1)$ MHz; attempting to recover the Doppler shift by subtracting the measured f_2 to f_0 is too imprecise as the difference is smaller than the error associated to the measurement of f_2. Earlier attempts to solve this problem involved combining the scattered light with a local optical oscillator and measuring the *beat signal* (Yeh & Cummings, 1964); *i.e.*, the frequency generated from the pattern of interference of the two combined wavelengths.

Improvements of this optical heterodyne detection method for single beams were attempted by Foreman *et al.* (1965) or Lewis *et al.* (1968), among others, but single beam LDV is not currently the standard approach. Instead, and taking advantage of the same optical heterodyne principle to eliminate the need to resolve the Doppler shift, the current most common LDV beam arrangement is the dual-beam mode, proposed by Lehmann (1968) and von Stein and Pfeifer (1969).

The standard configuration involves intersecting two laser beams of the same color, forming a measuring volume through which tracer particles travel. As they cross the measuring volume, they scatter light of both beams simultaneously (Figure 3.6.4). For each beam, Equation (3.6.4) holds. If the wavelength and the frequency of both beams is λ_0 and f_0, respectively, one has $f_0 - f_2^{(1)} = \boldsymbol{u}\cdot(\boldsymbol{n}_{sp}^{(1)} - \boldsymbol{n}_{pr})/\lambda_0$ and $f_0 - f_2^{(2)} = \boldsymbol{u}\cdot(\boldsymbol{n}_{sp}^{(2)} - \boldsymbol{n}_{pr})/\lambda_0$, where superscripts (1) and (2) stand for beams 1 and 2. The difference between frequencies $f_2^{(1)}$ and $f_2^{(2)}$ at the receiver (the "beat" frequency) is

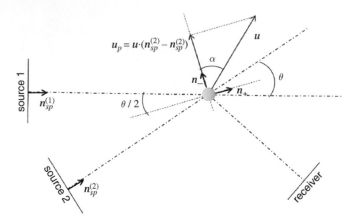

Figure 3.6.4 Geometry of the dual-beam LDV mode.

$$f_D = f_2^{(1)} - f_2^{(2)} = \frac{u \cdot \left(n_{sp}^{(2)} - n_{sp}^{(2)} \right)}{\lambda_0} = \frac{|u| |n_{sp}^{(2)} - n_{sp}^{(1)}|}{\lambda_0} \cos (\alpha) \qquad (3.6.5)$$

where α is the angle between the velocity u and direction $n_- = (n_{sp}^{(1)} - n_{sp}^{(2)})/|n_{sp}^{(1)} - n_{sp}^{(2)}|$, Figure 3.6.4. In this context, f_D is generally called the Doppler frequency (Durst *et al.*, 1976, 1981, see also Section 3.2.2.2), and can be formulated as a function of the geometry of the beams with simple trigonometric considerations. For this purpose note that vector $n_- = (n_{sp}^{(1)} - n_{sp}^{(2)})/|n_{sp}^{(1)} - n_{sp}^{(2)}|$ is orthogonal to the direction of the bisector of the angle θ formed by the beams, $n_+ = (n_{sp}^{(1)} + n_{sp}^{(2)})/|n_{sp}^{(1)} + n_{sp}^{(2)}|$. Hence, $u_p = |u| \cos (\alpha)$ is the component of the velocity of the tracer particle, in the plane defined by $n_{sp}^{(1)}$ and $n_{sp}^{(2)}$, that is perpendicular to the bi-sector of the angle defined by the beams (Figure 3.6.4). Since $|n_{sp}^{(2)} + n_{sp}^{(1)}| = 2\sin(\theta/2)$, Equation (3.6.5) becomes

$$u_p = f_D \frac{\lambda_0}{2\sin \left(\frac{\theta}{2} \right)} \qquad (3.6.6)$$

It is noteworthy that, in the dual-beam mode, the velocity of the tracer particle is independent of the location of the receiver. The receiver is a photodetector that can be located where it is most convenient for the user (see Sections 3.6.3 and 3.6.4). The accuracy of the method strongly depends on the precise specification of the beam arrangement, normally enclosed in a watertight probe with optical components (see Section 3.6.3.1). It is also important to note that the measured velocity depends only on the light wavelength and beam arrangement – LDV does not require calibration – and is a linear function of the Doppler frequency.

The velocity of a moving particle with the laser beam of the LDV system is supported with clear reasoning. The Doppler principle by which the dual-beam LDV operates is illustrated with great simplicity by the so-called *fringe model* based on the description of the interaction with geometrical optics (Durst & Stevenson, 1976; Miles & Witze, 1994). The fringe model best delineates two key considerations: the LDV measurement volume; and, the fringe shift used to eliminate the directional bias involved in LDV operation. The former consideration is detailed in Appendix 3.B.1 and the latter consideration in Appendix 3.B.2, respectively.

3.6.2.2 Multiple velocity components

As stated above, the dual-beam mode enables detection of the tracer velocity component perpendicular to the beam bisector and located in the plane defined by the directions of the intersecting beams of a given wavelength; *i.e.*, of a given color. The unit vector of the direction of the velocity detected by a pair of beams of that color is:

$$
n_{sp} = \frac{\left(n_{sp}^{(1)} - n_{sp}^{(2)} \right)}{\left| n_{sp}^{(1)} - n_{sp}^{(2)} \right|}
\tag{3.6.7}
$$

To determine the complete velocity vector it is necessary to detect particle velocities in two other independent directions. Therefore, two other pairs of beams must intersect at the same measuring volume. Their colors should differ so that, by optical filtering (see Section 3.6.3), the Doppler frequency of each can be detected independently. If the unit vectors of the detected velocity directions are designated as n_g, n_b, and n_v (as suggested by the common choice of emitted wavelengths green, blue and violet), and the corresponding velocities are u_g, u_b, and u_v (values may be positive or negative), retrieving the tracer particle velocity $u = (u_x, u_y, u_z)$, in an orthonormal direct Cartesian referential, requires the following transformation:

$$
u = M^{-1}v \Leftrightarrow \begin{bmatrix} u_x \\ u_y \\ u_z \end{bmatrix} = \begin{bmatrix} n_{gx} & n_{gy} & n_{gz} \\ n_{bx} & n_{by} & n_{bz} \\ n_{vx} & n_{vy} & n_{vz} \end{bmatrix}^{-1} \begin{bmatrix} u_g \\ u_b \\ u_v \end{bmatrix}
\tag{3.6.8}
$$

The matrix M is non-orthogonal, except if unit vectors n_g, n_b, and n_v form an orthonormal basis of 3D Euclidean space, which is a very difficult probe configuration to attain, even in the controlled conditions of a laboratory. The Reynolds stress tensor in the orthonormal referential, T, can be calculated from the transformed variables $u = (u_x, u_y, u_z)$ as $T = uu$ (using notation for dyadic product). It can also be calculated with the originally detected velocities in the non-orthonormal referential, $v = (u_g, u_b, u_v)$, and subsequently transformed. If $H = vv$ is the Reynolds stress tensor expressed in the non-orthonormal referential, Equation (3.6.8) yields $H = MuuM^{T}$. Given that M is an invertible matrix, the Reynolds stress tensor in the orthonormal referential becomes

$$
T = M^{-1}H(M^{-1})^{T}
\tag{3.6.9}
$$

Second-order tensors are transformed in accordance with Equation (3.6.9).

3.6.2.3 Summary of practical issues

Two practical issues are discussed here: beam positioning; and, variation in the speed of light passing through different media.

A) Beam positioning

Equation (3.6.6) reveals that the accuracy of LDV measurements strongly depends on precise positioning of the laser beams. The trigonometric quantity $\sin(\theta/2)$, derived from $n_{sp}^{(1)}$ and $n_{sp}^{(2)}$ (for each beam color and taking into account the propagating media) must be calculated with high accuracy. Specifications from LDV suppliers may not be precise enough and the user should, before a given experimental configuration, define the beam geometry. This task can be done by fixing the sources of the laser beams in a solidary frame (no relative motion) mounted in a traversing system with a high-accuracy movement step. A pinhole and a laser power meter are then used to determine the position of the measuring volume and to determine the points that establish the direction of each beam. The points belonging to the path of a given beam on arbitrary vertical planes (Figure 3.6.5). One position may be enough, though more positions increase precision. The procedure is repeated for both beams of a given color and for all colors. The unit vectors are calculated by

$$
\begin{aligned}
n_{sp}^{(1)} &= \frac{(x_A, y_A, z_0) - (x_C, y_C, z_C)}{|(x_A, y_A, z_0) - (x_C, y_C, z_C)|} \\
n_{sp}^{(2)} &= \frac{(x_B, y_B, z_0) - (x_C, y_C, z_C)}{|(x_B, y_B, z_0) - (x_C, y_C, z_C)|}
\end{aligned}
\tag{3.6.10}
$$

where (x_A, y_A, z_0) and (x_B, y_B, z_0) are the positions of an arbitrary point (fixed, relatively to the probes) when beam 1 and beam 2, respectively, pass the pinhole (Figure 3.6.5). Point (x_C, y_C, z_C) is the location of the same arbitrary fixed point when both beams pass the

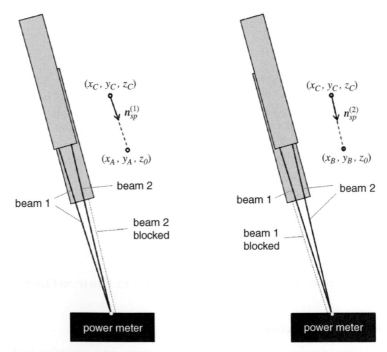

Figure 3.6.5 Procedure for the determination of the directions of the laser beams and the angle between beams. Example for two beams of one probe.

pinhole (Figure 3.6.5). Fine adjustments can be made using the power meter immediately behind the pinhole and maximizing the laser power through the pinhole for each beam and for the beam intersection. For simplicity of traversing operation x_0 is the same for both beams but this is not necessary. The angle θ is then calculated as

$$\theta = \arccos\left(n_{sp}^{(1)} \cdot n_{sp}^{(2)}\right) \tag{3.6.11}$$

B) Speed of light
In Equation (3.6.6) the wavelength of the emitting light corresponds to the medium through which it propagates. The fact that the speed of light changes with the medium must be taken into account. Equation (3.6.6) is thus best restated as

$$u_p = f_D \frac{\lambda_0^{(0)}}{2n^{(m)}\sin(\theta/2)} \tag{3.6.12}$$

where $\lambda_0^{(0)}$ is the wavelength of the emitted light in vacuum and $n^{(m)}$ is the refractive index of the medium, which should be known accurately (Resagk *et al.*, 2003; Huisman *et al.*, 2012).

3.6.3 Basic instrumentation configurations: lasers and optics

3.6.3.1 Typical components

A typical LDV system has multiple components: a continuous laser source; transmitting optical components; including beam splitters; frequency shifters and color splitters; receiving optical components; and, a photo-detector and signal processing components (Durst *et al.*, 1976). Figure 3.6.6 illustrates the layout and components of a typical LDV system.

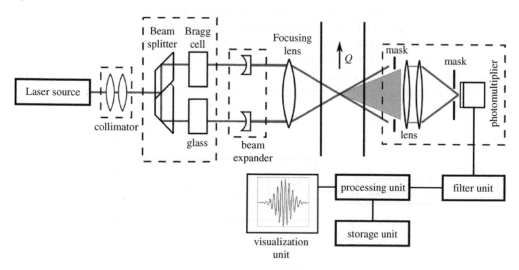

Figure 3.6.6 General layout and components of a LDV system.

Several lasers are commonly used for LDV: He-Ne (Helium-Neon), typically at 632.8 nm (red); Ar-Ion (ions of Argon), normally, in water, operated in the green, blue and deep-blue ranges (typically 514.5, 488.0 and 476.5 nm, respectively); and, Neodymium-doped Yttrium Aluminium garnet (Nd:Yag), operating mostly at 532, 518 and 412 nm (green, blue and violet, respectively). The first two lasers are gas-lasers; *i.e.*, coherent light beam is produced by the excitation of a gas by a strong electric current. The third laser is a solid-state laser, in which a crystalline medium is modified (doped) with small concentrations of impurities to change its electrical properties; the crystalline medium is optically excited (optically pumped, generally by diode lasers) to produce population inversion and, hence, emission of photons (Csele, 2004).

The laser beam is, in most LDV applications, a Gaussian beam (Siegman, 1986) although alternatives with lower divergence have been proposed (Voigt *et al.*, 2009). With all wavelengths mixed, it is split in two beams, one of which undergoes frequency shifting (Figure 3.6.7), to eliminate directional ambiguity in the detected velocities, as discussed in Appendix 3.B.2. Of the many ways tested to achieve frequency shifting (discussion in Durst & Zaré, 1974; Somerscales *et al.*, 1981), acoustic-optic modulation is the most used and is standard in commercial systems. The acoustic-optic modulator is a Bragg cell (see Chang, 1996) – it is composed of a glass crystal subjected to acoustic vibration, through which the laser beam is driven. The frequency of this beam is changed by an amount proportional to the frequency of the acoustic wave, ε_a. A typical value of the shifting frequency is 40 MHz; however, the signal is often down-mixed with a reference signal to reduce the actual frequency shift to values more appropriate to the hydrodynamic problem at hand, which may not require a high range of negative velocities (see example in Appendix 3.B.2).

Recent systems incorporate beam splitting and frequency shifting in the Bragg cell (Compton & Eaton, 1996). Beyond this component, the split beams pass through a color-splitter prism, which separates the colors that will be used for velocity detection. For three-dimensional velocities, six beams emit from the color-splitter: three beams phase-shifted; and three beams not phase-shifted (Figure 3.6.7). As mentioned above, the usual color choices are green, blue and deep blue or green, blue and violet.

In most contemporary LDV systems the beams are deflected by mirrors into single-mode optical fibers that conduct the light to emitting probes, preserving beam coherence and polarization. The alignment of the optical fibers and the beams is a sensitive issue – misalignment can cause fiber damage. Mechanical fiber manipulators (Figure 3.6.7) are normally employed to manually optimize the alignment.

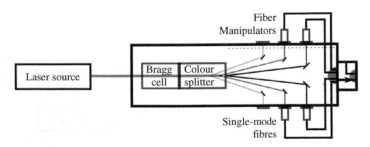

Figure 3.6.7 Scheme of the transmitter module.

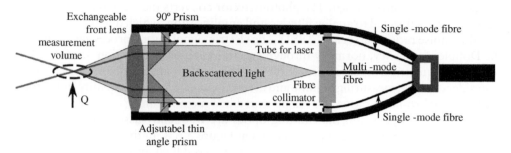

Figure 3.6.8 Scheme of emitting and receiving probe.

Inside the emitting probe, fiber collimators are installed at the end of the optical fibers, as schematized in Figure 3.6.8, to obtain the free-space collimated beams that will be projected out of the probe. Given that beam divergence is inversely proportional to beam radius, collimating optics are usually complemented with beam expanders which, by increasing beam radius, ensure low beam divergence and, thus, high values of the Rayleigh length (Siegman, 1986). This step ensures that the focal length of each beam is sufficiently large to avoid significant distortions of the fringe pattern (that can be eliminated through optical masking, Compton & Eaton, 1996; discussion in Zhang, 2010).

The beams are focused trough a lens, normally spherical, producing a measuring volume ideally at the location of the smallest beam radius – the *beam waist*. At this measuring volume, shifted and un-shifted beams of each monochromatic coherent light produce a pattern of moving fringes, as explained in Appendices 3.B.1 and 3.B.2.

Particles traveling through the measuring volume scatter the fringe pattern to the receiver optics. This is a system of lenses and pinholes that focus the light onto an optical fiber that carries it to the receiver module (Figure 3.6.9). LDV has relied mostly on single-mode optical fibers, to preserve coherence (Kaufmann & Fingerson, 1985). However, multimode fibers better handle larger output laser power and require less alignment effort. They have become standard for LDV, once signal degeneration issues were eliminated (Büttner & Czarske, 2004).

The receiver module is composed of a color-splitter, interference filters and photo-detectors for each color. The scattered light is separated again into the original colors and the use of filters ensures that only light from one wavelength is arriving to the

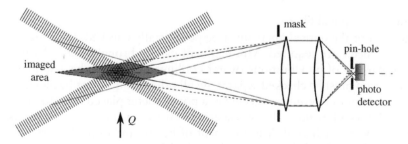

Figure 3.6.9 Scheme of receiving optics and receiver module.

corresponding photo-detector. The photo-detector converts the scattered light into an electric signal subjected to analog filtering and amplification, which does not reduce the signal-to-noise ratio (Durst *et al.*, 1981). The electric signal is then processed to identify the Doppler frequency, normally through spectral analysis (see Section 3.6.5.1). The resulting frequencies, other features of the detected tracer particle motion such as time-of-residence in the measuring volume and arrival instant, and signal quality indicators, such as signal-to-noise ratios, are finally stored in a memory unit.

The receiving optics and the emitting optics are frequently assembled in emitting probe (most commercial systems), as shown in Figure 3.6.8. In this case, the system is working in backscatter mode, *i.e.*, the paths of scattered light and emitting light are at zero degrees. As seen in Section 3.6.4 (and the associated Appendix 3.B.3) below, this may not be the most convenient arrangement as, in the Mie scattering regime, light intensity at angles larger than 90° is generally higher than light intensity scattered at small angles.

3.6.3.2 Summary of practical issues

Eight practical issues are considered here: quality of laser beam; alignment of optical fibers; beam focus; replacement of interference filters; positioning of receiving and emitting optics; alignment of measurement volume for 3-component LDV; determining the reference frame for LDV measurements; and, measuring flows with free-surface.

A) Quality of laser beam
Experimenters acquire either a commercial LDV system or assemble an LDV system based on the components shown in Figure 3.6.6. When selecting an LDV system, it is important to check the quality of the laser beam. This check is normally done through the M^2 parameter, the ratio of the actual and ideal beam parameter products (BPP). The actual BPP can be experimentally determined by measuring the minimum beam diameter (defined as the diameter for which the intensity drops to $1/e^2$ of the peak intensity, Siegman, 1986, if it is a Gaussian beam to a good approximation) and the beam diameter sufficiently far from the waist (Johnston, 1998). The ideal BPP of a Gaussian beam is the product of its waist diameter and divergence, determined theoretically. Values of M^2 closer to 1 configure good quality Gaussian beams which, when properly aligned, will generate more predictable fringe patterns. The user should also verify that the frequency shift introduced by the Bragg cell is enough to resolve the negative velocities expected in the phenomenon under investigation. An example of calculation is given in Appendix 3.B.2.

B) Alignment of optical fibers
The alignment of the optical fibers must be periodically checked, for which a pinhole and a power meter are required. Misalignment can be brought about by moving the emitting apparatus of by vibrations (water cooling systems can be a source of vibration). Each beam must be checked separately; all other beams must be blocked off. The intensity of the emitted beam is collected in a power meter placed just after the pinhole. The manipulators allow normally two degrees of freedom and should be tweaked until the highest power is achieved. Misalignment of beams and optical fibers can damage the latter. Mechanical polishing can prolong the life of the optical fibers but this must be done with great care.

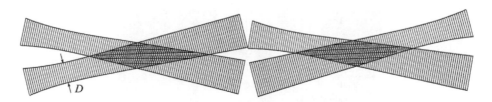

Figure 3.6.10 Two examples of laser beams not crossing at the focal distance, and thereby causing gradients in the interference pattern (fringe distortion).

C) Beam focus

The emitting probe is normally calibrated by the manufacturer in commercial LDVs ensuring that laser beams cross at the correct focal distance, *i.e.*, beams should cross at their waists. The calibration should be checked periodically, as poor beam focus due to lens misalignment causes spatial gradients in the fringe pattern (Figure 3.6.10). Inadequate focus may lead to poor estimates of the frequency f_D with the tools described in Section 3.6.5.1 or, if the signal is analyzed for different time windows, a non-accelerating particle will seem subjected to acceleration as different frequencies will be detected (Nobach, 2008).

If the fringe pattern is not uniform, the range of frequencies f_D that is detected is broadened, increasing the variance of the signal and, hence, overestimating turbulence (Zhang & Eisele, 1998). Note that fringe distortion leads to frequency broadening and, hence, to turbulence overestimation, but may be due to other causes (see section 3.6.4.2).

Note that the spatial fringe pattern of properly aligned beams of slightly and strategically out-of-focus beams can be used to determine the spatial distribution of velocities and velocity profiles (Pfister *et al.*, 2005; Yamanaka, *et al.*, 2006) or accelerations (Lehmann *et al.*, 2002; Nobach & Bodenschatz, 2011). In depth discussion of these issues is out of the scope of this book.

D) Replacement of interference filters

The interference filters must be replaced periodically. A faulty filter permits color leaking; *i.e.*, light scattered from two different pairs of beams arriving at the same photo-detector. To identify filter malfunctioning, the signal from each photo-detector should be analyzed with the remaining blocked off. If only one color is being emitted and more than one color is detected, the corresponding filters are probably deficient.

E) Positioning of receiving and emitting optics

The relative positioning of the receiving optics and emitting optics configures the functioning mode as backscattering (180° between emitting and receiving optics), forward scattering (0° between both directions) or 90° scattering. Backscattering is implemented in many commercial LDV systems, as it allows for compact probes incorporating receiving and emitting optics. This is a major advantage for a non-experienced user as there is no need to focus the receiving optics, a procedure done by the manufacturer and allows for measurements in challenging geometries. Figure 3.6.11 shows the usage of a 3-component back-scatter LDA/LDV in a free-surface flow with obstacles. 90° scattering has the advantage of maximizing spatial resolution for two-dimensional instruments. These configurations,

Figure 3.6.11 A 3-component LDA composed of two probes measuring velocity amidst randomly placed cylinders simulating vegetation. The LDA system is composed of a 5W Argon-Ion laser, a Dantec F80 flow processor, and two watertight probes with focal length of 198 mm in water mounted on an automated transverse system. The two-component probe transmits and receives two orthogonal pairs of beams (wavelengths 514.5 nm and 488 nm) while the one-component probe, positioned at 30° to the two-component probe, transmits the third pair of beams, with 476.5 nm wavelength (photo by Rui M.L. Ferreira, channel and LDA at the Leichtweiß-Institut für Wasserbau, Technische Universität Braunschweig, Germany).

especially the latter, are not efficient in terms of signal-to-noise ratio (SNR) since light is mostly scattered in the 0° direction (see Section 3.6.4 and Appendix 3.B.3).

The forward scatter configuration requires that the receiving optics is placed on the opposite side of the flow, relatively to the emitting optics, which may be a serious constraint given the frequent scarcity of laboratory space. Figure 3.6.12 shows a 2-component He-Ne LDV system operated in forward scattering mode where two pairs of a 632.8 nm laser beams intercept above a gravel bed of a water channel. In this example, a high inertia (to minimize vibrations) metal frame was built and installed over the channel to hold both the receiving and the emitting optics without relative movement. The major advantage of the forward scatter configuration is efficiency, since the largest possible scattered power reaches the receiver optics, however care must be taken to ensure that none of the beams hits the photodetector. This can be achieved by means of a blind system to be placed in front of the receiving optics. The need to focus the optics on the measuring volume every time the focal distance changes is frequently a time-consuming task (details in Ferreira, 2011).

F) Alignment of measurement volume for 3-component LDV
To obtain the three components of the velocity vector, two emitting probes are normally necessary, featuring three pairs of intersecting beams. Second and higher-order moments

Figure 3.6.12 A monochromatic 20 mW He-Ne Laser LDA (Dantec 55X Modular LDA), in forward scattering mode, measuring over a gravel bed (photo by Rui M.L. Ferreira, channel and LDA at the University of Aberdeen).

of turbulence velocity data require coincident measurements (see Section 3.6.5.3), *i.e.*, velocities of particles detected simultaneously in all three channels. To ensure that particles simultaneously cross the three measuring volumes, the probes must be perfectly aligned throughout the duration of the experimental test. Unlike two-dimensional systems, in which the alignment of measuring volumes is performed by the manufacturer, three-dimensional systems require the experimenter to perform the alignment of the third measuring volume. This requires a mounting frame for both probes in which one of the probes has at least two degrees of freedom, a power meter, a pinhole and optionally a fluid reservoir. The position of the mounting frame is first adjusted to maximize the light power through the pinhole for one of the probes (possibly inside the reservoir with the same fluid of the actual tests). The second probe is aligned by gradually adjusting the two or three degrees of freedom until the power reading attains a maximum for the same position of the pinhole. No relative motion between the mounting frame and the probes is allowed from this point onwards. The mounting frame with the probes can now be transported to the channel where actual experiments are to be performed.

G) Defining a reference frame for LDV measurements

The correct positioning of the LDV control volume inside a channel can be achieved using a pin-hole plate. The pin-hole plate location in the channel relative to the side walls will define the horizontal coordinates, whereas the pin-hole vertical location will define a reference height. The LDV should be mounted on a transversing system to enable ease of sweeping movement along the x-y-z coordinates (see Figure 3.6.5). The position of the control volume when it is located in the pinhole (measured, for instance, with a laser power meter) is stored and provides a reference for all other positions of the traversing system thus spatially referencing LDV measurements. It is often possible to have automatized solutions integrating the transversing system with the LDV acquisition program.

H) Measuring flows with free-surface

In some applications there may be the need to emit the laser across the free-surface. To avoid uncontrolled refraction of the laser beams by the free-surface and, in some cases, because the focal length is larger than the flow depth, it may be necessary to submerge the probes inside a watertight casing with a glass bottom (as in Figure 3.6.11). Medium continuity is ensured everywhere except the thin casing bottom which also simplifies the calculations to find the locus of the control volume (details in Ricardo *et al.*, 2014).

3.6.4 Tracer particles

3.6.4.1 Types of tracer particles

Laser Doppler techniques require the existence of tracer particles in fluid flow. Ideally, a turbulent flow should be uniformly seeded with tracers whose properties meet the following criteria (details on light scattering and the quantification of the tracking ability of tracers are given in Appendices 3.B.3 and 3.B.4, respectively):

– Particle density should be close to that of the fluid to reduce buoyancy effects (discussed in Appendix 3.B.4);
– Particles should possess a refractive index as different as possible from that of the fluid to maximize light scattering (see Appendix 3.B.3);
– Particles should be sufficiently large to scatter enough light intensity to be detected by the photo-detector with no need for signal amplification (see also Section 3.6.5);
– Particles should be large enough to avoid unpredictable kinematic behavior, such as Brownian motion (see Appendix 3.B.4);
– Particles should be small enough to follow fluid flow to the required level of detail (addressed in Appendix 3.B.4);
– Particles should be small enough to preserve a good signal modulation (Durst & Ruck, 1987, see Appendix 3.B.3).

Seeding particles used as tracers should have a reasonable cost to achieve ample seeding when using large volumes of fluid, as often required in hydraulics experiments. The tracers should be inert and non-toxic. For water flows, natural tracers, such as conifer pollen particles, have approximately zero relative density, high enough relative refractive index and adequate range of diameters for most hydraulics experiments. However, they eventually degrade and cause the appearance of bio-films in the experimental apparatus

Table 3.6.1 Common tracer particles for LDV.

	Mean diameter (µm)	Refractive index	Density (g/cm³)	Shape
Titanium dioxide	0.1–50.0	2.60	4.3	Irregular
Silicon carbide	1.5	2.65	3.2	Irregular
Nylon	4.0	1.53	1.14	Spherical
Metallic coated hollow glass spheres	14.0	0.21+i2.62	1.65	Spherical
Polystyrene latex	0.5	1.6	1.05	Spherical
Zirconium Oxide	30.0	2.2	5.7	Irregular
Polyurethane beads	5.0–150.0	1.5–2.1	1.05–1.3	Irregular
Air bubbles	0.1-...	1.0	1.3×10^{-3}	Spherical

and have other undesired effects. Air bubbles suspended in the flow or impurities, for instance resulting from pipe oxidation, are sometimes sufficient seeding in laboratory flumes. However, this tracer source is not recommended because the properties of impurities suspended in the water flow are usually not known: it will add uncertainty to velocity measurements, especially in turbulent flows, and lower the level of confidence of measurements as measurement accuracy depends on tracer size and density (see Appendix 3.B.4). Synthetic particles or mineral powders are normally more appropriate and have been successfully used in water flows even if their density is larger than that of the water. A list of some common tracers is given in Table 3.6.1. An electronic microscope image of one particle type, polyurethane beads, is shown in Figure 3.6.13.

For air flows, common techniques of generation of tracers include atomization of liquids and condensation of saturated vapor. Since this text is mostly concerned with water flows, these techniques are not be described here. A review of these and other techniques can be found in McKeon *et al.* (2007).

It should be highlighted that manipulation of substances capable of serving as tracers must obey to the general health and safety procedures associated to the manipulation of powders of micro-particles. For some of these substances there is no evidence on the long-term effects of its deposition in the human respiratory system.

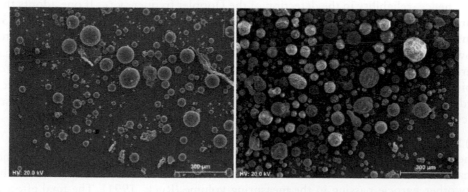

Figure 3.6.13 Examples of tracer particles: micrograph (100x) of polyurethane beads. Left: decosoft© 60 transparent beads (refractive index 1.5, specific gravity 1.05); right: decosoft© 60 white beads (refractive index 2.1, specific gravity 1.3).

3.6.4.2 Summary of practical issues

Briefly summarized below are the main practical issues associated with choosing appropriate tracer particles: tracking ability criteria for turbulent flows; use of non-spherical tracers; effects of tracer size on detection of the Doppler shift; particle-induced fringe distortion; laser power restrictions; and, choose inert tracers.

A) Tracking ability criteria for turbulent flows

The choice of appropriate tracer particles largely determines the quality of LDV measurements. It is useful to compile all available data on possible materials, including those in Table 3.6.1, and to consider cost and health hazard information. For turbulent flows, the criteria expressed in Appendix 3.B.4 (Figure 3.B.4.1), or similar appropriate criteria, should be applied in order to verify if the properties of the tracers are not limiting the resolution of turbulence in the flow phenomenon under investigation.

The tracking ability expressed by these criteria depends on tracer particle density and diameter. For small particles (*i.e.*, smaller than 10 μm), diameter is more relevant than density to determine particle tracking ability.

Note that an additional limiting factor for resolving small turbulence eddies is normally (notably for commercial systems) the dimension of the measuring volume; more advanced techniques may be required to obtain higher spatial resolution (Büttner & Czarske, 2004).

B) Use of non-spherical tracers

Perfectly spherical particles are normally more expensive than irregular particles, which makes the latter more attractive. However, the detailed knowledge of their dynamic interaction with fluid flow and their scattering patterns are generally unknown. Equation (3.B.4.3) in Appendix 3.B.4 and Mie's solution are only approximations for these non-spherical particles. The degree of approximation is not rigorously known for most types of tracers.

C) Effects of tracer size on the detection of the Doppler shift

Small tracer particles produce well modulated signals as they cross the fringe pattern. The use of particles larger than the fringe spacing affects negatively signal modulation and, hence, reliability in the detection of the Doppler frequency f_D. If one adopts this criterion, for example, when considering laser light with vacuum wavelength $\lambda = 532$ nm and an angle between beams $\theta = 5°$, then particle diameter to be used in water flows should ideally be smaller than 4.6 μm. However, it has been shown that particles whose diameter is larger than several times λ still produce good signal-to-noise ratios (Durst & Ruck, 1987). This signal, in this case, will exhibit a pronounced pedestal (or offset) that must be filtered out before attempting to determine f_D (see Section 3.6.5).

D) Particle-induced fringe distortion

The passage of a particle through a laser beam shortly before the measurement volume. That particle is undetected but the distortion it generates affects the burst generated by any other particle passing in the measuring volume (Ruck, 1991). The final effect is overestimation of flow turbulence, as in the case of beam astigmatism or inadequate focus (see Section 3.6.3.2).

E) Laser power restrictions

When the turbulence scales of interest are very small (*e.g.*, the order of magnitude of the laser wavelength), to be kept in mind is that the scattering area of particles, whose diameter is smaller than the laser wavelength, is proportional to d^4 (McKeon *et al.*, 2007). Hence, the required laser power is likely to be high. Also, these particles are likely to undergo Brownian motion, which decreases the confidence with which very small scales are measured. In this case, it would probably be advisable to change the dimensions of the experimental installation in order to obtain good quality measurements.

F) Choose inert tracers

Choose tracer particles that are biochemically inert. It is advisable to use hand and respiratory protection while handling powders of micro-particles. In the case of respiratory protection, masks with, for instance, European standard EN 143 P3 class particle filters must be used.

3.6.5 Signal processing and calculation of flow velocity statistics

3.6.5.1 Determination of the Doppler frequency

The unprocessed electric signal ensuing from the LDV photo-detector comprises irregular bursts; *i.e.*, the outcome of one or more particles crossing the measuring volume and scattering the fringe pattern, plus several forms of noise (Albrecht *et al.*, 2003; McKeon *et al.*, 2007; Nobach & Tropea, 2007). The Doppler frequency, f_D, must then be extracted from the signal. The fringe model for LDV is described in Appendix 3.B.1. The key issue for the LDV signal processing unit is to correctly estimate f_D from a signal that is contaminated with noise and modulated by the spatial shape of light beam intensity. Figure 3.6.14 shows a typical LDV signal as the sum of Doppler signal (upper left), noise (upper right) and the modulation introduced by the shape of light beam intensity, commonly referred to as the signal pedestal or the DC signal component (lower left).

The signal pedestal (Figure 3.6.14) is the imprint of the radial distribution of the beam light intensity for Gaussian beams. Note that large particles (larger than fringe spacing) exhibit a larger pedestal, because of poor modulation. There is no exact boundary of the laser beam; the beam's diameter is twice the radius for which light intensity has fallen to $1/e^2$ of the maximum registered at the axis. This relationship results in a modulation that can be removed, along with high wavelength noise by a high pass filter (Durst *et al.*, 1981). Burst detection (or validation) frequently involves an amplitude threshold on the pedestal signal.

Noise in the burst signal may be associated with the instrumentation, including: electric noise due to poor grounding; noise introduced by the photo-detector; and, noise introduced by the laser generation or by system electronics (see Albrecht *et al.*, 2003). It can also be generated by presence of several particles in the measuring volume; each particle generates a fringe pattern, in the terms laid out in Appendix 3.B.1. The superimposition of these fringe patters can increase the modulation of the signal (if the particles are in phase) but, if the particles are out of phase, can impair the determination of the correct f_D (in the limit, if the particles are π out of phase, it suppresses the Doppler signal, Durst *et al.*, 1981). In general, sources of noise are uncorrelated and produce white noise (except grounding noise, which is the easiest to eliminate).

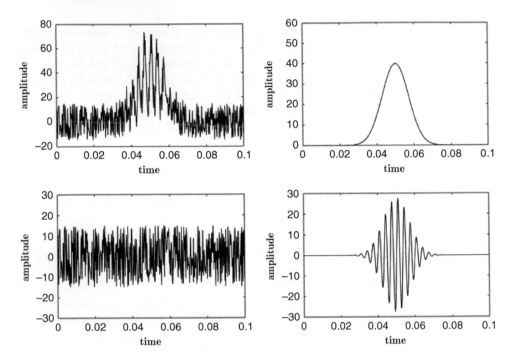

Figure 3.6.14 Components of the LDV signal as generated by a small particle (smaller than the fringe spacing) crossing the measuring volume. Top left: pre-filtered signal, composed of the superimposition of the actual Doppler burst, noise and pedestal; top right: pedestal (DC component); bottom left: noise; bottom right: Doppler burst (AC component).

Adding to limits imposed by particle size and density (see Appendix 3.B.3), noise in the LDV signal adds a new restriction to the resolution of small-scale turbulent motion. In the spectral domain, the level of noise will render less energetic velocity fluctuations non-assessable.

An important indicator of the quality of the signal is the signal-to-noise ratio (SNR), the ratio, in decibels, between the power of the signal and the power associated to noise. The latter power level can be estimated from the value of the auto-correlation function at the origin. Current signal processing units include a SNR threshold to validate burst detection (McKeon *et al.*, 2007). Noise contribution can be reduced by application of band-pass filters while searching for the Doppler frequency (Nobach & Tropea, 2007).

Although many techniques, including photon correlation (Vehrenkamp *et al.*, 1979), frequency tracking (Adrian, 1972; Li & Wang, 1994) and frequency analysis (Durst *et al.*, 1976), have been tried over the time span of LDV usage (Albrecht *et al.*, 2003), current commercial system almost exclusively use frequency analysis. The main procedure consists on the analysis of the energy spectrum of the signal, determined by Fourier techniques. An analysis of the resolution of the FFT was made by Alvarez *et al.* (1989). Today's estimators come very close to achieving Cramèr-Rao lower bound, *i.e.*, retrieving all the information from the signal (Nobach & Tropea, 2007). Figure 3.6.15 shows the energy spectral density of the signal shown in Figure 3.6.14, estimated as a periodogram employing FFT. The pedestal was removed and a

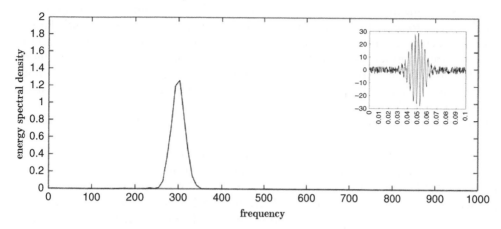

Figure 3.6.15 Doppler frequency of the signal shown in Figure 3.6.14 determined by analysis of the energy spectral density.

low-pass filter was applied before calculating the energy spectral distribution. The inset shows the actual signal subjected to analysis. The Doppler frequency corresponds to the peak at 302.7 Hz, a good approximation to the true value. A better result can be obtained with peak interpolation. The width of the peak depends on the spectral estimator employed.

It is noted that the optical arrangement of the beams influences the quality of the detection of the Doppler frequency. The link is the number of fringes in the measuring volume: naturally, the reliability in the estimation of f_D increases with the number of fringes, which depends on the optical arrangement through the geometrical relation

$$N_f = \frac{8F \tan\left(\frac{\theta}{2}\right)}{\pi D_L} \tag{3.6.13}$$

where F is the focal distance and D_L is the diameter of the beam just before the lens. Hence, number of fringes, and thus the reliability of the estimation of f_D increases with the angle between beams and with the focal distance. In contemporary commercial systems, these parameters are normally previously set; surely, the user should be aware of their values for a better understanding of the obtained results.

Other relevant signal parameters include arrival time and particle residence time in the measuring volume (or length of burst). Arrival time is particularly relevant to set up the coincidence criterion, for measurements of more than one velocity component. Given that the measuring volumes of the intersections of pairs of beams are not exactly superimposed, the arrival time for each color will not be exactly the same. The coincidence window method is generally employed to decide if bursts detected in different color channels belong to the scattering of the same particle. The maximum width of the coincidence window should be smaller than an appropriate statistic of the inter-arrival time. As the arrival times are not regularly distributed in time (see Section 3.6.5.2) and, in fact, several particles may arrive at the almost the same time at a measuring volume, the criterion for setting the width of the coincidence window is not founded on a rigorous optimization procedure. The general guidelines are simple: the width should small enough to prevent different bursts to be identified as one, but sufficiently large to correctly detect as one a slow particle arriving at a location where the measuring volumes are not exactly superimposed. A frequent option for the coincidence window width is to estimate it as a value an order of magnitude smaller than the mean inter-arrival time (Zhang, 2010). Other estimates make use of the residence time, for instance a percentage of the residence time (for instance 10% to 30%, McKeon *et al.*, 2007). Most of today's LDV systems include a "coincidence mode" operating option. However, the user can always decide on the coincidence criterion as a data post-processing step provided that particle arrival and transit times are also recorded.

The residence time is also relevant to minimize the positive bias to large velocities inherent to LDV measurements. As seen in Section 3.6.5.3 the correlation between particle velocity and particle rate (see Section 3.6.5.2) causes larger velocities to be overrepresented in a time series, generating a positive bias of the arithmetic mean as an estimator of the true mean particle velocity. Weighting the velocity with its own residence time is an effective (Buchhave *et al.*, 1979) way to reduce this bias. Another relevant use of the residence time is in burst validation: the registered time can be compared with the ratio between width of measuring volume and particle velocity; if they are different beyond a reasonable tolerance, the burst is rejected.

3.6.5.2 Spatial and temporal resolution

Spatial resolution is a topic closely linked with signal quality and noise, especially in turbulent flows. As a general principle, scales smaller than twice the size of the detecting volume may not be resolved. The energy associated with such small scales may be aliased into the resolved scales changing the slope in the high frequency range of the power spectral density function (Kirchner, 2005). The detecting volume is of the order of magnitude of the measuring volume. The volume can be smaller if the receiving optics have a short focal length and are positioned, by means of mirrors or other optical devices, close to the measuring volume and in the side-scatter mode (Orloff & Olson, 1982; Compton & Eaton, 1996), and a pinhole is used to truncate the region of optical inspection (Orloff & Olson, 1982). This is not a typical LDV arrangement, as, in the Mie regime, side-scattering is not a favorable arrangement as it needs much more power for the same light scattering intensity (see Appendix 3.B.3), and the optical arrangement is normally difficult to position and may be intrusive.

High resolution LDV data, typically to resolve the flow shear rate with high accuracy, has been attained inside the measuring volume by superimposing two wave-lengths that, using the heterodyne pattern, generate two fringe patterns. The fringe patterns are not parallel, which admits simultaneous determination of particle position and velocity (details in Büttner *et al.*, 2005), from which spatial gradients of the velocity can be computed. Note that this is not a standard technique and requires extra optical and signal processing complexity.

The detection volume is generally larger than the measuring volume since a particle can be detected outside the measuring volume, because the beams do not have sharp edges. In this case, they normally generate a bimodal pedestal and lower burst intensity (Nobach & Tropea, 2007), which makes it possible to filter them out.

The size of the measuring volume is most important when discussing the spatial resolution of LDV measurements. The measuring volume is an ellipsoid with length δ_x, width δ_y, and height δ_z:

$$\delta_x = \frac{D_w}{\sin(\theta/2)}$$
$$\delta_y = D_w \qquad\qquad (3.6.14)$$
$$\delta_z = \frac{D_w}{\cos(\theta/2)}$$

where

$$D_w = \frac{4F\lambda_0}{\pi D_L} \qquad\qquad (3.6.15)$$

is the diameter of the *beam waist*, focused at the measuring volume. As mentioned in Section 3.6.3.1 using a beam expander to increase the beam diameter D_L is a practical way of decreasing the measurement volume size. By decreasing the measurement volume, the laser power density increases thereby producing higher amplitude signals.

Turbulent motion whose spatial scales are smaller than the dimensions of the measuring volume are not resolved in the sense that particles belonging to different eddies may enter in different parts of the volume alternately. Consequently it can be impossible to distinguish the coherence of the turbulent motion. In fact, if δ_y equates to the length scale of the velocity gradients, the succession of particles with different velocities crossing the control volume is interpreted as turbulence even when the flow is laminar. Thus, even if particles were suitable to track small turbulence structures (see Appendix 3.B.4), a noisy signal is expected for scales smaller than the largest dimension of the measuring volume, especially if this direction is aligned with strong gradients in the mean flow.

Consider, as an example, a typical system with $F = 0.40$ m, $\theta = 30°$, $\lambda = 632.8$ nm and $D_L = 0.001$ m measuring in flowing water with a mean velocity of 0.12 ms^{-1}, and seeded with 1 µm (diameter) TiO$_2$ particles. The dimensions of the measuring volume are $\delta_x = 0.00065$ m, $\delta_y = 0.00032$ m, and $\delta_z = 0.00037$ m. According to Figure 3.B.4.1 in Appendix 3.B.4, the smallest turbulent scale resolved with high confidence (99%) using these particles would be 3×10^5 Hz. However, the largest size of the control volume can be translated, through the application of Taylor's frozen turbulence hypothesis (Taylor, 1938), as a frequency of 186 Hz. Beyond this frequency, noise is expected.

It is concluded that the LDV spatial resolution depends primarily on the optical arrangement, through the size of the measuring volume. However, it can also be strongly influenced by extra optical arrangements and influenced by signal processing technique.

The dimensions of the measuring volume influence also the time resolution of LDV. This issue must be understood together with the influence of seeding concentration. The mean (ensemble-averaged) number of particles in the measuring volume, $\{N_V\}$, depends on the number-concentration of tracer particles in the flow, $\{n_p\}$, and on its volume, $\forall_0 = 1/6\pi\delta_x\delta_y\delta_z$. Assuming that the tracer particles are homogeneously distributed, then

$$\{N_V\} = \{n_p\} \frac{1}{3} \pi \frac{D_w^3}{\sin(\theta)} \qquad\qquad (3.6.16)$$

Assuming also that the motion of the tracer particles is independent and that the spatial distribution of these particles is random and susceptible to be described by a Poisson distribution, the following regimes can be determined: $\{N_V\} < 0.1$, particles arriving in separate events (probability of 99.5%); $0.1 \leq \{N_V\} \leq 10$, multiple particles coexisting in the measuring volume and $\{N_V\} \geq 10$, quasi-continuous signal (McKeon *et al.*, 2007).

A quasi-continuous signal may not detect individual bursts and therefore may yield no signal. Coexistence of several particles in the control volume can enhance the signal, as seen in Section 3.6.5.1 , but most frequently reduces its quality. The optimal operating regime for LDV signal processing occurs when particles arrive to the measuring volume as separate events.

The mean data-rate, $\overline{G}$, in this regime can be calculated by dividing the mean number of particles in the measuring volume, Equation (3.6.16) by the mean time necessary to travel across it, $\delta_x/\overline{u}_p$. The resulting relationship is

$$\overline{G} = \{n_p\} \overline{u}_p \frac{1}{3} \pi \frac{D_w^2 \cos(\theta/2)}{\sin(\theta)} \qquad\qquad (3.6.17)$$

Time resolution of LDV, understood as mean rate of particle arrival, depends on the arrangement of laser beams, on the *waist beam* diameter, on the seeding concentration and on the particle velocity. This interdependence of particle rate and particle velocity is not specific of LDV and is the cause of positive bias to large velocities; such bias occurs in any technique employing single particles being detected in a finite measuring volume. In the case of LDV, given the high data-rates intended, the interdependence can pose a concern. A strategy to minimize bias in the LDV measurements is discussed in Section 3.6.5.3.

Evidently, particles arrive at the measuring volume in irregular sequences. Figure 3.6.16 shows an example of the probability distribution function of the inter-arrival time for a turbulent flow in a laboratory open channel flume. As mentioned above the arrival of tracer particles to the control volume may be idealized as a Poisson process, in which case the distribution of the interval between consecutive arrivals, $\varDelta\tau$ (herein inter-arrival time), can be approximated by a negative exponential distribution (Figure 3.6.16). The probability density of a given value of $\varDelta\tau$ is thus $p(\varDelta\tau) = \overline{G}e^{-\overline{G}\varDelta\tau}$.

The most probable inter-arrival time is zero, independent of the data rate. This tendency suggests that the mean data-rate does not fully reflect the temporal resolution of LDV and that high frequency information is included in the signal. It then becomes possible to estimate the distribution of signal power at frequencies much larger than the mean data-rate, as seen in Section 3.6.5.3.

As summary concerning time resolution, there is an optimal range of operation when considering particle concentration and measurement volume. For a given concentration of tracer particles, the measuring volume should not be too small to yield high data-rates. It should not be too large to avoid multiple-particle bursts and to maintain high spatial resolution. Because the optical arrangement that determines the measuring volume is normally fixed by the manufacturer, the preceding argument is normally cast in terms of concentration: tracer concentration should be optimized to work with high data-rates in the separate event regime.

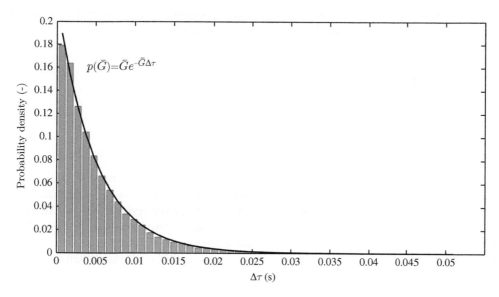

Figure 3.6.16 Examples of distribution of inter-arrival times for a mean data-rate $\dot{G} = 202$ Hz (mean interarrival time 4.1 ms). Black line stands for fitted negative exponential probability density function.

3.6.5.3 Calculation of statistics in turbulent flows

LDV produces time series of instantaneous velocities (determined from Equation (3.6.6)) acquired at high data-rates. The adjective "instantaneous" implies that velocities are obtained over extremely short time intervals (of the order of magnitude of tens of microseconds to tens of milliseconds) that ideally are smaller than the relevant turbulent time scales. This section addresses the calculation of statistics of turbulent flows, namely moments of the distribution of instantaneous velocities and estimates of the power spectral density function (see also Chapter 6, Volume I).

Calculation of moments from LDV data requires weighting-averaging techniques, because of the result expressed in Equation (3.6.17): particle count rate increases with particle velocity, resulting in larger velocities being overrepresented in a time series. The simple arithmetic mean of the velocity samples is a biased estimator of the true mean; *i.e.*, biased to larger velocities. The residence time, collected to characterize each burst, provides a suitable weighting factor to mitigate velocity bias: slower velocities weight more in the mean estimator than large velocities. The sample mean, $\bar{u}$, can be calculated as

$$\bar{u} = \frac{\sum\limits_{k=1}^{N} u_{p_k} \Delta t_k}{\sum\limits_{k=1}^{N} \Delta t_k} \qquad (3.6.18)$$

where N is the number of samples in the times series and Δt_k is the residence time associated to the crossing of particle with velocity u_{pk}. The equation is valid for each pair of beams of the same color. Second and higher-order centerd moments, including cross moments, can be obtained by the general formula

$$\sigma^n_{u_i^a u_j^b} = \frac{\sum\limits_{k=1}^{N} \left(u_{i_k} - \overline{u}_i\right)^a \Delta t_{i_k} \left(u_{j_k} - \overline{u}_j\right)^b \Delta t_{j_k}}{\sum\limits_{k=1}^{N} \Delta t_{i_k} \Delta t_{j_k}} \qquad (3.6.19)$$

where $\sigma^n_{u_i^a u_j^b}$ is a centered moment of order n, u_i and u_j are, respectively, the ith and j^{th} components of the velocity field, expressed in an orthogonal referential, $\overline{u}_i$ and $\overline{u}_j$ are their respective means, calculated by Equation (3.6.18), and $a + b = n$. Depending on the arrangement of the probes, velocities u_i and u_j may be the result of the transformation expressed in Equation (3.6.8). Alternatively, the centered moment can be calculated with the velocities of each color and then transformed through Equation (3.6.9). For instance, second-order moments between green and blue channels can be calculated as

$$\sigma^2_{u_g u_b} = \frac{\sum\limits_{k=1}^{N}\left(u_{g_k} - \overline{u}_g\right)\Delta t_{g_k}\left(u_{b_k} - \overline{u}_b\right)\Delta t_{b_k}}{\sum\limits_{k=1}^{N} \Delta t_{g_k} \Delta t_{b_k}} \qquad (3.6.20)$$

and then transformed through Equation (3.6.9) to contribute to Reynolds stresses.

Other possibilities for a weighting factor must respect the general rule that higher velocities must weigh less in the estimation of the mean. Hence, a possible weighting factor is the module of the velocity itself.

LDV data are frequently re-sampled into equally-spaced time series, greatly simplifying the procedure of calculating autocorrelation functions, structure function or power spectral density functions. Sample-and-hold (S+H) techniques (zero-order polynomial data reconstitution) are the most frequently employed but other polynomial reconstitution or other interpolation techniques can also be employed. Figure 3.6.17 shows an example of an actual LDV signal reconstituted into equally spaced data by S+H, linear interpolation and cubic interpolation. However, Adrian and Yao (1987) demonstrated that the power spectral density function of S+H reconstituted data is not reliable beyond the frequency $\dot{G}/(2\pi)$, where $\dot{G}$ is the mean data-rate, calculated from the series of inter-arrival times (and whose estimate is $\overline{G}$). This result is due to the combined effect of step-noise (noise introduced by the steps in the reconstitution procedure) and particle-rate filter, a low-pass filter with a cut-off frequency of $\dot{G}/(2\pi)$ (Adrian & Yao, 1987). Müller et al. (1994) showed that higher-order reconstituting techniques, particularly polynomial, would not change this result.

Nobach et al. (1996) devised a so-called refined sample-and-hold reconstitution technique (RS+H) to cancel the particle-rate noise. It employs a finite-impulse response filter parameterized using the mean particle rate. The autocorrelation function of the S+H reconstituted data is first obtained and then used to compute the refined estimation of the true autocorrelation function. Symbolically, the key step is

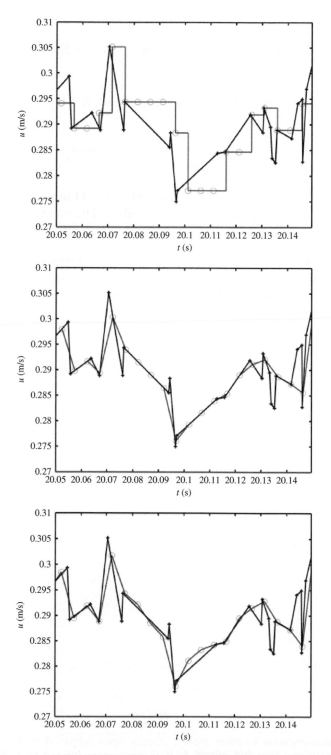

Figure 3.6.17 Examples of LDV data reconstitution. From top to bottom: Sample-and-hold reconstruction, linear interpolation, piecewise cubic interpolation.

$$R^{(RS+H)}(m\Delta T)$$
$$= (2c_G - 1)R^{(S+H)}(m\Delta T) - c_G\left(R^{(S+H)}\big((m-1)\Delta T\big) + R^{(S+H)}\big((m+1)\Delta T\big)\right)$$

$$(3.6.21)$$

where $R^{(RS+H)}$ is the refined auto-correlation function, $R^{(S+H)}$ is the autocorrelation function obtained from the S+H reconstruction, $c_G = e^{-\dot{G}\Delta T}/\left(1 - e^{-\dot{G}\Delta T}\right)^2$ and ΔT is the sampling interval (details in Nobach *et al.*, 1998a). Equation (3.6.21) is not valid at $m = 0$, for which $R^{(RS+H)}(0) = R^{(S+H)}(0)$. The power spectral density is obtained by a cosine transform of $R^{(RS+H)}$.

Figure 3.6.18 shows the results of both examples of S+H reconstruction techniques applied to a LDV velocity signal obtained in a rough wall boundary layer with Taylor's Reynolds number $Re_\lambda \approx 250$. For this particular signal, the differences are restricted to small scales.

Slotting techniques (Mayo *et al.*, 1974; Nobach *et al.*, 1998a, b) are other common techniques for estimating power distribution over turbulence frequencies of LDV data. They normally involve slot correlation for the autocorrelation function, followed by cosine transform for the power spectral density function.

The autocorrelation function can be calculated as

$$R^{(SC)}(m\Delta T) = \frac{\sum_{i=1}^{N}\sum_{j=i}^{N} u(t_i)u(t_j)b_m(t_j - t_i)}{\sum_{i=1}^{N}\sum_{j=i}^{N} b_m(t_j - t_i)} \qquad (3.6.22)$$

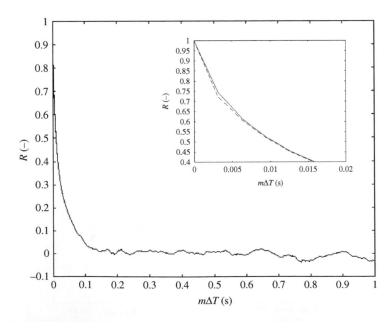

Figure 3.6.18 Autocorrelation functions from a velocity signal subjected to S+H reconstruction (continuous line) and refined S+H reconstruction (dashed line), $\Delta T = 2/\dot{G}$. LDV velocity signal obtained in a rough wall boundary layer with $Re_\lambda \approx 250$.

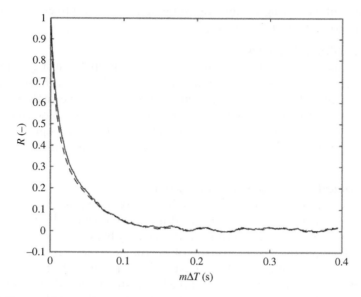

Figure 3.6.19 Autocorrelation functions obtained with slot correlation and RS+H with uniform $\Delta T = 0.5/\dot{G}$. Slot correlation: continuous line; RS+H: dashed line. Original velocity data as in Figure 3.6.18.

where normally $a = i$ or $a = i + 1$ (to avoid including variance noise) and kernel function $b_m(t_j - t_i)$ expressing a top-hat filter, a gaussian filter or other types of weighing functions. For instance, in the fuzzy slotting technique of Nobach *et al.* (1998b) it is a triangular window. The simplest form of kernel is

$$b_m(t_j - t_i) = \begin{cases} 1 - \left| \frac{t_j - t_i}{\Delta T} - m \right| & \text{for} & \left| \frac{t_j - t_i}{\Delta T} - m \right| < 1 \\ 0 & & \text{otherwise} \end{cases} \qquad (3.6.23)$$

A cosine transform is then applied to obtain the power spectral density function. Figure 3.6.19 shows the autocorrelation function and the power spectral density function corresponding to the slotting technique expressed in Equation (3.6.22) with $a = i + 1$ (estimate without variance noise) and kernel using Equation (3.6.23), and the RS+H technique.

The power spectral density of unevenly-spaced data can also be computed using a direct Fourier transform. However, such approach requires a high computational effort, which makes it less attractive. Benedict *et al.* (2000) reviewed several methods of computing the power spectral density function and arrived at the conclusion that the best options available consisted on cosine transformations of the autocorrelation functions obtained by Equations (3.6.21) or (3.6.22).

3.6.5.4 Summary of practical issues

Briefly summarized here are the main practical issues involved in the statistical analysis of LDV data: check coincidence data rates; defining coincidence at a post-processing step; alignment of measuring volume; re-sampling LDV data into equally spaced time series; and appropriate post-processing.

A) Check coincidence data rates

Working in coincidence mode, for which a crossing particle must generate a burst recognizable in all of the three wavelengths, may drastically reduce the frequency count if the measurement volumes are not perfectly aligned. A sensitive analysis to the mean data-rate is a straightforward way to find good alignment (*e.g.*, Derksen *et al.*, 1999). Coincidence data-rates should be checked periodically. If they are significantly smaller than non-coincidence data-rates for any of probes (*e.g.*, 3 times smaller), probe alignment should be corrected (see Section 3.6.3.2).

B) Defining coincidence at a post-processing step

Setting up the coincidence time can be done by the user as a post processing step. The key criterion is that bursting events detected in different channels should overlap as much as possible in time. Burst duration, as detected by burst processing hardware and software (see Figure 3.6.6), is an estimate of the time of residence of a particle in the measuring volume. The initiation of the burst is also determined by the processing unit. The initiation of the burst, or the center of the burst, should, for all pairs of beams, fall within a small radius; preferably one order of magnitude smaller than the mean interarrival time, and smaller than the residence time of the particle for that burst. Using this last criterion, take as an example a particle with 1.0 m/s crossing a measuring volume with 0.0005 m waist of a given color. The time of residence is 0.5 ms. The coincidence window should thus be smaller than 50.0 μs, *i.e.*, the center of the bursts detected for each color should fall within a 50.0 μs radius of each other.

C) Alignment of measuring volume

Align the largest direction of the measuring volume with the direction of the smallest gradients of the mean flow. For instance, in the case boundary layer flows, the largest dimension of the control volume to detect the longitudinal flow velocity should be aligned with the lateral flow direction, not with the wall-normal direction.

D) Re-sampling LDV data into equally spaced time series

The sampling interval is frequently the mean data-rate, calculated from the series of inter-arrival times (estimated as $\overline{G}$). This option may not be a good interpretation of Adrian and Yao (1987), as it represents a sub-optimal use of the information contained in the velocity time series, since it eliminates high-frequency components: equally spaced data is subjected to the sampling theorem; thus, frequencies above Nyquist frequency $1/(2\Delta T)$, where ΔT is the sampling interval, are lost in the resampling process. Also, noise is aliased back into the domain independently of the size of the sampling interval. Figure 3.6.20 shows the power spectral density function of the resampled signal with several ΔT and with the RS+H technique.

E) Appropriate post-processing

Many LDV commercial systems facilitate a significant amount of post-processing. It is possible to define user-defined functions to enhance the post-processing of the LDV data output (Aleixo *et al.*, 2016), *e.g.*, a despiking algorithm. Spiking is not typical of LDV data. However, errors may be introduced in the estimation of the Doppler frequency, f_D, for instance due to incorrect modulation originated by conflicting fringe patterns. In

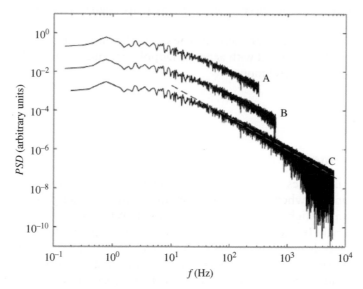

Figure 3.6.20 Power spectral density functions obtained from discrete cosine transforms of autocorrelation functions calculated with the S+H technique. A: $\Delta T = 1/\dot{G}$; B: $\Delta T = 0.5/\dot{G}$; C: $\Delta T = 0.05/\dot{G}$. Red dashed line represents Kolmogorov's $-5/3$ slope. Original velocity data as in Figure 3.6.18.

this case a despiking algorithm should be applied. The algorithm of Goring and Nikora (2002), although developed for ADV data, can be employed since it is compatible with unevenly spaced data, as it uses finite differences of velocity data.

Figure 3.6.20 shows that the results are unreliable for frequencies larger than about 1,000 Hz. Yet, assuming that the -5/3 slope is valid over a range of scales, valid results seem to be retrieved for $\Delta T = 0.5/\dot{G}$.

3.6.6 Final remarks

This section discusses the theoretical and practical principles associated with LDV. It includes on hardware components, key tracer requirements and the fundamentals of LDV data precessing. Practical issues considered include the description of the arrangement and positioning of probes to obtain referenced three-component measurements, beam and optical fiber alignment, maintenance of optical parts of the instrumentation, criteria for selecting tracers, based on tracing ability, and fundamental techniques of data processing to obtain turbulence quantities and autocorrelation and power spectral density functions.

LDV is a technique that should be considered for use in the following experiment situations:

–   Low intrusiveness is required. LDV is a relatively low intrusive technique. Given that the focal distances are large, no material pieces of instrumentation are normally required near the flow under measurement.
–   High temporal data rates are needed. Sampling frequencies obtained with LDV can be very high if an adequate amount of tracers is provided. In hydraulic flows, the

time discretization of three-component velocity signal is among the highest achievable.

- Small measuring volumes are needed. The size of the measuring volume is generally smaller than that obtained with acoustic Doppler techniques and other point-wise or profiling acoustic techniques.

3.B APPENDIX

3.B.1 Fringe model

Figure 3.B.1.1 illustrates the main geometric features of two laser beams. A fringe pattern is originated when two monochromatic (same wavelength) coherent (same phase for all waves) light beams are superimposed. Where the interfering light is in phase a peak (bright fringe) is generated; where it is out of phase, light is cancelled out and a trough (dark fringe) appears. In the LDV case this fringe pattern forms what is commonly designated the measurement volume. The fringe pattern is parallel to the axis of the transmitting optics (the bisector of the beams) and configures the optical heterodyne principle.

The spacing between fringes, δ, depends on the wavelength of the crossing beams, λ_0, and on the angle that they form, θ. Simple geometrical considerations (Figure 3.B.1.2) lead to the formula

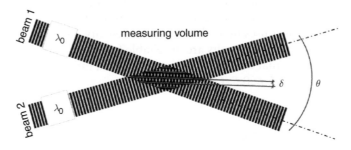

Figure 3.B.1.1 Basic geometry of the fringe model in the dual beam mode.

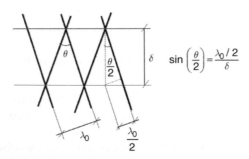

$$\sin\left(\frac{\theta}{2}\right) = \frac{\lambda_0/2}{\delta}$$

Figure 3.B.1.2 Determination of the fringe spacing from geometrical considerations.

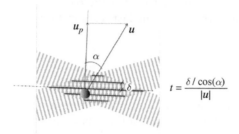

Figure 3.B.1.3 Particle crossing the fringe pattern and generating a wave pattern whose period is t.

$$\delta = \frac{\lambda_0}{2\sin(\theta/2)} \tag{3.B.1.1}$$

The time necessary for a particle to travel between two points with the same phase within the fringe pattern is

$$t = \frac{\left(\dfrac{\delta}{\cos(\alpha)}\right)}{|u|} = \frac{\delta}{|u_p|} \tag{3.B.1.2}$$

It is a function of the fringe wavelength and of the tracer velocity in the direction normal to the beam bisector (Figure 3.B.1.3). Hence, a tracer particle crossing the measuring volume with velocity u will scatter the fringe pattern generating a "burst" (see Section 3.6.5): a wave packet whose period is t and whose frequency is $t = 1/f_D$.

It is this frequency f_D, the "beat frequency", that the signal processor must be able to identify correctly. Introducing Equation (3.B.1.1) in Equation (3.B.1.2) one obtains the velocity of the particle that generated the burst as

$$u_p = \frac{\delta}{t} = f_D \frac{\lambda_0}{2\sin(\theta/2)} \tag{3.B.1.3}$$

which, evidently, is the same as that derived directly from the dual-beam mode Doppler shift (Equation (3.6.6)).

3.B.2 Directional ambiguity

To distinguish between positive and negative velocities, a frequency shift ε, in one of the beams is introduced (Durst & Zaré, 1974). In the layout of Figure 3.B.1.1, let the shift be applied to beam 2. For this beam, one has $(f_0 + \varepsilon) - f_2^{(2)} = u \cdot (n_{sp}^{(2)} - n_{pr})/\lambda_0^{(s)}$ where $\lambda_0^{(s)} = c/(f_0 + \varepsilon)$ is the shifted wavelength. Equation (3.6.5) becomes

$$f_R = f_2^{(1)} - f_2^{(2)} = \varepsilon + \frac{u \cdot \left(n_{sp}^{(2)} - n_{sp}^{(1)}\right)}{\lambda_0^{(s)}} = \varepsilon + f_0 \frac{u \cdot \left(n_{sp}^{(2)} - n_{sp}^{(1)}\right)}{c} + \varepsilon \frac{u \cdot \left(n_{sp}^{(2)} - n_{sp}^{(1)}\right)}{c} \tag{3.B.2.1}$$

where f_R is the frequency actually detected at the receiver. The shift frequency can be much smaller than f_0 and, therefore, the last term in Equation (3.B.2.1) is negligible, resulting in

$$f_R - \varepsilon = u \frac{2\sin(\theta/2)}{\lambda_0} \tag{3.B.2.2}$$

To better understand Equation (3.B.2.2) one defines a positive x-direction along n_- from the unshifted to the shifted beam (for this example, if beam 2 is shifted, positive x coincides with n_- in Figure 3.6.4. One notices that the frequency shift causes the fringe pattern to move with a velocity $v^{(s)} = \varepsilon\delta$ in the negative x-direction (shifted beam to unshifted beam). Hence when a tracer particle has zero x-velocity the detected frequency will be $f_R = \varepsilon$. If the x-velocity is smaller than $v^{(s)}$ the detected frequency is smaller than ε, as less fringe wavelengths are crossed by the tracer particle. On the contrary, if the tracer particle is faster than $v^{(s)} = \varepsilon\delta$ in the x-direction, more fringe wavelengths are crossed during the travel time across the measurement volume. This effect, mathematically expressed as Equation (3.B.2.2), is illustrated in Figure 3.B.2.1.

The shift frequency is, since early LDV systems, performed by acoustic vibration; the shifts thus attained are relatively small, frequently tens of MHz. This value should be enough to resolve the full spectrum of expected velocity values, which, in case of turbulent flows, may be difficult to determine. Figure 3.B.2.1 also illustrates this principle: the lowest value of the detected velocity is obtained when the detected frequency is near zero (more correctly equal to the measuring error). For instance, if $\theta = 5°$, $\lambda_0 = 532$ nm, and a frequency shift of 5 MHz, Equation (3.B.2.2) renders the limit $u_{min} \approx -30.5$ m/s, which, in open-channel flows, is unlikely to be reached. Note that the dynamic range is still limited by the receiver bandwidth and the frequency determination method used, frequency shift only affects the actual values of u_{max} and u_{min}, not its difference.

The velocity resolution of the LDV depends on the performance of the chosen method to determine the frequency. Nowadays the vast majority of LDV signal processors (see Section 3.6.5) rely on the Fast Fourier Transform (FFT) to measure the Doppler frequency. The frequency resolution of the FFT is

$$\Delta f = \frac{f_s}{N} \tag{3.B.2.3}$$

where f_s is the sampling frequency and N is the number of points. Since the number of points can be expressed as $N = f_s \Delta t$, where Δt is the acquisition period; *i.e.*, the duration

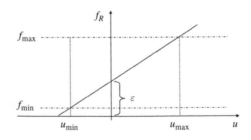

Figure 3.B.2.1 Eliminating directional ambiguity through frequency shifting.

of a burst. Therefore Equation (3.B.2.3) can be expressed as $\Delta f = 1/\Delta t$. The frequency resolution thus depends on the acquisition period, which is, of course, limited for each burst. A strategy to artificially increase the frequency resolution is by means of zero-padding the signal in the time-domain. This strategy requires filling the data series with a given number of zeros, often to obtain a number of samples that are a power of 2.

3.B.3 Tracer particles – light scattering

Light scattering is used here to name the process by which the direction and the intensity of a light beam are changed upon interacting with a tracer particle. The changes are due to combined effects of reflection, refraction and diffraction; details on these processes are out of the scope of this text and can be consulted in Albrecht *et al.* (2003). Light is scattered in all directions by particles but the angular distribution is rather complex. As a general rule, the amount of light scattered increases with the density of the material. The angular variation of the scattered light is mostly a function of the size of the particle.

If the diameter of the tracer particles is larger than the wavelength of the laser light, light scattering is thus said to be in the Mie regime (Albrecht *et al.*, 2003). Mie's solution of the Maxwell equations requires heavy computational effort (Grainger *et al.*, 2004) but can be considered exact if light is monochromatic and polarized into plane waves and if the particles are isotropic and spherical. In laser velocimetry, the conditions regarding the light are satisfied but the particles may not conform to these shape restrictions (see Table 3.6.1 and Figure 3.6.13).

Typical light scattering diagrams are shown in Figure 3.B.3.1 for particles with different refraction indexes subjected to the same power density of a monochromatic coherent light source. As shown in the figure, the forward direction (0°) is the most effective. LDV systems operating in the forward scatter mode are, however, not dominant, for constructive and operative reasons. Systems operating in backscattering systems have the receiving and the transmitting optics in one single probe, with the former optics focused by the manufacturer. In the forward scatter mode, the user is generally required to find the measuring volume and to focus the receiving optics, which may introduce measurement errors.

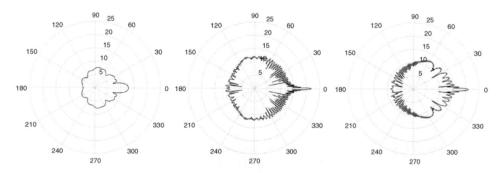

Figure 3.B.3.1 Power density scattered (in W/m²) by different types of particles when crossing a 6 W, 0.5 mm diameter green (514 nm) laser beam in water (receiver at 0.4 m from scattering particle). a) Titanium dioxide, diameter = 1 μm, refractive index = 2.6. b) Titanium dioxide, diameter = 10 μm, refractive index = 2.6. c) Polyamide, diameter = 10 μm, refractive index = 1.7. Note that power scale of a) is twice of the scale of b) and c).

The scattering cross-section in Mie regime depends on the particle diameter (d), the light wavelength and the relative refractive index. It may be taken, for constant light power density, a proxy for the amount of scattered light. For $d > \lambda_0$, the scattering cross-section depends on d^2, i.e., the decay of light power scattered to a receiver is quadratic. The signal associated to small particles will thus be much difficult to detect than the signal of large particles. Hence, tracers with wide grain-size distributions may lead to sub-representation of small turbulent scales: larger particles may introduce errors in the tracking small eddies (see Appendix 3.B.4) and small particles may not be detected or may be detected incorrectly. This shortcoming may lead to scale-dependent (or frequency dependent) noise, biased to small scales, in measurements of turbulent flows.

Regarding the advantages and disadvantages of the use of large particles, it should be noted that, although the light intensity increases the detection probability, the modulation of the signal can much reduced by particles equal or larger than the fringe spacing, δ, given by Equation (3.B.1.1) (Durst, 1973; Durst et al., 1981). In the limiting case f_D can become undetected if a particle scatters exactly the same number of bright and dark fringes, whose light intensities cancel each other. This outcome occurs if the particle is an exact multiple of δ.

Note that to avoid light scattering with angle dependence fluorescent particles may be used. Fluorescent light is isotropic which would eliminate the shortcomings of the backscatter mode. However, this possibility is generally not considered in LDV because the decay of light power scattered to a receiver is cubic. In other words, larger particles would be necessary to generate the same amount of brightness (McKeon et al., 2007).

3.B.4 Tracer particles – quantification of tracking ability

The dynamic behavior of the tracers in the fluid is one of the key aspects influencing the quality of the results of particle-based methods. To evaluate the performance of the many available materials susceptible to be used as tracers one needs to formulate their motion relatively to that of the fluid, including physically acceptable simplifications to render the analysis tractable. The equation of conservation of momentum of a spherical tracer particle transported by an unsteady non-uniform flow of a viscous incompressible fluid, is a widely employed model for that relative motion. According to Maxey and Riley (1983) or Crowe et al. (1998), this equation can be written for the velocity of the tracer particles relatively to that of the fluid motion as

$$
\frac{4}{3}\pi a^3 \left(\rho^{(s)} \frac{du_j^{(r)}}{dt} + \underbrace{\frac{1}{2}\rho^{(w)}\left(\frac{du_j^{(r)}}{dt} + \frac{1}{10}a^2 \frac{d}{dt}\left(\nabla^2 u_j^{(w)}\right) \right)}_{A} \right) + \underbrace{\frac{4}{3}\pi a^3 \rho^{(s)} \frac{du_j^{(w)}}{dt}}_{B}
$$

$$
+ \underbrace{6\pi\mu a^2 \int_0^t \frac{d}{d\tau}\left(u_j^{(r)} + \frac{a^2}{6}\nabla^2 u_j^{(w)} \right) \frac{1}{\sqrt{\nu(t-\tau)}} \, d\tau}_{C}
$$

$$
= \underbrace{-6\pi\mu a u_j^{(r)} + \pi\mu a^3 \nabla^2 u_j^{(w)}}_{D} + \underbrace{\frac{4}{3}\pi a^3 \rho^{(w)} \frac{Du_j^{(w)}}{Dt}}_{E} + \underbrace{\frac{4}{3}\pi a^3 \left(\rho^{(s)} - \rho^{(w)} \right) g_j}_{G}
$$

(3.B.4.1)

where $u_j^{(r)} = u_j^{(s)} - u_j^{(w)}$ is the jth component of the relative tracer velocity, $u_j^{(s)}$ and $u_j^{(w)}$ are the jth components of the velocity of tracers and fluid, respectively, $\rho^{(s)}$ and $\rho^{(w)}$ are the densities of the tracer grains and of the fluid, respectively, a is the radius of the tracer grains (assumed to be uniformly distributed), μ and v are the viscosity and the kinematic viscosity of the fluid, respectively and t stands for the independent variable time. The material derivatives d/dt and D/Dt differ in the velocities used in the convective operator, $u_j^{(s)}$ for the former and $u_j^{(w)}$ for the latter.

In addition to the inertia force associated to the accelerating mass of the tracer particle, the left-hand side of Equation (3.B.4.1) includes forces arising from flow unsteadiness, namely the "apparent" added mass

$$\frac{2}{3}\pi a^3 \rho^{(w)} \left(\frac{\mathrm{d}u_j^{(r)}}{\mathrm{d}t} + \frac{1}{10}a^2 \frac{\mathrm{d}}{\mathrm{d}t}\left(\nabla^2 u_j^{(w)} \right) \right) \tag{3.B.4.2}$$

(term A in Equation (3.B.4.1) multiplied by $\frac{4}{3}\pi a^3$) expressing the force necessary to accelerate the surrounding fluid (Crowe *et al.*, 1998), and the Basset force, term C in Equation (3.B.4.1), expressing the time delay in the development of the boundary layer around and accelerating particle. Both these unsteady forces include a Fáxen correction term depending on $a^2 \nabla^2 u_j^{(w)}$ (Fáxen 1922, cited in Maxey & Riley 1983) to account for non-uniformity of the flow field. The right-hand side of Equation (3.B.4.1) includes terms arising from the transfer of momentum between the particle and the undisturbed flow. These include Stokes (viscous) drag – term D in Equation (3.B.4.1) –, the effect of pressure and viscous stress gradients (except buoyancy) – term E in Equation (3.B.4.1) – and the net body force, normally the buoyancy – term G in Equation (3.B.4.1). Term –B in Equation (3.B.4.1) is a consequence of expressing particle inertia in terms of the relative velocity.

Equation (3.B.4.1) is valid for small values of the Reynolds number $Re^{(s)} = \rho^{(w)}Wa/\mu$, where W is a representative value of the relative tracer velocity. Thus, it is not valid for large deviations between tracer and fluid velocity such as those expected in sediment grains in saltation. It may be crude approximation for non-spherical particles, particles subjected to non-uniform shear (Saffman forces, see Zheng & Li, 2009) or spinning particles (Crowe *et al.*, 1998). It is also necessary that the tracer con-centration is small enough so that there are no collisional or frictional interactions among tracer particles.

Equation (3.B.4.1) may be applied to turbulent flows provided that: turbulence is locally homogeneous in the sense of Monim and Yaglom (1971) and stationary; the diameter of the tracer particle is much smaller than Kolmogorov length microscale, η, and $(a/\eta)a\zeta/v << 1$ where ζ is Kolmogorov velocity microscale. Notice that, under these conditions, Saffman forces are $O(a/\eta)$ and thus negligible relatively to Stokes drag. Under the same conditions and assuming that the Kolmogorov length scale is taken as the characteristic flow non-uniformity scale and that $W/\zeta \propto a/\eta$, Fáxen corrections are also negligible since they are $O(Re^{(s)}a/\eta)$ relatively to Stokes drag. As for the remaining forces, terms E and G are $O(Re^{(s)}\eta/a$ and $O(Re^{(s)}(s-1)ag/W^2)$, respectively, relatively to Stokes drag. Hence the former may not be negligible and the latter (buoyancy) will be relevant for denser particles whose diameters are not so small (Melling, 1997). Added mass (term A) and Basset forces (term C) are $O(Re^{(s)})$ and $O(Re^{(s)\ 0.5})$, respectively, relatively to Stokes drag and hence, should be unimportant in most applications.

As described in Hinze (1975), Hjelmfelt and Mockros (1966) idealized fluid and particle velocities as a superimposition of harmonic waves and solved Equation

(3.B.4.1) neglecting Fáxen terms and body forces. Introducing the effect of buoyancy as a first order perturbation term one obtains a simple method to determine a first approximation to the cut-off turbulent scale imposed by the tracers. The ratio of the amplitudes of velocities of tracer particles and fluid, $r_A = U^{(s)}/U^{(w)}$, where $U^{(s)}$ and $U^{(w)}$ are absolute values of tracer and fluid velocities, respectively, can be calculated analytically as

$$r_A = \left((1+\alpha_1)^2 + \alpha_2^2\right)^{\frac{1}{2}} \tag{3.B.4.3}$$

where

$$\alpha_1 = \frac{f\left(f + \dfrac{9\sqrt{vf}}{\left(s+\frac{1}{2}\right)}\right)\left(\dfrac{3}{2\left(s+\frac{1}{2}\right)} - 1\right)}{\left(\dfrac{18v}{4\left(s+\frac{1}{2}\right)a^2} + \dfrac{9\sqrt{vf}}{\left(s+\frac{1}{2}\right)}\right)^2 + \left(f + \dfrac{9\sqrt{vf}}{\left(s+\frac{1}{2}\right)}\right)^2} \tag{3.B.4.4}$$

$$\alpha_2 = \frac{f\left(\dfrac{18v}{4\left(s+\frac{1}{2}\right)a^2} + \dfrac{9\sqrt{vf}}{\left(s+\frac{1}{2}\right)}\right)\left(\dfrac{3}{2\left(s+\frac{1}{2}\right)} - 1\right)}{\left(\dfrac{18v}{4\left(s+\frac{1}{2}\right)a^2} + \dfrac{9\sqrt{vf}}{\left(s+\frac{1}{2}\right)}\right)^2 + \left(f + \dfrac{9\sqrt{vf}}{\left(s+\frac{1}{2}\right)}\right)^2} \tag{3.B.4.5}$$

and f is the frequency of the eddying motion. Results for three types of tracer particles are shown in Figure 3.B.4.1. The closer the ratio r_A is to unity, the better the tracking capability. Melling (1997) proposed a range of acceptability $0.95 < r_A < 1.05$ to quantify the adequacy of the tracer particles.

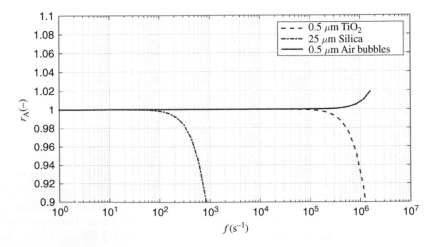

Figure 3.B.4.1 Amplitude ratio particle to fluid velocity in a turbulent flow of water at 20° C. Dashed line: titanium dioxide ($a = 0.5$ μm, $s = 4.20$). Dashed-dot line: silica powder ($a = 25$ μm, $s = 2.65$). Continuous line: air bubbles ($a = 0.5$ μm, $s = 0.0013$).

Large turbulent scales (small frequencies) are well represented by the motion of the tracer particles. In what concerns the small scales (large frequencies), Figure 3.B.4.1 shows that titanium dioxide particles of 1 μm diameter follow turbulent scales up to 2×10^5 Hz with 95% confidence. The tracking ability depends on the actual flow to be measured. Let the advection velocity be 0.2 ms^{-1} and Kolmogorov microscale 1×10^{-4} m. Using Taylor's frozen turbulence hypothesis (Taylor, 1938), the cut-off frequency of 2×10^5 Hz means that, from this theory, the smallest resolved turbulence scale would be 1×10^{-6} m. Such small scales would not be attainable since the size of the particle limits the resolution to scales twice the diameter of the particle, 2 μm. Larger particles, such as 50 μm silica powder of Figure 3.B.4.1 , although less dense, have a smaller cut-off frequency, approximately 2×10^2 Hz. For the same hypothetical flow, the smallest resolvable scale is 1×10^{-3} m, which means that a decade of turbulent scales, down to Kolmogorov length scale, would not be susceptible to be measured with less than 95% confidence, even if the interarrival time of tracers would be smaller than 5 ms. In some cases, dissolved air in the flow is the only readily available tracer. In the example of Figure 3.B.4.1, the velocity of 1 μm air bubbles is representative of fluid velocities (with 95% confidence) up to turbulent scales of 6×10^6 Hz. In the space domain, for the same hypothetical flow of the previous examples and from the same arguments, one finds that such tracer would be able to resolve eddies of 2 μm.

The above discussion indicates that particle diameter has a greater influence on tracking ability than density, for the range of values shown in Table 3.6.1. However, a more stringent limitation to resolution is associated to signal processing: several fringe wavelengths must be scattered to allow detection of f_D through FFT (see Section 3.6.5); thus, in practice, independently of the particle size, the smallest turbulence scales are of the order of magnitude of the dimensions of the measuring volume (Section 3.6.5). Hence, cheaper material such as Titanium dioxide can be used for common open-channel flow applications. The two main short-coming of this particular material are: that its grain-size distribution is generally far from uniform, which may reduce confidence at smaller scales; and, that it tends to deposit on the channel walls.

The lower bound of diameter range cannot be determined from the solution of Equation (3.B.4.1) expressed in Equation (3.B.4.3). Particles whose diameter is smaller than 1 μm are likely to interact with individual water molecules, thus be subjected to Brownian motion, a phenomenon not included in Equation (3.B.4.1). The order of magnitude of the velocities that arise due to this phenomenon is generally much lower than that of the mean convective flow but, in turbulent flows, the accuracy achieved in tracking the smallest scales may be reduced due to this phenomenon.

3.7 IMAGE-BASED VELOCIMETRY METHODS

3.7.1 Introduction

Image-based velocimetry methods include a wide variety of techniques that utilize images of a flow to obtain two- or three-dimensional measurements of the velocity field. Although the use of flow visualization to qualitatively understand flow behavior dates back to ancient times, contemporary image-based velocimetry techniques have only recently become viable with advancements that allow large quantities of image data to be collected and processed. The most common image-based techniques are referred to as particle-imaging techniques because they require the flow to be seeded with tracer particles which follow the motion of the fluid. For these techniques, the displacements of particles over a known time period are measured and used to calculate particle velocities. Adrian (1991) organizes particle-imaging techniques into three primary categories: laser speckle velocimetry (LSV), particle image velocimetry (PIV) and particle tracking velocimetry (PTV). These techniques are categorized based on the concentration of particles used to seed the flows that are being measured, PTV having the lowest seeding density and LSV having the highest.

PTV refers to the tracking of individual tracer particles to find their velocities. According to Adrian (1991), PTV is probably the oldest of the particle-imaging techniques because of its simplicity and ease of application without requiring advanced technology. For instance, one could easily manually track the displacements of small debris or pieces of ice floating in a river. Advances in PTV have come in the form of the ability to simultaneously and accurately track the movements of vast numbers of particles. LSV is associated with tracking the motion of laser speckle patterns produced by the illumination of high densities of tracer particles. Although LSV is a precursor to present-day PIV, it is not as widely used as PIV or PTV in hydraulic measurements and is only alluded to briefly here. For PIV applications, particle densities are generally low enough so that individual particles are identifiable but high enough so that groups of particles are tracked instead of individual particles.

Particle-imaging techniques are not the only image-based velocimetry methods. Image correlation methods, for instance, have been applied to assess velocity distributions in flows with continuous tracers like dyes and thermal patterns (*e.g.,* Gornowicz, 1997; Roux *et al.*, 2002). However, tracking the motion of continuous tracers is not as straightforward as tracking the motions of discrete particles, and particle-imaging techniques are definitely the most widely used image-based velocimetry methods. Of these, PIV and PTV are the most popular methods applied to hydraulic flows. Beginning with PIV, these two techniques will be the subject of this section.

3.7.2 Particle Image Velocimetry (PIV)

3.7.2.1 Introduction

PIV is an optical technique for measuring velocity fields that has revolutionized investigations in experimental fluid mechanics (Adrian, 2005; Westerweel *et al.*, 2013). The most basic implementation of PIV, depicted in Figure 3.7.1, allows two components of the velocity vector to be estimated at discrete points across a 2-dimensional slice of the flow field. This system is referred to as 2C-2D (two components – two dimensions), and it exclusively provides planar measurements of velocity. The key components of a typical PIV system are an illumination source, a digital camera to record the motion of small tracer particles added to the fluid, electronics to synchronize the camera and the light source, and

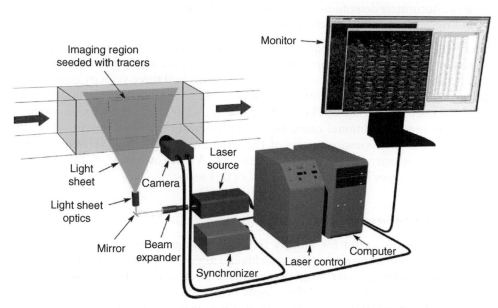

Figure 3.7.1 A typical two component, two dimension (2C-2D) Particle Image Velocimetry deployment. The dashed rectangle within the light sheet plane is the region of the flow captured in camera images.

software to analyze the digital images and estimate the displacements (velocities) of the tracer particles.

PIV measurements typically require:

1. Seeding the flow with small tracer particles that closely follow the motion of the fluid. Ideal tracer particles are small, have a density similar to that of the fluid, and scatter light well.
2. Illumination of a flow region of interest. A common source of illumination is the pulsed Nd:YAG laser because it provides very short (~nanoseconds) bursts of intense light, the light produced is easy to shape into a thin light sheet using readily available optics, and the light has a very narrow range of wavelengths (making it easy to filter and control).
3. Capturing images of the flow field. Pairs (or sequences) of images are collected for analysis. The time between the images, referred to as the separation time, is usually very short, and its selection depends on the flow velocity and image scale. The separation time can be accurately controlled by a synchronization device, which simultaneously controls the camera and triggers the laser to fire at appropriate instants. Typically, PIV images capture the positions of thousands of discrete tracers distributed over the region of interest. These particles move only small distances during the separation time. Thus, the second image of a PIV image pair is very similar to the first image, but with small displacements of all of the particles captured in the images.
4. Processing of captured images. Displacements of tracer particles during the separation time between pairs of PIV images are directly related to the velocity field of the

flow. Accurately determining the velocity field from the images requires an efficient but rigorous processing technique.

In this section, these four requirements will be covered in detail with illustrations focused on the basic 2C-2D implementation. However, the methods and many of the concepts that are introduced can be extended to resolve all three velocity components (stereoscopic PIV – Prasad, 2000) and even to measure velocity fields in three-dimensional volumes (tomographic PIV – Elsinga *et al.*, 2006). These extended methods require the use of additional cameras, precise calibration procedures to determine spatial positions and viewing angles of the different cameras, and revised software for extracting velocity fields from the raw images. A brief overview of different PIV-implementation types is provided at the end of this section.

3.7.2.2 PIV measurement process details

The PIV method is useful for quickly (relative to single point measurements) and nonintrusively estimating mean velocity fields and higher order moments of the velocity probability distribution over large spatial domains. It is also ideal for studying the instantaneous spatial structure of the velocity field, such as for the analysis of coherent structures. To successfully utilize the PIV method, users should have a basic understanding of the operating principles; it cannot realistically be treated as a 'black box' or 'plug and play' system. In the following sections, tracer particles, light sheet formation, image capture, and PIV software algorithms are briefly reviewed. Understanding measurement resolution and error is a critical part of data analysis. For PIV, the resolution and error depend on the interaction of many system and flow variables, these are examined in Sections 3.7.2.3 through 3.7.2.5.

3.7.2.2.1 Selection of tracer particles

Tracer particles for use with water flows are most often nearly neutrally buoyant glass or plastic microspheres, but metallic powders or natural particles (pollen) may also be used (*e.g.*, Raffel *et al.*, 2007). Hollow glass spheres are naturally hydrophilic and therefore disperse easily in water, whereas most plastics are hydrophobic and a surfactant may be required to help disperse the particles. Usually, tracer particle diameters should be as small as possible while still scattering sufficient light to properly expose an image (typically 10-100 microns in diameter for hydraulic experiments). If the particles are too large, or their concentration is too high, the particles may not follow the smallest flow scales accurately (*e.g.*, Mei, 1996; Melling, 1997), but this is less of a problem with water flows. Section 3.6.4 presents characteristics of LDV tracer particles in detail. Much of the discussion in Section 3.6.4 is also relevant for PIV tracer particles.

3.7.2.2.2 Light sheet formation and image capture

The typical light source used for contemporary PIV is a pulsed Nd:YAG laser with pulse energies of 10-200 mJ, pulse durations of 5-10 ns, and repetition rates of 10-10,000 Hz. Dual cavity lasers (essentially two lasers mounted on a common base plate) are often used to allow very short pulse separation times (Δ_t) that can be varied independently from the laser repetition rate. For laboratory PIV measurements, the pulse separation time is typically in the range 0.1-10 ms and is used to limit the displacements of particle images between exposures. In order to capture two separate images of the flow field, the laser must fire once in each of two sequential image frames. If the necessary pulse separation time is shorter than the shortest possible duration of a camera frame (due to camera hardware limitations), the synchronizing device must be used to trigger the first laser pulse near the end of the first image exposure and the second laser pulse near the start of the second image exposure. This strategy is called "frame straddling". There are cameras designed specifically to enable this type of frame straddling, but it can often also be accomplished with general purpose models. Higher repetition rate lasers (1-10 kHz) may be used in a single cavity configuration, but these lasers typically have lower pulse energies than dual cavity lasers. One drawback of dual cavity lasers is that small differences between the output light intensity distributions for each cavity, and any misalignment of the optics that combine the two laser beams onto a common axis will increase measurement error. Continuous wave lasers may also be used, but these have lower instantaneous power outputs than pulsed lasers, requiring longer exposure times. Consequently, care must be taken to limit particle motion during an image exposure in order to avoid motion blur. This requirement typically limits the application of continuous wave lasers to lower speed flows. Samples of pulsed and continuous lasers are provided in Appendix 2.A.

Laser light is formed into a thin sheet using a combination of spherical and cylindrical lenses. The example configuration shown in Figure 3.7.2 uses two spherical lenses to adjust the thickness of the light sheet and a negative cylindrical element to expand the beam in one direction to form a sheet. Several alternative configurations are given in Raffel *et al.* (2007). Due to diffraction, the thickness of the light sheet will not be constant, but will vary with distance from the laser. The minimum thickness of the light sheet (T) is given by:

$$T = 2\omega_0 \approx \frac{2M^2\lambda}{\pi\theta} \tag{3.7.1}$$

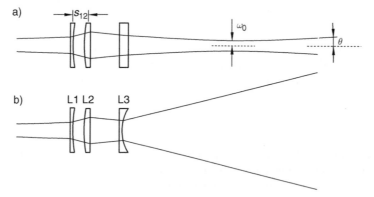

Figure 3.7.2 Light sheet formation optics: a) view parallel to the light sheet and b) view normal to the light sheet. The lenses for this configuration are: L1, plano-concave spherical; L2, plano-convex spherical; and L3, plano-concave cylindrical. The distance s_{12} controls the position of the beam waist.

where ω_0 is known as the waist radius, λ is the wavelength of the laser light, θ is the far field beam divergence angle, and $M^2 > 1$ is the beam quality factor. For the typical high power lasers used for PIV, the M^2 factor is often in the range 5 to 50. Equation (3.7.1) points to an inherent trade-off between sheet thickness and divergence angle. To achieve a thin light sheet, the divergence angle must be increased, which will limit the usable area of the light sheet (where its thickness is reasonably uniform).

The underlying optical process on which PIV is based is light scattering, as described in Appendix 3.C.1. The light scattered from tracer particles is focused by a lens onto the camera sensor which integrates the received light at discrete photosites (or sensor pixels) to form an image. The size and shape of particle images are important parameters affecting the accuracy of PIV and are a function of diffraction, lens aberrations, and the geometry (or physical size) of the seeding particles. If the influences of lens aberrations and geometry are small compared to that of diffraction, the image is said to be diffraction limited. Most lenses used for PIV will approach diffraction limited for lens f-numbers ($f_\#$) larger than about 8. This is the ideal condition for PIV because the size and shape of diffraction limited particle images are reasonably predictable and repeatable across different implementations.

3.7.2.2.3 Image processing to measure the velocity field

The central goal of the PIV image processing technique is to estimate the displacements (and ultimately the velocities) of groups of particles in small sub-regions of the flow (referred to as interrogation areas, windows or regions) by comparing the locations of the images of the particles in two or more consecutive images. The displacement of particles within an interrogation region is statistically determined by measuring a two-dimensional correlation of the interrogation region in two successive images. Thus, image processing usually begins with the division of the PIV images into a grid of interrogation areas as shown in Figure 3.7.3

The figure shows an original PIV image, the same image divided into a 20 by 15 cell grid of 32 by 32 pixel interrogation areas, and a magnified inset of one of the interrogation areas that shows particles from the first PIV image in green and particles from the second PIV image in red. In the inset, displacement of the particles is from left to right. Each of the interrogation areas is processed with the same algorithm, and as long as there are sufficient numbers of particles within each interrogation area and the algorithm is robust, the result will be one displacement measurement for each interrogation area. The velocity vector for each interrogation area is then simply determined by

$$\vec{V} = M \cdot \frac{\vec{d}}{\Delta_t} \tag{3.7.2}$$

where $\vec{d}$ is the displacement in image pixels, M is the magnification factor (the actual distance that each pixel represents in the flow field), and Δ_t is the image separation time. The final vector field is obtained by repeating the interrogation process on the grid of interrogation areas spanning the field of view.

During the evolution of PIV, several image processing algorithms were developed, driven by the technology available at the respective times. The first PIV systems used single frame/ multiple exposure recordings that required autocorrelation-based algorithms for

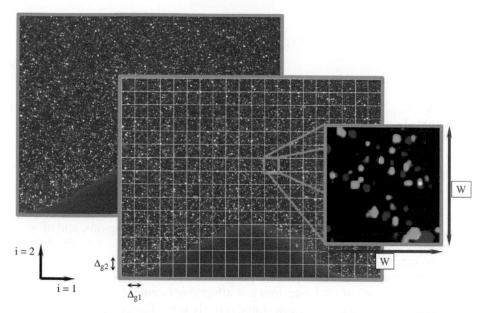

Figure 3.7.3 The PIV interrogation process showing original image with flow over a sand ripple, original image subdivided into a grid of interrogation areas, and an inset showing a close-up of one of the interrogation areas. Green and red particles are from matching pairs of images. Note the similar displacements of the particles between frames.

evaluation of the velocities (Adrian, 1991). Subsequently, with the increase in camera frame rates, double frame/single exposure recordings have emerged (Raffel *et al.*, 2007), leading to the development of cross-correlation methods for the evaluation of velocities (Raffel *et al.*, 2007). This latter system is the dominant contemporary interrogation technique, though some other correlation methods, such as the minimum quadratic difference method, have emerged for specific applications (Sveen, 2013; Gui & Merzkirch, 1996). Because of their prevalence, only cross-correlation algorithms are presented here for illustration purposes. In this approach, velocity vectors are extracted from a pair of PIV images by estimating the particle displacement from the cross correlation of image intensities computed between the matching interrogation windows of the first and second images. The interrogation window shown in Figure 3.7.3 has a side length W and is centered on a grid with spacing Δ_{g1} and Δ_{g2} in the $i=1$ and $i=2$ directions, respectively. Historically, interrogation windows have often been square with side lengths that are powers of two (*e.g.*, 16, 32, or 64 pixels, etc.) as analysis of windows with these dimensions is more efficient for some interrogation algorithms. In general, grid spacing is not required to equal the interrogation window side length – interrogation windows may be sparsely distributed or may overlap. In fact, the distribution of interrogation points is not restricted to any particular pattern, though a grid of points is usually selected. The cross-correlation function between a pair of interrogation regions $\phi(m', n')$ can be calculated as:

$$\phi(m', n') = \frac{A}{\sqrt{BC}} \tag{3.7.3}$$

with

$$A = \sum_{m,n=0}^{W-1} \eta(m,n)\Big(g(m,n) - \mu_g\Big)\eta(m+m',n+n')\Big(h(m+m',n+n') - \mu_h\Big)$$

$$(3.7.4)$$

$$B = \sum_{m,n=0}^{W-1} \Big[\eta(m,n)\Big(g(m,n) - \mu_g\Big)\Big]^2 \qquad\qquad (3.7.5)$$

$$C = \sum_{m,n=0}^{W-1} \Big[\eta(m,n)\Big(h(m,n) - \mu_h\Big)\Big]^2 \qquad\qquad (3.7.6)$$

where $g(m,n)$ and $h(m,n)$ are the gray level intensities of the first and second interrogation windows, μ_g and μ_h are the mean intensities of the respective regions, and m' and n' represent the shift of the second interrogation region with respect to the first. An optional weighting window (η) can be incorporated to in some cases reduce the measurement error. The term $(BC)^{0.5}$ in Equation (3.7.3) scales the correlation function to the range $-1 \le \phi(m', n') \le 1$ but does not affect displacement estimates. The cross correlation function is not usually calculated directly using Equation (3.7.3) because it is computationally expensive. Instead, a Fourier transform based method (Figure 3.7.4) can be used to significantly speed up the calculation. The Fourier transform method, can introduce some artifacts such as wrap-around aliasing and spectral leakage (Eckstein & Vlachos, 2009), but these can be mitigated with an appropriate choice of tapering window (η). Correlation values away from the origin ($m' = n' = 0$) will also be biased low due to reduced window overlap for larger window offsets. As suggested earlier, alternative formulations of the cross correlation function (Equation (3.7.3)) have also been suggested (e.g., Raffel et al., 2007) and may perform better in some situations.

Figure 3.7.5 shows a pair of matching interrogation areas and the two-dimensional cross correlation distribution that results from the process depicted in Figure 3.7.4. After the cross correlation function has been calculated, the highest peak is located and its position estimated to sub-pixel resolution by interpolation. A common interpolation method is the three-point Gaussian fit to the correlation maximum ϕ_c and its immediate neighbors ϕ_l, ϕ_r, ϕ_u, and ϕ_d (Figure 3.7.5):

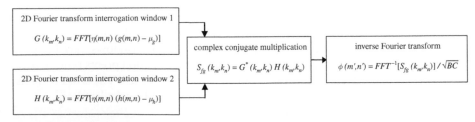

Figure 3.7.4 Calculation of the cross correlation function using fast Fourier transforms (*FFT*). G^* is the complex conjugate of G, *FFT*$^{-1}$ is the inverse fast Fourier transform, and k_m and k_n are the wavenumbers in the m and n directions respectively.

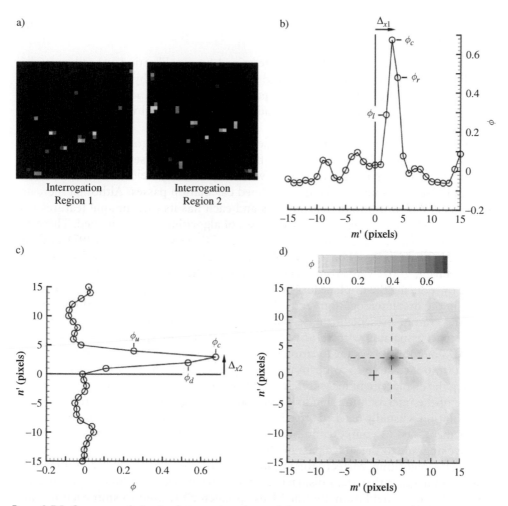

Figure 3.7.5 Cross correlation implementation: a) sample interrogation region pair; b), c), and d) cross correlation function. Subfigures b) and c) are cross sections through the correlation maximum marked by dashed lines in d).

$$\Delta_{x1} = m'_{\phi_c} + \frac{\ln\phi_l - \ln\phi_r}{2\ln\phi_l - 4\ln\phi_c + 2\ln\phi_r} \tag{3.7.7}$$

$$\Delta_{x2} = n'_{\phi_c} + \frac{\ln\phi_d - \ln\phi_u}{2\ln\phi_d - 4\ln\phi_c + 2\ln\phi_u} \tag{3.7.8}$$

where m'_{ϕ_c} and n'_{ϕ_c} are the discrete coordinates of the correlation peak, and Δ_{x1} and Δ_{x2} are estimates of the particle image displacement. Displacement estimates are related to velocity components (u_1, u_2) by:

$$u_1 = \frac{\Delta_{x1}}{S\Delta_t} \tag{3.7.9}$$

$$u_2 = \frac{\Delta_{x2}}{S\Delta_t} \qquad\qquad (3.7.10)$$

where S is a scale factor (pixels/mm) determined by a calibration procedure and is related to the image magnification (M) and pixel pitch (P) pitch by $S = M/P$, and Δ_t (ms) is the time separation between image exposures. Here u_1 and u_2 have units meters/second and would typically be the longitudinal and bed normal velocity components, respectively.

A number of improvements to the single pass correlation algorithm described thus far have been suggested over the years (Raffel *et al.*, 2007). These improvements come in the form of multi-pass algorithms, in which the velocity field obtained from one iteration is used as input to subsequent refined calculation passes. Although there are numerous PIV algorithm implementations and each has its own unique features and performance characteristics, two broad classes of algorithms have emerged. These are the 'iterative discrete shift' (IDS) method (*e.g.*, Westerweel *et al.*, 1997) and the 'iterative deformation method' (IDM; Nogueira *et al.*, 1999; Scarano & Riethmuller, 2000; Astarita, 2007; Cameron, 2011). Both algorithms improve on the single pass method by reducing the in-plane loss of particle images, that is, particle images that are present in the first interrogation region but due to their displacement in the x_1 and x_2 directions are not present in the second region. The IDM method can additionally reduce errors associated with velocity gradients and increase spatial resolution.

The IDS method allows the location of the first and second interrogation regions to be offset from each other to compensate for the particle motion. The offset is selected by rounding the displacement measured from the previous iteration to the nearest integer. The size of the interrogation window can also be changed with each iteration, and a stable solution is usually reached after 2 or 3 correlation passes. The IDS method is easy to implement, fast to compute, and the reduction of measurement error makes it an attractive upgrade to the single pass method.

The IDM algorithm corrects for in-plane loss of pairs and velocity gradients by deforming images rather than displacing interrogation grid coordinates. An interpolated displacement field estimate (the 'dense predictor') is used to shift each original image pixel to a new location, so that particle images in the first deformed image are ideally at the same location in the second deformed image. A low pass filtering operation, often implemented as a convolution of the velocity field with a kernel ξ is usually required after each iteration to ensure that the algorithm converges. Convergence is usually reached within 8–20 iterations depending on the selection of region weighting windows (η) and low pass filter kernel (ξ). The IDM approach is significantly more computationally expensive than IDS due to the larger number of iterations and the requirement for image interpolation. The benefits of IDM are that it can cope with larger velocity gradients than IDS, and that the shape of the frequency response function (Section 3.7.2.3) can be manipulated to achieve higher resolution.

3.7.2.3 Spatio-temporal resolution

The temporal resolution of a PIV system may be influenced by several configuration characteristics, including: exposure method (multi-exposure or multi-frame recording), camera sensitivity, camera frame rate, and the illumination source. Exposure method is important because the camera frame rate is only relevant if multiple frames must be exposed. However, for multi-frame methods, which constitute a majority of contemporary PIV applications in hydraulic engineering, the camera frame rate is often a limiting factor for temporal resolution. Most standard digital video cameras have a repetition rate of 30 frames per second, which limits the collection rate of PIV data to 15 Hz when PIV images are collected in pairs. For very high resolution cameras, data transfer rate limitations may restrict the frame rate to much lower values. In addition, slow hard drive data storage rates often limit the number of consecutive data ensembles that can be collected at the highest rate to that which can be stored in available RAM.

Often, the illumination source is also a limiting factor. For example, typical dual-cavity Nd:YAG lasers fire pairs of light sheets at about 15 Hz, which also limits the collection rate of PIV data to 15 Hz. On the other hand, if a continuous laser sheet is used to illuminate the flow, camera sensitivity and the necessary exposure time become important. The energy per unit time provided by a continuous laser is typically much less than that of the flash from a pulsed laser, so the time required to sufficiently illuminate tracers is substantially more. Thus, such configurations may only be practical for flows with very low velocities or very small imaging regions.

In fact, most of the above comments are system specific generalizations based on commonly used equipment. High resolution cameras that can collect images at very high frame rates are available, there are pulsed lasers that can illuminate the flow area thousands of times per second, and there are configurations that allow continuous lasers to be used to collect PIV images of high velocity flow fields. In other words, temporal resolution is case dependent and requires knowledge of available equipment and application conditions.

The spatial resolution of a PIV system is a measure of the size of the smallest flow scales that can be resolved (see Section 4.4.3 of Volume 1 for a discussion of turbulent flow scales). The hard limit to the spatial resolution of a PIV system is set by the Nyquist wavenumber $k_{Ni} = 1/(2\Delta_{gi})$ where Δ_{gi} is the grid spacing (or vector spacing) in the i ($i = 1, 2$) direction, and the wavenumber is defined as one over the wavelength. Only scales with wavenumbers less than k_{Ni} can be resolved by the PIV system. However, flow scales with wavenumbers smaller than k_{Ni} are likely attenuated as a result of spatial averaging from the interrogation window size, the light sheet thickness, and the displacement of particles between exposures. There may be additional contributions to the attenuation of small scales due to the spacing between particle images and the particle image diameter, but these factors are usually small for typical PIV deployments and may be neglected. The degree to which the amplitudes of velocity fluctuations are attenuated can be described as a function of wavenumber by a transfer function $T_{PIV}(k)$. The transfer function defines how much turbulence is attenuated at each wavelength and completely describes the resolution of a PIV system. Contributions to the PIV transfer function are evaluated in detail in the following paragraphs.

If the variation of particle displacements within an interrogation region is small compared to the particle image diameter, the PIV interrogation process can be approximated as a linear low pass filter (Theunissen, 2012). The frequency response function (Bendat & Piersol, 2000) of the filter is given by $T_{PIV}(k) = |T_{PIV}(k)|e^{-j\theta_{PIV}(k)}$ where $|T_{PIV}|$ is the system gain factor, θ_{PIV} is the system phase factor, $k = (k_1, k_2, k_3)$ is the wavenumber vector, $j = \sqrt{-1}$ and $|\ |$ is the complex modulus. The phase factor is near zero and may be neglected for IDS and IDM algorithms due to the symmetry of the window shifting and image deformation processes. The PIV system gain factor (often called the 'modulation transfer function', MTF, or simply the 'transfer function'), is the ratio of output (measured) to input (actual) amplitudes for sinusoidal velocity fluctuations (or Fourier components) of wavenumber k. For homogeneous turbulence, the measured 3-dimensional velocity spectrum ($\phi_{ii}{}^m$) is related to the actual spectrum (ϕ_{ii}) by $\phi_{ii}{}^m(k) = \phi_{ii}|T_{PIV}|^2$ and the measured 1-dimensional spectrum by $F_{ii}{}^m(k_1) = \int\int_{-\infty}^{\infty} \phi_{ii}|T_{PIV}|^2 dk_2 dk_3$. A cutoff wavenumber can be defined for each direction (k_{C1}, k_{C2}, k_{C3}) such that $|T_{PIV}(k_{Ci})| = 0.9$, and if $k_{Ci} < k_{Ni}$, $\lambda_{Ci} = k_{Ci}^{-1}$ may be referred to as the spatial resolution of the PIV system in the i direction ($i = 1, 2, 3$). Note that the case $k_{Ci} > k_{Ni}$ should be avoided as aliasing from the high wavenumber region will contaminate the resolved spatial scales. Ideally, the cutoff wavenumber k_{Ci} should be sufficiently large such that all scales present in the flow are resolved, i.e., $k_{Ci} > \zeta^{-1}$ where $\zeta = (\nu^3 \varepsilon_d{}^{-1})^{0.25}$ is the Kolmogorov length scale, ν is the kinematic viscosity of the fluid, and ε_d is the dissipation rate of turbulent kinetic energy. In practice this is rarely possible and measured velocity variance will inevitably be biased low by the lack of spatial resolution.

The PIV system gain factor can be written for scales that are large compared to the mean particle image spacing as $|T_{PIV}(k_1, k_2, k_3)| = |T_{CC}(k_1, k_2)| |T_{LS}(k_3)| |T_\Delta(k_1, k_2, k_3)|$, where $|T_{CC}|$, $|T_{LS}|$, $|T_\Delta|$ are contributions from the cross-correlation analysis, the light sheet thickness, and the particle displacement between image exposures, respectively. The gain factor associated with the light sheet intensity distribution ($L_S(x_3)$) may be estimated for camera viewing angles orthogonal to the light sheet as: $|T_{LS}(k_3)| = |FFT[L_s{}^2(x_3)]|/\overline{L_s{}^2(x_3)}$, where x_3 is the coordinate in the out-of-plane direction (orthogonal to the light sheet). The particle displacement transfer function ($|T_\Delta|$) is often approximated as a moving average filter (e.g., Lavoie et al., 2007; Atkinson et al., 2014) only in the dominant flow direction as:

$$|T_\Delta(k_1)| = \frac{\sin\left(\pi k_1 \overline{\Delta_{x_1}}\right)}{\pi k_1 \overline{\Delta_{x_1}}} \tag{3.7.11}$$

where $\overline{\Delta_{x_1}}$ is the time averaged particle displacement in the $i=1$ direction. For IDM algorithms $|T_{CC}|$ may be estimated as (Astarita, 2007):

$$|T_{CC}{}^{IDM}(k_1, k_2)| = \frac{|T_\eta|}{1 - |T_\xi||T_D| + |T_\eta||T_D|} \tag{3.7.12}$$

and for IDS algorithms $|T_{CC}|$ is given by:

$$|T_{CC}{}^{IDS}(k_1, k_2)| = |T_\eta| \tag{3.7.13}$$

where $|T_\eta|$, $|T_\xi|$, and $|T_D|$ are the gain factors associated with the interrogation region weighting windows, the low pass filtering of the velocity field within each IDM algorithm iteration, and the interpolation of the velocity field to obtain the dense predictor. These terms may be calculated as the normalized Fourier transforms of their respective convolution kernels (η, ξ, and D):

$$|T_\eta|(k_1,k_2) = \frac{|FFT[\eta^2(x_1,x_2)]|}{\overline{\eta^2(x_1,x_2)}} \tag{3.7.14}$$

$$|T_\xi|(k_1,k_2) = \frac{|FFT[\xi(x_1,x_2)]|}{\overline{\xi(x_1,x_2)}} \tag{3.7.15}$$

$$|T_D|(k_1,k_2) = \frac{|FFT[D(x_1,x_2)]|}{\overline{D(x_1,x_2)}} \tag{3.7.16}$$

Note that η^2 is used in Equation (3.7.14) rather than η because the interrogation region weighting window is applied to both the first and second interrogation regions used for calculating the cross correlation (*e.g.*, Equation (3.7.4)).

Example distributions of $|T_{CC}|$ for different algorithms are shown in Figure 3.7.6 where specific algorithms are referred to using the notation 'algorithm $_\eta^2\_\xi$ ', with algorithm either IDS or IDM, and η and ξ are the weighting windows described previously. The weighting windows are identified by their shape and size, such as BL64 which is a 64×64 pixel Blackman window, and TH32 which is a 32×32 pixel top-hat window. Note that the window ξ is not required for the IDS method.

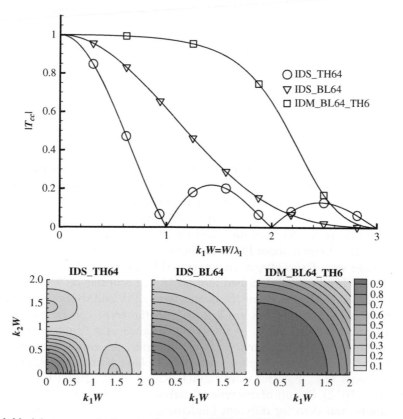

Figure 3.7.6 Modulation transfer function for selected PIV algorithms.

Figure 3.7.6 shows how resolution can be manipulated for a given base window size by implementing different algorithms. If measurements are to be used for estimating velocity spectra in the wavenumber domain, the extended range of near unity transfer function for IDM_BL64_TH6 may be particularly beneficial. Spatial resolution should, however, always be considered in conjunction with measurement noise as the two are inseparably linked. PIV measurement noise is examined in the following sections.

3.7.2.4 Measurement noise

Measurement error (ε_i) is defined as the difference between measured (u_i^m) and actual (u_i) velocity (or displacement) fluctuations: $\varepsilon_i = u_i^m - u_i$. The error can be decomposed into a mean component $\bar{\varepsilon}_i$ (also known as bias or systematic error) and a random component ε_i' (also known as the measurement noise) such that $\varepsilon_i = \bar{\varepsilon}_i + \varepsilon_i'$. PIV measurement errors can usually be considered to be normally distributed such that their probability density function is sufficiently characterized by the mean error $(\bar{\varepsilon}_i)$ and the standard deviation of the error $(\sigma_{\varepsilon i} = \sqrt{\overline{\varepsilon_i'^2}})$. The power spectrum of PIV measurement noise $(\phi_{\varepsilon i})$ for scales that are large compared to the mean particle spacing can be considered as a white noise multiplied by the squared PIV correlation algorithm gain factor $\phi_{\varepsilon i}(k_1, k_2) = S_{\varepsilon i}|T_{CC}(k_1, k_2)|^2$, where $S_{\varepsilon i}$ is the spectral noise saturation level for the ith displacement (velocity) component. The variance of the measurement noise is then:

$$\sigma_{\varepsilon i}^2 = \int\int\limits_{-\infty}^{\infty} \phi_{\varepsilon i}(k_1, k_2)dk_1 dk_2 = S_{\varepsilon i}E \tag{3.7.17}$$

where

$$E = \int\int\limits_{-\infty}^{\infty} |T_{CC}(k_1, k_2)|^2 dk_1 dk_2 \tag{3.7.18}$$

is known as the equivalent noise bandwidth. It is convenient to normalize the noise variance with the noise bandwidth $\phi_{\varepsilon i}^2/E = S_{\varepsilon i}$ when comparing the noise level of different PIV algorithms. In this way system noise can be assessed independently from spatial resolution. The noise bandwidths for IDS_TH64, IDS_BL64, and IDM_BL64_TH6 are $E = 64^{-2}$, 36.48^{-2} and 18.59^{-2} respectively.

Much of what is known about PIV measurement errors comes from the analysis of simulated image sets. These images are generated from randomly distributed particle coordinates which follow a prescribed motion (e.g., Lecordier & Westerweel, 2004; Raffel et al., 2007). After processing the images with a PIV code both the measured and actual (prescribed) velocity fields are known and the error can be calculated. It is common to present distributions of the error variance as a function of image and flow-field parameters such as: particle image diameter, the concentration of particle images, the image noise, the in-plane and out-of-plane particle displacements, and velocity gradients (e.g., Foucaut et al., 2004; Raffel et al., 2007; Cameron, 2011; Timmins et al., 2012). Results of this type of analysis can vary widely between different PIV algorithms, and even for different implementations of the same algorithm. It is therefore good practice to test every PIV code with simulated image sets rather than relying on published data. This is something that can be done even for commercial software where the details of the employed algorithms might not be available.

3.7.2.5 Sources of error

For uniform displacement fields (Δ_{x1} = constant), random measurement errors are composed of the contributions of two primary source errors; 1) the 'image aliasing' error which dominates for small particle image diameters, and 2) the 'change of brightness' error which dominates for large particle image diameters and large relative out-of-plane displacements (Δ_{x3}/T), where Δ_{x3} is the out-of-plane displacement component and T is the light sheet thickness). The superposition of these two errors, which have opposite slopes when plotted against particle image diameter, often leads to the formation of an 'optimum' diameter where the total error is minimized (Figure 3.7.7).

The image aliasing error is caused by under-sampling of the image plane light intensity field by the camera sensor (Appendix 3.C.1). The resulting aliased content of the image has incorrect phase which causes an error in measured particle displacements. Both mean ($\overline{\varepsilon}_i$) and random ($\sigma_{\varepsilon i}$) components of the image aliasing error are periodic functions of the particle image displacement and have idealized distributions given by $\overline{\varepsilon}_i = A\sin(2\pi\Delta_{xi})$ and $\sigma_{\varepsilon i}^2 = B[1 - \cos(2\pi\Delta_{xi})]$, where A and B are coefficients that depend on the particle image diameter and the correlation algorithm. The value of B additionally scales with the number of particles in an interrogation region (N) such that $B \propto 1/N$ reflecting an averaging of the random error associated with each particle image over N particles. For IDS algorithms, A and B are sensitive to the correlation peak interpolation function, whereas for IDM algorithms, A and B are

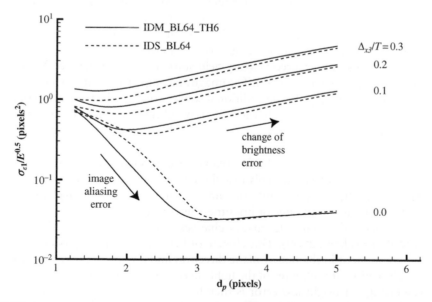

Figure 3.7.7 Random measurement error versus particle image diameter for different relative out-of-plane displacements (Δ_{x3}/T). The particle image concentration is 0.03 particles per image pixel, the fill factor is 1.0 and the error is averaged over in plane displacements $0 \le \Delta_{x1} \le 2.0$.

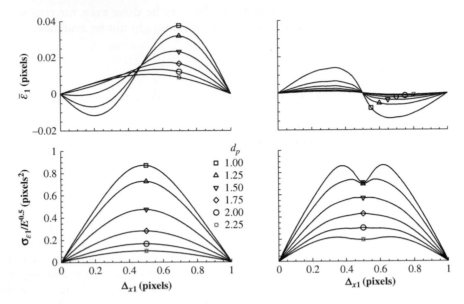

Figure 3.7.8 Mean and standard deviation of PIV measurement error versus in plane displacement for different particle image diameters. Left column: IDM_BL64_TH6; right column: IDS_BL64. The particle image concentration is 0.05 particles per image pixel and the fill factor is 1.0.

sensitive to the image interpolation function. Some example distributions of $\overline{\varepsilon_1}(\Delta_{x1})$ and $\sigma_{\varepsilon1}(\Delta_{x1})$ are given for IDS and IDM algorithms in Figure 3.7.8. It should be noted that A typically has opposite sign for IDS relative to IDM methods and that distortions from the idealized distributions are typical (including a well known longer wavelength contribution to the IDM mean error). PIV errors that are periodic with displacement are generally referred to as 'peak locking' errors, and they can cause significant distortion to distributions of measured velocity field statistics (*e.g.*, Angele & Muhammad-Klingmann, 2005; Cholemari, 2007). The image aliasing error is a major contributor to peak locking but there may also be additional sources such as the truncation of particle images at interrogation region borders if a weighting window is not used (*e.g.*, Nogueira *et al.*, 2001), and unique behavior of the imaging device (*e.g.*, Nogueira *et al.*, 2009).

The cross-correlation function $\phi(m', n')$ can be decomposed into the 'true peak' and a number of (usually smaller) randomly distributed 'false peaks'. Some of the false peaks inevitably overlap the true peak and cause an error in the measured displacement. This error is referred to as the 'change of brightness' error because it is only non-zero if the brightness of individual particle images changes from one image to the next (*e.g.*, Nobach & Bodenschatz, 2009). This change of brightness normally occurs if there is a velocity component in the out-of-plane direction and a particle moves, for example, from a bright region at the center of the light sheet toward a less bright region near the edge. The change of brightness error is therefore proportional to the relative out-of-plane displacement component (Δ_{x3}/T), and it is also proportional to the particle image diameter – reflecting the likelihood that a false peak will overlap the true peak. The change of brightness error is conveniently independent of the particle image con-

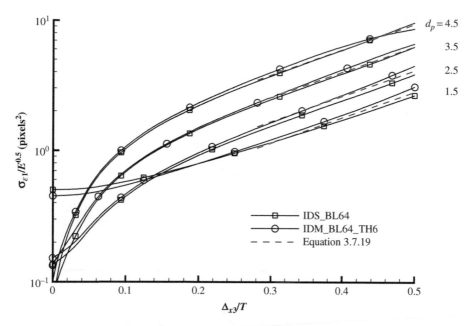

Figure 3.7.9 Standard deviation of measurement error versus out-of-plane displacement for several particle image diameters.

centration and reasonably independent of the analysis algorithm (Figure 3.7.9). For larger Δ_{x3}/T, the error approaches:

$$\frac{\sigma_{\varepsilon i}}{E^{0.5}} = C \exp\left(\frac{Q\Delta_{x3}}{T}\right) \tag{3.7.19}$$

where $Q = 4.530$ is an empirical constant and C is a function of particle image diameter approximated by $C = 0.046d_p^2 - 0.050d_p + 0.266$.

If the displacement of particle images within an interrogation region is not uniform, such as when velocity gradients are present in the flow, the IDS cross correlation peak is distorted and a measurement error proportional to the velocity gradient results. When the range of displacements is small relative to the particle image diameter, the IDS PIV algorithm continues to function as a linear filter and the variance of the measured displacement is proportional to the variance of the individual displacements within an interrogation region (σ_{Wi}^2) such that $\sigma_{\varepsilon i}^2 = \sigma_{Wi}^2/N$. If the range of displacements is large compared to the particle image diameter, the correlation peak can split into several peaks, each representing the displacement of different particles in the interrogation region. This case should be avoided as results are unpredictable. Simulation results for a constant gradient velocity field are shown in Figure 3.7.10. In this case the predicted trend

$$\frac{\sigma_{\varepsilon 1}\sqrt{N}}{d_p} = \frac{\sigma_{W1}}{d_p} = \frac{1}{\sqrt{12}}\frac{\partial\Delta_{x_1}}{\partial x_2}\frac{W}{d_p} \tag{3.7.20}$$

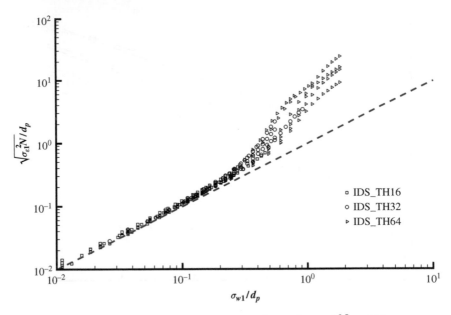

Figure 3.7.10 Effect of a constant displacement gradient $(\partial \Delta_{x1} / \partial x_2 = 12^{0.5}\sigma_{W1}/W)$ on measurement error for IDS algorithms. The zero-gradient measurement error has been subtracted from the total measurement error to isolate the effect of displacement gradients. Particle image concentrations vary between 0.01 and 0.09 particles per image pixel, and particle image diameters (d_p) vary between 1.5 and 3.5 pixels.

holds up to $\sigma_{W1}/d_p \approx 0.2$, after which the range of displacements in the interrogation regions becomes too large for the PIV algorithm to handle and the error rapidly increases. The IDM method in comparison to IDS iteratively corrects for varying displacement within an interrogation window and the measurement error is therefore reasonably insensitive to the gradient.

Simulation results such as those presented here are useful for understanding the baseline performance of the PIV method, however, it should be realized that for real implementations there may be many more contributing factors to the measurement error. These additional factors may include background reflections in the images, system vibrations, variations in the fluid refractive index, camera lens aberrations, non-uniform pixel response of the camera sensor, calibration errors, perspective errors, and probably many more. Fully accounting for all error sources is difficult, and due to the conflicting contributions of measurement errors and limited spatial resolution to the measured velocity variance, comparing PIV results to those obtained by other measurement techniques such as hotwire or laser Doppler anemometer may be misleading (*e.g.*, Atkinson *et al.*, 2014).

3.7.2.6 Practical PIV configurations

The PIV concepts described above can be assembled in a variety of practical systems. Irrespective of their configuration, the PIV systems are characterized by important sys-

Table 3.7.1 Key parameters describing a PIV system implementation and their typical values.

Parameter	Typical	Unit
image scale factor	10–100	pixels/mm
time between exposures	0.1–10	ms
lens f-number	2.8–22	(–)
camera pixel pitch	5–20	μm
camera fill factor	0.6–1.0	(–)
image field of view	50–500	mm
seeding diameter	10–50	μm
seeding density	0.9–2.0	g/cm^3
seeding concentration	10–50	mg/l
particle image diameter	1.5–5	pixels
particle image concentration	0.01–0.1	particles per image pixel
sampling frequency (time domain)	10–1000	Hz
grid spacing	0.1–5	mm
light sheet thickness	0.5–3	mm
details of correlation algorithm	IDS/IDM etc.	(–)
transfer function	Figure 3.7.6	(–)
equivalent noise bandwidth	$1/64^{0.5}$–$1/16^{0.5}$	pixels^2

tem parameters that define their performance. Typical values for these parameters are provided in Table 3.7.1. A comparison of contemporary PIV with other forms of velocimetry is provided in Appendix 3.C.2.

A brief overview summarizing the most common PIV-setups used in hydraulic engineering applications is provided below. These setups include the 2C-2D (two components, two dimensions) single camera configuration, the 3C-2D (three components, two dimensions) stereoscopic configuration, and the 3C-3D (three components, three dimensions) tomographic configuration. The configurations are shown in Figure 3.7.11. Detailed information on the individual configurations, calibration techniques, measurement procedures, and data processing can be found in textbooks and scientific literature (*e.g.*, Raffel *et al.*, 2007; Adrian & Westerweel, 2010). It is worth mentioning that there exist many more specific PIV setups such as Micro-PIV, dual-plane PIV, and Holographic PIV which are not described here. For details, the interested reader is directed to the literature (*e.g.*, Adrian & Westerweel, 2010; McKeon *et al.*, 2007).

3.7.2.6.1 Single camera PIV (2C-2D)

The 2C-2D PIV setup described extensively in the preceding section is the simplest and therefore most widely deployed form of particle image velocimetry. It allows for the measurement of two components of the velocity vector across a two-dimensional plane. A 2C-2D PIV system typically consists of a single camera, a double-pulsed laser with planar illumination optics and a synchronization unit (Figures 3.7.1 and 3.7.11a). The camera is placed with a viewing direction perpendicular to the plane of the laser sheet and the calibration procedure can be as simple as recording an image of known scale aligned with the illumination plane. Figures 3.7.3 and 3.7.17 demonstrate analysis and

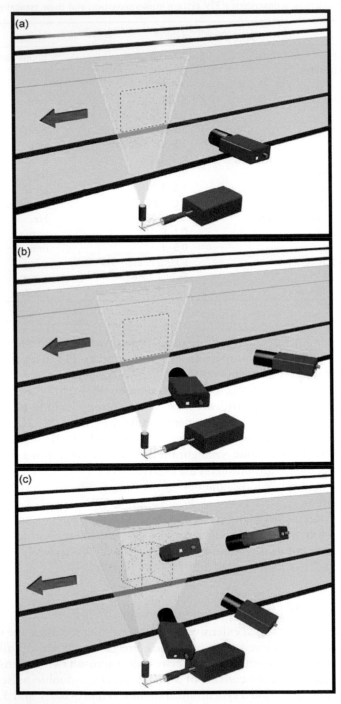

Figure 3.7.11 Examples of illuminated regions, measurement volumes (dashed lines) and camera orientations for (a) 2C-2D, (b) 3C-2D, and (c) 3C-3D camera configurations

results of a 2C-2D deployment. The inherent limitation of the 2C-2D method is that only two components of the velocity vector are resolved, potentially limiting the interpretation of turbulent flow measurements. The unresolved out-of-plane velocity component also leads to a 'perspective error' (Raffel *et al.*, 2007) in the resolved components, particularly for wide-angle camera lenses. This error is addressed by stereoscopic PIV systems.

3.7.2.6.2 Stereoscopic PIV (3C-2D)

Stereoscopic PIV systems enable the measurement of all three components of the velocity vector across a planar flow region. These systems require at least 2 cameras which are orientated such that they view the same flow region from different angles. If the camera viewing angle is large, the camera lens may be tilted with respect to the camera to re-align the plane of focus of the camera with the light sheet; *i.e.*, the so called 'Scheimpflug' condition (Prasad, 2000). The calibration procedure involves recording the positions of multiple points around the measurement plane and their corresponding (pixel) coordinates in the image plane. These data points are used to calculate a mapping function, usually either a polynomial equation or a model of the imaging system, which relates object and image coordinates. The mapping function allows interrogation regions from each camera to be co-aligned and the computation of the magnification and viewing angles at different points in the image. Images from each camera are typically processed separately following a similar correlation procedure to that used for single camera PIV. The resulting 2-component vectors from each camera are subsequently combined using the calibrated viewing angles to form the 3-component vector.

Figure 3.7.12 shows the results of a stereoscopic PIV measurement of the time-averaged velocity field over an armored gravel bed. The vectors show vertical and

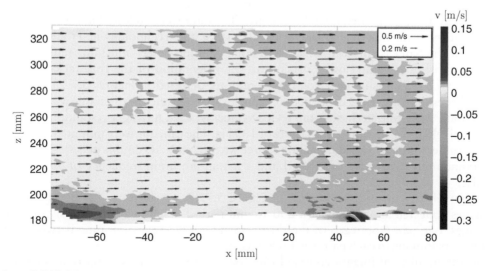

Figure 3.7.12 3C-2D (stereoscopic) measurements collected in a tilting flume (Image courtesy of C.U. Navaratnam)

streamwise velocity components whereas the color map represents the transverse velocity field. The data were collected using two CCD cameras, each with a resolution of four megapixels and a maximum frame rate of 180 frames per second. The illumination source was a dual-cavity Nd:YAG laser with a maximum energy output of 200 mJ per laser pulse and a maximum pulse rate of 50 Hz. The experiment was done in a 12.5 m long by 1 m wide tilting flume. For the results shown in Figure 3.7.12, the discharge in the flume was 73.4l/s, the slope was 0.17%, and the water depth was 159 mm above the mean bed elevation. The time-averaged velocity field shown in Figure 3.7.12 was measured for 150 seconds at a frequency of 20 Hz, corresponding to 6000 pairs of image captures (3000 pairs per camera). The separation time between the frames of each image pair was 1.5 ms. Figure 3.7.12 clearly shows the effects that bed features have on the transverse velocity distribution – information that 2C-2D PIV cannot provide.

3.7.2.6.3 Tomographic PIV (3C-3D)

Tomographic PIV (*e.g.*, Elsinga *et al.*, 2006) enables the measurement of all three components of the velocity vector in a volumetric domain. The hardware setup and calibration procedure are similar to stereoscopic PIV, but the thickness of the light sheet is significantly increased, and at least three cameras positioned with different viewing angles are required. The key to tomographic PIV is the process of reconstructing the three-dimensional intensity field from the projected views obtained by each camera; algorithms for which are the subject of ongoing research. Velocity vectors are extracted from the reconstructed intensity fields using a three-dimensional correlation procedure. The key challenges for tomographic PIV implementations are: increasing the thickness of the measurement volume (requiring more powerful illumination sources and more sensitive cameras); reducing measurement noise toward that of stereoscopic PIV; and reducing computation time.

Figure 3.7.13a shows velocity measurements collected downstream of a trashrack placed at an angle in a rectangular channel. The data were collected in the hydraulic laboratory at the Norwegian University of Science and Technology in Trondheim as part of the SafePass-project, which was done to improve eel and salmonid passage characteristics of hydropower structures. The orientation of the trashrack and the location of the measurement volume (in red) are depicted in Figures 3.7.13b and c. For the results shown, the depth in the channel was 50 cm and the discharge was 200 l/s. The data were collected using three CCD cameras, each having a resolution of four megapixels and a maximum frame rate of 180 Hz. The cameras were mounted on a TSI V3V camera mount. The illumination source was an Nd:YAG double pulse laser with a maximum energy output of 200 mJ and a maximum firing frequency of 50 Hz per pulse.

The velocity magnitudes shown in Figure 3.7.13 are averages for 3000 sets of images collected at 15 Hz over a 200 second period. Each image set consists of six images, one pair from each of the three cameras. The separation time between the frames belonging to each pair was 1 mS. Based on the resolution of the cameras (4MP) and the number of images collected ($3000 \times 6 = 18,000$ images), approximately 70 gigabytes of memory storage was required for this single experiment. Clearly, tomographic PIV experiments can require vast amounts of available hard drive storage.

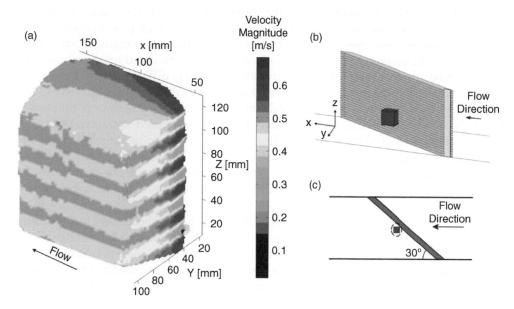

Figure 3.7.13 3C-3D (tomographic) measurements downstream of a grate in a rectangular channel. (a) Shows the measured velocity magnitudes, (b) shows the location of the measurement volume (in red) downstream of the grate, and (c) is a plan view of the channel showing the measurement volume (Image courtesy of C.U. Navaratnam and M. Szabo-Meszaros)

Nevertheless, as demonstrated by Figure 3.7.13a, which clearly shows the variations in velocity induced by the bars of the trashrack, the detail provided by tomographic PIV is impressive.

3.7.2.6.4 Special applications – field PIV

In recent years, field applications of PIV have become increasingly common. Large-scale PIV (LSPIV) is one method of applying PIV in the field, as described later in Section 3.7.4. This section, however, provides a brief overview of several examples of traditional PIV in field applications. These examples provide a sense of the challenges and possibilities associated with field PIV.

The first applications of PIV in the field were largely associated with measurements of velocity fields and turbulence in coastal boundary layers. One example of a field PIV system that was developed and extensively tested used a fiber optic to transmit laser pulses to a submerged imaging system (Bertuccioli *et al.*, 1999; Doron *et al.*, 2001; Nimmo Smith *et al.*, 2002, 2005; Luznik *et al.*, 2007). In the experiments that employed this system, the laser was a dual-cavity flashlamp pumped dye laser. A dye laser was selected instead of an Nd:YAG laser because the discharge energy of dye lasers is distributed over a longer duration pulse, reducing the possibility that the optical fiber would be damaged by the laser. The system could transmit 120 mJ pulses via fiber optic to the submerged imaging region. The high energy output was necessary to visualize the small tracer particles. Continuous lasers were ruled out because they deliver significantly less energy during the short pulse durations necessary for PIV. It was not practical or safe to submerge the laser because of its high power requirements. Various cameras were used with the system with resolutions ranging from 1 to 4 megapixels. The imaging system was attached to a frame mounted on the ocean floor that could be raised or lowered to measure velocity fields in different parts of the water column to better understand boundary layer flow and turbulence characteristics. The system was deployed on the bed in water depths of as high as 25 m in some of the studies listed above. Naturally occurring particles in the ocean were used as tracers; their sizes and concentrations could not be controlled but were found to be satisfactory for the reported tests. Incidentally, the same system was also used to measure the flow field above an irrigated corn field (Zhu *et al.*, 2006), but the air above the corn had to be seeded with an oil based fog in order to measure the velocity and turbulence field above the corn.

There have also been field applications of PIV in which light sheets were formed using continuous lasers. Tritico *et al.* (2007) used a 90 mW Diode Pumped Solid State (DPSS) laser and a pair of cylindrical lenses to form the light sheet. A 1 megapixel camera was used to visualize the flow field. A chopper wheel was used to synchronize the light sheet with the camera and to control the duration and separation time of the light sheet. In their tests, they were able to measure velocities of about 11 cm/s in the Huron River. Natural particles were sufficient as tracer particles, but pulse separation and duration times were both about 7 ms, which is rather long. With such a low energy light source, blurring and out-of-plane particle motion would likely be an issue for higher flow velocities. Katija and Dabiri (2008) used a slightly more powerful 300 mW laser sheet to illuminate the flow field. Timing was controlled by the high speed camera that they used, which could collect 1 megapixel images at up to 3000 frames per second. The system was entirely self-contained and was small enough to be deployed as a hand held system so that divers could study the flow field around aquatic organisms, such as jellyfish. It relied on naturally occuring particles in the water as tracers. The non-stationary reference frame of the system resulted in some velocity ambiguity problems.

In order to increase the energy reflected by particles in the water, Liao *et al.* (2009, 2015) and Wang *et al.* (2012, 2013) scanned a laser beam across the imaging plane using a galvanometer instead of relying on a light sheet formed with cylindrical lenses. Liao *et al.* (2009) synchronized a galvanometer and a 4 megapixel camera so that a 375 mW laser beam was scanned across the imaging area of the camera each time the shutter was opened. Thus, the separation time was based on the frame rate of the camera. The camera also employed a band pass filter to reduce the effects of ambient light on the system, which is a concern for all field techniques that are employed in daylight. Liao *et al.* (2009) were able to utilize their system in Lake Michigan to study the flow above Quagga Mussels. Naturally occurring phytoplankton and suspended sediment were suitable as flow tracers. Velocities higher than 20 cm/s required faster laser scanning rates to reduce out-of-plane motion of particles, but the higher scanning rates also meant less energy was reflected from each particle, leading to more noise. Liao *et al.* (2009) also noted that changes in field conditions (*e.g.*, the direction of the current) during the course of an experiment resulted in loss of data.

Liao *et al.* (2015) scanned a 1 W laser beam across the flow field while imaging it with a 1.3 megapixel camera operating at 30 frames per second. They used their system to investigate shear stresses induced on a harbor bed by propeller jet wash in order to determine the critical shear stress required to suspend sediment from the bed. For some propeller speeds, they were unable to measure the velocity with the PIV system because the turbidity of the water was too high. Nevertheless, at lower velocities, the PIV system could record velocity and turbulence information useful for bed shear stress estimates – information not provided by point measurement devices.

Using a galvanometer, Wang et al. (2012, 2013) simultaneously scanned two lasers of different wavelength and at an offset angle across the imaging area. Aside from the offset angle, the lasers were closely aligned. Two cameras were also closely aligned to view the same imaging region. Each of the cameras had a different filter that allowed them to view only one of the two laser wavelengths. Thus, the separation time between the pictures collected by the two cameras was dependent on the angle between the two lasers and not on the scan rate of the system. Using this approach, the separation time could be reduced without increasing the scanning speed of the lasers (which would reduce the energy scattered from particles in the imaging region). Consequently, the system would be able to measure a wider range of flow velocities than the system described by Liao et al. (2009, 2015). The most difficult aspect of using the system was aligning the lasers and the cameras – misalignment of either resulted in velocity measurement error. Wang et al. (2012) describe different solutions to timing and alignment issues. Wang et al. (2013) used the system to investigate flow and turbulence conditions below waves using an upward-looking free-floating PIV system mounted above a beach on Lake Michigan.

Cameron et al. (2013) used a stereoscopic PIV system to study velocity and turbulence behavior associated with flow through submerged aquatic vegetation in a small river in Scotland. They utilized a 532 nm, 50 Hz, 100 mJ dual-cavity Nd:YAG laser to generate a light sheet in the flow. Two cameras, each with a resolution of four-megapixels, were used to visualize the flow field. Each of the cameras also had a 532 nm bandpass filter to reduce noise caused by ambient light. The entire system was mounted above the river surface, and the laser sheet and cameras were directed downward through a specially designed 'boat-like' window that prevented distortion of the imaging region by undulations in the water surface. Unlike all of the other hydraulic PIV field studies described in this section, the flow in this case was artificially seeded with 60-80 μm conifer pollen that was pumped in as a concentrated mixture approximately 5 m upstream of the study site.

The operational and processing constraints described for PIV in previous sections are no less applicable for field PIV, and the same level of data quality is required if similar flow and turbulence variables are to be measured. It is clearly more difficult to apply traditional PIV techniques in field conditions than in the laboratory. Added complexities include: the sizes and distribution of the flow tracers are not easily controlled, and if artificial tracers are added, they must be biodegradable; commonly used lasers are expensive and very susceptible to damage in field applications; and it is more difficult to confine the intense light produced by the illumination source, resulting in a potential safety hazard. Finally, establishing an appropriate frame of reference can be more difficult in the field, and flow conditions are more apt to change with time, adding additional challenges to PIV field applications. Nevertheless, the ability to apply standard PIV techniques in the field is clearly useful because many field conditions are difficult or impossible to reproduce in the laboratory, and it is sometimes questionable whether the artificial conditions produced in the laboratory adequately represent actual conditions.

3.7.3 Particle Tracking Velocimetry (PTV)

3.7.3.1 Introduction

Particle Tracking Velocimetry (PTV) and Particle Image Velocimetry (PIV) are closely related, both belonging to the more general classification of particle-imaging techniques. Both methods analyze pairs or sequences of images of a tracer particle field, using the tracer particles as surrogates to measure fluid displacements in the flow field. Knowing the time between images, the image-to-image displacements are then used to determine the velocity field. The key difference between PTV and PIV is that in PTV, velocities are determined from discretely tracked particles, whereas in PIV, velocities are determined from the motions of groups of particles. This difference between PTV and PIV has implications for seeding concentrations, spatial resolution, and field averaging of results.

PTV generally requires significantly lower tracer concentrations than PIV (Adrian, 1991). PIV requires relatively higher concentrations of particles so that an ample number of particles appear in the overlapping interrogation regions of image pairs. In PIV applications, image fields with low particle concentrations inevitably lead to a higher number of erroneous vector calculations. Conversely, for PTV, high tracer concentrations lead to increased difficulty in identifying tracer matches between image frames. Unfortunately, the need for low tracer concentrations in PTV generally reduces the number of available measurements for each interrogation of the velocity field.

In PIV, the spatial resolution of velocity measurements is closely tied to the sizes of interrogation regions. For instance, a simple 32 pixel by 32 pixel interrogation region with an image-object scale of 64 pixels per cm will have an area of 0.25 cm^2 and a corresponding spatial resolution of 0.5 cm. The travel distances of the tracer particles within the interrogation region are limited between frames so that a good correlation can be found for the velocity measurement. Furthermore, it is best that the velocity distribution within the interrogation area be as uniform as possible or the measured velocity will be biased low (Adrian, 1991) because slower moving particles stay within the interrogation region for a longer duration, biasing the correlation. In contrast, in PTV, the displacement of tracers is less restrictive. As long as tracer particle matches are correctly identified between adjacent frames, and as long as the tracers have velocity fidelity (the velocities of the tracers match the velocity of the flow), uniformity of the velocities of tracers in close proximity is not as critical. One obvious implication of this is that PTV can be used to obtain finer resolution and more accurate results in regions with strong velocity gradients.

PIV involves some incidental filtering of velocity results since interrogation methods are applied to groups of tracers. The velocity obtained for an interrogation region is a representation of the velocities of the tracers in the interrogation region (note that for cross-correlation-based PIV methods, the velocity is not an average of the particle velocities within the interrogation region). PTV measures the velocities of the individual tracers, and a single velocity measurement is only subject to variations in the velocity of the flow over the trajectory of the particle between PTV images. Ultimately, PTV also requires filtering of results, since individual velocity measurements are usually combined to form results that are more useful, but the user has some control over selection of data combination methods and the resulting filtering effects.

3.7.3.2 Methodology

Figure 3.7.14 provides a flow chart that describes a typical PTV methodology. To the right of the flow chart, progression of the methodology is demonstrated using subsections extracted from a pair of PTV images. As shown in Figure 3.7.14, the PTV process begins with the collection of a pair of particle images. Alternatively, a sequence of multiple images may be collected if multi-frame tracking will be applied.

As with PIV, selection of appropriate tracer particles is important. Selection criteria include velocity fidelity (how well the particles follow the flow), visibility in images, environmental considerations (*e.g.*, biodegradable particles may be necessary for field measurements), and application specific criteria. Mei (1996) and Melling (1997) both offer good information for assessing how well tracer particles track the flow, and the reader should refer to these sources for additional information on this topic. In terms of

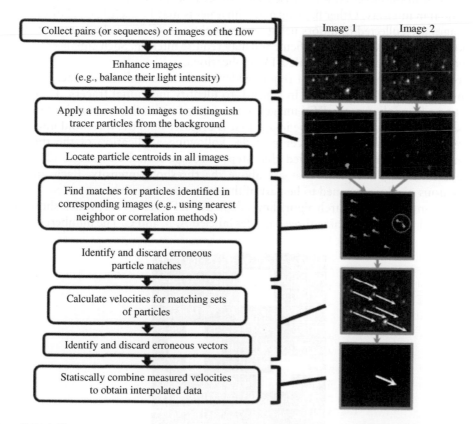

Figure 3.7.14 Flow chart showing a typical particle tracking velocimetry methodology. Images shown on the right demonstrate the progression of the procedure given by the flow chart.

visibility of the tracer particles in images, Cowen and Monismith (1997) suggest that tracer particles with particle-image diameters of between 1.8 and 2.4 pixels are optimal for reducing positioning error. In comparison, for PIV, research has suggested particle-image diameters of 1.0 (Westerweel, 1993), 2.0 (Prasad *et al.*, 1992) or 2.0 to 4.0 (Cowen & Monismith, 1997) pixels for optimizing accuracy. In any case, tracer particle-images should be comprised of more than one pixel to allow location of particle centroids to subpixel accuracy. Furthermore, large particle-image diameters should be avoided to optimize fidelity of the velocity measurements in locations with strong temporal or spatial velocity gradients. These general considerations are equally valid for both PTV and PIV.

The pair of images shown in Figure 3.7.14 was collected using a thin light sheet generated with a dual-head Nd:YAG laser, but other light sources can be used to illuminate the particles, as well. For example, continuous lasers, strobe lights, or even natural lighting may be used to illuminate a flow. As with PIV, the choice of an appropriate light source depends on the required sizes of the tracer particles, how fast the tracer particles are moving, and any other factors that influence the needed light intensity and how the light is applied.

After the PTV images are collected, some initial processing may be required to make sequences of images more uniform. In the example shown in Figure 3.7.14, the brightness of one of the images was increased slightly because the two laser head outputs were not equal in intensity. Other filtering and adjustments may be useful also. For example, if images are collected outdoors using natural lighting, effects of shadows and clouds make image processing more difficult and may require filtering. Such enhancements are especially beneficial to Large-scale PTV applications.

Processing of the images then requires identification of individual tracer particles within images of the flow field. In general, to identify individual tracers, a threshold must first be applied to the PTV images. For tracer particles that are brighter than the background flow, anything in the image that is brighter than the threshold intensity is considered to be part of a tracer particle. Following application of the threshold intensity, an algorithm is applied that steps through each PTV image, identifying groups of contiguous pixels that are above the threshold intensity. Each group of contiguous pixels is assumed to be part of the same tracer particle.

An example particle search algorithm is depicted in Figure 3.7.15, which shows a 10 by 10 pixel matrix containing three particles labeled A, B, and C. The numbered cells in

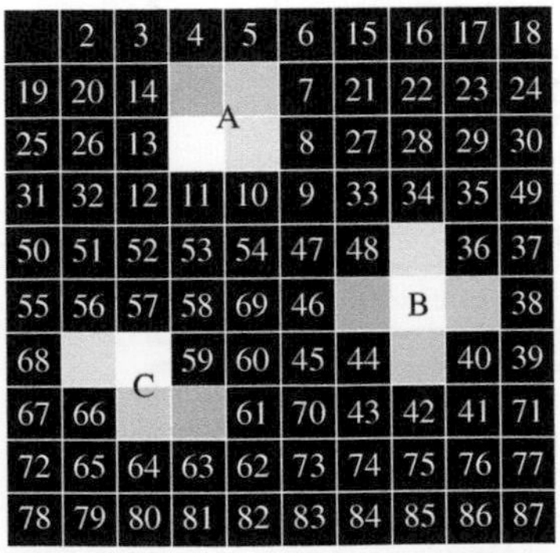

Figure 3.7.15 Sample search algorithm that finds three particles within an image.

Figure 3.7.15 are pixels that are below the threshold intensity. The algorithm begins at cell 1 in the upper left corner of the image and searches to the right and down, one row at a time, as shown in the figure by increasing cell numbers. When the algorithm encounters a pixel on its right that is above the threshold intensity, it searches clockwise around the particle until it reaches a previously analyzed pixel. The algorithm then tags all of the pixels in the interior of the clockwise search area and identifies them as belonging to the same particle. After identifying a particle, the algorithm continues searching the remainder of the matrix where it left off. Once a pixel has been analyzed, it is ignored in subsequent row searches. In Figure 3.7.15, the search that identifies particle A begins at cell 3 and ends at cell 14. The algorithm resumes searching for additional particles at cell 15.

Once individual tracer particles are found, the centroids of the tracer particle-images are calculated based on the contiguous pixels that form the tracer. Centroid estimators are discussed in many articles, including Ouellette *et al.* (2006), Cowen and Monismith (1997), Westerweel (1993), and Willert and Gharib (1991). Ouellette *et al.* (2006) suggest that selection of an estimator depends on several criteria: (1) it should yield subpixel accuracy – the more accurate, the better, (2) it should be fast and efficient (particularly for large PTV tasks), (3) It should be robust if it is applied to noisy images, and (4) it should be able to handle overlapping particles if the PTV application is three-dimensional.

Table 3.7.2 shows six centroid estimators. In Table 3.7.2, x_c and y_c are the centroid image-coordinates determined for a tracer particle-image – these coordinates are what we want to find; x_p and y_p are x and y image-coordinates of a particular pixel (pixel p) belonging to the tracer particle-image; N is the number of pixels that make up the tracer particle-image, and I_p is the intensity of pixel p. Curve and surface fit estimators like the parabolic and Gaussian 1-D fits and the Gaussian 2-D fit, yield centroid image-coordinates as the result of a parametric estimation procedure. For 1-D fit estimators, two curve fits are applied to the pixel intensities of each tracer particle-image, one for each of the two principle directions, and centered on the pixel with the highest intensity. For the parabolic 1-D fits, $I_{A,x}$, $I_{B,x}$, and $I_{C,x}$ are regression coefficients for the x equation, and $I_{A,y}$, $I_{B,y}$, and $I_{C,y}$ are regression coefficients for the y equation. Similarly, for the Gaussian 1-D fits, $I_{M,x}$ and σ_x are fitted parameters for the x equation, and $I_{M,y}$ and σ_y are fitted parameters for the y equation. The Gaussian 2-D fit is a surface fit of the pixel intensity signature of each tracer particle. Because the fit is a function of both x and y, only one equation is necessary for each tracer particle. The Gaussian 2-D fit requires estimation of the fitted parameters I_M, σ_x, and σ_y in addition to the centroid coordinates x_c and y_c. Such two-dimensional fits can be very accurate but are also computationally intensive (Ouellette *et al.*, 2006).

In addition to the functional equations of the methods shown in Table 3.7.2, the table provides three-point solutions for three of the methods. To use a three-point solution for a particular particle-image, the highest intensity pixel of the particle-image must first be identified. The coordinates of this pixel represent the origin for the three point solutions given in Table 3.7.2, and the intensity of the pixel is denoted I_0. The pixels to the left and right of the highest intensity pixel thus have the coordinates $(-1, 0)$ and $(1, 0)$, respectively, and the intensities I_{x-} and I_{x+}, respectively. Similarly, the pixels above and below the highest intensity pixel have the coordinates $(0, 1)$ and $(0, -1)$, respectively, and the intensities I_{y+} and I_{y-}, respectively. Therefore, knowing the pixel

Table 3.7.2 Common centroid estimators

Method	Functional Equations	Corresponding 3-pt Estimator	References
Unweighted Center-of-Mass	$x_c = \dfrac{\Sigma_p x_p}{N}$ and $y_c = \dfrac{\Sigma_p y_p}{N}$	This estimator is very inaccurate for small particles	Lloyd et al. (1995) Admiraal et al. (2004)
Weighted Center-of-Mass	$x_c = \dfrac{\Sigma_p x_p I_p}{\Sigma_p I_p}$ and $y_c = \dfrac{\Sigma_p y_p I_p}{\Sigma_p I_p}$		Ouellette et al. (2006)
Parabolic 1-D Fit	$I_p = I_{A,x}(x-x_c)^2 + I_{B,x}(x-x_c) + I_{C,x}$ and $I_p = I_{A,y}(y-y_c)^2 + I_{B,y}(y-y_c) + I_{C,y}$	$x_c = \dfrac{I_{x-} - I_{x+}}{2(I_{x-} + I_{x+} - 2I_0)}$ $y_c = \dfrac{I_{y-} - I_{y+}}{2(I_{y-} + I_{y+} - 2I_0)}$	Westerweel (1993)
Gaussian 1-D Fit	$I_p = \dfrac{I_{M,x}}{\sqrt{2\pi}\sigma_x} exp\left\{-\dfrac{1}{2}\left[\left(\dfrac{x-x_c}{\sigma_x}\right)^2\right]\right\}$ and $I_p = \dfrac{I_{M,y}}{\sqrt{2\pi}\sigma_y} exp\left\{-\dfrac{1}{2}\left[\left(\dfrac{y-y_c}{\sigma_y}\right)^2\right]\right\}$	$x_c = \dfrac{\ln(I_{x-}) - \ln(I_{x+})}{2\left(\ln(I_{x-}) + \ln(I_{x+}) - 2\ln(I_0)\right)}$ $y_c = \dfrac{\ln(I_{y-}) - \ln(I_{y+})}{2\left(\ln(I_{y-}) + \ln(I_{y+}) - 2\ln(I_0)\right)}$	Ouellette et al. (2006) Cowen and Monismith (1997) Westerweel (1993)
Gaussian 2-D Fit	$I_p = \dfrac{I_M}{2\pi\sigma_x\sigma_y} exp\left\{-\dfrac{1}{2}\left[\left(\dfrac{x-x_c}{\sigma_x}\right)^2 + \left(\dfrac{y-y_c}{\sigma_y}\right)^2\right]\right\}$	Requires a five parameter fit to the pixel intensity data	Ouellette et al. (2006)
Neural Networks	Procedure described by Ouellette et al. (2006)	Requires training of neural network	Ouellette et al. (2006)

intensity distribution of a particle-image, any of the three point solutions can be easily and efficiently used to estimate the centroid coordinates of the particle-image with respect to the position of the highest intensity pixel. When using one of these three-point solutions to determine the displacement of a particle, the centroid locations found for an image-particle before and after its displacement must be converted to a common reference frame. Note that the three point solutions given in the table are also used for sub-pixel estimation of the location of the peak of PIV cross-correlation functions (see Westerweel, 1993).

The simplest way shown in Table 3.7.2 to find the centroid of a particle is to algebraically determine the center-of-mass of its pixels. This method may be improved upon by finding the center-of-mass of the pixels weighted by their intensities. Center-of-mass methods are biased toward integer positions (Cowen & Monismith, 1997), so fits to the image-particle intensity signatures have also been explored. For optimally sized image-particles (~2.0 pixels), research has shown that a three-point Gaussian fit (or a similar fit) to the intensity signature of the particle is optimal for determining its centroid (Cowen & Monismith, 1997). For large-scale PTV, it is not always possible to use optimally-sized tracer particles, and when larger particles are necessary, a center-of-mass method may be sufficient and easier to apply than other methods (Admiraal *et al.*, 2004).

After particles and their centroids are found in two or more consecutive images, an attempt is made to match pairs of particles in adjacent images. There are various methods of identifying corresponding tracer particles, including correlation functions and nearest neighbor algorithms. For example, Lloyd *et al.* (1995) applied PTV to study the velocity field in the wake of a model island. After finding the centroids of each tracer particle, Lloyd *et al.* (1995) represented each particle with identically sized virtual particles and applied a cross-correlation method to determine the most likely particle matches between each image pair. Their correlation method found the displacement of tracer particles between images by determining the particle match that resulted in the maximum overlapping area of the virtual particles.

The simplest method of finding tracer matches between images is the nearest neighbor (NN) algorithm, which searches for a particle in the second image that is closest to a particle in the first image. This method works well for disperse tracer quantities but often results in erroneous vectors when two tracer particles are in the same general vicinity. A number of "predictor" methods can be used to improve the nearest neighbor algorithm. For example, Cowen and Monismith (1997) developed a hybrid method that combines the benefits of PIV and PTV; this method utilizes PIV to get an initial estimate of the velocity field. The estimate is then used to provide an offset for the search location of the matching particle, reducing the necessary search area for a match and more reliably identifying a correct match. One advantage of such "hybrid" PIV/PTV methods is that they allow for higher tracer concentrations and thus finer spatial resolution. The matches found for the example shown in Figure 3.7.14 are based on a nearest neighbor algorithm. The algorithm does a fairly good job in the present case, with only one erroneous match (circled in the image in Figure 3.7.14). The error occurs because the initial particle position is closer to an incorrect match than to the correct match in the second image. Using PIV to find an offset for the nearest neighbor search location may have eliminated this error.

3.7.3.3 Sources of error

Many of the error sources identified for PIV in Section 3.7.2 are directly applicable to PTV. In addition to these error sources, the importance of finding accurate particle matches is unique to PTV. There are a number of image-particle match identification errors that can result when applying PTV. Different possible tracer particle matching scenarios are demonstrated in Figure 3.7.16. Possible matching errors include:

1. No match is found for a tracer particle identified in the first or second image.
2. Two tracer particles in the first image have the same end point in the second image (*e.g.*, see the circled matches on the right side of Figure 3.7.14.
3. A tracer particle identified in the first image has two possible end points in the second image.
4. A wrong match is found in the second image for a tracer particle identified in the first image, and the resulting velocity vector is not possible or is wrong.

These errors may be the result of imperfect lighting of the flow, out-of-plane motion of tracer particles, or multiple tracer particles in close proximity. The matching errors described in 1-3 are easily identified with robust software, and tracer particle matches with these errors are discarded. Such errors can be managed before calculating velocities. Matching errors described in 4 are often more subjective and require the user to fully consider the physical characteristics of the flow. The errors described in 4 must be managed after the velocity field is determined. Usually, a set of criteria must be developed that delineate the reasonable range of velocities of the flow. These criteria may include minimum and maximum velocity, a range of possible velocity directions, criteria that describe how much the measured velocity is allowed to deviate from other velocities measured in close proximity, strength of the correlation function used to identify a particle match, etc. Many of these criteria can be programed so that

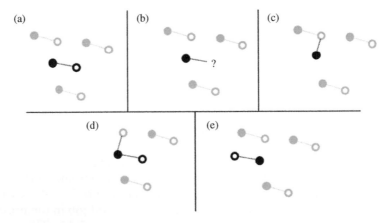

Figure 3.7.16 Examples of possible tracer particle matches. Closed circles represent Image 1 particles and open circles represent Image 2 particles. Cases include: (a) Correct tracer match found, (b) no match found in second image, (c) two matches found for a tracer particle in Image 2, (d) two matches found for a tracer particle in Image 1, and (e) the match found is inconsistent with surrounding matches.

unreasonable vectors are automatically removed, but all filtering must be done carefully and with good justification so as not to bias measurement results.

3.7.3.4 Averaging of velocity data

The last step shown in Figure 3.7.14 is to combine PTV velocity measurements to produce meaningful results. Unlike for PIV, velocity vectors collected using PTV are randomly located throughout the flow field based on the positions of the tracer particles when the images are collected. It is usually desirable to know the velocities at selected points in the flow, so the PTV vector data must be combined to form weighted average vectors at the selected points. Such vectors may be time averages of data collected from many PTV realizations or instantaneous averages of all of the tracers within a known radius of the point of interest. A very basic method of combining individual velocity measurements is to calculate an unweighted average of all of the vectors that are within a specific radius of the location where the average velocity is required. Such a simplistic approach has the potential to bias the resulting average velocity since it presumes that the velocity distribution over the sampled area is uniform. Consequently, methods of computing vector averages have been developed that give more weight to vectors that are closer to the location of averaging.

Agüí and Jiménez (1987) and Spedding and Rignot (1993) examined various methods of interpolating randomly positioned velocity data. Spedding and Rignot (1993) investigated two techniques of interpolating velocimetry data, the thin-shell spline (STS) and the adaptive Gaussian window (AGW). To test the methods, data were interpolated to grid points that were more suitable for data analysis than the original measurement locations. Spedding and Rignot (1993) found that the STS method significantly outperformed the AGW method in terms of mean and local error in velocity and vorticity in most cases that they tested, though they also indicated that AGW is more efficient for large numbers of data. Agüí and Jiménez (1987) tested polynomial interpolation, kriging, polynomial least squares, and convolutions. Various forms of polynomial interpolations and kriging were the most accurate, but the improvement over less complex convolution methods was not substantial, and they suggested that an AGW convolution method was suitable and efficient. As a demonstration of interpolation methods, the AGW method will be described, assuming that a dataset of randomly located PTV measurements has been collected. Each datum is a velocity vector (u_i) measured at a known position vector (x_i). To compute an interpolated velocity (u) at any specified position (x), the vectors are combined using the averaging formula:

$$u = \frac{\sum_i \alpha_i u_i}{\sum_i \alpha_i} \qquad (3.7.21)$$

Here, α_i is a weighting coefficient. For a simple mean, α_i is 1. For the AGW method, the weighting coefficients are given by:

$$\alpha_i = \exp\left(-\left(\frac{h_i}{H}\right)^2\right) \qquad (3.7.22)$$

where H is the width of the averaging window and h_i is the distance from the interpolation position to the position of each individual velocity measurement used in the average (*i.e.*, $h_i = |x - x_i|$). Agüí and Jiménez (1987) found the optimum size of the averaging window to be ~1.24δ, where δ is the average distance between PTV tracer particles (*i.e.*, between velocity data). AGW weighting coefficients are very small except for particles in close proximity to the interpolation position, and when calculating the velocity at a particular location, it is not necessary or practical to apply the formula to every measured PTV velocity. It is generally better to limit the number of vectors included in the calculation to those in the vicinity of the interpolation position, though the exact range to include in each interpolation varies with flow and tracer seeding conditions.

Figure 3.7.17 compares PTV and PIV applied to the same pair of images of flow over a vortex ripple. The PTV analysis produces many more vectors than the PIV analysis because there is one vector for each identified particle match. The PTV results can be locally averaged as described in this section to produce results that are similar in form to the results given for PIV (Figure 3.7.17c). When compared for this flow, it is clear that PTV results differ from PIV results, particularly in regions with strong velocity gradients, like near the bed on the leeward side of the vortex ripple. This does not necessarily imply that PIV results are inferior to PTV results or vice-versa, rather, that the experimentalist must fully understand the implications of all data collection procedures, image processing, and averaging algorithms to ensure accurate results when applying either PIV or PTV.

3.7.3.5 Multi-frame tracking

Multi-image tracking can be used to improve the performance of particle matching (Hassan & Canaan, 1991; Cierpka *et al.*, 2013). To apply multi-image tracking, a sequence of more than two PTV images is collected for each interrogation. The additional images provide confirmation of a correct particle match, assuming that the trajectory of the particle is consistent from frame to frame. Figure 3.7.18 illustrates the concept. Figure 3.7.18(a) shows the trajectories of two particles, Particle A and Particle B, which were determined from three images, Images 1, 2, and 3. The positions of Particle A from the three images are denoted A1, A2, and A3. Similarly, the positions of Particle B are given by B1, B2, and B3. In Figure 3.7.18(b), the second image is searched to find the nearest neighbor to particle A1 in the first image. The search area is indicated by the dashed circle. Two possible matches are found in the second image, A2 and B2. Particle B2 is closer to the position of particle A1, and the nearest neighbor algorithm yields an erroneous result. As shown in Figure 3.7.18(c), multi-frame tracking will reveal this error, because when the erroneous trajectory of A1 is extended to the third frame (as shown by the dashed line), no particle is found near the end of the predicted trajectory. However, if the trajectory predicted by the A1 and A2 match is extended to a third frame, as shown in Figure 3.7.18(d), particle A3 is found very near the end of the trajectory, indicating that A1 and A2 is a better match than A1 and B2.

There are numerous multi-image tracking algorithms (*e.g.*, see Ouellette, 2006). As an example, a multi-image tracking algorithm might search the second frame to find all possible matches to a particle found in the first frame. The trajectories of those matches are then used to identify particles in subsequent frames that are nearest to the trajectories. The trajectory which most accurately predicts the presence of particles in all

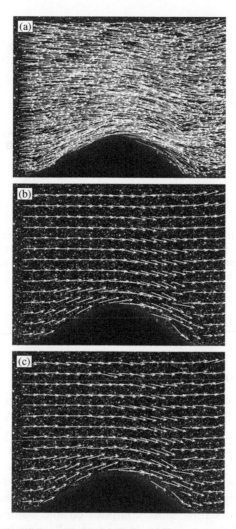

Figure 3.7.17 Comparison of PTV and PIV analysis of flow over a vortex ripple: (a) PTV vector field (using a NN hybrid PIV/PTV algorithm), (b) PIV vector field (32 by 32 pixel cross-correlation), and (c) locally averaged PTV results using AGW averages on the same grid used for PIV. Units are consistent but arbitrary.

subsequent frames (based on an objective function) is identified as the correct trajectory. It is common for multi-image tracking algorithms to utilize four or more consecutive images of the flow that are equally spaced in time.

Multi-image tracking methods are more robust than traditional PTV, resulting in more successful identification of tracer particle matches in flows with high tracer particle densities. Consequently, error rates can be reduced when multi-image tracking is applied. A disadvantage of multi-image tracking methods is that they require a sequence of PTV images instead of a pair. For many flows, this necessitates the use of sophisticated cameras and lasers because necessary sampling rates may substantially

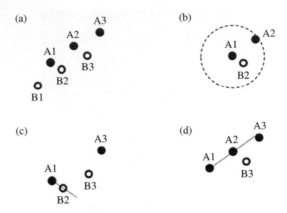

Figure 3.7.18 Illustration of multi-image tracking showing: (a) the trajectories of two particles (A and B) collected in three images (1, 2, and 3), (b) Particles found in the second frame (A2 and B2) that are nearest neighbors to particle A1 from the first frame, (c) the projected trajectory of particle A1 if B2 is erroneously found to be its match, and (d) the projected trajectory of particle A1 if A2 is correctly found to be its match.

exceed standard video rates of 30 frames per second or pulsed laser flash rates of 30 strobes per second.

3.7.3.6 3-D Particle tracking (for complex flows)

A number of researchers, including Maas *et al.* (1993), Malik *et al.* (1993), Guezennec *et al.* (1994), and Ouellette *et al.* (2006), have extended PTV to three-dimensional flows. In this section, a brief introduction to the topic is provided. To obtain more information on the subject, the reader is encouraged to explore the sources provided.

Conceptually, the extension of traditional PTV to three-dimensional flows is relatively straight-forward. In fact, the sequence of steps is very similar to the sequence shown in Figure 3.7.14. However, several of the processes listed in Figure 3.7.14 are significantly more complicated and computationally intensive. For three-dimensional flow volumes, instead of collecting images of thin slices of the flow field, sets of images must be collected that capture particles distributed over the entire volume of interest, sets that can resolve particle positions in three dimensions. This requires the use of multiple cameras, often three or more, that image the flow field at different angles.

Resolving particle positions in three dimensions from a set of two-dimensional images is not trivial, and requires an understanding of photogrammetry. In basic terms, each imaged particle lies along the same line as the corresponding object particle ("image" referring to what is output by the camera and "object" referring to the true physical space). A ray tracing concept is applied to determine the three-dimensional image/object line on which each imaged particle falls. The equations for such lines can be determined from optical theory but are more accurately found by calibrating each camera with a fixed array of particles with known positions. For a PTV interrogation, image/object lines are found for every particle identified in the images collected from all cameras. If each camera captures the entire measurement volume, and none of the tracer particles overlap in the images, each object particle within the volume will have a corresponding particle-image in each camera frame. Furthermore, each of the particle-images will have an image/object line. The image/object lines associated with a particular object particle all intersect (within a small tolerance). An optimization routine is used to determine the image/object lines that intersect, and thus the set of particle-images from each camera that correspond to the same object particle. The intersecting lines are then used to determine the location of the object particle in three dimensions. Increasing the number of cameras reduces the impact of overlapping particles on flow measurements.

Once particle locations are identified in two (or more) sequential sets of images, particle matches in the sequential sets must be found. Ouellette *et al.* (2006) describe four methods of determining matches from sequential sets of images. The nearest neighbor algorithm is easily extended from 2-D to 3-D PTV measurements. For 3-D measurements, however, the search region for particle matches is spherical instead of circular. The other three methods described by Ouellette *et al.* (2006) require sequences of multiple sets of images and are similar in form to the multi-frame tracking algorithm described earlier. However the matches are identified, they are used to calculate frame-to-frame velocities throughout the measurement volume just as in 2-D PTV.

Thus far, applications utilizing 3-D PTV have been somewhat limited for several reasons. First, 3-D PTV techniques require relatively sparse seeding densities and relatively low velocities. Otherwise, identifying particle-image sets and particle matches is quite difficult. Second, a relatively intense source of light is required for flows with moderate to high velocities because a volume of fluid must be illuminated instead of a thin sheet of fluid. Third, implementing 3-D PTV can be complicated and time consuming compared to simpler 2-D PIV and PTV methods.

3.7.4 Image velocimetry applied to large scales

3.7.4.1 Introduction

Visual investigation was the first way of understanding flow features, and it continues to be highly useful. The intricate flow patterns depicted in Leonardo da Vinci's sketches suggest that the human eye can sense important qualitative aspects of a river surface flow. Converting these visual impressions into quantitative river flow information, however, has only recently become possible. Developments over the last three decades in optics, lasers, electronics, and computer- related technologies have facilitated implementation of image-based techniques for flow visualization and quantitative measurements as discussed in the preceding sections. Yet instruments and methods developed for intricate laboratory experiments on relatively small-scale flow phenomena are difficult to apply directly to large-scale natural phenomena such as flood flows. For example, seeding of particles for flow visualization is virtually impossible under natural river flow conditions. Therefore, as an extension of conventional PIV applicable to river flow investigation, large-scale PIV (LSPIV) has been developed since the mid 1990's (Fujita & Komura, 1994; Aya *et al.*, 1995; Fujita *et al.*, 1997).

The essential steps of LSPIV are the same as those described for PIV in Section 3.7.2 of this chapter with adaptations that are imposed by imaging of large scales, typically acquired with conventional video recordings illuminated by natural or artificial lighting. The scopes of PIV and LSPIV measurements differ, as the former measures internal flow structures while the second non-intrusively measures velocities mainly at the free surface of a water body. LSPIV involves tracking surface features such as water surface ripples, drifting material (e.g., vegetation) or differences in water surface color generated by small- or large-scale river turbulence. Basic LSPIV data are instantaneous water surface velocity fields, spanning flow areas of up to hundreds of square meters. The technique has undergone continuous development and testing in anticipation of various hydraulic applications (Muste et al., 2004b). An extension of LSPIV is mobile LSPIV that allows the instantaneous onsite measurement with an all-in-one set of imaging instruments (Kim et al., 2008; Muste et al., 2008; Dramais et al., 2011). Another version of LSPIV is aerial LSPIV, which utilizes airborne images obtained from an aircraft or a helicopter (Fujita & Hino, 2003; Fujita & Kunita 2011). Aerial LSPIV can cover a measurement area much larger than LSPIV, but stricter meteorological conditions are required for successful measurements, primarily associated with a clear view of the flow. In LSPIV, video images viewed from an oblique angle are used to analyze surface velocities, and images taken from smaller oblique angles tend to reduce LSPIV accuracy due to larger image distortion. For solving this problem, space-time image velocimetry (STIV) has been developed by Fujita et al. (2007a).

3.7.4.2 Basics of Large-Scale PIV (LSPIV)

In LSPIV, sequential images of river surface flow viewed from a riverbank or a bridge are sampled at a certain time interval and utilized to extract quantitative flow information; i.e., two-dimensional surface velocity distributions in screen coordinates. The images thus obtained are first ortho-rectified using a mapping relation between the image coordinates and the physical coordinates with the additional information of water surface level. Note that the spatial resolution of rectified images and the time separation between the images have to be determined to yield an accurate result. Once a sequence of rectified images is generated, conventional PIV analysis can be applied to them to determine two-dimensional velocity distributions at the water surface.

3.7.4.2.1 Mapping relation

The key factor for a successful LSPIV measurement is to generate accurate ortho-rectified images. Therefore, the mapping relation between the screen coordinates (x, y) and the physical coordinates (X, Y, Z) has to be established with great care. Conventionally, the direct linear transformation (DLT) (Chen et al., 1994) expressed by the following relation has been used to relate the above coordinates (Aya et al., 2002; Fujita & Aya, 2000; Meselhe et al., 2004; Le Coz et al., 2010):

$$x = \frac{A_1 X + A_2 Y + A_3 Z + A_4}{C_1 X + C_2 Y + C_3 Z + 1} \tag{3.7.23}$$

$$y = \frac{B_1 X + B_2 Y + B_3 Z + B_4}{C_1 X + C_2 Y + C_3 Z + 1} \tag{3.7.24}$$

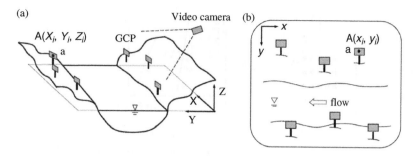

Figure 3.7.19 The relationship between screen and physical coordinates; a) physical coordinates (X, Y, Z); b) screen coordinates (x, y) (Muste et al., 2008).

The eleven coefficients from A_1 to C_3 can be determined if at least six corresponding sets of screen and physical coordinates are identified. These coordinate sets are referred to as ground control points (GCPs) or mark points as shown in Figure 3.7.19 and are usually identified on both sides of a river as panels. A rule of thumb when installing GCPs is to set them in a zigzag arrangement both vertically and horizontally as shown in Figure 3.7.19.

An alternative mapping relation is the two-dimensional collinearity equation expressed as

$$x - x_o = -f\left[\frac{m_{11}(X - X_0) + m_{12}(Y - Y_0) + m_{13}(Z - Z_0)}{m_{31}(X - X_0) + m_{32}(Y - Y_0) + m_{33}(Z - Z_0)}\right] \tag{3.7.25}$$

$$y - y_o = -f\left[\frac{m_{21}(X - X_0) + m_{22}(Y - Y_0) + m_{23}(Z - Z_0)}{m_{31}(X - X_0) + m_{32}(Y - Y_0) + m_{33}(Z - Z_0)}\right] \tag{3.7.26}$$

where (x_0, y_0) is the projection center, (X_0, Y_0, Z_0) are the camera coordinates, f is the focal distance, and m_{ij} are coefficients mainly related to the viewing angles. Once the mapping relationship between (x, y) and (X, Y, Z) is established, the water level equation is substituted into Equations (3.7.23) and (3.7.24) or Equations (3.7.25) and (3.7.26) to obtain a relation between the screen coordinates (x, y) and the physical coordinates on the water surface (Xs, Ys). The water surface is usually assumed to be a flat plane and is expressed as

$$Z = D_1 Xs + D_2 Ys + D_3 \tag{3.7.27}$$

Hence the relations for Xs and Ys simplify to the following equations:

$$Xs = \frac{c_1 x + c_2 y + c_3}{p_1 x + p_2 y + p_3} \tag{3.7.28}$$

$$Ys = \frac{q_1 x + q_2 y + q_3}{p_1 x + p_2 y + p_3} \tag{3.7.29}$$

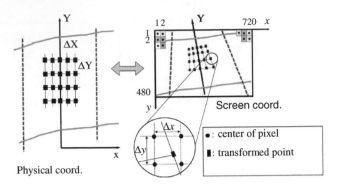

Figure 3.7.20 Generation of ortho-rectified image (Fujita *et al.*, 1997, with permission from IAHR).

where the coefficients have the following relations when Equations (3.7.23) and (3.7.24) are used:

$$
\begin{aligned}
c_1 &= (B_2 C_3 - B_3 C_2) D_3 + (B_3 - B_4 C_3) D_2 + (B_2 - B_4 C_2) \\
c_2 &= -\{(A_2 C_3 - A_3 C_2) D_3 + (A_3 - A_4 C_3) D_2 + (A_2 - A_4 C_2)\} \\
c_3 &= (A_3 B_4 - A_4 B_3) D_2 + (A_2 B_3 - A_3 B_2) D_3 + (A_2 B_4 - A_4 B_2) \\
p_1 &= (B_3 C_2 - B_2 C_3) D_1 + (B_1 C_3 - B_3 C_1) D_2 + (B_1 C_2 - B_2 C_1) \\
p_2 &= (A_2 C_3 - A_3 C_2) D_1 + (A_3 C_1 - A_1 C_3) D_2 - (A_1 C_2 - A_2 C_1) \\
p_3 &= (A_3 B_2 - A_2 B_3) D_1 + (A_1 B_3 - A_3 B_1) D_2 + (A_1 B_2 - A_2 B_1) \\
q_1 &= -\{(B_1 C_3 - B_3 C_1) D_3 + (B_3 - B_4 C_3) D_1 + (B_1 - B_4 C_1)\} \\
q_2 &= (A_1 C_3 - A_3 C_1) D_3 + (A_3 - A_4 C_3) D_1 + (A_1 - A_4 C_1) \\
q_3 &= -\{(A_3 B_4 - A_4 B_3) D_1 + (A_1 B_3 - A_3 B_1) D_3 + (A_1 B_4 - A_4 B_1)\}.
\end{aligned}
\tag{3.7.30}
$$

It should be noted that Equation (3.7.30) holds only for the water surface and not for ground area above the water level. Ortho-rectified images can be generated using the above equations with arbitrary spatial resolution as depicted in Figure 3.7.20. More specifically, image intensity at a point (X, Y) in the ortho-rectified image is spatially interpolated from the image intensities of the nearest corresponding points in the oblique image. Since a grid composed of (X, Y) coordinate points can take arbitrary spacing ΔX and ΔY, an ortho-rectified image becomes larger for smaller grid spacing and vice versa. Usually, the ortho-rectification is executed so that the transformed image has an image size comparable to the original image. However, when a largely distorted image is generated for a considerably oblique viewing angle, the resolution of velocity becomes extremely low farther from the camera location.

3.7.4.2.2 Image processing

Once the ortho-rectified images are generated, conventional PIV analysis (see Section 3.7.2) can be applied directly to the images. As a similarity index for a template pattern matching, the cross-correlation coefficient (R_{ab}) is usually used, defined as:

$$R_{ab} = \frac{\sum\limits_{i=1}^{M}\sum\limits_{j=1}^{N}\left\{\left(a_{ij} - \overline{a_{ij}}\right)\left(b_{ij} - \overline{b_{ij}}\right)\right\}}{\sqrt{\sum\limits_{i=1}^{M}\sum\limits_{j=1}^{N}\left(a_{ij} - \overline{a_{ij}}\right)^2 \sum\limits_{i=1}^{M}\sum\limits_{j=1}^{N}\left(b_{ij} - \overline{b_{ij}}\right)^2}} \qquad (3.7.31)$$

where M and N are the size of the template or the interrogation area (IA), a_{ij} and b_{ij} are the distribution of image intensities in two templates separated by a time interval Δt. The over bar stands for the average value of the intensities within the template. A Gaussian or parabolic sub-pixel fit is applied to the resulting cross-correlation distribution to improve measurement accuracy (Raffel *et al.*, 2007), which is about five times more accurate than a simple search of the maximum correlation. The measurement accuracy Δu is, therefore, roughly estimated by the following relation:

$$\Delta u \approx 0.2 \frac{\Delta X}{\Delta t} \qquad (3.7.32)$$

where ΔX is the physical pixel size of the ortho-rectified image. Therefore, improvement of measurement accuracy can be achieved by taking smaller ΔX or larger Δt, with great care not to allow large deformation of surface features during Δt. For example, when ΔX is 0.1m, the minimum time separation of an NTSC signal, 1/30sec, yields the resolvable velocity of about 0.3m/s, which is fairly coarse; however, for a time separation of 0.1 seconds Δu becomes 0.1m/s, which is a better setting of PIV. It should be noted that the fundamental accuracy further depends on the image distortion.

3.7.4.2.3 Sample measurements

Sample 1: Velocity distributions. This sample uses for illustration purposes a set of laboratory LSPIV measurements. The laboratory study was carried out in a large-scale laboratory model encompassing a reach of a river illustrated in Figure 3.7.21a (Muste *et al.*, 2008). The raw LSPIV measurement outcomes are instantaneous vector fields as plotted in Figure 3.7.21b. LSPIV is the only available technique that provides instantaneous two-component velocity measurements on a plane. The LSPIV vector field makes it possible to conduct Lagrangian and Eulerian analyses for determining spatial and temporal flow features such as the mean velocity field, streamlines, and vorticity (see Figure 3.7.21c, d and e) as well as other velocity-derived quantities (strain rates, fluxes, dispersion coefficients due to shear, etc).

The field implementation of LSPIV illustrates that the technique can be readily applied to river reaches spanning scales as large as illustrated in Figure 3.7.22 (Fujita *et al.*, 2007b). These images were acquired from a helicopter hovering over the Yodo River (Japan) during a flood. Since the position of the camera changes at every instant during the flight, the field of view varies gradually from frame to frame. Consequently, an additional step was added to the typical LSPIV protocol to extract only the flow-related information (Fujita *et al.*, 2007b; Fujita & Kunita, 2011; Detert & Weitbrecht 2015). As a result, velocity distributions for an area about 800m long and 270m wide were obtained by tracking only the naturally-occuring features conveyed at the free

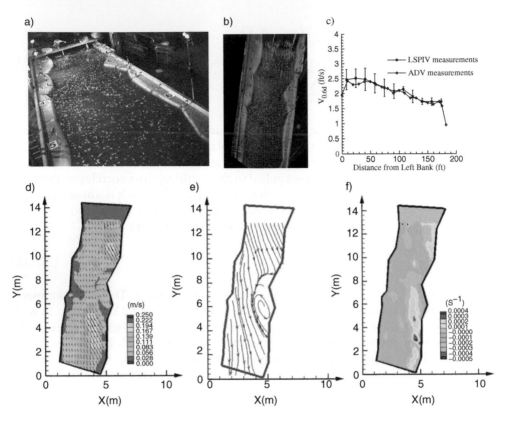

Figure 3.7.21   LSPIV results obtained in a laboratory physical model (Muste *et al.*, 2008): a) video frame of the upstream reach of a 5 m x 40 m hydraulic model; b) instantaneous vector field superimposed on an undistorted video frame; c) comparison of a LSPIV velocities with ADV velocities in a cross-section; d) mean vector field; e) streamlines established on the mean vector field; f) vorticity field established from the mean vector field.

surface. It should be noted that aerial LSPIV yields successful data only when surface features are clearly visualized, such as reflections of sunlight, numerous floating drift material (*e.g.*, vegetation) or foam, or recognizable differences in water surface color due to the presence of suspended sediments (Hauet *et al.*, 2008).

Sample 2: Discharge estimation. LSPIV surface velocity in conjunction with bathymetry can provide flow rates in streams. The principal of the method for estimation of the discharge stems from the velocity-area method, as illustrated in Figure 3.7.23a. The channel bathymetry can be obtained independently from direct surveys using specialized instruments (*e.g.*, sonar or ADCP measurements). The channel bathymetry can be surveyed at the time of the LSPIV measurements or prior to them under the assumption that bathymetry is not changing in the time interval between the bed and free-surface measurements. Surface velocities at several points along the surveyed cross section (V_i in Figure 3.7.23a) are computed by linear interpolation from neighboring grid points of the PIV-estimated surface velocity vector field (V_s). Assuming that the shape of the

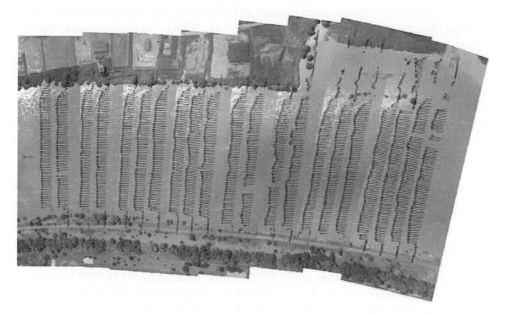

Figure 3.7.22 Surface velocity distributions of the 2006 Yodo River Flood measured using aerial LSPIV (Fujita *et al.*, 2007b, with permission from IAHR).

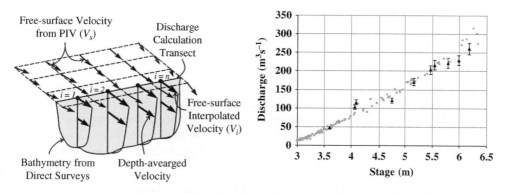

Figure 3.7.23 Discharge estimation with LSPIV: a) discharge estimation protocol (Muste *et al.*, 2008); b) LSPIV discharge estimates (black triangles) are compared with the calibration measurements for the stage-discharge relationship existing at the same location (gray squares). Error bars indicate the sensitivity of the LSPIV results to the correction coefficient applied to the surface velocity to derive the depth-averaged velocity Creutin *et al.*, 2003. Republished with permission of Elsevier; permission conveyed through Copyright Clearance Center, Inc.

vertical velocity profile is the same at each point i, the depth-averaged velocity at each i vertical is related to the free-surface velocity by a velocity index. The discharge for each river subsection $(i, i+1)$ is computed following the classical velocity-area procedure (*e.g.*, Rantz, 1982).

Sample discharge estimates from a series of LSPIV measurements in the Iowa River (USA) are illustrated in Figure 3.7.23b (Creutin *et al.*, 2003). For these measurements, a value of $k = 0.85$ for the index velocity is used. This value is generally accepted for river flows by the hydraulic community and used in conjunction with other measurement techniques (Costa *et al.*, 2000). The impact of varying the index velocity from 0.8 to 0.9 is also shown in Figure 3.7.23b by error intervals. The index velocity value has been, and continues to be, a subject of research (Polatel, 2005). This index depends on the shape of the vertical velocity profile, which is affected by the flow aspect ratio, Froude and Reynolds number, micro and macro bed roughness, and relative submergence of the large-scale roughness elements.

3.7.4.2.4 Error sources in LSPIV

Errors are generated in all steps of the measurement process: illumination, seeding, recording, transformation, processing, and post-processing. Many of the errors associated with image processing are aligned with those described in Section 3.7.3.5. They are reiterated here in the context of large-scale image velocimetry. Kim (2006) identified twenty-seven errors sources that might affect LSPIV measurements in field conditions. Some of the error sources discussed in Section 3.7.3.5 are not included in this discussion as their effect for LSPIV measurements is in most cases negligible or pertains to procedures that do not apply to LSPIV protocol. In addition to conventional terminology (*i.e.*, bias or precision), error sources are identified below in terms of their effect on the LSPIV results. Specifically, the global error type is constant over the whole image area and affects all calculated velocities. The local error type varies from one grid point to another over an image. A short description of each elemental error is given next.

1. Illumination
   – *Global illumination* (Bias, Global): field LSPIV is deployed under natural conditions. The overall intensity in the recorded images is affected by sunlight brightness, with conditions commonly known as sunny, partly cloudy, and cloudy.
   – *Local illumination non-uniformity* (Precision, Local): Local differences in the light intensity in images may occur due to stationary reflections (*e.g.*, from trees, buildings) on the water surface and affect estimates of local velocity.

2. Seeding
   – *Clustering of seeding* (Bias, Local): The presence of inter-particle forces (*e.g.*, electrostatic surface tension) may cause tracer particles to cluster. This alters the appearance of particle patterns in successive images and might induce flow-independent velocities in the measurement.
   – *Tracing error* (Bias, Precision, Local): Denotes the difference between the real velocity at the water surface and the velocity of the tracer particles. Seeded particles float on or just below the water surface depending on particle specific gravity, and may be affected by wind or other free-surface processes.

3. Recording
   – *Image resolution* (Bias, Global): An important parameter for accurately recording the image patterns on the free surface. The higher the resolution, the more detailed information images contain about water surface area.

- *Time interval between images* (Bias, Precision, Global): This is a crucial parameter in image recordings with direct impact on the velocity calculation. Attention is needed to carefully assess this important factor as contemporary cameras use various proprietary algorithms for compressing and transferring images such that in some cases the time step between the images is not constant throughout the recordings.
- *Sampling time or number of images* (Bias, Precision, Global): When the overall recording time is increased, more images are available for the analysis, providing higher quality results.
- *Camera movement* (Bias, Precision, Global): Unexpected causes (*e.g.*, wind) move or vibrate the camera support, and will be mistakenly interpreted as movement of the patterns in the images, and hence inaccurate velocities.

4. Transformation
- *Distance error* (Bias, Precision, Global): This error is induced while conducting the geodetic survey and is due to instrument accuracy and operator manipulation of the instrument.
- *Marker point identification* (Precision, Global): This error occurs during the identification of the CRT coordinates of marker points on the recorded image required to calculate the transformation coefficients between the CRT coordinates and the physical coordinates.
- *Brightness allocation* (Bias, Precision, Local): When the image transformation is applied, the brightness of each pixel of an undistorted image is calculated using the cubic convolution interpolation (Fujita *et al.*, 1998). The calculated brightness of each mesh point might be different from the brightness of the point in the distorted image.
- *Camera positioning and orientation* (Bias, N/A): This error is related to the implicit image transformation method, in which case the geodetic survey is not necessary. The camera position and orientation parameters are called extrinsic parameters in the implicit transformation.
- *Magnification* (Bias, N/A): Magnification is the ratio between the physical coordinates and image coordinates of the objects. Scaling the image coordinates to real-world dimensions is one of the critical sources of uncertainty in conventional PIV and LSPIV.

5. Processing
- *Peak finding error* (Bias, Precision, Global): This error is related to the sub-pixel peak finding algorithm. There are several peak finding algorithms at the sub-pixel level (Lourenco & Krothapalli, 1995).
- *Background noise* (Precision, Global): In practice, the image background is not perfectly uniform as there is always some unavoidable noise in the recording process (Joseph, 2003). The background non-uniformity may be induced by shot noise associated with the camera sensor array or noise due to light sources other than the particles of interest and results in the deterioration of image contrast.
- *Velocity gradient error* (Precision, Local): The effect of displacement gradients within the interrogation region must be considered in the evaluation of the uncertainty of velocity measurements (Forliti *et al.*, 2000).

- *Interrogation area size* (Precision, Global): The size of interrogation area used in the image processing step is commensurate with particle or pattern size. If the interrogation area size is not large enough to recognize the particle's form or the pattern from a group of particles, velocity calculation error will occur.
- *Searching area* (Precision, N/A): The selected search area during image processing is proportional to expected particle displacements to calculate correct velocity vectors. If the displacements of particles exceed the scope of the search area, the peak finding algorithm cannot determine the peak cross-correlation between image patterns.
- *Window offset* (Precision, N/A): For cross-correlation analysis of a pair of single exposed images, it is relatively simple to offset the interrogation windows by the integer part of the particle-image displacement. Hence, the residual displacement is only a fraction of the particle-image displacement (in pixel units), which is always smaller than 1/2 pixel, and subsequently would yield a more accurate result compared to the original analysis without the window offset.
- *Out of plane motion* (Precision, N/A): This error occurs in highly three-dimensional flows. The river free surface can usually be considered as two-dimensional.
- *Image displacement error* (Precision, Global): For most displacements, the uncertainty due to the in-plane pair losses is nearly constant except for displacements of less than 0.5 pixels where a linear dependency can be observed. Image displacement error is different for various interrogation window sizes up to 0.05 pixels (Raffel *et al.*, 2007).
- Image quantization error (Precision; Global): This error is generated by the variation of image quantization levels. Image quantization represents the bit/pixels ratio. In case of an 8 bit image which is used in the current LSPIV program, the maximum uncertainty due to it is 0.03 pixel (Raffel *et al.*, 2007)
- *Seeding concentration* (Bias, Precision, Global): Seed particles floating at the water surface form image patterns in the recordings. These patterns are used to calculate the cross correlation coefficients between two successive images in determining velocity vectors. Seeding concentration, an important factor in the creation of the image patterns, is defined as the ratio between the number of pixels occupied by particles and the total number of pixels in the reference area.
- *Seed size* (Bias, Precision, Global): The size of seeds in images is related to the image resolution and the image magnification factor. Therefore the seeding size in images for the same physical size would impact the accuracy of the measurements. According to Raffel *et al.* (2007), when the seed size is 3~ 6 pixel, a measurement uncertainty up to 0.07 pixel might occur.
- *Seed brightness and contrast* (Bias, Precision, Global): Image contrast is defined as the difference in the intensity of brightness in images and it is directly dependent on the color and brightness of particles with respect to the background. High contrast of an image pattern favors image pattern recognition during image processing, while poor contrast induces errors due to the difficulty of recognizing patterns in successive images (Hauet *et al.*, 2008).

6. Post-processing
 - *Velocity filtering* (Precision): Given the multitude of factors involved in the measurement process (not all of them accountable in the analysis), the calculated vector fields might contain erroneous vectors. Velocity filtering applies various criteria to fill in stray vectors. The accuracy of various algorithms needs to be taken into account in the overall analysis.
 - *Minimum correlation* (Precision): An averaged velocity field is calculated from a number of instantaneous velocity fields. In averaging instantaneous velocities, the velocities calculated from low correlation coefficients can be excluded from the average velocity calculation by setting the minimum correlation.

Assessment of this extensive list of elemental errors has to make use of the best available information obtained from prior estimations or through the development of specially-designed experiments and/or numerical simulations. Currently, information on elemental uncertainties is scarce, as the typical measurement-based investigations rarely afford the extensive resources needed to estimate such errors. However, the use of standardized methodologies for assessing uncertainties allowed Kim *et al.* (2007) to estimate uncertainty in a manner that addresses scientific and legal scrutiny. In particular, their assessment of uncertainty for LSPIV measurements acquired in a small stream with a basic LSPIV system and minimum preparation resulted in uncertainties of the free-surface velocities ranging between 10% and 35% (at 95% confidence level). The uncertainty of the spatial distribution depends on the location of the velocity vector over the imaged area with most of the centrally located velocities being at 10% uncertainty. Note however, that many of the sources of errors could have been reduced with additional system optimization.

3.7.4.3 Basics of Space-Time Image Velocimetry (STIV)

As mentioned in the previous section, the accuracy of LSPIV depends on a combination of PIV parameters such as ΔX or Δt with lower velocity resolution farther from a camera position when the image is taken at a more oblique angle. STIV was developed to achieve a uniform velocity resolution irrespective of the vertical viewing angle and with minimum parameters that control the accuracy.

3.7.4.3.1 Generation of space-time image (STI)

In STIV, the evolution of brightness distribution in a line segment embedded in the free surface is the main image element that is subsequently subjected to analysis to determine velocity vectors in a cross section of the river. In this respect, the line segment is equivalent to the interrogation area used in LSPIV for estimating the local velocity at a grid point (see Figure 3.7.3). Once a line segment parallel to the river banks is selected (see Figure 3.7.24a) its evolution in terms of brightness is tracked over time as image recording progresses. By vertically stacking the evolution of image intensity along the line segment, a synthetic space-time image (STI) is created as depicted in Figure 3.7.24b. The inclined patterns in this image indicate the advection of image intensity along the line segment, *i.e.*, the larger the inclination of the segment, the higher the local

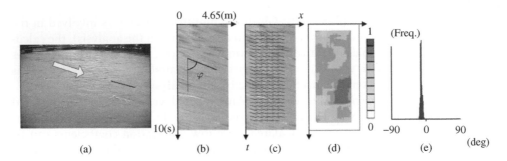

Figure 3.7.24 Outline of STIV procedure for the Chigusa River flood: (a) set line segment, (b) generate STI, (c) calculate orientation angles, (d) calculate coherency, and (e) generate histogram of orientation angles (Fujita *et al.*, 2009, with permission from IAHR).

velocity. Each STI is generated such that distortion of the image will be canceled out. The mean gradient of the pattern over the STI image is attributed to the local mean velocity along the line segment as will be explained in detail later. The STI vertical axis is constructed using the images of the selected segment in all the recorded frames, hence the finer unit of the time axis for video-recordings is 1/30 seconds for the NTSC standard and 1/25 seconds for the PAL standard. The horizontal axis of an STI corresponds to the physical length of the selected search line, which can take an arbitrary value. In the case of Figure 3.7.24b its length is 4.65m.

Another example of STI is provided in Figure 3.7.25 for the Uono River (Japan) during flood conditions. The STIs for several search lines are shown in Figure 3.7.26, in which the horizontal scale is 17.6m and the vertical scale is ten seconds generated by 300 image frames. The histogram equalization applied to the images reveals the variation of the velocity gradient (slope of dotted red line in the figure) across the river. The image quality for STI can alternatively be improved by applying a two-dimensional fast Fourier transformation (FFT) with a band pass filter (Fujita *et al.*, 2009).

Figure 3.7.25 Arrangement of search lines for STIV measurements with a length of 17.6m and a spacing of 5.6m: (a) original image and (b) ortho-rectified image (Fujita *et al.*, 2013, with permission from IAHR).

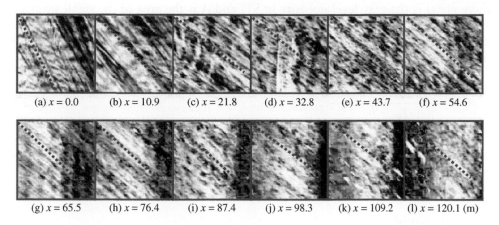

(a) $x = 0.0$ (b) $x = 10.9$ (c) $x = 21.8$ (d) $x = 32.8$ (e) $x = 43.7$ (f) $x = 54.6$

(g) $x = 65.5$ (h) $x = 76.4$ (i) $x = 87.4$ (j) $x = 98.3$ (k) $x = 109.2$ (l) $x = 120.1$ (m)

Figure 3.7.26 STIs for search lines indicated in Figure 3.7.25; the length is 17.6m and vertical scale is ten seconds.

A limitation of STIV and LSPIV is that images cannot be acquired in low visibility conditions or during night time. Use of far-infrared cameras was tested for overcoming this limitation. The customized experiment conducted by Fujita *et al.* (2012), whereby alternative velocity measurement techniques (*i.e.*, radar, ADCP, luminescent floats) were simultaneously employed along with STIV applied to infrared camera images, showed good agreement between all the engaged measurement alternatives.

3.7.4.3.2 STIV algorithm

STIV is computationally efficient compared with LSPIV; once an STI is generated for a search line the corresponding velocity can be calculated manually without recourse to intensive analysis such as the estimation of the cross-correlation coefficient. The mean STI gradients can be calculated by a fast algorithm developed by Fujita *et al.* (2007a). In this algorithm, STI is divided into a number of small square segments allowing overlaps, and for each segment a gradient of the image orientation is calculated with the following equation:

$$\tan 2\varphi = \frac{2J_{xt}}{J_{tt} - J_{xx}} \qquad (3.7.33)$$

where

$$J_{xx} = \int_A \frac{\partial g}{\partial x}\frac{\partial g}{\partial x}dxdt \qquad (3.7.34)$$

$$J_{xt} = \int_A \frac{\partial g}{\partial x}\frac{\partial g}{\partial t}dxdt \qquad (3.7.35)$$

$$J_{tt} = \int_A \frac{\partial g}{\partial t}\frac{\partial g}{\partial t}dxdt \qquad (3.7.36)$$

where $g(x,t)$ is the gray level intensity in STI and A is the area of the small segment. Figure 3.7.24c provides image orientation vectors for each segment, indicating the vectors follow the image pattern well. Figure 3.7.24d represents the distribution of the coherency defined as:

$$C = \frac{\sqrt{(J_{tt} - J_{xx})^2 + 4J^2_{xt}}}{J_{xx} + J_{tt}} \tag{3.7.37}$$

which is a measure of image-pattern coherence that takes a value of 1 for ideal local orientation and 0 for an isotropic gray image. Therefore it is possible to calculate the mean orientation angle by picking up clearer orientation information using the coherency level as a weighting function as:

$$\overline{\varphi} = \frac{\int \varphi C(\varphi) d\varphi}{\int C(\varphi) d\varphi} \tag{3.7.38}$$

Figure 3.7.24e is the weighted histogram of the orientation angle, showing a sharp peak distribution. Finally, since the length scale and the time scale of STI are given, the mean velocity along the line segment can be calculated directly from the mean orientation angle by using the following relationship:

$$U = \frac{S_x}{S_t} \tan \overline{\varphi} \tag{3.7.39}$$

where S_x (m/pixel) is a unit length scale for a horizontal axis of STI and S_t (s/pixel) is a unit time scale for a vertical axis of STI. The velocity thus obtained represents a mean velocity averaged both spatially and temporally along a search line segment.

3.7.4.4 Traceability of surface features

An issue that often arises when utilizing surface features for estimating flow velocity concerns the accuracy of the tracing of the surface flow velocities by means of naturally-occurring image patterns evident on the surface of the moving water body. To investigate this issue and its underlying assumption, the STI trajectory of driftwood floating on the free surface at the time of the video recordings was compared with the STI images created by natural patterns moving at the free surface. Figure 3.7.27 contains STI for the Uono River in flood, and comprises images of both tracers. The pair of images in the figure indicates that the trajectory of the driftwood is parallel to images generated by surface ripples with the speed of the driftwood at 3.2m/s and the advection speed of surface features at 3.1m/s. Therefore the assumption held for this case. This illustration confirms that, in the absence of wind, naturally-occuring free-surface patterns can be reliable tracers for image-based techniques. The study of Plant and Wright (1980) found that for situations where wind-induced velocity at the water surface is less than 3% of the wind speed, the tracing provided by naturally occurring tracers is appropriate. However, further field studies are needed to confirm this suggestion.

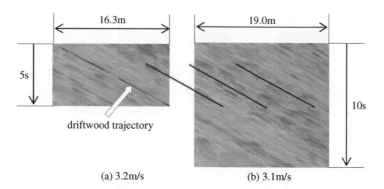

16.3m

19.0m

5s

10s

driftwood trajectory

(a) 3.2m/s   (b) 3.1m/s

Figure 3.7.27 Comparison of STIs and measured velocity of driftwood and surface features (Fujita *et al.*, 2013, with permission from IAHR).

3.7.4.5 Performance of LSPIV and STIV

3.7.4.5.1 Comparison between LSPIV and STIV

To compare STIV and LSPIV, measurements were used from a flood flow along the Ibo River Flood in Japan (Fujita *et al.*, 2014). Figure 3.7.28 shows the original images with search lines for STIV, and the ortho-rectified images for LSPIV. Both methods yielded close results for viewing angle A, in which surface features are clearly visible in the entire river width. For viewing angles B and C, however, LSPIV underestimates velocities. For angle B, surface features in areas farther from the camera are vague, making pattern matching in LSPIV difficult. As for angle C, since the image distortion is very large, LSPIV failed to yield an appropriate velocity distribution. On the other hand, STIV yielded results that were almost as stable and reliable as for angle B and angle C, even under deteriorated image shooting conditions.

3.7.4.5.2 Comparison with other velocity measurement methods

During the field measurements of flood flow along the Uono River (Fujita *et al.*, 2012), other measurement methods were also used; *i.e.*, an acoustic Doppler current profiler (ADCP) and a radio wave velocity meter. In terms of the imaging techniques, video cameras were installed on both sides of the river, and additionally a far-infrared ray camera was used for another comparison to be shown later. Figure 3.7.29 compares the surface velocity distribution measured by various methods. The data collected with an ADCP are for the layer about 30cm below the water surface. The radio wave velocity meter was shifted step by step manually in the transverse direction along the bridge. Allowing for scatter between the respective measurement techniques, the overall measurements were in good agreement with each other, verifying the accuracy of image-based techniques.

3.7.4.5.3 Comparison with discharge measurements

Figure 3.7.30 compares a discharge directly measured by ADCP and estimated discharges from surface velocity distributions by STIV and a radio wave velocity meter.

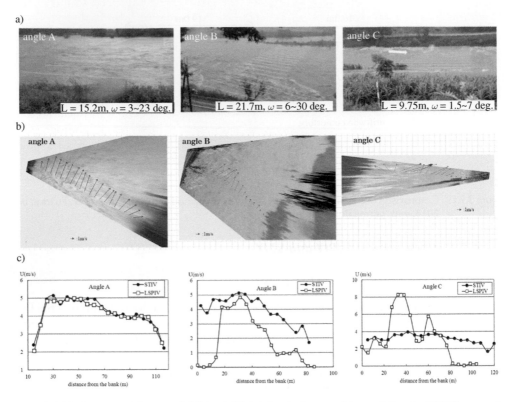

Figure 3.7.28 Comparison between LSPIV and STIV; a) Original image with search lines for STIV (L = search line length, ω = range of viewing angles); b) ortho-rectified image with LSPIV result; c) comparison of velocity distribution between LSPIV and STIV (Fujita *et al.*, 2014. Reprinted by permission of the publisher, Taylor & Francis Ltd, http://www.tandfonline.com).

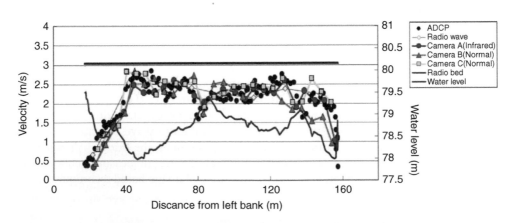

Figure 3.7.29 Comparison of surface velocity distributions at the downstream section of the Negoya Bridge of the Uono River; 2012 Flood on April 21, 10:00AM (Fujita *et al.*, 2013, with permission from IAHR).

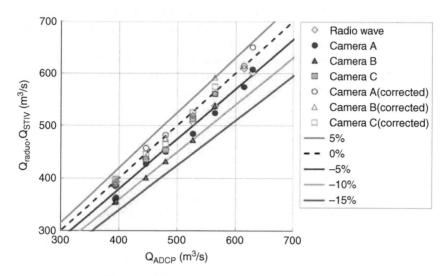

Figure 3.7.30 Comparison of discharges by various methods for the 2012 Uono River Flood (Fujita et al., 2013, with permission from IAHR).

In integrating the velocity distribution measured with the ADCP, the surface velocity was assumed to be the same as the data 30cm below the water surface. Generally, the discharge estimated by STIV by using the conventional velocity index of 0.85 for the transformation coefficient from surface to mean velocity results in a mean velocity that is somewhat lower than the value found using the ADCP, but the difference is less than ten percent. The STIV data, adjusted with a correction coefficient to account for the different location of the measurements (*i.e.*, STIV at the surface while ADCP close to the surface), along with the ADCP discharge measured at the same location shown in Figure 3.7.30 show differences of less than 5%. Regarding the comparison of discharges, Sun *et al.* (2010) compared LSPIV data with an ADCP and showed the validity of imaging methods. Hauet *et al.* (2008) utilized the LSPIV results as validation of a numerical simulation. This is a good example of the utilization of imaging techniques for evaluating simulated results.

3.7.4.6 Application of PTV to large-scale measurements (LPTV)

PTV methods are particularly well-suited for moderately sized large-scale free surface flows, such as those that are typically employed to measure water surface velocity distributions in physical models. Tracer concentrations in large-scale measurements are often low for a number of reasons, including the limited practicality of seeding the flow with large numbers of tracers and limiting interactions between proximate tracers (Admiraal *et al.*, 2004). Low tracer counts make PTV an appropriate measurement method in many cases. Furthermore, since the tracers used in large-scale measurements are buoyant, they only appear on a single curved surface (the water surface). Thus, there is no tracer particle overlap or out-of-plane motion of the tracer particles.

Many of the concepts described in the previous PTV sections are also applicable to large-scale particle tracking velocimetry (LPTV). In fact, the flow chart shown in Figure 3.7.14 is directly applicable to LPTV with one exception. Images captured in large-scale settings such as physical models and field measurements are often captured at oblique angles (*i.e.*, the camera axis is not perpendicular to the flow surface). Prior to analyzing images of the free surface, the oblique images must be orthorectified. Converting oblique images into ortho-rectified images was discussed in detail in the previous section and will not be discussed here.

Although LPTV is very similar to traditional PTV, it does have a few additional idiosyncrasies that the reader should be aware of:

-    The LPTV object field is often large, and the spatial resolution of LPTV images is usually more limited than in traditional PTV. Consequently, larger tracer particles are necessary, resulting in reduced ability to track small scale flow structures.
-    In field measurements, natural light is often used to illuminate the tracer particles. Reflections, shadows, and variations in lighting intensity throughout the image make consistent identification of tracer particles difficult. Generally, a diverse set of filters – both optical and digital – is useful.
-    Standard video cameras are often used for LPTV since they are easier to use in the field than machine vision cameras. Most current commercial video cameras have CMOS sensors which employ a rolling shutter instead of a global shutter. Images collected with these cameras are not collected at an instant but are rapidly scanned across the sensor. For some applications, this design is not optimal. Furthermore, most standard video cameras collect images at fixed rates of 25 or 30 frames per second. Lower frame rates can be obtained by skipping frames, but high speed flows may require higher frame rates and non-standard cameras.
-    When using LPTV in field measurements, the wind and surface waves can affect the flow field and the actual and perceived motion of tracer particles. Tsanis (1989) provides a review of the observed effects of wind on water surface velocity. Based on a literature review, Tsanis (1989) suggests that the wind-induced water surface drift velocity is about 3% of the wind velocity.

While most large-scale PTV measurements have been applied to flows that have surfaces that are nearly planar, it is also practical to extend PTV measurements to three-dimensional surfaces. For example, Baud *et al.* (2005) used PTV to measure the flow over a drop structure. Since the PTV images for their study were of a three dimensional surface, this application required multiple cameras, as discussed by Maas *et al.* (1993), so that the horizontal and vertical positions of tracer particles could be determined through photogrammetry. As a result, the method provided both the water surface profile and the water surface velocity distribution over the drop structure.

A common application for large-scale particle-imaging techniques is to estimate discharge, but usually only the water surface velocity is measured because the tracers float. For many large-scale PIV and PTV studies, it is often assumed that the velocity distribution below the free surface is described by the log-law or a power law, and surface velocity measurements are extended to the bed by way of a curve fit. This works well for shallow flows with relatively uniform depth (Weitbrecht *et al.*, 2002;

Creutin *et al.*, 2003; Meselhe *et al.*, 2004), but for rapidly varied flows like the one described by Baud *et al.* (2005), a different approach is necessary to evaluate the subsurface velocity distribution. Baud *et al.* (2005) used a k-ε turbulence model to estimate the velocity distribution of the flow. In their case, the water surface velocities were used to optimize the model to improve its performance. The resulting discharge predictions were estimated to have a relative error of 2 to 3% with a standard deviation of 1%.

3.C APPENDIX

3.C.1 Light scattering in PIV

The shape of a diffraction limited particle image can be approximated by a two-dimensional Gaussian distribution integrated over the light sensitive area of a sensor pixel:

$$I(m,n) = I_0 \int_{x_1 = m_{x_1} - 0.5fP}^{x_1 = m_{x_1} + 0.5fP} \exp\left(\frac{-(x_1 - p_{x_1})^2}{\frac{1}{8}d_p^2}\right) dx_1 \int_{x_2 = n_{x_2} - 0.5fP}^{x_2 = n_{x_2} + 0.5fP} \exp\left(\frac{-(x_2 - p_{x_2})^2}{\frac{1}{8}d_p^2}\right) dx_2$$

$$(3.C.1.1)$$

where $I(m,n)$ is the grayscale value of the image pixel at the discrete (m,n) coordinate of the digital image, f^2 is the area fill factor of the camera sensor (or proportion of a sensor pixel that is sensitive to light), d_p is the diameter of the diffraction limited particle image defined at the $1/e^2$ level of the Gaussian curve, I_0 is the brightness of the particle, the image plane coordinate system (x_1, x_2) is aligned with the rows and columns of the digital image, (m_{x_1}, n_{x_2}) is the image plane coordinate of the (m,n) photosite, (p_{x_1}, p_{x_2}) is the image plane coordinate of the particle, and P is the sensor pixel pitch (center to center spacing of photosites). Image plane axis directions are denoted using the subscript $i = 1,2,3$ where $i = 1$ is the image plane horizontal direction, $i = 2$ is the image plane vertical direction, and $i = 3$ is orthogonal to the image plane. The image plane coordinate system often has the same orientation as the 'world' coordinate system such that the $i=1$ axis is aligned with the longitudinal flow direction and the $i = 2$ axis is aligned with the bed normal direction, but in general, the mapping from image to world coordinates depends on the positioning and orientation of the camera and the light sheet. It is convenient to normalize image plane lengths using the sensor pixel pitch (P). The resulting dimensionless length is often referred to as a 'pixel'. It is important to distinguish between 'pixel' – the dimensionless length, 'image pixel' – the discrete elements making up a digital image, and 'sensor pixel' – the photosites of a CCD or CMOS camera sensor.

The diffraction limited particle image diameter is a function of the camera lens f-number, the system magnification (M), the wavelength (λ) of the light source, and the pixel pitch P according to $d_p = 1.80f_\# (M + 1)\lambda/P$. Small particle image diameters ($d_p <\sim 3$) will be under-sampled by the camera sensor, that is, they contain information at wavenumbers larger than the Nyquist limit of the sampling process. The normalized Fourier transform of a continuous particle image, $S(k_1,k_2)/S(0,0)$, is approximated by:

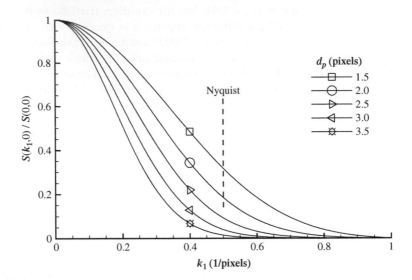

Figure 3.C.1.1 Normalized Fourier transform of a particle image for different particle image diameters (d_p) with a fill factor of 1.

$$\frac{S(k_1, k_2)}{S(0,0)} = \exp\left(-\frac{1}{8}\left(\pi d_p k_1\right)^2\right) \frac{\sin(\pi k_1 f)}{\pi k_1 f} \exp\left(-\frac{1}{8}\left(\pi d_p k_2\right)^2\right) \frac{\sin(\pi k_2 f)}{\pi k_2 f}$$

$$(3.C.1.2)$$

where $k_i = 1/\lambda_i$ is the wavenumber and λ_i is the wavelength in the i-direction ($i = 1, 2$). Equation (3.C.1.2) is plotted in Figure 3.C.1.1 for different values of d_p and for a fill factor $f^2 = 1$. The aliased content of PIV images (from wavenumbers larger than 0.5 per pixel) is a contributor to measurement error and should be minimized by appropriate selection of camera $f_\#$ if sufficient light is available. If there is insufficient light available, as is often the case, smaller $f_\#$'s (larger apertures) will be needed and lens aberrations will contribute to the particle image shape. This can be problematic for PIV because lens aberrations will be a function of the radial distance from the lens axis and measurement errors will inevitably vary from the center of an image to the corners. The geometric particle image size is given by DM/P, where D is the physical diameter of the tracer particles. Unless the field of view is very small (large magnification), the geometric particle image size is typically much smaller than the diffraction particle image size and its contribution to the total particle image size can be ignored.

3.C.2 Velocimeter comparison

Table 3.C.2.1, provided below, serves two purposes: first, it highlights criteria that the experimentalist should consider when selecting a velocimeter, and second, it compares capabilities and limitations of several generations of instruments, starting with Hot-Wire Anemometry (HWA), which was developed at the beginning of the 20th century.

Table 3.C.2.1 Comparison of hot-wire anemometry (HWA), LDA, PIV, and ADV specifications

Specifications	HWA	LDA	PIV	ADV
Operating principle	thermal	optical	optical	acoustic
Measurement volume	point	point	area/volume	point
Flow interference	yes	no	no	some
Requires seeding	no	yes	yes	yes
Instrument response	non-linear	linear	linear	linear
Single-input response[1]	no (varies with temperature)	yes	yes	no (varies with sound speed)
Sampling type[2]	continuous	random in time	random in space	quasi-continuous
Output data type	continuous analog	discrete digital	discrete digital	discrete digital
Frequency response	excellent	very good	good	good

[1] Single-input response sensors respond only to changes in one flow variable (*i.e.*, flow velocity) and are not affected by changes in other flow variables (*e.g.*, temperature)
[2] Sensors that require seeding record data only when a seeding particle passes through the measurement volume of the instrument

3.8 HIGH-FREQUENCY RADAR

Albeit a relatively expensive option in terms of investment costs, ultra-high frequency (UHF) radar offers an alternative means for continuous monitoring of surface flow in rivers and streams. Contrary to PIV, radars can operate day and night, and under extreme conditions of bad weather. The term radar abbreviates radio detection and ranging. Ultra-high frequency refers to an operating frequency of the instrument between 300 MHz and 3 GHz. Common radars for river applications operate at a frequency around 400 MHz, yielding radial velocity measurements over a sector of a circle with the antenna in its center. Within this sector, radial velocity components at the water surface are measured along rays from the antenna (Figure 3.8.1).

The theory of surface velocity measurements from UHF radar relies on Bragg scattering theory and Doppler processing. On a specular water surface, none of the transmitted electromagnetic energy returns, and surface velocity cannot be monitored. Surface gravity waves, turbulent boils or other irregularities at the water surface are needed for backscattering toward the receiver. Provided the water surface is slightly rough; *i.e.*, the root-mean-square surface elevation is small compared to the electromagnetic wave length, the backscatter is dominated by Bragg scattering, resulting from coherent contributions from a series of scattering surfaces regularly spaced apart. The condition for first-order Bragg scattering reads (Costa *et al.*, 2006):

$$2\lambda_b \sin \theta = \lambda \tag{3.8.1}$$

where λ_b is the wave length of the Bragg-resonant water wave at the water surface, θ is the incidence angle between the wave propagation direction and the normal vector perpendicular to the water surface, determined by the elevation of the transmitter above the water surface and the distance to the target area, and λ is the wave length of the emitted radar signal. With the antennas little elevated above the water surface, θ

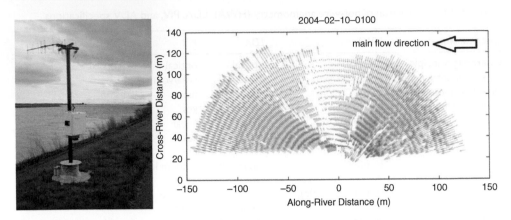

Figure 3.8.1 Left: application of a UHF radar to monitor river flow with three antennas connected to the transceiver in the enclosure attached to the mast. Right: example of an instantaneous radial velocity field. Adapted from Cheng et al. (2007).

is close to 90 degrees and surface waves causing resonance have about half the wave length of the emitted signal. For a 435 MHz radar, λ is 0.70 m and thus λ_b is 0.35 m (Cheng *et al.*, 2007). In fact, two distinct water waves will satisfy the resonance condition. Both have the required wave length. One is moving directly toward the radar, while the other is moving directly away from the radar.

Using the frequency of the emitted (known) and received (determined) radar signal, a Doppler spectrum can be determined that is associated with the radial surface velocity at the point of contact with the water surface. Assuming the surface wave amplitude is much smaller than the UHF radar wavelength, first-order scattering theory is valid and predicts two discrete lines (or Bragg lines) in the Doppler spectrum (Sassi & Hoitink, 2014). If no surface flow is present, these two lines lie symmetrically in the Doppler spectrum, on either side of the carrier frequency and the mean Doppler shift (averaged over both lines) is zero. However, if the surface velocity has a non-zero component along the line of sight of the radar, the lines will no longer be symmetrical. Both Bragg lines will be shifted over a range proportional to the flow velocity. Then, the mean Doppler shift of both lines is no longer zero and represents the Doppler frequency solely originating from the flow. Consequently, the velocity of the surface current along the line of sight from the radar can readily be determined from the average Doppler frequency shift.

When only one Bragg line is visible in the Doppler spectrum, the situation is less straightforward. Then the moving Bragg wave will result in a Doppler frequency shift that is proportional to the sum of its phase velocity and the velocity of the underlying current. For this surface gravity wave, the phase velocity is known from the resonant condition combined with the linear-dispersion relation between wavelength, frequency and depth.

Using a dual-radar deployment, the two-dimensional surface velocity field can be resolved for the overlap of the two segments on the free surface covered by the UHF radars. Figure 3.8.2 shows a comparison between two UHF radars and simultaneous ADCP measurements acquired at a river bifurcation. The difference between the radar

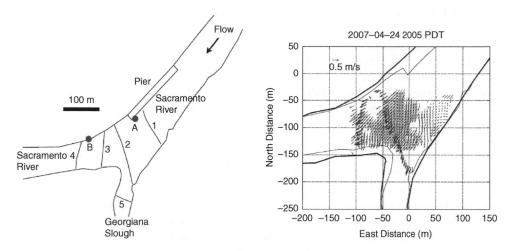

Figure 3.8.2 Comparison of the surface velocity field retrieved from a dual UHF radar deployment with near-surface ADCP measurements at a river bifurcation. The left panel indicates the position of the two radars (red dots) and the ADCP transects (blue lines). Right panel compares surface velocity vectors measured by radars (red vectors) with near-surface velocity measured by the ADCP (blue vectors). © 2008 IEEE. Reprinted, with permission, from Teague *et al.* (2008).

measurement location and the location of the first measured ADCP cell limits somewhat the degree in which such a comparison is meaningful, as the first cell is typically located about 1m below the surface (due to the immersion depth and a blanking distance related to transducer ringing as discussed in Section 3.3.2.3.2 in Volume II). The radar measurements are acquired within a thin film below the surface. The radar penetration depth can be estimated as $\lambda b/(4\pi)$ (Stewart & Joy, 1974), which is less than 0.03 m for a 0.35m Bragg-resonant surface wave.

The need for the presence of Bragg resonant waves is among the main limitations of measuring surface velocity with dual UHF radars. In addition, the fact that the penetration depth is very small makes the measurement susceptible to wind-induced surface flow, which may have little to do with the stream flow at larger depths. Despite these limitations, comparisons of velocity estimates from dual UHF radars with traditional propellers (Wang *et al.*, 2007) and ADCPs (Teague *et al.*, 2008) have shown to be favorable. The method is well-suited for measurements of the free-surface velocity in coastal areas and streams for situations where the deployment of permanent gaging stations is costly or during extreme flow events (such as floods).

3.9 DRIFTERS AND DROGUES

Drifters are the oldest form of velocimetry. The classic method of tossing an object (an orange works well) into a river and tracking its trajectory has likely been used since prehistoric times to measure flow path and velocity. A detailed review of modern drifter techniques was provided in Thomson and Emery (2014). Today, surface drifters with attached drogues are outfitted with on-board Global Positioning System (GPS) and telemetry for real-time Lagrangian measurement of the translation of fluid parcels. A typical drifter is a surface buoy tracked by GPS, although neutrally buoyant sub-surface drifters that are tracked acoustically are also used. Drifters often carry instrumentation for measurement of other parameters such as temperature. A drogue is a drag device attached below the drifter and designed such that the drifter tracks the current at the elevation of the drogue. Even if the objective is to track surface currents, a drogue is required to minimize the influence of wind and wave drag on the drifter. It should be recognized, however, that the drifter trajectory is a function of drag forces on both the drifter and the drogue. Thomson and Emery (2014) reviewed various drogue designs, and Niiler *et al.* (1995) tested drifter performance as a function of drogue design under various wind conditions. A typical drogue is a holey sock, essentially a perforated nylon cloth tube. Drogues designed as kites are also used. The drogue is tethered to the drifter, sometimes with an intermediate subsurface float that reduces risk of loss of the drogue by minimizing momentary high tensile loads along the tether due to wind and wave action on the drifter. Strain gauges can also be incorporated in the tether to monitor for loss (detachment) of the drogue. Drifters are most often used in ocean and lacustrine environments to track surface or near-surface currents (*e.g.*, Limeburner *et al.*, 1995; Hurdoware Castro *et al.*, 2007; Olson *et al.*, 2007), but a few researchers have used drifters in riverine and estuarine environments (*e.g.*, Tsanis *et al.*, 1996; Stieglitz & Reid, 2001; Schacht & Lambert 2002). Recently, small drifters for river measurements have been designed and tested in field conditions as illustrated in Figure 3.9.1 (Christian Noss – personal communication).

Figure 3.9.1 Surface flow, turbulence, and wave measurement in a small stream by a tiny drifter (2.5 cm diameter). The arrow points to the custom-made battery-powered device, which logs data using an accelerometer, gyroscope, and compass sensor at up to 100 Hz. The device includes a micro-controller and memory for managing data storage, communication, and settings (photos contributed by C. Noss).

3.10 DILUTION METHOD

The dilution method provides a reach averaged streamwise water velocity. The method involves injection of a conservative, neutrally buoyant tracer into a river. The tracer advects downstream as a plume. Measurements of the tracer concentration at a downstream location are used to estimate the plume advection, from which velocity and discharge are calculated. While it may be utilized in various riverine environments, the tracer method is particularly effective for measurement of low velocities and discharge in small, shallow rivers. Errors on the order of 2% compared to volumetric discharge measurement have been reported (Tazioli, 2011).

The dilution method has been used to measure stream discharge for decades. Fischer *et al.* (1979) provided a description of the technique. In the middle of the 20th century, radioactive tritium was utilized as the tracer. Nowadays, less toxic tracers are used, such as sodium chloride salt (*e.g.*, Weijs *et al.*, 2013) or the fluorescent dye rhodamine (*e.g.*, Tazioli, 2011). Salt concentration is measured indirectly via electrical conductivity. Rhodamine concentration is measured using a fluorometer. The amount of required tracer depends upon the stream discharge (which sets the fully mixed concentration for a given injected mass of tracer) and the sensitivity of the instruments used to measure electrical conductivity or fluorescence. Tazioli (2011) compared salt versus rhodamine tracing, finding that rhodamine was less conservative due to absorption to suspended sediments, but rhodamine measurement is more sensitive thus smaller tracer concentrations are required. The traditional sampling method for rhodamine involved collection of a time series of bottle samples for laboratory measurement of fluorescence. However, fluorometers are now available for *in situ* continuous measurement (*e.g.*, Pilechi *et al.*, 2014).

Fischer *et al.* (1979) emphasized the importance of achieving complete cross-sectional mixing at the measurement section, which ensures that the advection of the tracer plume represents the reach averaged velocity. They presented two versions of the method: slug release and continuous release. The slug release uses less tracer but requires continuous measurement of the tracer concentration. On the other hand, continuous release requires large quantities of tracer but the tracer concentration sampling is made only once.

The slug release method involves release of a known mass (M) of tracer upstream of the measurement reach. If the release is distributed across the channel section, less stream length is required to achieve full cross-sectional mixing. The tracer concentration (C_s) is measured quasi-continuously at the measurement section within the zone of full cross-sectional mixing. The total mass of tracer at the measurement section is

$$M = \int_0^{T_s} C_s Q \, dt \tag{3.10.1}$$

where Q is the stream discharge, T_s is the total time duration of sampling of the tracer plume. Fischer *et al.* (1979) showed that Equation (3.10.1) can be rearranged as:

$$Q = \frac{M}{C_{avg} T_s} \tag{3.10.2}$$

Thus, if the entire plume is sampled, the average concentration of the measured samples (C_{avg}) can be used to estimate the discharge. However, depending on the length of the plume at the measurement section, this could require a long sampling effort T_s.

An alternative slug method approach is to set up two sampling sections at the upstream and downstream ends of the sampling reach within the zone of full cross-sectional mixing. By continuously sampling at each section, the passage of the centroid of the tracer plume at the measurement section is calculated, from which the travel time of the tracer is determined (see Figure 3.10.1). The sampling reach length divided by the travel time yields the reach averaged velocity. The discharge can then be calculated from the product of the reach averaged velocity and the average channel cross-sectional area of the sampling reach. For river reaches with large roughness elements and low relative submergences (*i.e.*, in reaches where the flow is spatially heterogeneous such as mountain streams with step-pool systems) such measurements can be used to estimate the averaged cross-sectional area in the sampling reach based on the determined velocity and the discharge determined through Equation (3.10.2).

The continuous release method involves continuous injection of tracer with concentration C_i at a rate Q_i. Once equilibrium is attained, the tracer concentration (C_s) is

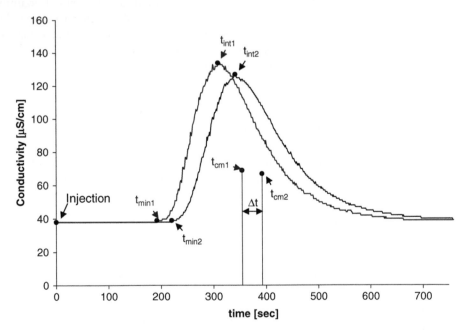

Figure 3.10.1 Velocity and discharge measurement using slug tracer (salt), sampled with 1 Hz frequency at two measurement sections. The time t_{min} indicates the travel time of the tracer to the probes and the time t_{cm} represents the time of passage of the centroid of the plume at the particular measurement section. The measurement section separation distance (10.9 m for this case) divided by Δt yields the reach averaged velocity ($U = 0.29$ m/s). Alternatively, the reach averaged velocity could be determined using the peak values of the curves, t_{int}. The discharge in this case was determined through Equation (3.10.2) after conversion of the measured conductivity to tracer concentration ($Q=93$ L/s). (plots contributed by J. Aberle).

sampled in the zone of full cross-sectional mixing. Fischer *et al.* (1979) demonstrated that the discharge can then be calculated as:

$$Q = Q_i \frac{C_i}{C_s} \tag{3.10.3}$$

The downside of this approach is that depending upon the distance required to obtain full mixing, large quantities of tracer may be required.

REFERENCES

Aberle, J., Koll, K., & Dittrich, A. (2008) Form induced stresses over rough gravel-beds. *Acta Geophys* 56(3), 584–600, doi:510.2478/s11600–11008-10018-x.

Admiraal, D.M., Stansbury, J.S., & Haberman, C.J. (2004) Case study: Particle image velocimetry in a model of Lake Ogallala. *J Hydraul Eng*, 130(7), 599–607.

Adrian, R.J. (1972) Statistics of laser Doppler velocimeter signals: Frequency measurement. *J Phys E*, 5, 91–95.

Adrian, R.J. (1991). Particle-imaging techniques for experimental fluid mechanics. *Annu Rev Fluid Mech*, 23, 261–304.

Adrian, R.J. (2005) Twenty years of particle image velocimetry. *Exp Fluids*, 39(2), 159–169.

Adrian, R.J. & Goldstein, R.J. (1971) Analysis of a laser Doppler anemometer. *J Phys E*, 4(7), 505–511.

Adrian, R.J. & Westerweel, J. (2010) *Particle Image Velocimetry*. Cambridge University Press.

Adrian, R.J. & Yao, C.S. (1987) Power spectra of fluid velocities measured by laser Doppler velocimetry. *Exp Fluids*, 5(1), 17–28.

Agüí, J.C. & Jiménez, J. (1987) On the performance of particle tracking. *J Fluid Mech*, 185, 447–468.

Albrecht, H.-E, Borys, M., Damaschke, N., & Tropea, C. (2003) *Laser Doppler and Phase Doppler Measurement Techniques*. Berlin, Springer-Verlag.

Aleixo, R., Carvalho, E., & Ferreira R. (2016) A toolbox for laser Doppler velocimetry data post-processing. In: Constantinescu, G., Garcia, M., & Hanes, D. (eds.) *River Flow 2016, 11.14 July 2016*, St. Louis, USA. Boca Raton, FL, CRC Press, pp. 339–344.

Alvarez, J.M., Rebolledo, M.A., & Tornos, J. (1989) Velocity resolution in laser Doppler velocimetry experiments by measuring the Fourier transform of the time-interval probability. *Opt Laser Technol* 21(5), 325–327. doi:10.1016/0030-3992(89)90065-0.

Angele, K.P. & Muhammad-Klingmann, B. (2005) A simple model for the effect of peak-locking on the accuracy of boundary layer turbulence statistics in digital PIV. *Exp Fluids*, 38(3), 341–347.

Arndt, R.E.A., Kawakami, D., Wosnik, M., Perlin, M., Duncan, J.H., Admiraal, D.M., & García, M.H. (2007) Hydraulics. In: Tropea, C., Yarin, A.L., & Foss, J.F. (eds.) *Springer Handbook of Experimental Fluid Mechanics*. Berlin, Springer-Verlag. pp. 959–1042.

Astarita, T. (2007) Analysis of weighting windows for image deformation methods in PIV. *Exp Fluids*, 43(6), 859–872.

Atkinson C., Buchmann, N.A., Amili, O., & Soria, J. (2014) On the appropriate filtering of PIV measurements of turbulent shear flows. *Exp Fluids*, 55(1), 1654. doi:10.1007/s00348-013-1654-8.

Aya, S., Fujita, I., & Yagyu, M. (1995) Field observation of flood in a river by video image analysis. In: *Proceedings of Hydraulic Engineering JSCE*, 39, pp. 447–452 (in Japanese).

Aya, S., Kakinoki, S., Aburaya, T., & Fujita, I. (2002) Velocity and turbulence measurement of river flow by LSPIV. In: Ninokata, H., Wada, A., & Tanaka, N. (eds.) *Advances in Fluid Modeling and Turbulence Measurements*. Singapore, World Scientific Publishing, pp. 177–184.

Baker, D.W. & Watkins, D.W. (1967) A phase coherent pulsed Doppler system for cardiovascular measurement. In: *Proceedings of the 20th Alliance for Engineering in Medicine and Biology, Stockholm*, 27, pp. 2–12.

Baud, O., Hager, W.H., & Minor, H.E. (2005) Toward nonintrusive flood discharge measurement. *J Hydraul Eng*, 131(12), 1031–1035.

Bendat, J.S. & Piersol, A.G. (2000) *Random Data: Analysis and Measurement Procedures*. New York, John Wiley & Sons.

Benedict L.H., Nobach H., & Tropea C. (2000) Estimation of turbulent velocity spectra from laser Doppler data. *Meas Sci Technol*, 11, 089. doi:10.1088/0957-0233/11/8/301.

Bertuccioli, L. Roth, G., Katz, J., & Osborn, T. (1999) A submersible particle image velocimetry system for turbulence measurements in the bottom boundary layer. *J Atmos Oceanic Technol*, 16, 1635–1646.

Blanckaert, K. & Lemmin, U. (2006) Means of noise reduction in acoustic turbulence measurements. *J Hydraul Res*, 44(1), 1–37.

Brumley, B.H., Dienes, K.L., Cabrera, R.G., & Terray, E.A. (1996) Broadband Acoustic Doppler Current Profiler. *United States Patent*. Patent Number 5483499. January 9.

Buchhave, P., George Jr., W.K., & Lumley, J.L. (1979) The measurement of turbulence with the laser-Doppler anemometer. *Annu Rev Fluid Mech*, 11, 443–503, doi:10.1146/annurev.fl.11.010179.002303.

Buffin-Bélanger, T. & Roy, G.A. (2005) 1 min in the life of a river: Selecting the optimal record length for the measurement of turbulence in fluvial boundary layers. *Geomorphology*, 68(1–2), 77–94.

Büttner, L. & Czarske, J. (2004) Multi-mode fibre laser Doppler anemometer (LDA) with high spatial resolution for the investigation of boundary layers. *Exp Fluids*, 36(1), 214–216.

Büttner, L., Czarske, J., & Knuppertz, H. (2005) Laser-Doppler velocity profile sensor with submicrometer spatial resolution that employs fiber optics and a diffractive lens. *Appl Opt*, 44, 2274–2280.

Cabrera, R., Deines, K., Brumley, B., & Terray, E. (1987) Development of a practical coherent Doppler current profiler. In: *Proceedings of the Oceans'87. The ocean, an international workplace*. Halifax, NS, Canada, IEEE, pp. 93–97.

Cameron, S.M. (2011) PIV algorithms for open-channel turbulence research: Accuracy, resolution and limitations. *J Hydro Environ Res*, 5(4), 247–262.

Cameron, S., Nikora, V., Albayrak, I., Miler, O., Stewart, M., & Siniscalchi, F. (2013) Interactions between aquatic plants and turbulent flow: A field study using stereoscopic PIV. *J Fluid Mech*, 732, 345–372.

Carr, M.L., Rehmann, C.R., & Gonzalez, J.A. (2005) Comparison between dispersion coefficients estimated from a tracer study and ADCP measurements. In: *Proceedings of the World Water and Environmental Resources Congress 2005* (CD-ROM), ASCE, Reston, VA.

Carter, R.W. & Anderson, I.E. (1963) Accuracy of current meter measurements. *J Hydr Div*, 89 (4), 105–115.

Cea, L., Puertas, J., & Pena, L. (2007) Velocity measurements in highly turbulent free surface flow using ADV. *Exp Fluids*, 42(3), 333–348.

Chang, I.C. (1996) Acousto-Optic Bragg Cell. US patent 5,576,880, Nov. 1996.

Chen, L., Armstrong, C.W., & Raftopoulos, D.D. (1994) An investigation on the accuracy of three-dimensional space reconstruction using the direct linear transformation technique. *J Biomech* 27(4), 493–500.

Cheng, R.T., Burau, J.R., DeRose, J., Barrick, D.E., Teague, C.C., & Lilleboe, P.M. (2007) Measuring two-dimensional surface velocity distribution using two River Sondes. In: *Proceedings of the Hydraulics Measurement and Experimental Methods 2007*, Lake Placid, NY.

Cholemari, M.R. (2007) Modeling and correction of peak-locking in digital PIV. *Exp Fluids*, 42 (6), 913–922.

Cierpka, C., Lütke, B., & Kähler, C.J. (2013) Higher order multi-frame particle tracking velocimetry. *Exp Fluids* 54, 1533.

Cisse, M., Homann, H., & Bec, J. (2013) Slipping motion of large neutrally buoyant particles in turbulence. *J Fluid Mech.*, *735*, RI, doi:10.1017/jfm.2013.490.

Compton, D.A. & Eaton, J.K. (1996) A high-resolution laser Doppler anemometer for three-dimensional turbulent boundary layers. *Exp Fluids*, 22, 111–117.

Comte-Bellot, G. (1976) Hot-Wire Anemometry. *Annu Rev Fluid Mech*, 8, 202–231.

Costa, J.E., Spicer, K.R., Cheng, R.T., Haeni, P., Melcher, N.B., & Thurman, E.M. (2000) Measuring stream discharge by non-contact methods: A Proof-of-concept experiment. *Geophys Res Lett*, 27(4), 553–556.

Costa, J.E., Cheng, R.T., Haeni, F.P., Melcher, N., Spicer, K.R., Hayes, E., Plant, W., Hayes, K., Teague, C., & Barrick, D. (2006) Use of radars to monitor stream discharge by noncontact methods. *Water Resour Res*, 42(7), W07422, doi:10.1029/2005WR004430.

Coulter, R.L. & Kallistratova, M.A. (2004) Two decades of progress in sodar techniques: A review of 11 ISARS proceedings. *Meteorol Atmos Phys*, 85(1), 3–19.

Cowen, E.A. & Monismith, S.G. (1997) A hybrid digital particle tracking velocimetry technique. *Exp Fluids*, 22(3), 199–211.

Creutin, J.D., Muste, M., Bradley, A.A., Kim, S.C., & Kruger, A. (2003). River Gauging Using PIV Technique: Proof of Concept Experiment on the Iowa River. *J Hydrol*, 277(3–4), 182–194.

Crowe, C., Sommerfeld, M., & Tsuji, Y. (1998) *Multiphase Flows with Droplets and Particles*. Boca Raton, FL, CRC-Press.

Csele, M. (2004) *Fundamentals of Light Sources and Lasers*. New York, John Wiley & Sons. ISBN 0-471-47660-9.

Davies, A.G. & Thorne, P.D. (2005) Modeling and measurement of sediment transport by waves in the vortex ripple regime. *J Geophys Res*, 110, C05017, doi:10.1029/2004JC002468.

Derksen, J.J., Doelman, M.S., & Van den Akker, H.E.A. (1999) Three-dimensional LDA measurements in the impeller region of a turbulently stirred tank. *Exp Fluids*, 27, 522–532 doi:10.1007/s003480050376.

Detert, M. & Weitbrecht, V. (2015) A Low-Cost Airborne Velocimetry System: Proof of Concept. *J Hydr Res*, 53(4), 532–539.

Dinehart, R.L. & Burau, J.R. (2005) Averaged indicators of secondary flow in repeated acoustic Doppler current profiler crossing of bends. *Water Resour Res*, W09405, doi:10.1029/2005WR004050.

Dombroski, D.E. & Crimaldi, J.P. (2007) The accuracy of acoustic Doppler velocimeter (ADV) measurements in turbulent boundary-layer flows over a smooth bed. *Limnol Oceanogr Methods*, 5, 23–33.

Doron, P., Bertuccioli, L, Katz, J., & Osborn, T. (2001) Turbulence characteristics and dissipation estimates in the coastal ocean bottom boundary layer from PIV data. *J Phys Oceanogr*, 31, 2108–2134.

Doroudian, B., Bagherimiyab, F., & Lemmin, U. (2010) Improving the accuracy of four-receiver acoustic Doppler velocimeter (ADV) measurements in turbulent boundary layer flows. *Limnol Oceanogr Methods*, 8, 575–591.

Dramais, G., Le Coz, J., Camenen, B., & Hauet, A. (2011) Advantages of a mobile LSPIV method for measuring flood discharges and improving stage–discharge curves. *J Hydro Env Res*, 5(4), 301–312.

Durst, F. (1973) Scattering phenomena and their applications in laser-Doppler anemometry. *Z Angew Math Phys* 24, 619–643.

Durst, F. & Ruck, B. (1987) Effective particle size range in laser-Doppler anemometry. *Exp Fluids*, 5(5), 305–314. doi:10.1007/BF00277709.

Durst, F. & Stevenson, H. (1976) Visual modeling of laser Doppler anemometer signals by Moiré fringes. *Appl Opt*, 15(1), 137–144. doi:10.1364/AO.15.000137.

Durst, F. & Zaré, M. (1974) Removal of pedestals and directional ambiguity of optical anemometer signals. *Appl Opt*, 13(11), 2562–2579. doi:10.1364/AO.13.002562.

Durst, F., Melling, A., & Whitelaw, J.H. (1976) *Principles and Practice of Laser Doppler Anemometry*. London, Academic Press.

Durst, F; Melling, A., & Whitelaw, J.H. (1981) Principles and Practice of Laser Doppler anemometry. 2nd Ed. London, Academic Press.

Dushaw, B.D. & Colosi, J.A. (1998) Ray Tracing for Ocean Acoustic Tomography (No. APL-UW-TM-3-98). Applied Physics Laboratory, Seattle, University of Washington.

Eckstein, A. & Vlachos, P.P. (2009) Assessment of advanced windowing techniques for digital particle image Velocimetry (DPIV). *Meas Sci Technol*, 20, 075402.

Elkins C.J. & Alley, M.T. (2007) Magnetic resonance velocimetry: Applications of magnetic resonance imaging in the measurement of fluid motion. *Exp Fluids*, 43(6), 823–858.

Elgar, S., Raubenheimer, B., & Guza, R.T. (2005) Quality control of acoustic Doppler velocimeter data in the surf zone. *Meas Sci Technol*, 16(10), 1889–1893.

Elsinga, G.E., Scarano, F., Wieneke, B., & van Oudheusden, B.W. (2006) Tomographic particle image velocimetry. *Exp Fluids*, 41(6), 933–947.

Engel, P. (1999). Current Meter Calibration Strategy. *J Hydraul Eng*, 125(12), 1306–1308.

Essen, H.H. (1994) On the capability of an upward-looking ADCP deployed in the Iceland-Faeroe frontal area. *Ocean Dyn*, 46(3), 211–228.

Farmer, W.M. (1976) Sample space for particle size and velocity measuring interferometers. *Appl Opt* 15(8), 1984–1989. doi:10.1364/AO.15.001984.

Faxén, H. (1922) Der Widerstand gegen die Bewegung einer starren Kugel in einer zähen Flüssigkeit, die zwischen zwei parallelen ebenen Wänden eingeschlossen ist. *Annalen der Physik*, 373(10), 89–119. doi:10.1002/andp.19223731003.

Fernando, H.J.S., Princevac, M., & Calhoun, R.J. (2007) Atmospheric measurements. In: Tropea, C., Yarin, A.L., & Foss, J.F. (eds.) *Springer Handbook of Experimental Fluid Mechanics*. Berlin, Springer. pp. 1157–1178.

Ferreira, R.M.L. (2011) Turbulent flow hydrodynamics and sediment transport. Laboratory research with LDA and PIV. In Rowinski, P. (ed.). *Experimental Methods in Hydraulic Research, GeoPlanet-Earth and Planetary Sciences*. Berlin, Springer Verlag. pp. 67–111. ISBN: 978-3-642-17474-2, doi:10.1007/978-3-642-17475-9_4

Ferreira R.M.L., Franca, M.J., Leal, J.G.A.B., & Cardoso, A.H. (2012) Flow over rough mobile beds: Friction factor and vertical distribution of the longitudinal mean velocity. *Water Resour Res*, 48, W05529, doi:10.1029/2011WR011126.

Fischer, H.B., List, E.J., Koh, R.C.Y, Imberger, J., & Brooks, N.H. (1979). *Mixing in Inland and Coastal Waters*. New York, Academic Press.

Foreman, J.W., George, E.W., & Lewis, R.D. (1965) Measurement of localized flow velocities in gases with a laser Doppler flowmeter. *Appl Phys Lett*, 7(4), 77–78. doi:10.1063/1.1754319.

Forliti, D.J., Strykowski, P.J., & Debatin, K. (2000) Bias and precision errors of digital particle image velocimetry. *Exp Fluids*, 28, 436–447.

Foucaut, J.M., Miliat, B., Perenne, N., & Stanislas, M. (2004) Characterization of different PIV algorithms using the EUROPIV synthetic image generator and real images from a turbulent boundary layer. In: *Particle Image Velocimetry: Recent Improvements. Proceedings of the EUROPIV 2 Workshop, 31. March 31–1 April 2003, Zaragoza, Spain*. Berlin, Springer-Verlag.

Fouras, A., Dusting, J., Lewis, R., Hourigan, K. (2007) Three-dimensional synchrotron x-ray particle image velocimetry. *J Appl Phys*, 102, 064916.

Franca, M., & Lemmin, U. (2006) Eliminating velocity aliasing in acoustic Doppler velocity profiler data. *Meas Sci Technol*, 17(2), 313–322.

Fujita, I., & Komura, S. (1994) Application of video image analysis for measurements of river-surface flows. *Annual J Hydraul Eng JSCE*, 38, 733–738, (in Japanese).

Fujita, I. & Aya, S. (2000) Refinement of LSPIV Technique for Monitoring River Surface Flows. *Building Partnerships*, 1–9, doi:10.1061/40517(2000)312.

Fujita, I. & Hino, T. (2003) Unseeded and seeded PIV measurements of river flows videotaped from a helicopter. *J Visual*, 6(3), 245–252.

Fujita, I. & Kunita, Y. (2011) Application of aerial LSPIV to the 2002 flood of the Yodo River using a helicopter mounted high density video camera. *J Hydro Env Res*, 5(4), 323–331.

Fujita, I., Aya, S., & Deguchi, T. (1997) Surface velocity measurement of river flow using video images of an oblique angle. In: *Proceedings of the 27th Congress of IAHR, Theme B, Vol.1, San Francisco, CA*, pp. 227–232.

Fujita, I., Muste, M., & Kruger, A. (1998) Large-scale particle image velocimetry for flow analysis in hydraulic engineering applications. *J Hydraul Res*, 36(3), 397–414.

Fujita, I., Watanabe, H., & Tsubaki, R. (2007a) Development of a non-intrusive and efficient flow monitoring technique: The Space Time Image Velocimetry (STIV). *Int J River Basin Man*, 5(2), 105–114.

Fujita, I., Tsubaki, R., & Deguchi, T. (2007b) PIV measurement of large-scale river surface flow during flood by using a high resolution video camera from a helicopter. In: *Proceedings of the Hydraulic Measurements and Experimental Methods*, ASCE-IAHR, Lake Placid, NY. pp. 344–349.

Fujita, I., Ando, T., Tsutsumi, S., & Hara, H. (2009) Efficient space-time image analysis of river surface pattern using two dimensional fast Fourier transformation. In: *Proceedings of the 33rd IAHR Congress*, pp. 2272–2279.

Fujita, I., Kosaka, Y., Honda, M., & Yorozuya, A. (2012) Tracking of river surface features by space time imaging. In: *Proceedings of the 15th International Symposium on Flow Visualization (ISFV15)*, ISFV15–045 S23.

Fujita, I., Kosaka, Y., Honda, M., Yorozuya, A., & Motonaga, Y. (2013) Day and night measurements of snow melt floods by STIV with a far infrared camera. In: *Proceedings of the 35th IAHR Congress*, A10458.

Fujita, I., Asami, K., & Kumano, G. (2014) Evaluation of 2D river flow simulation with the aid of image-based field velocity measurement techniques. In: Schleiss, A., De Cesare, G., Franca, M. J., & Pfister, M. (eds.) *Proceedings of the River Flow 2014, 3–5 Sept. 2014*, Lausanne, Switzerland, pp. 1969–1977.

Fulford, J.M. (1995) Effects of pulsating flow on current meter performance. In: *Proceedings of the 1st International Conference Water Resources Engineering*, ASCE, Reston, Va.

Fulford, J.M. (2001) Accuracy and consistency of water-current meters. *J Am Water Resour Assoc*, 37(5), 1215–1224.

Fulford, J.M., Thibodeaux, K.G., & Kaehrle, W.R. (1994) Comparison of current meters used for stream gaging. In: *Proceedings of the Symposium on Fundamentals and Advancements in Hydraulic Measurements and Experimentation*, Buffalo NY USA, August 1994.

Ganju, N.K., Dickhudt, P.J., Thomas, J.A., Borden, J., Sherwood, C.R., Montgomery, E.T., Twomey, E.R., & Martini, M.A. (2011) Summary of oceanographic and water-quality measurements in West Falmouth Harbor and Buzzards Bay, Massachusetts. U.S. Geological Survey, Open-File Report 2011–1113. [Online] Available from: http://pubs.usgs.gov/of/2011/1113/ [Accessed 21st February 2017].

Garbini, J.L., Forster, F.K., & Jorgensen, J.E. (1982) Measurement of fluid turbulence based on pulsed ultrasound technique (part I and part II). *J Fluid Mech*, 118, 445–505.

García, C.M., Oberg, K., & García, M. (2007a) ADCP Measurements of gravity currents in the Chicago River, Illinois. *J Hydraul Eng*, 133(12), 1356–1366.

García, C.M., Mariano, I.C., Nino, Y., & Garcia, M.H. (2007b) Closure to 'Turbulence measurements with acoustic Doppler velocimeters', by Garcia, C.M., Mariano, I.C., Nino. Y. & Garcia, M.H. *J Hydraul Eng*, 133(11), 1289–1292.

García, C.M., Tarrab L., Oberg, K., Szupiany R., & Cantero, M.I. (2012) Variance of discharge estimates sampled using acoustic Doppler current profilers from moving platforms. *J Hydraul Eng*, 138(8), 684–694.

George, W.K. & Lumley J.L. (1973) The laser-Doppler velocimeter and its application to the measurement of turbulence. *J Fluid Mech*, 60(2), 321–362, doi:10.1017/S0022112073000194.

Goldstein, R.J. (1996) *Fluid Mechanics Measurements*. 2nd Edition. Washington DC, Taylor & Francis.

Gonzalez-Castro, J.A. & Muste, M. (2007) Framework for estimating uncertainty of adcp measurements from a moving boat by standardized uncertainty analysis. *J Hydraul Eng*, 133(12), 1390–1410.

Goring, D.G. & Nikora, V.I. (2002) Despiking acoustic Doppler velocimeter data. *J Hydraul Eng*, 128(1), 117–126.

Gornowicz, G.G. (1997) Continuous-field image-correlation velocimetry and its application to unsteady flow over an airfoil. *Eng thesis*. California Institute of Technology [Online] Available from: http://thesis.library.caltech.edu/3021/ [Accessed 21st February 2017].

Grainger, R.G., Lucas, J., Thomas, G.E., & Ewan, G. (2004) The calculation of Mie derivatives. *Appl Opt*, 43(28), 5386–5393.

Grant, G.R. & Orloff, K.L. (1973) Two-color dual-beam backscatter laser Doppler velocimeter. *Appl Opt*, 12(12), 2913–2916, doi:10.1364/AO.12.002913.

Guerrero, M. & Lamberti, A. (2011) Flow field and morphology mapping using ADCP and multibeam techniques: Survey in the Po River. *J Hydraul Eng*, 137(12), 1576–1587. doi:10.1061/(ASCE)HY.1943-7900.0000464.

Guerrero M., Szupiany R.N., & Amsler M. (2011) Comparison of acoustic backscattering techniques for suspended sediments investigation. *Flow Meas Inst* 22, 5, 392–401. doi:10.1016/j.flowmeasinst.2011.06.003.

Guezennec, Y.G., Brodkey, R.S., Trigui, N., & Kent, J.C. (1994) Algorithms for fully automated three-dimensional particle tracking velocimetry. *Exp Fluids*, 17(4), 209–219.

Gui, L.C. & Merzkirch, W. (1996) A method of tracking ensembles of particle images. *Exp Fluids*, 21(6), 465–468.

Hassan, Y.A. & Canaan, R.E. (1991) Full-field bubbly flow velocity measurements using a multiframe particle tracking technique. *Exp Fluids*, 12, 49–60.

Hauet, A., Creutin, J.-D., & Belleudy, P. (2008) Sensitivity study of large-scale particle image velocimetry measurement of river discharge using numerical simulations, *J Hydrol*, 349(1–2), 178–190. doi:10.1016/j.jhydrol.2007.10.062.

Hay, A.E. (1991) Sound scattering from a particle laden turbulent jet. *J Acoust Soc Am*, 78, 2055–2074.

Hay, A.E. & Sheng, J. (1992) Vertical profiles of suspended sand concentration and size from multifrequency acoustic backscatter. *J Geophys Res*, 97(C10), 15661–15677.

Heindel, T.J. (2011) A review of x-ray flow visualization with applications to multiphase flows. *J Fluids Eng*, 133(7), 074001.

Hinze, J.O. (1975) *Turbulence*. 2nd Edition. New York, McGraw-Hill.

Hjelmfelt, A.T. & Mockros, L.F. (1996) Motion of discrete particles in a turbulent fluid. *Appl Sci Res*, 16(1), 149–161.

Hoitink, A.J.F., Buschman, F.A., & Vermeulen, B. (2009) Continuous measurements of discharge from a horizontal acoustic Doppler current profiler in a tidal river. *Water Res Res*, 45, W11406, doi:10.1029/2009WR007791.

Howarth, M.J. (2002) Estimates of Reynolds and bottom stress from fast sample ADCPs deployed in continental shelf seas. In: *Proceedings of the Hydraulic Measurements & Experimental Methods* (CD-ROM), Reston, Va. ASCE.

Hubbard, E.F., Schwarz, G.E., Thibodeaux, K.G., & Turcios, L.M. (2001) Price current-meter standard rating development by the U.S. Geological Survey. *J Hydraul Eng*, 127(4), 250–257.

Huisman, S.G., van Gils, D.P.M., & Sun, C. (2012) Applying laser Doppler anemometry inside a Taylor–Couette geometry using a ray-tracer to correct for curvature effects. *Eur J Mech B-Fluids*, 36, Nov–Dec 2012, 115–119. doi:10.1016/j.euromechflu.2012.03.013.

Hurdoware-Castro, D., Tsanis, I., & Simanovskis, I. (2007) Application of a three-dimensional wind driven circulation model to assess the locations of new drinking water intakes in Lake Ontario. *J Great Lakes Res*, 33, 232–252.

Hurther, D. & Lemmin, U. (1998) A constant beam width transducer for 3D acoustic Doppler profile measurements in open-channel flows. *Meas Sci Technol*, 9(10), 1706–1714.

Hurther, D. & Lemmin, U. (2001) A correlation method for turbulence measurements with a 3D acoustic Doppler velocity profiler. *J Atmos Oceanic Technol*, 18(3), 446–458.

Hurther, D. & Lemmin, U. (2003) Turbulent particle and momentum flux statistics in suspension flow. *Water Resour Res*, 39, 1139, doi:10.1029/2001WR001113, 5.

Hurther, D. & Lemmin, U. (2008) Improved turbulence profiling with field-adapted acoustic Doppler velocimeters using a bifrequency Doppler noise suppression method. *J Atmos Oceanic Technol*, 25(3), 452–463.

Hurther, D., Thorne, P.D., Bricault, M., Lemmin, U., & Barnoud, J-M. (2011) A multi-frequency acoustic concentration and velocity profiler (ACVP) for boundary layer measurements of fine-scale flow and sediment transport processes. *Coast Eng*, 58(7), 594–605.

Ishigaki, T., Shiono, K., & Rameshwaran, P. (2002) PIV and LDA measurements of secondary flow in a meandering channel for overbank flow. *J Visual*, 5(2), 153–159.

Jackson, P., García, C.M., Oberg, K., Johnson K., & García, M. (2008) Density currents in the Chicago River: Characterization, effects on water quality, and potential sources. *Sci Total Environ*, 401(1–3), 130–143, doi:10.1016/j.scitotenv.2008.04.011.

Jamieson, E.C., Rennie, C.D., Jacobson, R.B., & Townsend, R.D. (2011a) 3-D flow and scour near a submerged wing dike: ADCP measurements on the Missouri River. *Water Resour Res*, 47, W07544, doi:10.1029/2010WR010043.

Jamieson, E.C., Rennie, C.D., Jacobson, R.B., & Townsend, R.D. (2011b) Evaluation of ADCP apparent bed load velocity in a large sand-bed river: Moving versus stationary boat conditions. *J Hydraul Eng*, 137(9), 1064–1071, doi:10.1061/(ASCE)HY.1943-7900.0000373.

Johnston, T.F. (1998) Beam propagation (M²) measurement made as easy as it gets: The four cuts method. *Appl Opt*, 37(21), 4840–4850.

Joseph, S. (2003). *Uncertainty Analysis of a particle tracking algorithm: Developed for super-resolution Particle Image Velocimetry*. [Master of Science Thesis], Canada, The University of Saskatchewan.

Katija, K. & Dabiri, J. (2008) In situ field measurements of aquatic animal-fluid interactions using a self-contained underwater velocimetry apparatus (SCUVA). *Limonol Oceanogr: Methods*, 6(4), 162–171.

Kaufmann, S.L. & Fingerson, L.M. (1985) Fiber optics in LDV applications. In: *Proceedings of the International Conference on Laser Anemometry—Advances and Applications 16–18 December 1985. Manchester, UK*

Kawanisi, K. (2004) Structure of turbulent flow in a shallow tidal estuary. *J Hydraul Eng*, 130(4), 360–370.

Kawanisi, K., Watanabe, S., Kaneko, A., & Abe, T. (2009) River acoustic to-mography for continuous measurement of water discharge. In: *Proceedings of the 3rd International Conference and Exhibition on Underwater Acoustic Measurements: Technologies and Results. Hellas Foundation for Research and Technology Nafplion*. 2, pp. 613–620.

Kawanisi, K., Razaz, M., Kaneko, A., & Watanabe, S. (2010) Long-term measurement of stream flow and salinity in a tidal river by the use of the fluvial acoustic tomography system. *J Hydrol*, 380(1), 74–81.

Kim, S.C., Friedrichs, C.T., Maa, J.P.Y., & Wright, L.D (2000) Estimating bottom stress in a tidal boundary layer from acoustic Doppler velocimeter data. *J Hydraul Eng*, 126(6), 399–406.

Kim, Y., Muste, M., Hauet, A., Krajewski, W., Kruger, A., & Bradley, A. (2008) Real-time stream monitoring using mobile large-scale particle image velocimetry. *Water Resour Res*, 44, W09502, doi:10.1029/2006 WR005441.

Kim, Y. (2006). Uncertainty analysis for non-intrusive measurement of river discharge using image velocimetry. [PhD Thesis] Iowa City, IA, The University of Iowa.

Kim, Y., Muste, M., Hauet, A., Bradley, A., Weber, L., & Koh, D. (2007) Image velocimetry for discharge measurements in streams. In: *Proceedings of the XXXII IAHR Congress, Venice, Italy*.

Kirchner, J.W. (2005) Aliasing in 1/f(alpha) noise spectra: Origins, consequences, and remedies. *Phys Rev E Stat Nonlin Soft Matter Phys*, 71(6 Pt 2):066110.

Koca, K., Noss, C., Anlager, C., Brand, A., & Lorke, A. (2017) Performance of the Vectrino Profiler at the sediment–water interface. *J Hydraul Res*, doi:10.1080/00221686.2016.1275049.

Kostaschuk, R., Villard, P., & Best, J. (2004) Measuring velocity and shear stress over dunes with acoustic Doppler profiler. *J Hydraul Eng*, 130(9), 932–936.

Lading, L. (1971) Differential Doppler Heterodyning Technique. *Appl Opt*, 10(8), 1943–1949, doi:10.1364/AO.10.001943.

Lane, S.N., Biron, P.M., Bradbrook, K.F., Butler, J.B., Chandler, J.H., Crowell, M.D., McLelland, S.J., Richards, K.S., & Roy, A.G. (1998) Three-dimensional measurement of river channel flow processes using acoustic Doppler velocimetry. *Earth Surf Proc Land*, 23 (13), 1247–1267.

Lang, S. & McKeogh, E. (2011) LIDAR and SODAR measurements of wind speed and direction in upland terrain for wind energy purposes. *Remote Sens*, 3, 1871–1901; doi:10.3390/rs3091871.

Latosinski, F.G., Szupiany, R.N., García C.M., Guerrero, M., & Amsler, M.L. (2014) Estimation of concentration and load of suspended bed sediment in a large river by means of acoustic Doppler technology. *J Hydraul Eng*, 140(7), doi:10.1061/(ASCE)HY.1943-7900.0000859.

Latosinski, F., Szupiany, R., Guerrero, M., Amsler, M., & Vionnet, C. (2017) The ADCP's bottom track capability for bedload prediction: Evidence on method reliability from sandy river applications. *Flow Meas Instrum*, 54, 124–135. doi; 10.1016/j.flowmeasinst.2017.01.005.

Lavoie, P., Avallone, G., De Gregorio, F., Romano, G.P., & Antonia, R.A. (2007) Spatial resolution of PIV for the measurement of turbulence. *Exp Fluids*, 43(1), 39–51.

Le Coz, J., Hauet, A., Pierrefeu, G., Dramais, G., & Camenen, B. (2010) Performance of image-based velocimetry (LSPIV) applied to flash-flood discharge measurements in Mediterranean rivers. *J Hydrol*, 394(1–2), 42–52.

Lecordier, B. & Westerweel, J. (2004) The EUROPIV Synthetic Image Generator (S.I.G.). In: *Particle Image Velocimetry: Recent Improvements. Proceedings of the EUROPIV 2 Workshop, 31. March 31–1 April 2003, Zaragoza, Spain*. Berlin, Springer-Verlag.

Lehmann, B. (1968) Geschwindigkeitsmessung mit Laser- Doppler-Anemometer Verfahren. *Wissenschaftlicher Bericht AEG-Telefunken*, 41, 141–145.

Lehmann, B., Nobach, H., & Tropea, C. (2002) Measurement of acceleration using the laser Doppler technique. *Meas Sci Technol*, 13(9), 1367–1381.

Lemmin, U. & Rolland, T. (1997) Acoustic velocity profiler for laboratory and field studies. *J Hydraul Eng*, 123(12), 1089–1098.

Lewis, R.D., Foreman, J.W. Jr., Watson, H.J., & Thornton, J.R (1968) Laser Doppler velocimeter for measuring flow-velocity fluctuations. *Phys Fluid*, 11, 433–435.

Lhermitte, R. (1999) Tidal flow velocity and turbulence measurements. Internal publication, Rosenstiel School of Marine and Atmospheric Science (RSMAS), University of Miami, USA.

Lhermitte, R. & Lemmin, U. (1990) Probing water turbulence by high frequency Doppler sonar. *Geophys Res Letters*, *17*(10), 1549–1552.

Lhermitte, R. & Lemmin, U. (1993) Turbulent microstructures observed by sonar. *Geophys Res Letters*, *20*(9), 823–826.

Lhermitte, R. & Lemmin, U. (1994) Open-channel flow and turbulence measurement by high resolution Doppler sonar. *J Atmos Oceanic Technol*, *11*(5), 1295–1308.

Lhermitte, R. & Serafin, R. (1984) Pulse-to-pulse coherent Doppler signal processing techniques. *J Atmos Oceanic Technol*, *1*(4), 293–308.

Li, E.-B. & Wang, S.-K. (1994) LDA system with both tracking and counting. *Flow Meas Instrum*, *5*(1), 59–60. doi:10.1016/0955-5986(94)90009-4

Liao, Q., Bootsma, H., Xiao, J., Klump, J., Hume, A., Long, M., & Berg, P. (2009) Development of an in situ underwater particle image velocimetry (UWPIV) system. *Limnol Oceanogr Methods*, *7*(2), 169–184.

Liao, Q., Wang, B., & Wang, P. (2015) In situ measurement of sediment resuspension caused by propeller wash with an underwater particle image velocimetry and an acoustic Doppler velocimeter. *Flow Meas Instrum*, *41*(3), 1–9.

Limeburner, R., Beardsley, R.C., Soares, I.D., Lentz, S.J., Candela, J. (1995). Lagrangian flow observations of the Amazon River discharge into the North Atlantic. *J Geophys Res—Oceans*, *100*(C2), 2401–2415.

Lipscomb, S.W. (1995) Quality assurance plan for discharge measurements using broad-band acoustic Doppler profilers. *U.S. Geological Survey Open-File Report 95–701*, 7.

Lloyd, P.M., Ball, D.J., & Stansby, P.K. (1995) Unsteady surface-velocity field measurement using particle tracking velocimetry. *J Hydraul Res*, *33*(4), 519–534.

Lohrmann, A., Cabrera, R., & Kraus, N.C. (1994) Acoustic Doppler velocimeter (ADV) for laboratory use. In: Pugh, C.A: (ed.) *Proceedings of the Symposium on fundamentals and advancements in hydraulic measurements and experimentation*. Reston, VA, ASCE. pp. 351–365.

Lohrmann, A., Cabrera, R., & Kraus, N.C. (1995) Direct measurements of Reynolds stress with an acoustic Doppler velocimeter. In: *Proceedings of the IEEE 5th Conference On current measurements*. St. Petersburg, FL., USA, IEEE Oeng. Soc., pp. 205–210.

López, F. & Garcia, M.H. (1999) Wall similarity in turbulent open-channel flow. *J Eng Mech*, *125*(7), 789–796.

Lourenco, A. & Krothapalli, A. (1995) On the accuracy of velocity and vorticity measurements with PIV. *Exp Fluids*, *18*, 421–428.

Luznik, L., Gurka, R., Nimmo Smith, W., Zhu, W., Katz, J., & Osborn, T. (2007) Distribution of energy spectra, Reynolds stresses, turbulence production, and dissipation in a tidally driven bottom boundary layer. *J Phys Oceanogr*, *37*, 1527–1550.

Ma, Y., Varadan, V.K., & Varadan, V.V. (1987) Acoustic response of sediment particles in the nearfield of high-frequency transducers. *IEEEE Trans Ultrason Ferroelctr Frequ Contr*, *34*(1), 3–7.

Maas, H.G., Gruen, A., & Papantoniou, D. (1993) Particle Tracking Velocimetry in three-dimensional flows. Part 1. Photogrammetric determination of particle coordinates. *Exp Fluids*, *15*(2), 133–146.

MacVicar, B.J., Beaulieu, E, Champagne, V., & Roy, A.G. (2007) Measuring water velocity in highly turbulent flows: Field tests of an electromagnetic current meter (ECM) and an acoustic Doppler velocimeter (ADV). *Earth Surf Proc Land*, *32*(9), 1412–1432.

Malik, N.A., Drakos, T., & Papantoniou, D. (1993) Particle Tracking Velocimetry in three-dimensional flows. Part 2. Particle tracking. *Exp Fluids*, *15*(4), 279–294.

Maxey, M.R. & Riley, J.J. (1983) Equation of motion for a small rigid sphere in a nonuniform flow. *Phys Fluids*, *26*(4), 883–889, doi:10.1063/1.864230.

Mayo W.T.Jr, Shay, M.T., & Ritter S. (1974) Digital estimation of turbulence power spectra from burst counter LDV data. In: *Proceedings of the 2nd International Workshop on Laser Velocimetry (Purdue University, 1974)* pp 16–26.

McKeon, B.J., Comte-Bellot, G., Foss, J.F., et al. (2007) Velocity, vorticity, and Mach number. In: Tropea, C., Yarin, A.L., & Foss, J.F. (eds.) *Springer Handbook of Experimental Fluid Mechanics*. Berlin, Springer-Verlag. pp. 215–472.

McLelland, S.J. & Nicholas, A.P. (2000) A new method for evaluating errors in high-frequency ADV measurements. *Hydrol Processes, 14*(2), 351–366.

Medwin, H. (1975) Speed of sound in water: A simple equation for realistic parameters. *J Acoust Soc Am, 58*(6), 1318–1319.

Mei, R. (1996) Velocity fidelity of flow tracer particles. *Exp Fluids, 22*(1), 1–13.

Melling, A. (1997) Tracer particle and seeding for particle image velocimetry. *Meas Sci Technol, 8*(12), 1406–1416.

Meselhe, E., Peeva, T., & Muste, M. (2004) Large scale particle image velocimetry for low velocity and shallow water flows. *J Hydraul Eng, 130*(9), 937–940.

Miles, P.C. & Witze P.O. (1994) Fringe field quantification in an LDV probe volume by use of a magnified image. *Exp Fluids, 16*, 330–335.

Monin, A.S. & Yaglom, A. (1971) *Statistical Fluid Mechanics: Mechanics of Turbulence.* Vol I. M. Lumley, J. (Ed.) MIT Press.

Moore, S.A., Le Coz, J., Hurther, D., & Paquier, A. (2012) On the application of horizontal ADCPs to suspended sediment transport surveys in rivers. *Cont Shelf Res, 46*(1), 50–63.

Mueller, D.S. (2002) Use of acoustic Doppler instruments for measuring discharge in streams with appreciable sediment transport. In: *Proceedings of the Hydraulic Measurements & Experimental Methods* (CD-ROM), Reston, Va., ASCE.

Mueller, D.S. & Oberg, K.A. (2011) Discussion of Near-Transducer Errors in ADCP Measurements: Experimental Findings. *J Hydraul Eng, 137*(8), 863–866.

Mueller, D., Abad, J., García C.M., Gartner, J., García, M., & Oberg, K. (2007). Errors in acoustic Doppler profiler velocity measurements caused by flow disturbance. *J Hydraul Eng, 133*(12), 1411–1420.

Mueller, D.S., & Wagner, C.R. (2009) Measuring discharge with acoustic Doppler current profilers from a moving boat: U.S. Geological Survey Techniques and Methods 3A–22, 72. [Online] Available from http://pubs.water.usgs.gov/tm3a22 [Accessed February 4, 2017].

Mueller, D.S., Wagner, C.R., Rehmel, M.S., Oberg, K.A., & Rainville, F. (2013) Measuring discharge with acoustic Doppler current profilers from a moving boat. (ver. 2.0, December 2013) U.S. Geological Survey Techniques and Methods 3A–22, 95. [Online] Available from http://pubs.water.usgs.gov/tm3a22 [Accessed 15th February 2017].

Müller, E., Nobach, H., & Tropea, C. (1994) LDA signal reconstruction: Application to moment and spectral estimation. In: *Proceedings of the 7th International Symposium on Application of Laser Techniques to Fluid Mechanics, July 11th–14th, Lisbon, Portugal*, paper 23.2.

Muste, M., Yu, K., & Spasojevic, M. (2004a) Practical aspects of ADCP data use for quantification of mean river flow characteristics; Part I: Moving-vessel measurements. *Flow Meas Instrum, 15*(1), 17–28.

Muste, M., Xiong, Z., Schöne, J., & Li, Z. (2004b) Validation and extension of image velocimetry capabilities for flow diagnostics in hydraulic modeling. *J Hydraul Eng, 130*(3), 175–185.

Muste, M., Fujita, I., & Hauet, A. (2008) Large-scale particle image velocimetry for measurements in riverine environments, *Water Resour Res, 40*(4), doi:10.1029/2008WR006950.

Muste, M., Kim, D., & González-Castro, J. (2010) Near-transducer errors in ADCP measurements: Experimental findings. *J Hydraul Eng, 136*(5), 275–289.

Muste, M., Kim, D., & Merwade, V. (2012) Modern digital instruments and techniques for hydrodynamic and morphologic characterization of streams, Chapter 24 in Gravel-bed rivers:

Processes, tools, environments, Church, M., Biron, P., Roy, A.G. (Eds.), John Wiley & Sons, LTD., ISBN 978-0-470-68890-8, Chichester, UK, 315–342.

Neti, S. & Clark, W. (1979) One-axis velocity component measurement with laser velocimeters. *AIAA J*, 17(9), 1013–1015, doi:10.2514/3.61267.

Nezu, I., & Nakagawa, H. (1993) *Turbulence in Open-Channel Flows*. IAHR Monograph, Balkema.

Nezu I. & Rodi W. (1985) Experimental study on secondary currents in open channel flows. In: *Proceedings of the 21st IAHR congress*, Melbourne, A125–32.

Nezu, I., Kadota, A., & Nakagawa, H. (1997) Turbulent structure in unsteady depth-varying open-channel flows. *J Hydraul Eng*, 123(9), 752–763. doi:10.1061/(ASCE)0733–9429 (1997)123:9(752).

Niiler, P.P., Sybrandy, A.S., Bi, K., Poulain, P.M., & Bitterman, D. (1995). Measurements of the water-following capability of holey-sock and TRISTAR drifters. *Deep-Sea Research*, 42(11/ 12):1951–1964.

Nikora, V.I. & Goring, D.G. (1998). ADV measurements of turbulence: Can we improve their interpretation. *J Hydraul Eng*, 124(6), 630–634.

Nikora, V.I. & Goring, D.G. (2000) Flow turbulence over fixed and weakly mobile gravel beds. *J Hydr Eng*, 126(9), 679–690.

Nimmo Smith, W., Atsavapranee, P., Katz, J., & Osborn, T. (2002) PIV measurements in the boundary layer of the coastal ocean. *Exp Fluids*, 33, 962–971.

Nimmo Smith, W., Katz, J., & Osborn, T. (2005) On the structure of turbulence in the bottom boundary layer of the coastal ocean. *J Phys Oceanogr*, 35, 72–93.

Nobach, H. (2008) Messung von Teilchenbeschleunigungen mit dem Laser-Doppler-Anemometer. In: *Proceedings of the Messtechnisches Symposium des Arbeitskreises der Hochschullehrer für Messtechnik e.V. Dresden*, 264–275.

Nobach, H. & Bodenschatz, E. (2009) Limitations of accuracy in PIV due to individual variation of particle image intensities. *Exp Fluids*, 47(1), 27–38.

Nobach, H. & Bodenschatz, E. (2011) Acceleration measurement with a laser Doppler system. In: *Proceedings of the 13th European Turbulence Conference, September 12–15, 2011, Warsaw*, Poland.

Nobach, H., Müller, E., & Tropea, C. (1996) Refined reconstruction techniques for LDA analysis. In: *Proceedings of the 7th International Symposium on Application of Laser Techniques to Fluid Mechanics, Lisbon*.

Nobach, H., Müller, E., & Tropea, C. (1998a) Efficient estimation of power spectral density from laser Doppler anemometer data. *Exp Fluids*, 24, 5–6, 499–509.

Nobach, H., Müller, E., & Tropea, C. (1998b) Correlation estimator for two-channel, non-coincidence Laser-Doppler-Anemometer. In: *Proceedings of the 9th International Symposium on Application of Laser Techniques to Fluid Mechanics, July 13–16, 1998, Lisbon, Portugal*.

Nobach, H. & Tropea, C. (2007) Fundamentals of Data Processing, In: Tropea, C., Yarin, A.L. & Foss, J.F. (eds.) *Springer Handbook of Experimental Fluid Mechanics*. Berlin, Springer-Verlag. pp. 1339–1418

Nogueira, J., Lecuona, A., & Rodríguez, P.A. (1999) Local field correction PIV: On the increase of accuracy of digital PIV systems. *Exp Fluids*, 27(2), 107–116.

Nogueira, J., Lecuona, A., & Rodríguez, P.A. (2001) Identification of a new source of peak locking, analysis and its removal in conventional and super-resolution PIV techniques. *Exp Fluids*, 30(3), 309–316.

Nogueira, J., Lecuona, A., Nauri, S., Legrand, M., & Rodríguez, P.A. (2009) Multiple Δt strategy for particle image velocimetry (PIV) error correction, applied to a hot propulsive jet. *Meas Sci Technol*, 20(7), 074001.

Nystrom, E.A., Oberg, K.A., & Rehmann, C.R. (2002) Measurement of turbulence with acoustic Doppler current profilers – Sources of error and laboratory results. In: *Proceedings of the Hydraulic Measurements & Experimental Methods 2002* (CD-ROM), Reston, Va., ASCE.

Oberg, K. & Mueller, D. (2007) Validation of streamflow measurements made with acoustic Doppler current profilers. *J Hydraul Eng*, *133*(12), 1421–1432.

Oberg, K., Morlock, S., & Caldwell, W. (2005) Quality-assurance plan for discharge measurements using acoustic Doppler current profilers. *U.S. Geological Survey Scientific Investigations Report 2005–5183*.

Olson, D.B., Kourafalou, V.H., Johns, W.E., Samuels, G., & Veneziani, M. (2007) Aegean surface circulation from a satellite-tracked drifter array. *J Phys Ocean*, *37*, 1898–1017.

Ooij, P., Guédon, A., Poelma C., Schneiders, J., Rutten, M.C.M., Marquering, H.A., Majoie, C. B., Vanbavel, E., & Nederveen, A.J. (2011) Complex flow patterns in a real-size intracranial aneurysm phantom: Phase contrast MRI compared with particle image velocimetry and computational fluid dynamics. *NMR Biomed*, *25*(1), 14–26.

Orloff, K.L. & Olson, L.E. (1982) High-resolution LDA measurements of Reynolds stress in boundary layers and wakes. *AIAA J*, *20*(5), 624–631. doi:10.2514/3.51120

Ouellette, N.T., Xu, H., & Bodenschatz, E. (2006) A quantitative study of three-dimensional Lagrangian particle tracking algorithms. *Exp Fluids*, *40*(2), 301–313. doi:10.1007/s00348-005-0068-7

Parsons, D.R., Best, J.L., Lane, S.N., Orfeo, O., Hardy, R.J., & Kostaschuk, R. (2007) Form roughness and the absence of secondary flow in a large confluence-diffluence, Rio Paraná, Argentina. *Earth Surf Proc Land*, *32*(1), 155–162, doi:10.1002/esp.1457.

Pedocchi, F. & Garcia, M.H. (2012) Acoustic measurement of suspended sediment concentration profiles in an oscillatory boundary layer. *Cont Shelf Res*, *46*, 87–95.

Pfister, T., Büttner, L., Shirai, K., & Czarske, J. (2005) Monochromatic heterodyne fiber-optic profile sensor for spatially resolved velocity measurements with frequency division multiplexing. *Appl Opt 44*(13), 2501–2510, doi:10.1364/AO.44.002501.

Pierce, C.H. (1941) Investigations of methods and equipment used in stream gaging; Part 1, Performance of current meters in water of shallow depth. *U.S. Geol. Survey Water-Supply Paper*, *868-A*, 35.

Pilechi, A., Rennie, C.D., Mohammadian, M., & Zhu, D. (2014) In situ field measurements of transverse dispersion of a wastewater effluent in an extended natural meandering river reach. *J Hydraul Res*, *53*(1), 20–35, doi:10.1080/00221686.2014.950611

Pinkel, R. (1979) Observations of non-linear motions in the open sea using a range-gated Doppler sonar. *J Phys Oceanogr*, *9*, 675–680.

Plant, W.J. & Wright, J.W. (1980) Phase speeds of upwind and downwind traveling short gravity waves, *J Geophys Res*, *85*(C6), 3304–3310.

Poelma, C. (2016) Ultrasound Imaging Velocimetry: A review. *Exp Fluids*, *58*(3), doi:10.1007/s00348-016-2283-9.

Polatel, C. (2005) Indexing free-surface velocity: A prospect for remote discharge estimation. In: *Proceedings of the XXXI IAHR Congress, Seoul, Korea*.

Prasad, A.K. (2000) Stereoscopic particle image velocimetry. *Exp Fluids*, *29*(2), 103–116.

Prasad, A.K., Adrian, R.J., Landreth, C.C., & Offutt, P.W. (1992) Effect of resolution on the speed and accuracy of particle image velocimetry interrogation. *Exp Fluids*, *13*(2), 105–116.

Raffel, M., Willert, C., Werely, S., & Kompenhans, J. (2007) *Particle Image Velocimetry: A Practical Guide*. Berlin, Springer-Verlag.

Ramooz, R. & Rennie, C.D. (2010) Laboratory measurement of bedload with an ADCP. In: Bedload-surrogate monitoring technologies: United States Geological Survey Scientific Investigations Rep. 2010–5091, Reston, VA. [Online] Available from: http://pubs.usgs.gov/sir/2010/5091/papers/Ramooz.pdf [Accessed 15th February 2017].

Rantz, S.E. *et al.* (1982) Measurement and computation of streamflow: 2. Computation of discharge. *US Geological Survey Water Supply Paper*, *2175*, 373. [Online] Available from https://pubs.usgs.gov/wsp/wsp2175/html/wsp2175_vol2_pdf.html [Accessed 1st March 2017].

Rehmel, M. (2007) Application of acoustic Doppler velocimeters for streamflow meaurements. *J Hydraul Eng, 133*(12), 1433–1438.

Rennie, C.D. & Rainville, F. (2006) Case study of precision of GPS differential correction strategies: Influence on aDcp velocity and discharge estimates. *J Hydraul Eng, 132*(3), 225–234.

Rennie, C.D. & Church, M. (2010) Mapping spatial distributions and uncertainty of water and sediment flux in a large gravel bed river reach using an acoustic Doppler current profiler. *J Geophys Res, 115*, F03035, doi:10.1029/2009JF001556.

Rennie, C.D., Millar, R.G., & Church, M.A. (2002) Measurement of bedload velocity using an acoustic Doppler current profiler. *J Hydraul Eng, 128*(5), 473–483.

Resagk, C., du Puits, R., & Thess, A. (2003) Error estimation of laser-Doppler anemometry measurements in fluids with spatial inhomogeneities of the refractive index. *Exp Fluids, 35*, 357–363. doi:10.1007/s00348-003-0679-9

Ricardo, A.M., Koll, K., Franca, M.J., Schleiss, A.J., & Ferreira, R.M.L. (2014) The terms of turbulent kinetic energy budget within random arrays of emergent cylinders. *Water Resour Res, 50*(5), 4131–4148. doi:10.1002/2013WR014596

Rolland, T. (1994) Développement d'une instrumentation Doppler ultrasonore adaptée à l'étude hydraulique de la turbulence dans les canaux. PhD dissertation No 1281, Swiss Federal Institute of Technology (EPFL), Lausanne, Switzerland.

Rolland, T. & Lemmin, U. (1997) A two-component acoustic velocity profiler for use in turbulent open- channel flow. *J Hydraul Res, 35*(4), 545–561.

Roux, S., Hild, F., & Berthaud, Y. (2002) Correlation image velocimetry: A spectral approach. *Appl Opt, 41*(1), 108–115.

Ruck, B. (1991) Distortion of LDV fringe patterns by tracer particles. experiments in fluids. *Exp Fluids, 10*(6), 349–354.

Rudd, M.J. (1969) A new theoretical model for the laser Dopplermeter. *J Phys E Sci Instrum, 2*(1), 55–58.

Samimy, M. & Wernet, M.P. (2000) Review of planar multiple-component velocimetry in high-speed flows. *AIAA J, 38*(4), 553–574.

Sassi, M.G. & Hoitink, A.J.F. (2014) Assessment of approaches to validate surface velocity measurements from HF coastal radars. Technical report, Hydrology and Quantitative Water Management Group, Wageningen University.

Sassi, M.G., Hoitink, A.J.F., & Vermeulen B. (2012) Impact of sound attenuation by suspended sediment on ADCP backscatter calibrations. *Water Resour Res, 48*, W09520, doi:10.1029/2012WR012008, 1–14.

Sathe, A. & Mann, J. (2013) A review of turbulence measurements using ground-based wind lidars. *Atmos Meas Tech, 6*, 3147–3167, doi:10.5194/amt-6-3147-2013.

Scarano, F. & Riethmuller, M.L. (2000) Advances in iterative multigrid PIV image processing. *Exp Fluids, 29*(7), S51–S60.

Schacht, C. & Lambert, C. (2004). Drogue-based measurements of estuarine suspended sediment characteristics: A new approach. *J Coast Res, SI 41*, 124–129.

Schemper, T.J. & Admiraal, D.M. (2002) An examination of the application of acoustic Doppler current profiler measurements in a wide channel of uniform depth for turbulence calculations. In: *Proceedings of the Hydraulic Measurements & Experimental Methods* (CD-ROM), Reston, Va., ASCE.

Schroeder, A. & Willert, C.E. (2008) *Particle Image Velocimetry*. Berlin, Springer-Verlag.

Seim, H.E. & Edwards, C.R. (2007) Comparison of buoy-mounted and bottom-moored ADCP performance at Gray's Reef. *J Atm Oceanic Technol, 24*, doi:10.1175/JTECH 1972.1.

Shen, C. & Lemmin, U. (1997a) Ultrasonic scattering in highly turbulent clear water flow. *Ultrasonics, 35*(1), 57–64.

Shen, C., Niu J., Anderson E.J., & Phanikumar M.S. (2010) Estimating longitudinal dispersion in rivers using Acoustic Doppler Current Profilers. *Adv Water Resour*, *33*(6), 615–623.

Sheng, J.Y. & Hay, A.E. (1988) An examination of the spherical scattering approximation in aqueous suspensions of sand. *J Acoust Soc Am*, *83*(2), 598–610.

Shercliff, J.A. (1962) *The Theory of Electromagnetic Flow-Measurement*. Cambridge, Cambridge University Press.

Siegman, A.E. (1986). *Lasers*. Mill Valley, CA, USA, University Science Books.

Simpson, M.R. (2001) Discharge measurements using a broadband acoustic Doppler current profiler. [Online] U.S. Geological Survey Open-File Report 01–01, 123. [Online] Available from: http://pubs.usgs.gov/of/2001/ofr0101/ [Accessed 15th February, 2017].

Smoot, G.F. & Carter, R.W. (1968) Are individual current meter ratings necessary? *J Hydr Div*, *94*(2), 391–397.

Smoot, G.F. & Novak, C.E. (1977) Calibration and maintenance of vertical-axis typecurrent meters. Techniques of water-resources investigations, Book 8. Reston Va., U.S. Geological Survey, Chapter 2.

Somerscales, E.F.C., Papyrin, A.N., & Soloukhin, R.I. (1981) Tracer Methods. In: *Methods of Experimental Physics*, *18A*, pp. 1–240, Section 1.1 of Chap. 1 - Measurement of Velocity. Academic Press Inc. ISBN 0–12-475960-2.

SonTek/YSI (2000) *Acoustic Doppler profiler principles of operation*. San Diego, CA, SonTek/YSI, 28.

Spedding, G.R. & Rignot, E.J.M. (1993) Performance analysis and application of grid interpolation techniques for fluid flows. *Exp Fluids*, *15*, 417–430.

Stacey, M.T., Monismith, S.G., & Burau, J.R. (1999) Observations of turbulence in partially stratified estuary. *J Phys Oceanogr*, *29*(8), 1950–1970.

Steffler, P., Rajaratnam, N., & Peterson, A. (1985) LDA measurements in open channel. *J Hydraul Eng*, *111*(1), 119–130, doi:10.1061/(ASCE)0733-9429(1985)111:1(119)

Stewart, R.H., & Joy, J.W. (1974). HF radio measurements of surface currents. In: *Deep Sea Research and Oceanographic Abstracts*. *21*(12), pp. 1039–1049.

Stieglitz, T. & Reid, P.V. (2001). Trapping of mangrove propagules due to density-driven secondary circulation in the Normanby River estuary, NE Australia. *Mar Ecol Prog Ser*, *211*, 131–142.

Sun, X., Shiono, K., Chandler, J.H., Rameshwaran, P., Sellin, R.H.J., & Fujita, I. (2010) Discharge estimation in small irregular river using LSPIV. *Water Manag*, *163*(WM5), 247–254.

Sveen, J.K. (2013) Laser Doppler anemometry (LDA) and particle image velocimetry (PIV) for marine environments. In: Watson, J.E., & Zielinski, O. (eds.) *Subsea Optics and Imaging*. Cambridge, Woodhead Publishing Ltd., pp. 353–378.

Szupiany, R.N., Amsler, M.L., Best, J.L., & Parson, D.R. (2007) A Comparison of fixed- and moving-vessel measurements with an acoustic Doppler profiler (ADP) in a large river. *J Hydraul Eng*, *133*(12), 1299–1310.

Szupiany R.N., Amsler M.L., Parsons D.R., & Best J.L. (2009) Morphology, flow structure, and suspended bed sediment transport at two large braid-bar confluences. *Water Resour Res*, *45*, W05415, doi:10.1029/2008WR007428, 2009.

Szupiany, R.N., Amsler, M.L., Hernandez, J., Parsons, D.R., Best, J., Fornari, E., & Trento, A.E. (2012) Flow fields, bed shear stresses and suspended bed sediment dynamics in bifurcations of a large river. *Water Resour Res*, *48*, W11515, doi:10.1029/2011WR011677.

Tarrab, L., García, C.M., Cantero, M.I., & Oberg, K. (2012) Role of turbulence fluctuations on uncertainties of acoustic Doppler current profiler discharge measurements. *Water Resour Res*, *48*(W06507).

Tatarski, V.I. (1961) *Wave propagation in a turbulent medium*. New York, USA, Dover Publications.

Taylor, G. (1938) The spectrum of turbulence. *Proceedings of the Royal Society London. Series A-Mathematical and Physical Sciences.* The Royal Society, *164*, 476–490,. doi:10.1098/rspa.1938.0032.

Tazioli, A. (2011). Experimental methods for river discharge measurements: Comparison among tracers and current meter. *Hydrol Sci J*, *56*(7):1314–1324, doi:10.1080/02626667.2011.607822

Teague, C.C., Barrick, D.E., Lilleboe, P.M., Cheng, R.T., Stumpner, P., & Burau, J.R. (2008) Dual-RiverSonde measurements of two-dimensional river flow patterns. In: *Proceedings of the 9th Working Conference on Current Measurement Technology, CMTC 2008. IEEE/OES.* IEEE, pp. 258–263.

Teledyne RD Instruments (1992) *User's Manual for RD Instruments Transect Program.* San Diego, CA, RD Instruments.

Teledyne RD Instruments (1996) *Principals of operation – A practical primer for broadband acoustic Doppler current profilers.* 2nd ed., San Diego, CA, Teledyne RD Instruments, 57.

Teledyne RD Instruments (2010) *ADCP coordinate transformation: Formulas and transformations.* Teledyne RD Instruments, P/N 951–6079-00, 36 p.

Terzi, R.A. (1981) Hydrometric Field Manual – Measurement of streamflow. *Water Resources Branch, Environment Canada*, Ottawa, Canada.

Theunissen, R. (2012) Theoretical analysis of direct and phase-filtered cross-correlation response to a sinusoidal displacement for PIV image processing. *Meas Sci Technol*, *23*(6), 065302.

Thibodeaux, K.G. (1994) Review of literature on the testing of point-velocity current meters. *U.S. Geological Survey Open-File Report 94–123*, 44.

Thomson, R.E. & Emery, W.J. (2014). Data analysis methods in physical oceanography. 3rd edition, Amsterdam, Elsevier.

Thorne, P.D. & Hardcastle, P.J. (1997) Acoustic measurements of suspended sediments in turbulent currents and comparison with in-situ samples. *J Acoust Soc Am*, *101*, 2603–2614.

Thorne, P.D. & Hanes, D.M. (2002) A review of acoustic measurement of small-scale sediment processes. *Cont Shelf Res*, *22*(4), 603–632.

Thorne, P.D., Vincent, E.C., Hardcastle, P.J., Rehman, S., & Pearson, N. (1991) Measuring suspended sediment concentrations using acoustic backscattering devices. *Mar Geol*, *98*(1), 7–16.

Timmins, B.H., Wilson, B.W., Smith, B.L., & Vlachos, P.P. (2012) A Method for automatic estimation of instantaneous local uncertainty in particle image velocimetry measurements. *Exp Fluids*, *53*(4), 1133–1147.

Tominaga, A. & Nezu, I. (1991) Turbulent structure in compound open-channel flows. *J Hydraul Eng*, *117*(1), 21–41. doi:10.1061/(ASCE)0733-9429(1991)117:1(21).

Tritico, H., Cotel, A. & Clarke, J. (2007) Development, testing and demonstration of a portable submersible miniature particle imaging velocimetry device. *Meas Sci Technol*, *18*, 2555–2562.

Tropea, C., Yarin, A.L., & Foss, J. (eds.) (2007) *Springer handbook of experimental fluid mechanics.* Berlin, Springer.

Tsanis, I.K. (1989) Simulation of wind-induced water currents. *J Hydraul Eng*, *115*(8), 1113–1134.

Tsanis, I.K., Shen, H., & Venkatesh, S. (1996). Water currents in the St. Clair and Detroit Rivers, *J Great Lakes Res*, *22*(2), 213–223.

Turnipseed, D.P., & Sauer, V.B. (2010) Discharge measurements at gaging stations: U.S. Geological Survey Techniques and Methods book 3, chap. A8, 87. [Online] Available from http://pubs.usgs.gov/tm/tm3-a8/; [Accessed 15th February 2017].

USGS (2002a) Configuration of acoustic profilers (RD Instruments) for measurement of streamflow. OSW Tech. Memo. 2002.01. Washington, D.C., U.S. Geological Survey.

USGS (2002b) Policy and technical guidance on discharge measurements using acoustic Doppler current profilers. OSW Tech. Memo. 2002.02. Washington, D.C., U.S. Geological Survey.

USGS (2006) Availability of the report "Application of the Loop Method for Correcting Acoustic Doppler Current Profiler Discharge Measurements Biased by Sediment Transport" by David S. Mueller and Chad R. Wagner (Scientific Investigations Report 2006–5079) and guidance on the application of the Loop Method. USGS, Office of Surface Water Technical Memorandum 2006.04. [Online] Available from https://pubs.usgs.gov/sir/2006/5079/pdf/SIR2006-5079.pdf [Accessed 4th February 2017].

USGS (2011) Exposure time for ADCP moving-boat discharge measurements made during steady flow conditions. USGS, Office of Surface Water Technical Memorandum 2011.08. [Online] Available from https://water.usgs.gov/admin/memo/SW/sw11.08.pdf [Accessed 4th February 2017].

Vehrenkamp, R., Schätzel, K., Pfister, G., Fedders, B.S., & Schulz-DuBois, E.O. (1979) A comparison between analog LDA, photon correlation LDA and rate correlation technique. *Phys Scr*, *19*, 379. doi:10.1088/0031-8949/19/4/016

Visbeck, M. & Fischer, J. (1995) Sea surface conditions remotely sensed by upward-looking ADCPs. *J. Atmos Oceanic Technol*, *12*, 141–149.

Voigt, A., Heitkam, S., Büttner, L., & Czarske, J. (2009) A Bessel beam laser Doppler velocimeter. *Opt Comm*, *282*(9), 1874–1878.

von Stein H.D. & Pfeifer H.J. (1969) A Doppler difference method for velocity measurements. *Metrologia*, *59*(5), 59–61, doi:10.1088/0026-1394/5/2/006.

Voulgaris, G. & Trowbridge, J.H. (1998) Evaluation of the acoustic Doppler velocimeter for turbulence measurements. *J Atmos Oceanic Technol*, *15*, 272–289.

Wahl, T. (2003) Discussion of Despiking acoustic Doppler velocimeter data, by D.G. Goring and V.I. Nikora. *J Hydraul Eng*, *129*(6), 484–487.

Wang, B., Liao, Q., Bootsma, H., & Wang, P. (2012) A dual-beam dual-camera method for a battery powered underwater miniature PIV (UWMPIV) system. *Exp Fluids*, *52*(6), 1401–1414.

Wang, B., Liao, Q., Xiao, J., & Bootsma, H. (2013). A free-floating PIV system: Measurements of small-scale turbulence under the wind wave surface. *J Atmos Oceanic Technol*, *30*(7), 1494–1510.

Wang, C.J., Wen, B.Y., Ma, Z.G., Yan, W.D., & Huang, X.J. (2007). Measurement of river surface currents with UHF FMCW radar systems. *J Electrom Waves Appl*, *21*(3), 375–386.

Weijs, S.V., Mutzner, R., & Parlange, M.B. (2013). Could electrical conductivity replace water level in rating curves for alpine streams? *Water Resour Res*, *49*, 343–351, doi:10.1029/2012WR012181

Weitbrecht, V., Kühn, G., & Jirka, G.H. (2002) Large scale PIV-measurements at the surface of shallow water flows. *Flow Meas Instrum*, *13*, 237–245.

Westerweel, J. (1993) Digital particle image velocimetry - theory and application. Ph.D. Dissertation, Delft University Press, Delft, The Netherlands. 237.

Westerweel, J., Dabiri, D., & Gharib, M. (1997) The effect of a discrete window offset on the accuracy of cross-correlation analysis of digital PIV recordings. *Exp Fluids*, *23*(1), 20–28.

Westerweel, J., Elsinga, G.E., & Adrian, R.J. (2013) Particle image velocimetry for complex and turbulent flows. *Annu Rev Fluid Mech*, *45*, 409–436.

Willert, C.E. & Gharib, M. (1991) Digital particle image velocimetry. *Exp Fluids*, *10*(4), 181–193.

Yamanaka, G., Becker, S., & Durst, F. (2006) Limitations of Gaussian beam property based LDA-velocity profile measurement. In: *Proceedings of the 13th International Symposium on Applications of Laser Techniques to Fluid Mechanics, 26–29 June 2006, Lisbon, Portugal.* Paper 17.3.

Yanta, W.J. & Ausherman, D.W. (1981) A 3-D laser Doppler velocimeter for use in high-speed flows. In: *Proceedings of the 7th Biennial Symposium on Turbulence*. 21–23 September 1981, University of Missouri-Rolla.

Yeh, Y. & Cummins, H.Z. (1964) Localized fluid flow measurements with an He-Ne laser spectrometer. *Appl Phys Lett*, 4(10), 176–178, doi:10.1063/1.1753925.

Zappa, C.J., Raymond, P.A., Terray, E.A., & McGillis, W.R. (2003) Variation in surface turbulence and the gas transfer velocity over a tidal cycle in a macro-tidal estuary. *Estuaries*, 26(6), 1401–1415.

Zedel, L., Hay, A.E., & Lohrmann, A. (1996) Performance of a single beam pulse-to-pulse coherent Doppler profiler. *IEEE J Oceanic Technol*, 21(3), 290–297.

Zhang, Zh., & Eisele, K. (1998) On the overestimation of the flow turbulence due to fringe distortion in LDA measurement volumes. *Exp Fluids*, 2, 371–374.

Zhang, Z. (2010) LDA Application Methods. Laser Doppler anemometry for fluid dynamics. *Experimental Fluid Mechanics*, Springer-Verlag, Berlin, Heidelberg. 272. ISBN: 978–3-642-13513-2. doi:10.1007/978-3-642-13514-9_16.

Zheng, X. & Li, Z.S. (2009) The influence of Saffman lift force on nanoparticle concentration distribution near a wall. *Appl Phys Lett*, 95(12), 124105. doi:10.1063/1.3237159

Zhu, W., van Hout, R., Luznik, L., Kang, H., Katz, J., & Meneveau, C. (2006) A comparison of PIV measurements of canopy turbulence performed in the field and in a wind tunnel model. *Exp Fluids*, 41, 309–318.

Zrnic, D.S. (1977) Spectral moment estimate from correlated pulse pair. *IEEE Trans AES-13*, 344–354.

Chapter 4

Topography and Bathymetry

4.1 INTRODUCTION

This chapter describes instrumentation and methods to collect data on topography, bathymetry, and bed roughness. Such data are required for the generation of digital elevation models, to provide knowledge on the composition of the bed material, and to determine roughness coefficients. There exists a variety of different instruments and methods that can be used for the collection of such data. However, compared to instruments and methods for flow measurements, which represent a core element in hydraulic engineering and fluid mechanics studies and which have been developed in collaboration with researchers in these fields, the development of the instruments described in this chapter has mainly been driven by disciplines other than hydraulic engineering such as physics (optics and acoustics), geodesy, construction, or by industrial applications (*e.g.*, quality control). Therefore, in contrast to Chapter 3 (Volume II), this chapter provides less background information on the physical measurement principles. However, links to further reading on these principles are given in pertinent sub-sections.

Table 4.1.1 presents an overview of the instruments and methods described in this chapter. The chapter is organized with regard to the measurement of topography and bathymetry from larger to smaller spatial scales and hence also from 2.5D to 1D measurements. The latter measurements are generally point based and represent traditional methods to determine topography and bathymetry. However, due to technical advances, the corresponding information can be obtained from remote sensing applications across larger spatial scales. Methods other than those described in this chapter exist, such as airborne and satellite surveying. However, it may be necessary to consult specialists in these fields, as additional equipment and expertise is needed to apply these methods. This chapter includes also a sub-section focusing on traditional sieve-analysis and an advanced optical method to get automated information about bed surface grain-size distributions.

4.2 TERRESTRIAL LASER SCANNING: TOPOGRAPHIC MEASUREMENT AND MODELING

4.2.1 Introduction

Accurate digital elevation models (DEM) of channel and floodplain topography are fundamental tools for river science and engineering. They provide the boundary

Table 4.1.1 Methods and instruments for measurements of topography, bathymetry and roughness

Method	Abbrev.	Section	Technique	Range	Measurement coverage
Terrestrial Laser Scanning	TLS	4.2	optical	<1 m - > 6 km	volume
Mobile Laser Scanners	MLS	4.2	optical	<1 m - > 6 km	volume
Multibeam Echo Sounder	Sonar	4.3	acoustic	< 1 m - > 11 km	point to area
Side-scan Sonar	Sonar	4.3	Acoustic	< 1 m - > 1 km	point to area
Single Beam Echo Sounder	Sonar	4.3	acoustic	< 1m - > 10 km	point
Photogrammetry		4.4	optical	<1 m - > 6 km	area to volume
Displacement meters		4.5 4.3	optical/ acoustic	< 1 mm - > 6 km	point to profile
Mechanical profilers		4.5	mechanic	<1 mm – 1 m	point to profile
Basic surveying tools		4.5	optical	1 mm - > 6 km	Point

conditions for numerical hydrodynamic simulations, enable quantification of erosion and deposition by DEM differencing, and under limiting assumptions can be used to support indirect estimates of sediment transport (Ashmore & Church, 1998; Bates *et al.*, 2003; Brasington *et al.*, 2003; see also Section 5.2.4 in Volume II).

In the last decade, improvements in the resolution and precision of survey data have led to the development of a new generation of DEMs characterized by 10s to 1000s of observations per square meter. Such "hyper-resolution" models record information at the particle-scale building blocks of the terrain and thus provide insights well beyond low frequency surface topography (Brasington *et al.*, 2012). These new datasets have extended the scope of traditional terrain modeling and are now used increasingly to estimate surface roughness and facies characteristics (Heritage & Milan, 2009; Hodge *et al.*, 2009a, 2009b), vegetation cover type and density (Antonarakis *et al.*, 2009, 2010), and to incorporate microscale features such as kerbs and road camber into urban flood simulations (Sampson *et al.*, 2013).

This next-generation of terrain models has been developed through recently established geomatics methods that include laser scanning (airborne and particularly terrestrial) and image-based methods such as paired- and multi-view stereo photogrammetry. Standard deployment and data acquisition strategies enable dense data collection over a wide range of spatial scales so that, as illustrated in Figure 4.2.1, these methods enable quantification of the fluvial system from the particle to the reach scale seamlessly. Such broad capability facilitates applications that stretch from geometric measurements in laboratory scale models to catchment scale measurements such as detection of fluvial hazards or morphological changes.

The adoption of terrestrial laser scanning (TLS) has accelerated rapidly over the last five years, driven in part by the falling cost of hardware and simplification of field-to-product workflows. While image-based methods, such as structure-from-motion photogrammetry (Appendix 4.A.1; see James & Robson, 2012; Javernick *et al.*, 2014; Westoby *et al.*, 2014; and Section 4.4) offer extremely low-cost alternative data acquisition strategies, TLS offers the ability to realize results rapidly (even in real-time), requires comparatively minor post-processing, and has predictable data quality and well-understood sources of error. In this Section, standard approaches to terrestrial laser scanning are described, including the range of available instrumentation, the physical principles of

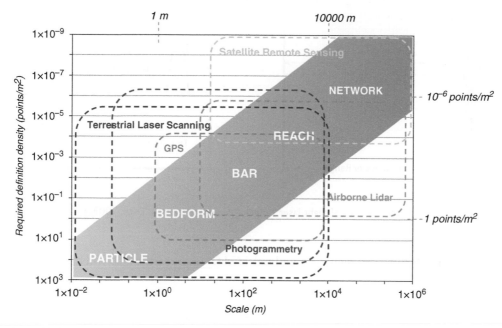

Figure 4.2.1 Relationship between data acquisition density and spatial extent for geomatics methods applied to fluvial problems (from Brasington *et al.*, 2012).

laser scanning, data acquisition and modeling strategies, and illustrations of the different scales of application.

4.2.2 Principles of measurement

Terrestrial laser scanners are active laser imaging systems that combine high frequency range observations with precision angular sampling to generate a dense 3D point cloud of xyz observations. In general, instruments comprise three components: a laser generator; an opto-mechanical reflector to divert the beam; and a receiver that records the reflected light. This review outlines the basic physical principles and for further information the reader is directed elsewhere (Wehr & Lohr, 1999; Petrie & Toth, 2009a, 2009b).

4.2.2.1 Laser ranging

All laser scanning operations involve some measurement of the line of sight or slant distance also referred to as the range (R), by the time of flight (ToF) principle. Several approaches to the estimation of range have been proposed and the two most commonly applied in terrestrial laser scanners are the timed pulse and phase difference methods (for information on triangulation methods see Beraldin *et al.*, 2000).

Timed pulse range estimates are based on recording the time elapsed between the transmission and receipt of a very short burst of laser radiation that is directed toward and reflected from a target. Thus, for the simple case shown in Figure 4.2.2, a laser

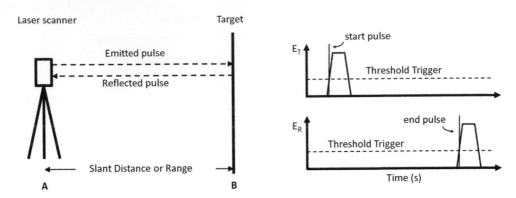

Figure 4.2.2 Operation of a time-of-flight laser ranger (based on Petrie & Toth, 2009b).

pulse emitted by an instrument at A and reflected from the target at B will have a range given as:

$$R = \frac{1}{2}c\varDelta t \qquad (4.2.1)$$

where R is the slant distance or range (m), c is the celerity of the emitted light wave (m s^{-1}) which is known and $\varDelta t$ is the time elapsed (s). Given Equation (4.2.1), it follows that as a time interval of just 1 ns corresponds to 0.3 m (or a range of 0.15 m) precise determination of range requires very short, precise laser pulses and very accurate, high frequency timing clocks. This is achieved by diverting a small fraction of the emitted pulse radiation onto a photodiode that generates a signal to start the timing clock when an energy threshold (photons) is exceeded. The timing cycle is terminated and $\varDelta t$ estimated when the reflected laser pulse returns to the receiving photodiode, exceeding a second threshold (Figure 4.2.2).

The second popular ranging method involves the transmission of a continuous beam of laser radiation instead of discrete pulses of laser energy. In this approach, the range is determined by calculating the phase difference (or phase angle, φ) between the transmitted and reflected beam, in an approach similar to that used in electronic distance measurement or precise ranging with Global Navigation Satellite Systems (GNSS). Given the shortwave nature of the laser energy, the fundamental carrier wave is modulated with a lower frequency sinusoidal measurement wave. In the standard amplitude-modulation approach, the range is then estimated by relating the phase difference of the measurement wave to the travel time:

$$\varDelta t = \left(\frac{\varphi}{2\pi} + n\right)\frac{1}{f} \qquad (4.2.2)$$

$$R = \frac{1}{2}c\varDelta t = \left(\frac{\varphi}{2\pi} + n\right)\frac{c}{2f} \qquad (4.2.3)$$

where, φ is the phase angle (rad), n is the integer number of wavelengths, f is the frequency of the measuring wave (Hz) and c is the celerity of light. The maximum unambiguous range (or range interval) that can be measured is determined by the wavelength (λ) of the modulated signal where, recalling that $f = 1/\lambda$:

$$R_{max} = \frac{2\pi c}{4\pi f} = \frac{c}{2f} = \frac{\lambda}{2}$$ (4.2.4)

However, the resolution of the range measurement (δR) is in turn inversely proportional to the wavelength of the measurement wave and also the resolution of the phase difference measurement:

$$\delta R = \frac{\pi \delta \varphi}{4\pi f}$$ (4.2.5)

where, $\delta \varphi$ is the resolution of the phase difference calculation (determined by the sampling frequency of the waveform). Paraphrasing Equations (4.2.4) and (4.2.5), the longer the measurement wave, the greater the range achievable but at the expense of the resolution of the estimate. A solution to this problem is found by calculating the integer number of wavelengths between the scanner and the target (n in Equation (4.2.3)), a process termed ambiguity resolution. Typically, this is achieved by modulating the carrier wave with multiple frequency waves in succession, the highest frequency (shortest) determines the resolution of the measurement, while the lowest frequency (longest) dictates the maximum range obtainable. This leads to a set of simultaneous range equations that can be solved to provide a final range. The Faro 880 scanner (see Table 4.2.1), for example, employs three wavelengths, scaled by a factor of 8: 1.2 m, 9.6 m and 78 m (Lichti, 2007) and offers measurements over a range from 0.6 – 76 m at a resolution of 0.6 mm.

4.2.2.2 Beam divergence and reflectivity

It is important to note that the precision and accuracy of the range estimates obtained by either method described above are subject to two further fundamentals of laser behavior, beam divergence and reflectivity. Even a focused and collimated laser beam will diverge with distance due to diffraction. Consequently, on incidence with a target, the beam will illuminate a circular (for normally incident rays) or elliptical (for oblique rays) region from which photons will be reflected. Even for perfectly flat targets, this introduces variability in the travel time of the reflected light and when the surface is irregular or incoherent, a complex set of reflections is likely that must be averaged or filtered to return either a single representative or multiple set of ranges. In well-constrained cases, the spread of the beam introduces a minimum target size that can be ranged uniquely, with the footprint determined by:

$$AI = \frac{\pi(\theta R + d)}{2}$$ (4.2.6)

where AI is the illuminated area of the footprint (m^2), θ is the beam divergence angle (rad), R is the range and d (m) is the aperture of the laser collimator. A significant problem associated with beam divergence occurs when the footprint is split across targets separated in space (for example, due to partial occlusion). In this situation, a complex backscatter function is returned and when averaged simply, can lead to range estimates that lie anomalously between targets, giving rise to so-called 'mixed pixels'. A growing number of scanners now employ more complex signal processing, digitally sampling the time-series of backscatter to identify multiple reflection signatures associated with 'porous' footprints (such as vegetation).

The backscatter function is also governed by the laws of reflection and the optical properties of the target surface, in particular the nature of reflection (diffuse or specular) and the wavelength specific pattern of absorption or transmission. These properties dictate the intensity of the reflected light and will differ for diffuse (Lambertian) and specular reflectors. For typical diffuse reflectors, very oblique viewing angles are likely to lead to low levels of reflection toward the scanner and ultimately, no return being logged. By contrast, specular reflectors (*e.g.*, mirrors, metal) may result in full deflection of the beam, and light recorded at the scanner may have followed a complex path that involves reflections from a number of targets; a phenomenon termed *multipathing* that is associated with erroneous overestimation of slant range.

The reflectivity of the target is also a function of the wavelength of the laser which may be strongly absorbed by the target material. Estimates of surface albedo are less well known for coherent laser energy at specific wavelengths (Petrie & Tooth, 2009a, 2009b) but tabulated data from Wehr and Lohr (1999) provide useful insight into the influence of albedo on reflectivity. Similar data are also published by manufacturers for specific materials. In a fluvial context, TLS with long optical wavelengths (*e.g.*, 780–900 nm) are not water penetrating and cannot be used either in the field or laboratory to obtain bathymetric information. By contrast, scanners employing shortwave 532 nm Nd:Yag lasers, such as the Leica ScanStation and C10 (see Table 4.2.1), have been used to penetrate shallow water successfully at angles of incidence close to nadir and over relatively short ranges. Under specific circumstances (low turbidity, depths < 0.5 m, and ranges of < 20 m) these scanners have been used to derive high quality sub-aqueous DEMs (*e.g.*, Smith & Vericat, 2014).

Finally, most scanners record the intensity of the backscatter. Such measurements are generally uncalibrated so that unique spectral classification of reflections are confounded by compensatory relationships with slant angle and range.

4.2.2.3 Laser scanning

To convert a laser ranger into an automated 3D scene reconstruction tool, a scanning mechanism is required. For static TLS, this is typically provided by a sensor head comprising rotating mirrors, servo-motors and accurate encoders that enable fine angular sampling at micro- to milli-radian resolution across a wide field of view. As technology has advanced over the last decade, the field of view of scanning systems has increased from the 40° x 40° window used in the Cyrax 2500 scanner, to full "dome" scans that acquire data through 360° in the horizontal and 270° in the vertical (see Table 4.2.1). Alternatively, scanners can be locked to acquire profile data in only a single axis (vertical), with spatially-distributed data provided by moving the scanner along a given trajectory (see discussion of mobile scanning below). Ultimately, the density of the point cloud obtained is a function of the radial sampling interval and target range. For example, a 50 μrad sampling interval at a range of 100 m, equates to

5 mm point spacing and a density of 40,000 pts/m^2, with point densities on nearer or more distant targets scaled proportionally.

4.2.3 Available instrumentation

Current commercially available instruments operate over ranges of less than a meter to over 6 km and at measurement frequencies of between 1–1000 kHz. Most scanners also record the intensity of the reflected laser beam, which although strongly influenced by distance, incidence angle and surface moisture, may also provide information on mineralogy, moisture content and roughness (Lichti & Jamtshoo, 2006; Franchesci *et al.*, 2009; Nield *et al.*, 2011, 2014). Additionally, some instruments incorporate or can be augmented with high-resolution digital cameras, enabling the true-color pixel values to be remapped directly onto each survey observation to produce photo-realistic 3D renderings. Furthermore, a number of instruments now also facilitate data-feeds from positioning instrumentation (GNSS and inertial navigation sensors, INS) and temporal correlation of these measurements by time-stamping the backscatter data.

As described in Section 4.2.2.1, a fundamental distinction between instruments lies in the laser ranging technique employed, which can be classified as either 'timed pulse' or 'continuous wave'. Pulse scanners typically operate over longer distances (100-1000+ m) but at reduced measurement frequencies (2–100 kHz). By contrast, continuous-wave (also referred to as "phase-based") scanners can acquire data at very fast rates and therefore at high densities, but generally over only shorter ranges (typically less than 170 m). Additionally, as range increases, errors associated with divergence of the laser beam and verticality (or attitude) of the sensor head are magnified, so that point accuracy ultimately deteriorates with distance (Lichti & Jamtsho, 2006). In effect, a trade-off between data quality (spatial density and point accuracy) and range emerges that must be tailored to the particular application and field logistics. This trade-off is typically governed by available time and resources.

Major manufacturers of survey grade TLS include Leica Geosystems, Reigl, and Faro. A number of other survey equipment manufacturers also offer instruments, including Trimble and Topcon and there is considerable re-selling of instruments under OEM (Original Equipment Manufacturer) agreements, such as the Zoller-Frohlich (Z&F) and Maptek scanners resold by Leica Geosystems. The selection of a laser scanner should be based largely on "fitness for purpose," although availability and cost are common limiting factors. Considerable variability exists within the range of instruments currently available and the pace of technological development is fierce. In the last five years, a wide range of scanners have come to market, with a broad range of capabilities. A full review of this rapidly evolving instrumentation is beyond the scope of this section and for an up-to-date listing of commercially available TLS the reader is recommended to review the product listings on the Geo-Matching website which provide useful tools to filter scanners according to functionality and logistics (*e.g.*, laser class and wavelength, range measurement method, max/min range, weight etc. http://www.geo-matching.com/category/id46-terrestrial-laser-scanners.html; accessed 30th January 2017). Table 4.2.1 provides an illustrative overview of the functionality of three scanners that span the spectrum of available technology. Figure 4.2.3 illustrates a popular long-range laser scanner, the Reigl VZ-1000 being used in the field to acquire a dense 3D point cloud of a recent debris flow.

Table 4.2.1 Characteristics of three popular survey grade TLS

	Faro Focus[3D]	Leica C10	Riegl VZ6000
Year of release	2014	2009	2012
Ranging principle	continuous Wave	timed Pulse	timed Pulse
Wavelength	905 nm	532 nm	1064 nm
Laser safety classification[1]	3R	3R	3b
Min. range	0.6 m	0.1 m	5 m
Max. range[2]	130 m	300 m	3000–6000 m
Beam diameter at exit	3 mm	4.5 mm	15 mm
Beam divergence	0.19 mrad	0.24 mrad	0.12 mrad
Camera[3]	integrated (72 MP)	integrated (4 MP)	integrated (5 MP)
Acquisition rates (pts/s)[4]	122,000–976,000	50,000	23,000–222,000
Max vertical field of view	300°	270°	60°
Waveform digitization	No	No	Yes
Horizontal field of view	350°	360°	360°
Additional sensors	compensator, compass	compensator, GNSS	compensator, GNSS, compass
Weight	5.0 kg	13 kg	14.5 kg

1 Laser safety classification based on ANSI Z136.1
2 Range data are based on manufacturers estimates for targets with a (very high) 90% albedo
3 Image resolution from manufacturers specification, may be based on composite (mosaic) imaged (Faro) or individual images (C10 and VZ6000)
4 Acquisition rate varies according to settings based on desired range for Focus3D and VZ6000

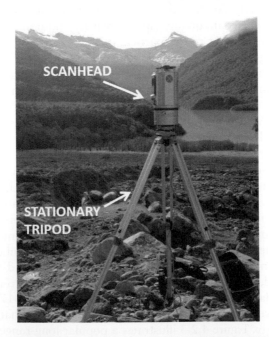

Figure 4.2.3 A long-range (>1000 m) Reigl VZ-1000 TLS deployed in the field to measure the three-dimensional morphology of a recent debris flow. Photo: Brasington, May 2014.

4.2.4 Deployment methods

Terrestrial laser scanners are generally deployed from a tripod and held stationary for the duration of measurements at a fixed position. The most popular scanners have rotating sensor heads that enable scans over a 360° horizontal field of view and a wide, often overhead, vertical field of view. Despite this large sampling window, data acquisition is limited by line-of-sight, so that surveys of complex objects typically require multiple scanner setups to ensure adequate overlap and coverage (Figure 4.2.4). The data acquired at each position are a set of measurements of range, horizontal and vertical angles and intensity. These are converted directly to Cartesian coordinates (x, y, z) where the origin $(0, 0, 0)$ and orientation are defined relative to an internal scanner coordinate system (so-called scanner space). Two important steps are thus required for the production of data deliverables; registration and geo-referencing. Registration refers to the merging of multiple scans into a common unified, but relative coordinate system. Geo-referencing then involves transformation of the point cloud onto a national or global coordinate system. These two processes are often treated as synonymous, but represent distinct processes which incorporate different sets of errors.

Registration of multiple scans can be achieved using one or both of two approaches; (i) a rigid body similarity transformation from scanner-space to object-space based on common, observed control points or targets (*e.g.*, Horn, 1987); or (ii) cloud-to-cloud registration. The latter involves matching overlapping point clouds on the basis of local

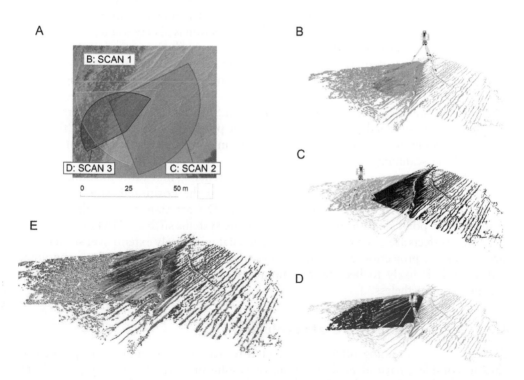

Figure 4.2.4 Registration of overlapping scans (B, C, D) to generated a single, unified point cloud (after Brasington *et al.*, 2012).

geometry and adjusting the position and orientation of clouds iteratively to minimize differences in these common regions. Cloud-registration techniques offer the possibility to co-register datasets without the need for targets, however success is strongly dependent on the geometry and 'texture' of the objects scanned. A variety of automated algorithms have been developed to solve this complex problem (see Lichti & Skaloud, 2010) although most require a degree of user intervention to identify initial seed-points to orientate the point clouds *a priori* (*e.g.*, the iterative closest point, ICP, algorithm of Besl & McKay, 1992).

These two methods are often best used in combination. In this hybrid approach, a 3D similarity transformation based on known targets provides the initial orientation that is then refined using cloud-registration to minimize the volumetric error in the overlap region and tighten the overall fit. Registration is also best undertaken as a global transformation problem, in which multiple overlapping scans are orientated simultaneously (as opposed to a progressive survey traverse) and global least-squares minimization used to ensure the homogeneity of the solution.

While registration of scan data to a common, relative coordinate system may be sufficient to generate products suitable for characterizing a particular object, it is often desirable to tie these data to a recognized frame of reference. This step facilitates the integration of additional data, such as airborne lidar or ground-observations positioned by GNSS and importantly, also enables the comparison of datasets over time. Geo-referencing is usually undertaken during post-processing as part of registration, where control points or targets with observed coordinates based on a national or global mapping system are used to define the primary similarity transformation.

Direct (forward) geo-referencing during data collection is also possible, in which the location and orientation of the sensor head are prescribed *a priori* and the scan observables (range and angles) are transformed on the fly. This approach has become more popular, as the latest generation of instruments incorporates positioning instrumentation such as GNSS and attitude sensors. More recently, this has led to the emergence of mobile laser scanning systems (MLS) in which scanners operating in profile mode are supplied with information on the position and orientation of the scan head by GNSS-informed inertial navigation units in a similar approach to that used in airborne laser scanning (see discussion below).

Finally, it is important to recognize that geo-referencing, inevitably introduces additional uncertainties associated with the accuracy of the positional data used. Care should also be taken to work within consistent 3D Cartesian frameworks prior to projecting the point cloud into a planar coordinate systems such as UTM or a national grid. These effects are often neglected as, to date, datasets rarely extend over sufficiently large areas for projection errors to become significant. However, as the technology matures this is likely to become commonplace and best-practice should always be encouraged nonetheless.

4.2.5 Spatial and temporal resolution

Terrestrial laser scanning offers the potential to acquire high quality morphometric data in complex, natural field situations at resolutions previously only possible in controlled laboratory studies. This is achieved through high sampling frequency, precise non-contact ranging and accurate angular measurements that, in combination, present a new scope of applications as summarized in Table 4.2.2. Recent use of TLS within fluvial and broader hydrodynamic research has focused largely on: a) detailed characterization of surface morphology and land-cover, for example: a) automating the analysis of fluvial gravel facies (*e.g.*, Hodge *et al.*, 2009a,b) or quantifying vegetative flow resistance (*e.g.*, Antonarakis *et al.*, 2009, 2010); b) high resolution change detection studies based on differencing sequential terrain models (*e.g.*, Milan *et al.*, 2007; Brasington *et al.*, 2012); and c) detailed parameterization of the topographic boundary condition to support high resolution geophysical fluid modeling (*e.g.*, Sampson *et al.*, 2013; Williams *et al.*, 2013).

The physical principles of laser scanning imply a series of trade-offs between target range, achievable data resolution (density) and the reliability of individual observations. Historically, this has led to applications focused at specific spatial and temporal scales. For example, studies of fluvial sediments have typically been based on high fidelity measurements of patches or small samples of bedforms, while data models of channel change have largely been restricted to small channels or bar-scale studies (*e.g.*, 0.3–2 channel widths). Attempts to upscale data acquisition (to encompass river reaches of 10–50+ channel widths) while maintaining consistent resolution and quality, have been confounded by line-of-sight losses, particularly in the shallow relief typical of fluvial systems. Comprehensive data coverage therefore requires observations from multiple locations, resulting in long and complex field data collection strategies (see for example, Brasington *et al.*, 2012, who used 18 scan locations to derive a detailed terrain model of a 1 km reach of the River Feshie in northern Scotland).

More recently, this restriction has been relaxed through the development of mobile deployment and data acquisition strategies. At their simplest, this can involve deploying the sensor head on a stable, mobile platform to facilitate rapid movements between survey stations. Scans are then obtained in "stop-and-go" mode, while the scanner is held stationary and then moved rapidly between scans. This approach was employed by Williams *et al.* (2011, 2014) to enable a high density topographic survey of the braided gravel-bedded Rees River in New Zealand. In their experimental setup, a high speed phase-based Leica 6200 scanner was mounted on a ruggedized, amphibious, all-terrain ARGO vehicle, which also incorporated GPS positioning, a panoramic camera and continuous AC power and data storage. The short-range scanner used provided very high frequency scans (less than 3 minutes for a full dome scan), but only over ranges of 40–60 m. Coverage of a 2.5 x 0.7 km reach involved over 300 individual scans, that were co-registered and georeferenced using a mobile network of GNSS-located targets, requiring between 3–5 days of field sampling per survey. Using this approach they were able to acquire a detailed time-series of surveys with typical vertical errors of 0.03 m that captured the evolution of this large braided river through a consecutive set of competent flood events (Figure 4.2.5).

More complex deployment strategies involve fully mobile laser scanning (MLS), in which integrated navigation sensors are used to record the position and orientation of the sensor platform carrying the scanner in 3D space. Accurate time synchronization of the instruments then enables on-the-fly geo-referencing of scan data that can be acquired continuously as the platform moves along a trajectory. Until recently, MLS systems were restricted to research prototypes, such as the ROAMER system developed by the Finnish Geodetic Institute (Vaaja *et al.*, 2011). However, they are now coming to market as integrated instrument packages such as the Leica Pegasus I and II (released

Table 4.2.2 Defining characteristics of terrestrial laser scanning and the relevant implications for geomorphological data applications.

Characteristic	Definition	Potential Research Applications
Sampling Rate	1000–500,000 observations per second, enabling 360° x 270° field of view scans acquired by continuous wave scanners in < 5 minutes over ranges of up to 80 m. Longer range (100–300 m) scans possible with slower time-of-flight scanners.	Rapid data acquisition permits multiple scans covering large spatial extents. Rapid reoccupation of survey areas possible to measure change in highly dynamic environments, e.g., tidal environments and slow geophysical flows.
Spatial Density	Typical point resolutions of 5–50 mm. Trade-off between acquisition time and data density. Limited ultimately by precision of encoders, beam divergence and focusing.	'Scale-free landscape models'; information acquired at the particle scale but over reach or landscape scales enables simultaneous representation of micro- meso- and macro-morphologies.
Point Quality	Point quality intermediate between total station and RTK GPS. Varies with range, but manufacturer quoted estimates of 2–6 mm are typical. Individual point quality dependent on beam footprint, angle of incidence, instrument verticality and surface roughness.	High fidelity individual measurements enable the detection of very small changes in object geometry, enabling accurate monitoring of erosion, deposition or deformation. Both close range and landscape scale applications are possible with a multi-purpose time-of-flight scanner.
Non-Intrusive Measurement	'Reflectorless' point observations without the need for prisms or targets. Effective instrument range is dependent on laser wavelength, power and albedo. Strongly influenced by water content for most short-wave lasers.	Remote survey of dynamic, deformable and inaccessible targets (e.g., mudflats, soft sands, slow geophysical flows). High quality representations of complex, fragile objects such as bedforms, speleothems and living objects.
Oblique, Multi-View Sampling	Data can be acquired from multiple, oblique positions and co-registered to generate fully-3d object models, as opposed to the single look-angle provided by vertical airborne sensors.	Possible to quantify objects in 3d; capturing re-entrant curves (overhangs), internal geometries (inside caves, buildings). Comprehensive object coverage is possible, but dependent on survey strategy (line of sight) and field logistics.
Intensity and Passive Optical Retrievals	3d point cloud geometry supported by additional information on the laser return intensity and remapped RGB color data from accompanying optical sensors.	High quality, photo-realistic renderings with minimal post-processing. Possible extraction of additional information on surface composition and micro-roughness from calibrated backscatter.
Unselective Sampling	Untargeted point selection. All observations are collected in uniform increments of arc, resulting in complete scene capture.	Data modeling is required to generate desired products, free from extraneous reflectors within the field of view. Substantial automated and manual edited often necessary to 'decimate' the point cloud to the appropriate selection and resolution.

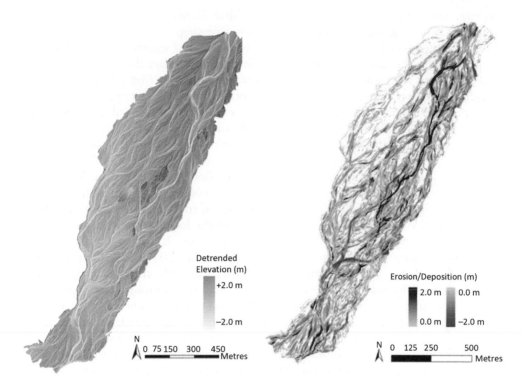

Figure 4.2.5 A) 0.25 m Digital Elevation Model of a braided reach of the Rees River, NZ derived from mobile TLS survey; B) DEM difference model derived by comparison of before and after flood terrain models.

2014). This generic system architecture can be configured for a range of platforms, including boats, trolleys and even mounted on a backpack (see the review by Alho *et al.*, 2011). Such systems integration creates exciting new opportunities to deploy multiple mobile sensors targeting different components of the environment, such as boat-mounted mobile TLS and sonar systems (either acoustic Doppler current profilers or multibeam sonars) that enable simultaneous quantification of exposed and sub-aqueous morphology (*e.g.*, Dix *et al.*, 2012).

4.2.6 Post-processing and data modeling

Unlike traditional terrestrial survey data acquired using GNSS or tacheometry, laser scanning generates voluminous quantities of xyz observations unselectively. Large area scans, such as those derived for the Rees River by Williams *et al.* (2014) discussed above, frequently incorporate 10^6–10^9 individual xyz observations, resulting in data volumes well beyond the capacity of standard desktop GIS, CAD or visualization software. Moreover, the unselective sampling incorporates significant image noise associated with components of the scene that are in motion (*e.g.*, vegetation, people and cars) and confound simple interpretation of the "bare-earth" topography. The extraction of primary terrain products, such as digital elevation models, typically therefore, necessitates a systematic workflow that incorporates a combination of manual data cleaning and automated data post-processing.

Following geo-registration, the removal of blunders and significant scene noise by manual editing, scan data are generally modeled through either the construction of a full 3D mesh or are resampled to a 2.5D terrain model function ($z = f(x, y)$) and represented as either a raster or triangular network. While this latter approach involves significant data reduction, intelligent filtering of points can be applied to segment the point cloud into key classes, such as ground and off-ground points. Brasington *et al.* (2012) presented a computationally-efficient method that enables such intelligent decimation of large point cloud datasets and is designed to process large TLS datasets on standard desktop computers. The Topographic Point Cloud Analysis Toolkit or ToPCAT, segments the scene into a regular grid and computes local properties of the elevation distribution in each gridcell. A locally fitted triangulation of the neighborhood topography is then used to detrend estimates the local standard deviation and higher order moments of the elevations. This algorithm has been encoded in the Geomorphic Change Detection software (Wheaton *et al.*, 2010; gcd.joewheaton.org [accessed 30th January 2017]) and has been used not only to process bare earth topographic models, but also to derive information on surface roughness, facies mapping, grain sorting and flow resistance from TLS and photogrammetric datasets (*e.g.*, Williams *et al.*, 2014; Javernick *et al.*, 2014; Bangen *et al.*, 2014; Storz-Peretz & Laronne, 2015). This approach has also been used in three-dimensions, generalizing the point cloud as a lower resolution set of voxels to characterize the three-dimensional form and flow resistance associated with riparian vegetation (Antonarakis *et al.*, 2009).

4.3 ULTRASONIC SENSING

4.3.1 Introduction

Ultrasonic transducers are devices that convert ultrasound waves to electrical signals, as well as the reverse. Ultrasonic waves carry information about their source and the environment that they encounter as they propagate through different transmission media (gases, fluids and solids). Data collected with ultrasonic transducers can thus offer insights into many different physical processes. Therefore, such instruments are used in many different scientific disciplines as well as for many practical and industrial applications as pointed out by Bradley (2007) and Dowling and Sabra (2015).

There are two basic types of ultrasonic devices. Passive devices (also called "listening" devices) detect sounds emitted by other sources. Active devices (ultrasound transceivers), on the other hand, are instruments that transmit and receive ultrasonic waves. There is a long history of the use of ultrasonic sensing in the aquatic environment (hydroacoustics). In hydraulic engineering, such instruments are used for flow velocity measurements (Chapter 3, Volume II), sediment transport measurements (Chapter 5, Volume II), discharge measurements (Chapter 7, Volume II), and remote sensing to determine water depth (see also Section 6.2) and to scan underwater topography (or bathymetry). Although the focus of this section is on field measurements of bathymetry, ultrasonic sensors are also widely used in laboratory applications to measure water levels and bed elevations. The measurement principles of laboratory sensors are the same as for field-instruments. The versatility of ultrasonic devices is due to the wide range of available signal frequencies (10^{-2} to 10^7 Hz) and transmission distance capabilities (10^{-2} to 10^7 m) (Dowling & Sabra, 2015).

4.3.2 Principles of sonar measurements

The technique to measure and analyze sound propagation is called "sound navigation and ranging" (SONAR or sonar). There exists a large body of literature related to this topic (*e.g.*, Singal, 1997; Kuperman & Roux, 2007; Lurton 2010; Ainslie, 2010; Dowling & Sabra, 2015; and references therein) and this sub-section therefore provides only a brief overview on active sonar devices (often referred to as sonars).

Active sonar devices emit sound waves at specific, controlled frequencies and listen for their echoes returned from objects in the water and on the ground. Distances are calculated from the measured travel time of the echo-return of a generated pulse (or ping) from an acoustic reflector (*e.g.*, bottom or suspended particles). This process is called echo sounding and therefore sonars are often called echo sounders.

A sonar device produces a sound wave using a projector. For bathymetric sonars it is important that acoustic pulses be precise, controllable, and repeatable. This can be achieved with projectors constructed of piezoelectric ceramic (L3- communications, 2000). The sound pulse generated from the instrument expands spherically from its source, forming an acoustic cone with dimensions that depend on instrument characteristics. The acoustic energy disperses over an increasing area as it travels away from the projector, causing a drop in energy per unit area; this is known as spreading loss. Moreover, pulse energy is reduced by attenuation loss (due to absorption of energy by the transmission medium). The combination of spreading and attenuation loss is called transmission loss (Christ & Wernli, 2014). This loss is directly correlated to the travel distance.

A target or the portion of the bed of a water body that reflects a sound wave (corresponding to the acoustic footprint of the pulse; *i.e.*, the cone-area at the bed) is said to be ensonified. During ensonification, a portion of the acoustic energy is absorbed by the bed. The amount of energy absorbed by the bed depends on the bed material. For example, sand and silt absorb energy fairly easily whereas rocks and metal objects absorb minimal acoustic energy (L3-communications, 2000). Energy that is not absorbed by the ensonified target is reflected or scattered back into the water. The fraction of incident energy per unit area directed back to the projector determines the backscattering strength of the bed. The instrument detects the reflected part of the sound wave using a hydrophone, which converts the impinging sound waves into voltages. The projector of a sonar is often also used as the hydrophone since the behavior of piezoelectric materials is reciprocal. Thus, the projector/hydrophone sensor is called a transducer (L3-communications, 2000). Note that the strength of the return signal can be used to infer sediment types and the height of sub-bottom sediment layers, but this application is not discussed in detail here as it is mainly used in Oceanography (*e.g.*, Saleh & Rabah, 2016).

In every stage of the above process the signal is affected by noise; *i.e.*, by sources of ambient sound that add to the received signal, limiting the maximum range of the instrument. At some range, the echo becomes too weak to be distinguished from noise. The quality of a measurement can be evaluated on the basis of the signal-to-noise ratio (SNR: the ratio of the received signal strength to the noise level). *The minimal SNR required for a signal to be detectable depends on the specific application* (L3-communications, 2000).

The strength of the measured echo-return in terms of signal excess can be evaluated based on the sonar equation, which represents an energy budget between transmitted, received, and processed sonar signals based on the parameters described above. More details on the sonar equation can be found in Urick (1983), Lurton (2010) and Lurton (2016).

4.3.3 Types of sonar devices

As indicated in the foregoing sections, sonars are used for a wide range of applications which are partly described in different sections of this book. For bathymetric surveys or distance measurements, three types of sonars are common: single beam, multibeam, and side scanning sonars (Figure 4.3.1).

Single beam sonars (echo sounders) make discrete local depth measurements. Miniature versions of such instruments are often used in laboratory investigations to monitor bed and water surface elevations (see Section 6.2, Volume II) at local positions. When attached to a traversing system, these sensors can be used to map the bathymetries of large areas by simultaneously recording sensor position and depth reading as the sensor is moved horizontally across the region of interest. In field applications, single beam depth sounders usually collect individual depth measurements from a moving vessel at many locations. Using a GPS-system, the corresponding location of the sensor is known so that maps can be constructed as shown in Figure 4.3.2. The spatial resolution can be enhanced by utilizing multiple single beam sonars in a sensor array; this is often done for surveying vessels.

A multibeam sonar is a device that collects multiple depth measurements at the same time by emitting sound waves in a fan shape (Figure 4.3.1b). The technique has a high level of accuracy and a high data point density. A disadvantage of this method is its relatively high cost.

Side-scan sonar efficiently covers a large area on the bottom of the water body by taking advantage of the differing sound reflecting and absorbing characteristics of different materials (L3-communications, 2000). The side-scan technique uses a sonar device which is either towed from a vessel or mounted to the hull of the vessel. The

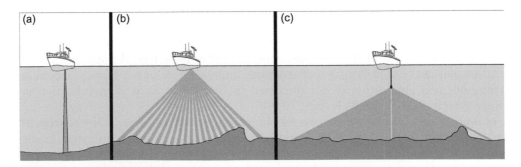

Figure 4.3.1 Acoustic signatures of vessel based a) single beam, b) multibeam, and c) side-scan sonars.

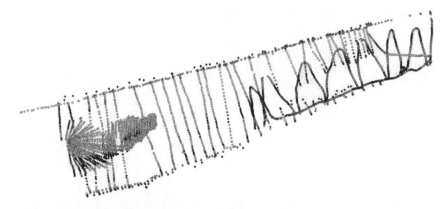

Figure 4.3.2 Single beam echo-sounding in a riverine application. The continuous line in the right part of the picture represents data recorded with a single beam echo sounder. The other points were recorded using a differential GPS-system.

device emits conical or fan-shaped pulses across a wide angle perpendicular to the path of the sensor. The resulting acoustic beam is wide in the across-track direction and narrow in the along-track direction (Johnson & Helferty, 1990). The strengths of acoustic reflections from the bottom are recorded in a series of cross-track slices. When these slices are combined they form an image within the coverage width of the beam (swath). The sound frequencies used in side-scan sonar usually range from 50 to 500 kHz; higher frequencies yield better resolution but with less range. This technique can be used to detect debris and other obstacles in the water body. Side-scan sonar is also used for fisheries research, dredging operations and environmental studies. Figure 4.3.3 illustrates the resulting bathymetry map of a 250 kHz side-scan sonar of several river arms in a freshwater delta in the south of Norway.

4.3.4 Spatial and temporal resolution

The described sonar types are characterized by different spatial and temporal resolutions. A single beam echo sounder makes only one depth measurement at a time; *i.e.*, the ensonified bottom area is the only distance recorded by a ping. In order to get a complete picture of the bottom with such an instrument, the sensor must be relocated with a high spatial resolution. This process is time consuming but can be hastened by using sensor-arrays as mentioned earlier. The enhanced spatial resolution of multibeam and side-scan sonars allows large areas to be scanned in a shorter amount of time.

One restriction for the movement of the sensor (and hence also for the temporal resolution) results from the requirement that the ping echo must return before the next ping can be transmitted; *i.e.*, the sounding depth and the speed of sound influence the maximum temporal resolution. Sonars with narrow beams can provide measurements with higher spatial resolution due to the smaller footprint areas of their beams (*i.e.*, the spatial resolution of a sensor is governed by the footprint size of the beam). However, compared to instruments with wider beams, the use of narrow beams requires longer scanning times to fully map the same region of the bed. Using instruments with wider

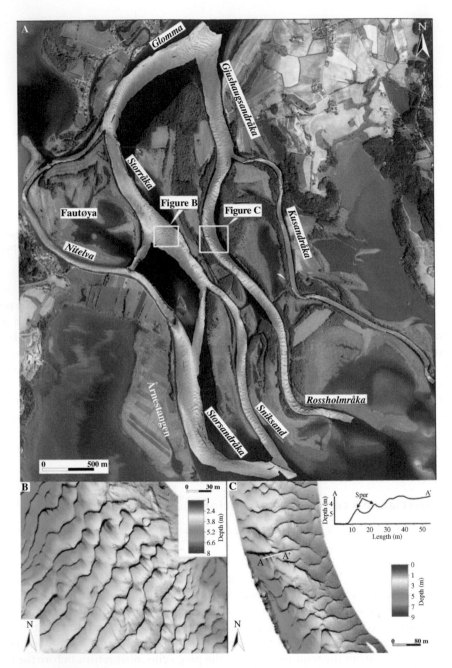

Figure 4.3.3 250 kHz interferometric side-scan sonar (GeoAcoustics) measurements at Øyeren lake, Norway. A) Overview of the lake Øyeren delta plain with measured data from October 2007. The aerial photo is from www.norgedigitalt.no; B) Lee side-scours in the Storråka channel; C) Lee-side scours with spurs in the Striksand channel. (Figures from Eilertsen *et al.*, 2013).

beams, a larger area can be mapped at the expense of spatial resolution. Note also that narrow beams are the result of larger transducer faces, resulting in higher costs (L3-communications).

4.3.5 Error sources

Besides error introduced by noise, there exist various other error sources with regard to sonar measurements such as errors in instrument calibration and sound velocity (Wu *et al.*, 2010) which are briefly discussed here.

Noise can be introduced by various sources related to footprint area, overlapping beams, bed characteristics and vessel movement. In fact, depth sounding is based on the return signal from the bottom which has the spatial resolution according to the acoustic footprint. However, the surface covered by the acoustic beam is often not completely flat and therefore larger individual objects within the footprint can bias the measurement. The extent of this bias can partly be determined in specifically designed experiments under controlled conditions (*e.g.*, by inserting objects of known sizes in the footprint area). Moreover, uneven surfaces also affect backscattering strength. The same applies to bed surfaces composed of different materials - muddy surfaces are harder to detect (depending on the consolidation stage) compared to sandy and rocky surfaces.

When using arrays of sensors, care must be taken so that the individual beams do not overlap as this can cause flawed measurements and low signal quality. If large areas of mobile beds are mapped with moving sonars (*e.g.*, dune fields), the resulting bathymetry may be prone to additional uncertainties dependent on the propagation speed of bed forms and the time required for the mapping. Although such scans are often considered as "instantaneous scans" the resulting bathymetry of large scale scans will contain some noise as the bed forms will move during the measurement. Note also that fish interference (or interference caused by other obstacles) can introduce measurement errors.

If the instrument is not stationary and attached to a vessel, further noise can be introduced by water surface conditions as the vessel will continuously pitch and roll. Vessel movement will result in a scattered (or unstabilized) beam, requiring a correction. This error can become even larger in case the sensors are poorly calibrated prior to their use. Such a calibration is, in general, required to determine the errors inherent in the timing and triggering circuits and the accuracy of the positioning system (GPS). For multibeam systems, it is important to know the static offsets of the sensors (the distance between sensors and a reference point on the vessel), transducer draft (the depth of the transducer below the waterline of the vessel), and the time delay between the positioning system, sonar measurement, and heave-pitch roll sensor (which in turn is required for data processing) (Mann, 1998). An accurate estimate of the speed sound in the water must also be known; this depends on the water temperature and salinity.

4.4 PHOTOGRAMMETRY

4.4.1 Introduction

Photogrammetry is the science of obtaining reliable information about physical objects and the environment through recording, measuring, and interpreting photographic images and patterns of electromagnetic radiant energy and other phenomena (ASPRS,

2013). It can be classified into aerial and terrestrial (close-range) photogrammetry, depending on the location of the camera during photography (Schenk, 2005). Aerial photogrammetry makes use of aircraft mounted cameras, and therefore the distance between the camera and the target objects is large. In close-range photogrammetry, the camera is closer to the object as the photographs are taken from the ground (*i.e.,* terrestrial) or close to the ground, with distance much shorter than for typical aerial photogrammetry (Atkinson, 1996). In hydraulic applications, both methods are used to obtain geometric and temporal information about objects and surfaces.

Geometric information obtained through photogrammetry is used to determine the spatial positions and shapes of objects, which is of particular importance for many different purposes, including topography and bathymetry measurements in the field and laboratory, generation of digital elevation models (*e.g.,* Butler *et al.,* 1998; Lane, 2000; Lane *et al.,* 2000, 2001; James & Robson, 2012; see also Section 4.2), determination of grain size distribution and bed roughness (*e.g.,* Detert & Weitbrecht, 2012; Bertin & Friedrich, 2014; see also Section 4.6) or the determination of the area of flexible vegetation elements (*e.g.,* Sagnes, 2010) to name just a few applications.

Temporal information enables the determination of the change of an object in time and is obtained by comparing images recorded at different times, *i.e.,* the temporal resolution depends on the time-scale separating the records. For example, using large time scales, changes in river channels can be monitored (Chandler *et al.,* 2002) while at smaller time scales it becomes possible to obtain data on the evolution of sediment beds (*e.g.,* Bouratsis *et al.,* 2013) or the movement of individual grains or bed forms through particle- or bed form tracking (*e.g.,* Henning, 2013; Heays *et al.,* 2014). The use of Structure from Motion (SfM) with MultiView Stereo (MVS) applications (see Appendix 4.A.1) can provide terrestrial topography (Westoby *et al.,* 2012; James & Robson, 2012; Javernick *et al.,* 2014; Smith *et al.,* 2016), marine bathymetry (*e.g.,* Leon *et al.,* 2015), through-water bathymetry of shallow rivers (Woodget *et al.,* 2015), sediment size statistics of dry river beds (Detert *et al.,* 2016a), and river surface flow image velocimetry (Detert *et al.,* 2016b, 2017, see also Section 3.7.4, Volume II). Using high-speed photogrammetry it is also possible to relate the movement of objects to the turbulent flow field (*e.g.,* Cameron *et al.,* 2013) or to measure flow velocities (*e.g.,* Particle Image Velocimetry (PIV), large-scale PIV (LSPIV), or space-time image velocimetry (STIV); see Section 3.7.4, Volume II).

These examples show that photogrammetry is a well-established and widely used method in hydraulic research which has been triggered by technical advances in digital methods and automated image recordings. However, a complete description of photogrammetric methods is beyond the scope of this handbook as photogrammetry is in itself an academic discipline. Fundamental principles of photogrammetry can be found in various text books (*e.g.,* Kraus, 2011; Brinker & Minnick, 2012; ASPRS, 2013; Luhmann *et al.,* 2014). The following sub-sections provide a brief overview of photogrammetry, addressing the measurement principles, available instrumentation, spatial and temporal resolution, and error sources with respect to hydraulic experiments.

4.4.2 Principles of measurement

Photogrammetry is accomplished through the analysis of photographs; *i.e.,* without physical contact with the target-objects. Photography converts 3D objects into "flat"

2D-images. In doing so, some information from one dimension is lost, primarily the depth. Photogrammetry reverses the process, converting 2D-images back into three dimensions (Luhmann *et al.*, 2014).

Photographs can be taken with both analog (film based) and digital cameras, although nowadays, the use of digital cameras (charge coupled devices; CCD or complementary metal–oxide–semiconductor devices; CMOS) is more common, as the photographs are instantly available for further processing and analysis. In fact, digital image processing techniques (not discussed here) enhance the potential of photogrammetric measurements and analyses (*e.g.*, Schenk, 2005).

A camera represents a spatial system consisting of the image plane (film or image sensor) and front lens. A photographic image is based upon central perspective ("one point perspective"). This means that each light ray reaching the image plane during exposure passes through the camera lens. The lens is mathematically considered as a single point (perspective center O' in Figure 4.4.1) and defines the origin of the camera coordinate-system, while the image (photograph) coordinates (x', y', z') define the locations of the object points on the image plane. Based on the collinearity equations, it can be demonstrated that each object point is projected onto a unique image point if its view is not obstructed by other object points (Luhmann *et al.*, 2014).

The focal length, position of the perspective center (X_0, Y_0, Z_0 in Figure 4.4.1), and lens distortion define the interior geometry of a camera, which is also known as interior orientation. The interior orientation parameters define the deviation of the camera's perspective center from the ideal central perspective mapping and must be known to accurately take measurements from photographs by reconstructing the ray-bundles which reach the image plane. For professional photogrammetry cameras, the interior orientation is known. However, for amateur cameras the interior orientation needs to be determined from camera calibration (see Section 4.4.3).

The exterior orientation of the camera defines the spatial location and orientation of the camera coordinate-system with respect to the global object coordinate system. It is defined by the space coordinates X_0, Y_0, Z_0 (*i.e.*, by the position vector $\mathbf{X_0}$ to the perspective center) and the angular orientation angles ω, φ, and κ which define the orthogonal rotation matrix; Figure 4.4.1). The exterior orientation is determined based on known XYZ reference points of objects whose image coordinates can be measured. The use of bar encoded reference points allows for automated orientation (*e.g.*, Godding *et al.*, 2003). A minimum of three XYZ reference points are required to apply the method of space resection (calculation of exterior orientation based on collinearity equations; *i.e.*, the establishment of a connection line between an object point P and its image point P' through the perspective center O') while at least five XYZ reference points are required for the use of projective relations (see Schenk, 2005; Luhmann *et al.*, 2014 for details).

The parameters of both the interior and exterior orientation are required to establish the relationship between the image information and object geometry through triangulation, the main principle used by photogrammetry. In order to reconstruct the object coordinates from triangulation, existing procedures require information about objects (reference points, known distances, geometric elements), the measurement of image points to determine the orientation (image coordinates), the calculation of orientation parameters (interior and exterior orientation), and object reconstruction from oriented images (new points, geometric elements) (Schenk, 2005; Luhmann *et al.*, 2014).

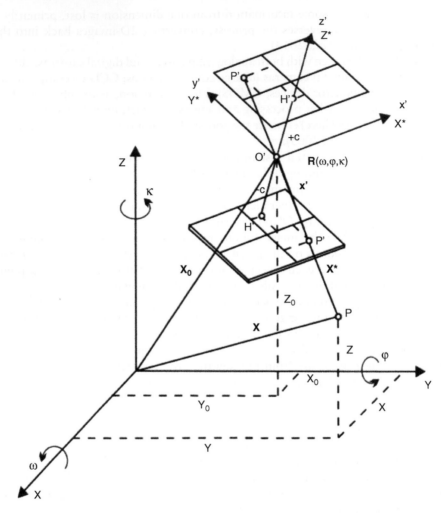

Figure 4.4.1 Sketch for the central perspective transformation (modified from Luhmann *et al.*, 2014). O': perspective center; $\mathbf{R}(\omega, \varphi, \kappa)$ = orthogonal rotation matrix; $\mathbf{X_0}$ position vector to perspective center (X_0, Y_0, Z_0); $\mathbf{x'}$ position vector to image point; $\mathbf{X}^*$ vector from perspective center to object point P.

Measurements from a single camera are only useful for plane (2D) objects, as obliquely photographed plane objects show perspective deformations. These need to be rectified for correct measurements. Rectification can only be neglected if the object is flat and the image plane is parallel to the object. In this special case, the photograph has a unique scale factor which can be determined from a known distance at the object. However, in general it is not possible to determine 3D-coordinates from a single photograph unless the scale factor of every ray is known. Corresponding photogrammetric procedures require information about image orientation and object geometry such as reference points, reference elements and digital surface model (Luhmann *et al.*, 2014). In hydraulic experiments, typical single camera applications are, *e.g.*, Particle

Tracking Velocimetry (PTV) of particles floating on a water surface plane and PIV, where a measuring plane is usually defined by a laser (see Section 3.7, Volume II).

The application of single photographs in photogrammetry is limited because they cannot be used for reconstructing the object space (*e.g.*, Schenk, 2005). Two photographs made at different positions with the same camera showing the same area (or parts of it) are called a stereopair. These can be generated by stereometric cameras or by using one camera and changing the position for taking the photograph. If both photographs have parallel viewing directions which are at a 90° angle to the base (the line between the two projection centers), the overlapping area of the two photographs can be seen in 3D. Using principles of stereoscopic image viewing and analysis (correspondence of homologous points lying in an epipolar plane) the coordinates of the visible parts of the objects in both photographs can thus be determined (*e.g.*, Luhmann *et al.*, 2014).

To reconstruct a 3D-object, more photographs are required that show the object from different perspectives. These need to be analyzed in a multi-image process using bundle triangulation (*a method for the simultaneous fit of an unlimited number of spatially distributed images;* Luhmann *et al.*, 2014). The most important criterion within a multi-image analysis based on bundle triangulation is that corresponding image rays should intersect in their corresponding object point with minimum inconsistency. As an example, Figure 4.4.2 shows a photogrammetric system used at the

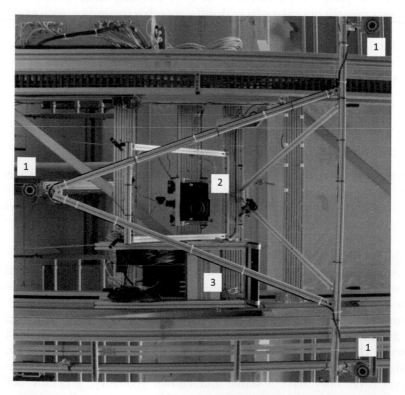

Figure 4.4.2 Photogrammetric system mounted to a traveling crane at the BAW; view from measurement area toward ceiling: 1) Camera with LED ring light; 2) Projector for grid projection (see Figure 4.4.3); 3) Computer and control unit.

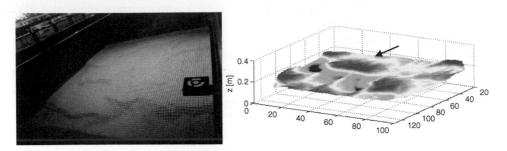

Figure 4.4.3 Photograph of an experimental setup for measurement of dune field bathymetry through water using photogrammetry and a corresponding digital elevation model (Henning, 2013). The grid projected on the bed allows for the reconstruction of the bed elevations using a multi-media approach. The retro-reflexive bar-code (14 bit-encoded) acts as a reference point and can be automatically identified by the system (see Figure 4.4.2).

Federal Waterways Engineering and Research Institute (BAW) in Karlsruhe, Germany which was used by *e.g.*, Henning (2013).

The standard photogrammetric methods outlined above assume collinearity of an object point, the perspective center and the corresponding image point. However, if light rays pass through optical media with different refractive indices such as glass walls or water, they no longer follow a straight line. This needs to be accounted for through additional considerations in both the experimental setup and analysis as the location of the interface and its thickness needs to be known (see, for example, Henning *et al.*, 2008 and Figure 4.4.3). In aerial photogrammetry correction for refraction of the atmosphere, earth curvature and image motion may also be necessary (*e.g.*, Schenk, 2005).

4.4.3 Available instrumentation

There exist many different photogrammetric instruments and systems which are commercially available. Compared to traditional instruments (mechanical or optical), digital systems are composed of both software and hardware components which are constantly being improved. A broad overview of different instruments and systems is provided in ASPRS (2013), but many more individual systems are on the market. Besides the commercially available systems, there exist also a large number of "self-made" systems as the basic instrumentation for digital photogrammetry is primarily the camera, and the software for data processing and analysis may be coded independently if no access to commercial software packages is available.

Contemporary cameras for photogrammetry are mainly digital. The classification of cameras as digital or analog is, however, insufficient for photogrammetry, and cameras must also be classified as metric or non-metric (*e.g.*, Luhmann *et al.*, 2014; Kraus, 2011). Metric cameras are particularly designed for photogrammetry and are therefore rather expensive. They have a stable and known geometry, generally defined by reference points, and very low lens distortion. Thus, images collected using metric cameras have minimal geometric distortion and camera characteristics do not change from photograph to photograph. A stereometric camera consists of two metric cameras mounted at both ends of a base of known length, permitting photographs of an object from two different positions. Off-the-shelf non-metric cameras (amateur cameras) are less-expensive than metric cameras but have an unknown and non-stable geometry. The geometry of such cameras can, for a given setup, be calibrated by photographing a test field with many control points (*e.g.*, Henning *et al.*, 2008) and at a repeatable fixed distance setting and using the four corners of the camera frame function as fiducials. Such cameras will never reach the precision of metric cameras (Luhmann *et al.*, 2014), but satisfactory results are possible in many cases (ASPRS, 2013).

4.4.4 Spatial and temporal resolution

The spatial and temporal resolution of photogrammetric methods cannot be generally given as it varies significantly dependent on the application type and the specifications of the individual system. For example, using aerial photogrammetry it becomes possible to cover large spatial scales, while close-range photogrammetry can resolve small scale structures (see also Figure 4.2.1). The same applies to the temporal resolution, which may be very high using high-speed cameras (*e.g.*, velocimetry or object-tracking applications) or rather low (*e.g.*, the generation of digital elevation models for static surfaces).

The achievable accuracy of photogrammetric methods strongly depends on the type of camera (*e.g.*, metric or non-metric), imaging configuration, camera stability, and experimental setup (*e.g.*, measurement through the water surface). For an optimally configured system, it is possible to achieve digital point measurements with sub-pixel accuracy.

4.4.5 Error sources

There exist many different unknown factors that can affect the accuracy of photogrammetric measurements which are related to different components of the system, the setup of the system, and the application type (close-range or aerial photogrammetry). Besides these systematic errors there is always the danger of random errors due to human factors as well as software bugs. The following paragraphs will provide a short overview of instrument-related systematic errors. Random errors will not be discussed.

Some error sources are directly related to the camera. Dai *et al.* (2014) categorized the systematic error due to camera imperfection into the aspects i) lens distortion, ii) approximated principal distance, and iii) resolution. The first two errors can be minimized through the use of metric cameras or using off-the-shelf cameras and camera calibration (see also Wackrow & Chandler, 2008, 2011). As a rule of thumb one can assume that the accuracy of the measurements increases with higher end camera models. The spatial resolution of the camera determines the smallest object that can be resolved by one pixel on the sensed image, providing a theoretical limit of accuracy for digital photogrammetric methods. In a time-series of images, the temporal resolution of the camera determines the time scales that can be represented in successive images. In general, cameras with resolutions better than the required spatial and temporal resolutions of the processes under investigation should be used for photogrammetric applications.

Another error source is related to the planning of the measurements, *i.e.*, the setup of the system in terms of shooting distance, baselines, percentage of photo overlaps, number of overlapping photos, camera intersection angles, and angles of incidence. In their review-paper on photogrammetric error sources, Dai *et al.* (2014) stated that the accuracy of a photogrammetric system increases with increases in the cameras' convergence angles, the number of intersection rays to an object point, the number of measured points, and the orthogonality of roll angles (at least two images should have a viewing angle of ± 90°, *i.e.*, portrait and landscape). Moreover, they pointed out that a strong correlation exists between measurement error and shooting distance.

Furthermore, good lighting is important for successful measurement from photographs. In order to enable the (automated) identification of objects and surfaces of interest, strong contrast is necessary between the objects or retroreflective targets and the background, and uniform illumination of the recording area is helpful. Where applicable, this can be achieved by, *e.g.*, lighting, reflectors, background coloring, housing, etc. In this context, the use of infrared cameras or color filters with corresponding illumination may be useful in some applications. Digital images, recorded directly or produced indirectly by digitizing, can be enhanced through methods of digital image processing (see *e.g.*, Kraus, 2011; ASPRS, 2013 for details).

It is not uncommon in various applications (*e.g.*, flow-vegetation interaction) to extract information on geometric properties of objects from 2D-images (*e.g.*, frontal projected area). However, if the object is 3D, corresponding estimates may be biased due to image distortion effects. A method to prevent these effects has been presented by Sagnes (2010).

4.5 OTHER SURFACE PROFILING METHODS

4.5.1 Introduction

The preceding sections outlined a range of optical and acoustic methods and instruments for the measurement of topography and bathymetry. However, these methods and instruments may not always be suitable or applicable for experimental conditions or data needs associated with particular field or laboratory experiments. In such cases, other methods and instruments are available. For example, there are numerous methods and instruments available for point measurements of bed elevations. The corresponding point measurements must be spatially referenced to an (arbitrary) coordinate system which may, dependent on the application, range from the Gauss-Krüger coordinate system to a local flume-based coordinate system. With an adequate reference system it is then possible to determine profiles (*e.g.*, longitudinal and cross-sectional) or to construct digital elevation models (DEMs).

4.5.2 Optical displacement meters

Optical displacement meters such as laser displacement meters are predominantly used in laboratory investigations for the measurement of surface properties. They are heavily used in industrial applications and are therefore commercially available at a reasonable price.

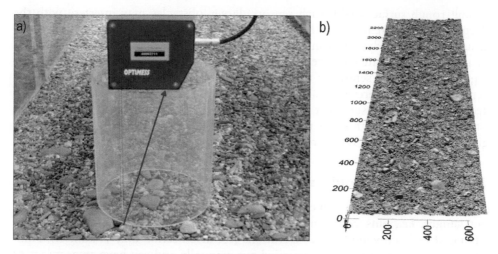

Figure 4.5.1 a) Laser displacement meter (the perspex cylinder was placed into the picture to enable visualization of the light); the red arrow indicates the reflected light toward the PSD (Photo: B. Neumann); b) Digital elevation model of a gravel surface measured with a the laser displacement meter shown in a); units in mm.

Laser displacement sensors detect the distance to an object by emitting focused light toward the surface (see Figure 4.5.1). The surface reflects the light back to the instrument where it is focused on an integrated position-sensitive detector (PSD) which enables the determination of the displacement via triangulation. Besides "point-based" sensors (*e.g.*, the sensor shown in Figure 4.5.1 has an optical footprint of 4 mm x 0.2 mm; Aberle & Nikora, 2006), there exist also "profile-based" sensors. Such sensors project a line of visible laser light on a target surface thus enabling more elevation readings in a single ensemble of data compared to "point-based" sensors. Although such sensors increase the spatial resolution of the measurements in one dimension, care needs to be taken in regard to distortion effects for measurements points located toward the outer ends of the projected line.

In order to measure surface properties, the sensor must be moved over the target surface. This can be easily achieved by attaching the sensor to a traversing system. Within the direction of movement, the spatial sampling interval is dependent on the speed of the traverse system and the sampling frequency of the displacement meter. For point based measurements, or for very wide surfaces, DEMs can only be generated by merging individually measured profiles or surface strips. While it is generally possible to merge profiles based solely on information from the traverse-positioning system, experience has shown that identifiable objects with known geometry help to better align the profiles and reduce measurement noise. For example, a cross-bar installed across the flume allows the accurate determination of the start-point of individual profiles in the merging process, and thus helps to reduce noise in the DEM which may arise due to the combined errors associated with traverse positioning and sensor signal. Such a strategy has been used in the study by Aberle and Nikora (2006).

Laser displacement meters have a high accuracy and cover different measurement ranges. Especially the latter aspect should be considered when purchasing such sensor.

Profiling rough surfaces with such sensors, larger obstacles may interfere with the reflected light so that it cannot be picked up by the PSD, resulting in an erroneous measurement. Similarly, edges on the target surface or surfaces that poorly reflect light may result in inaccurate measurements. These factors can partly be accounted for by the positioning of the sensor on the traverse system and/or by measuring the surface several times with different sensor orientations. Lastly, note that void space cannot be detected using this measurement principle as the light cannot penetrate through solid obstacles. This means that the roughness density function (porosity) in the lower regions of the rough bed is underestimated by such scans (Aberle, 2007).

4.5.3 Mechanical profilers

Mechanical profiling can be achieved through both the use of point or contour gauges. Point gauges, *i.e.*, pins with a needle tip, are a classical laboratory instrument for profiling of bed and water surface elevations (see also Section 6.2, Volume II). Point gauge measurements are generally carried out manually and can have an accuracy of up to 0.1 mm. However, profiling surfaces with point gauges is rather time consuming and care needs to be taken with regard to the accuracy of the spatial position, especially if the point gauge is manually positioned.

A contour gauge consists of many pins which are tightly set together in a frame at a given spacing to keep them parallel and in the same plane. The pins can move independently perpendicular to the frame and, when being lowered to the surface, the contour gage provides information about the bed profile (Figure 4.5.2), allowing for the measurement of local bed elevations. These devices, also known as "thousand-footers" when being used in the field (de Jong, 1995; Smart *et al.*, 2004), have a high spatial resolution along the measured profile with the frame position defining the

Figure 4.5.2 Contour gauge (thousand-footer) used by Smart *et al.* (2004) with 200 pointed 6 mm diameter rods and 10 mm spacing between centers of adjoining rods.

reference system. However, the spatial extension of the profiles is restricted by the frame length, and to cover longer distances, the frame must be moved and adequately positioned so that the profiles can be merged during post-processing. Measurements with contour gauges are time consuming and require significant effort due to the potentially large dimensions of the device.

4.5.4 Surveying

Standard survey equipment is commonly used to obtain information about topography and bathymetry (Figure 4.5.3). Surveying techniques have constantly evolved over time and provide the basis for modern remote sensing and satellite imagery. There exist many different surveying instruments and methods including level and rod, steel band, theodolites, total stations, photogrammetry, and surveying GPS-systems. Providing information on these instruments and their application is beyond the scope of the present section and can be found in the geodetic literature (*e.g.*, Brinker & Minnick, 2012).

However, one issue worth consideration is that the use of standard survey equipment may result in inconsistent bed elevation measurements of rough beds due to the leveling rod base diameter. If this diameter is smaller than the particle diameter, the leveling rod may be arbitrarily placed either on top of a particle or penetrate gaps between adjacent particles. DeVries and Goold (1999) therefore recommended the use of an adequately sized rod base plate which is in contact with the top of at least three of the more elevated particles and should not "settle substantially into the interstices between them". Further information on the sizing of the rod base plate can be found in DeVries and Goold (1999).

Figure 4.5.3 Surveying of a waterfall in winter conditions using a Topcon Imaging Station equipped with advanced auto-tracking technology. Photo: Rüther.

4.6 GRAIN SIZE DISTRIBUTION

4.6.1 Introduction

The preceding sections provided an overview on instruments that can be used to determine topography and bathymetry in the field or laboratory in order to construct digital elevation models or to derive roughness characteristics (*e.g.*, Coleman *et al.*, 2011). Another important characteristic of fluvial systems, one which cannot be directly obtained by applying these methods, is the grain size distribution of the bed material. Such information is important for several reasons, including: to obtain grain roughness, perform sediment transport calculations, assess channel stability, predict the morphodynamic regime, classify aquatic habitats, and evaluate geological deposits. The objective of this section is to present an overview of instruments and methodologies being available for the determination of grain-sizes and hence the grain-size distribution. For this purpose, sampling and sieving methods like laboratory sieving or line-sampling as well as image based techniques will be summarized.

Sand-bed rivers, in contrast to gravel-bed rivers, have relatively narrow grain-size distributions (within a 1 mm range), and thus determining grain-size variability is less important. Therefore, the following section mainly focusses on data analysis related to gravel-bed rivers. The appendix to this section provides the classification of grain-sizes (4.A.2) and the spatial variability (4.A.3) of grain-sizes in a river bed.

4.6.2 Principles of sampling

The spatial variability of the sediments, as discussed in more detail in Appendix 4.A.3, has a direct effect on the required sampling method in order to determine characteristic grain-sizes of the river bed surface or subsurface layer. For this purpose, two sampling strategies can be applied. They comprise direct methods – sampling of the subsurface material, and indirect methods – sampling the surface layer to get an indication to the grain-size distribution of the subsurface layer. Technically, two sampling procedures are used: volumetric and surface sampling, as explained in the following.

In general, volumetric sampling and subsequent sieving is the standard method to determine the grain-size distributions of both gravel beds and sand bed rivers. For this kind of sampling, a preselected sediment volume is taken from a predefined sedimentary layer. Laboratory sieving to classify sediment samples requires substantial effort, and the process of digging, transporting and sieving the samples is time-consuming and cost-intensive. This is especially true for coarse sediment mixtures, as the required sampling volume is typically governed by the maximum grain-diameter d_{max}, leading to sample weights of up to 1000 kg (see Section 4.6.3).

Alternatively, numerous *in situ* surface sampling methods have been developed to obtain grading curves, especially for coarse sediment. These procedures do not necessarily require complete withdrawal of material. Bunte and Abt (2001) summarize three different types of surface sampling methods, where a preselected number of surface particles within a predefined sampling area are assessed: 1) pebble counts, like heel-to-toe walks (Wolman, 1954) or line-by-number sampling (Fehr, 1986, 1987a), 2) grid counts (Kellerhals & Bray, 1971), and 3) areal samples, like manual sampling (Billi & Paris, 1992), adhesive sampling (*e.g.*, Little & Mayer, 1976; Diplas, 1992) and photographic areal sampling (*e.g.*, Graham *et al.*, 2005a, 2005b; Buscombe *et al.*, 2010; Detert & Weitbrecht, 2012). Depending on the sampling method, the obtained data set must be converted to be comparable to the analyses of other sampling methods. For instance, the statistics of a volumetric sampling are different from sampling along a line. Conversion between the various combinations of sampling methods and analysis is discussed in detail in Wolman (1954), Kellerhals and Bray (1971), Fehr (1987b), and Graham *et al.* (2012).

4.6.3 Volumetric sampling and sieving

4.6.3.1 Sampling

Several procedures and equipment for taking volumetric samples have been developed (Bunte & Abt, 2001). Sample volumes can be collected using a dredger or manually with a shovel. Depending on site conditions, samples may also be gathered as grab samples, using mesh bag scoops, or by collecting freeze or resin cores. The latter two often tend to overestimate the finer material whereas the former tends to overestimate the larger particles due to wash-out effects of fines. The main challenge during the sampling process is to get an undisturbed sample.

Furthermore, different empirical relations can be found in the literature for the required mass of an representative sample. The required sampling volume V usually is expressed as a function of the maximum particle size diameter as $V = i\, d_{\max}{}^k + j$, with i, j being factors and k ranging between 1 and 3. A rule of thumb to estimate a sufficiently large representative sampling volume V was given by Huber (1966):

$$V\ [\mathrm{m}^3] = 2.5\ d_{\max}\ [\mathrm{m}] \tag{4.6.1}$$

Based on an approach originally given by Church *et al.* (1987), Bunte and Abt (2001) proposed the following relationship for the required mass of a sample:

$$m\ [\mathrm{kg}] = 2881.6\ d_{\max}\ [\mathrm{m}] - 47.56 \tag{4.6.2}$$

without limit restrictions to $d_{\max}$.

4.6.3.2 Sieving

During laboratory sieving a volumetric sample is sorted using sieves with decreasing mesh sizes. The resulting ranges of sorted sizes may then be, e.g., classified based on the ϕ-scale (Krumbein, 1938) (see Appendix 4.A.2). The mass m_i retained in each sieve is weighed and its fraction is determined in relation to the total sample mass m_{tot}. The grading curve typically is documented as a plot of the cumulative sum of the retained mass against sieve aperture size, the aperture size being a proxy of sediment diameter d_i (Figure 4.6.1; d denotes the particle diameter). The resulting cumulative frequency distribution $p_i(d_i)$ represents the mass fraction of sediment that is smaller than diameter d_i.

In most cases, square-hole sieves are used for laboratory sieving. For such sieves, the grain area formed by the b and c axis determines if a grain is retained or not (a, b and c define the longest, intermediate and shortest axis of a sediment grain, respectively; see appendix). The square sieve opening d_s classifies the size-fractions. The geometric relation between d_s and the intermediate axis b reads (Church *et al.*, 1987)

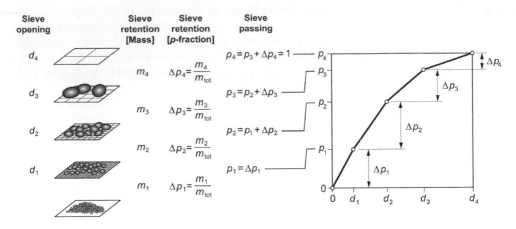

Figure 4.6.1 Illustration of the laboratory sieving procedure.

$$d_s = b \left(\frac{1}{\sqrt{2}} \sqrt{1 + \left(\frac{c}{b}\right)^2} \right)$$ (4.6.3)

with values of c/b = [0.0–1.0]. For infinitesimally flat particles to spherical particles d_s = [$\sqrt{0.5}$–1.0] b, and for natural river beds d_s = [0.8–0.9] b. Thus, the results obtained by square hole sieving typically give a smaller value than the 'real' length of the intermediate b-axis.

4.6.4 Surface sampling and sieving

4.6.4.1 Line-by-number sampling

The line-by-number sampling or transect-by-number method, a typical surface sampling procedure, is a pebble count method commonly used for gravel-bed rivers (Fehr, 1987a, 1987b). The procedure is a user-friendly *in situ* sampling method to analyze a gravel surface and to estimate the grain-size distribution of the subsurface bed material without complex utilities. For this method, a string is spanned over the dry (or nearly dry) bed parallel to the main flow direction (Figure 4.6.2). The b-axes of all grains that lie below the line and that are larger than a threshold value of 4.1 mm (Fehr, 1987a) are

Figure 4.6.2 Sketch of line-by-number sampling (top-view). For the subsequent analysis, the b-axes of the gray-shaded grains lying along the line are measured.

measured. For practical purposes a threshold value of 10 mm is recommended in field applications (Fehr, 1987b).

The sampling data are classified into fractions defined in the sampling log. To ensure the representativeness of the analysis, at least 150 grains should be counted, and at least 30 grains should belong to medium grain size fractions. The transfer calculation of a (truncated) line-sample of a surface layer to an equivalent full grading curve of the surface or subsurface layer comprises three steps; only after applying these steps a result equivalent to a volumetric analysis is obtained.

In the first step, the distribution by "number along the line on the surface" is transferred to a distribution by "mass fraction in the subsurface layer", *i.e.*, a quasi-sieve passing, by using

$$\Delta p_i = \frac{\Delta q_i \, d_{mi}{}^{\alpha}}{\sum\limits_{1}^{n} \Delta q_i \, d_{mi}{}^{\alpha}} \qquad (4.6.4)$$

with Δp_i = mass fraction i (in the subsurface layer, without fines), Δq_i = (number of grains in fraction i)/(number of grains in entire sample), d_{mi} = characteristic (mean) grain-diameter of fraction i, n = number of fractions, and α = fitting parameter. Fehr (1987a) found a value of $\alpha = 0.8$ by curve fitting to an extensive data set.

As finer grains are neglected within the sampling process, their cumulative frequency has to be corrected in a second step. To account for the mass fraction of finer grains, p_i has to be corrected toward p_{iC} via

$$p_{iC} = f + (1-f) \sum\limits_{1}^{i} \Delta p_i \qquad (4.6.5)$$

with a pre-estimation of percentages for fine material of $f = 0.25$ for the subsurface layer (Fehr, 1987a, 1987b). However, a closer inspection of the original data of Fehr (1987a) shows, that $f = \sim0.4$ should be chosen for a subsurface layer with $[d_m, d_{90}] = [29-43, 68-124]$ (mm), if a pebble-count threshold of 10 mm is taken instead of 4.1 mm. Figure 4.6.3 summarizes the methodology by example plots of the results gained during the conversion procedure. Note that compared to Fehr (1986) the first and the second step are permuted.

To close the gap due to the missing finest fractions in the sample, Fehr (1986, 1987a, 1987b) suggests using a Fuller distribution (Fuller & Thomson, 1907), *i.e.*, a grain-size distribution with minimal pore volume, via

$$p_f = \left(\frac{d_i}{d_{max\,FU}} \right)^{0.5} \qquad (4.6.6)$$

with d_{maxFU} being the maximal (virtual) grain-diameter of a Fuller distribution. After choosing an appropriate overlapping region of p_{iC} and p_f, *i.e.*, a region of similar $\Delta p/\Delta d$, the finer fractions of p_{iC} are accordingly replaced by p_f. As a side effect, the pre-estimation of f gets slightly corrected for fractions smaller than the overlapping region.

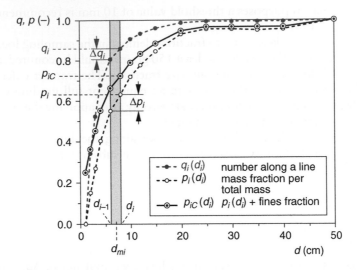

Figure 4.6.3 Basic data $q_i(d_i)$ obtained by line sampling, converted data $p_i(d_i)$ and corrected results $p_{iC}(d_i)$. The latter represents the estimated grain size distribution of the subsurface layer for fraction of larger grains.

An explicit square-hole sieve correction (Church *et al.*, 1987) is disregarded during the line sampling analysis. However, in a simplified approach it is implicitly incorporated as the Fehr's three-step methodology was calibrated against laboratory (square-hole) sieving data.

4.6.4.2 Photo sieving

Image based sieving is time- and cost-efficient. Several automatic approaches have been developed over the past years to obtain areal surface information from digital photography or laser scanning with the goal of achieving only a single characteristic grain size parameter. Successful procedures originate from *e.g.*, Carbonneau *et al.* (2004, texture analysis of airborne photographs), Heritage and Milan (2009, texture analysis of terrestrial laser-scanning), or Buscombe *et al.* (2010, frequency analysis of digital micro-photographs). Beyond this, techniques to detect and measure single grain-areas in digital photographs allow for classifying grain-sizes at the uppermost layer of a gravel bed. Weichert *et al.* (2004) used a simple grayscale threshold approach to determine a binary image where single grain elements are separated from interstices. Graham *et al.* (2005a, 2005b) applied a more complex approach based on a double grayscale threshold, bottom-hat filtering and a simple watershed algorithm. In-line with the research of Graham and co-workers, similar procedures have been applied by *e.g.*, Maerz *et al.* (1996), Gislao (2009), Strom *et al.* (2010), ASTM E1382-97 (2010) and Kozakiewicz (2013).

Detert and Weitbrecht (2012) developed the software BASEGRAIN. It is a free software-tool with a graphical user interface that enables non-intrusive gravelometric analysis of non-cohesive river beds. Furthermore, if geotagged photographs are analyzed, georeferencing of the results is done automatically and analysis reports can be exported to common file formats. Compared with the approach of Graham and co-workers, the results are more accurate, especially due to an improved separation methodology and image filtering for determination of areas of single grains. However, typically the results of every image based grain-size analysis suffer from the truncation at the fine end of the grain-size distribution. BASEGRAIN computes the truncated grain-size distribution of the top view layer, but also gives estimates of the subsurface layer according to the approach of Fehr (1987a, 1987b).

The core of the implemented five-step methodology in BASEGRAIN involves MATLAB-based object detection techniques applied to analyze digital top-view photographs of gravel layer surfaces. The steps are illustrated in Figure 4.6.4 and are explained in the following. Figure 4.6.5 gives typical snapshot results. In the first step, interstices are detected using a double grayscale threshold approach. Single grayscale threshold values are determined for the initial grayscale gravel image in a rasterized block structure of typically 16×16 px^2. The resulting threshold matrix is applied twice to the grayscale image to form binary images which show possible interstice areas, whereby during the second application the threshold matrix gets attenuated by multiplication with a positive factor <1. Interstice areas on both binary images are object detected. A feature-AND operation is applied to exclude possible interstices on the first binary image if they are not confirmed by the existence of smaller interstices on the second binary image. The second step applies a bottom-hat transformation technique to determine further interstices. Here, possible interstices are

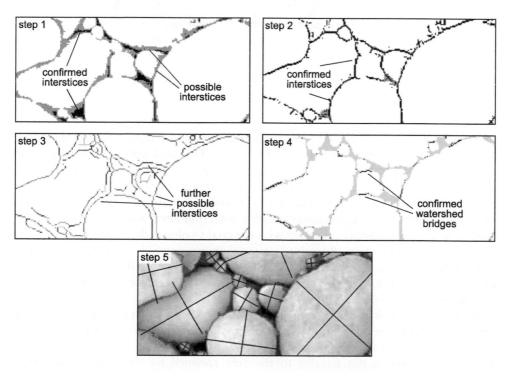

Figure 4.6.4 Illustration of the five-step methodology as applied in BASEGRAIN.

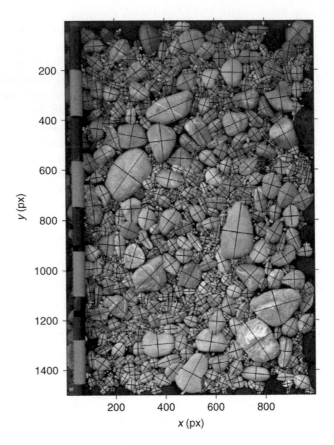

Figure 4.6.5 Automatic object detection by free software BASEGRAIN showing the final result with detected *a*- and *b*-axis on a top view photograph.

confirmed if they are connected by at least a small percentage (*e.g.*, ≥ 5%) of their area to confirmed extended interstices from the first step. The third step uses information of edge-detection and further gradient filter techniques to reveal further possible interstices that are confirmed or excluded in a similar way as in the first step. Within the fourth step the focus changes from detection of interstices to separation of single grain-areas. This is done by the combination of edge-detection filters and watershed bridges. Step five is needed for final operations with the goal to obtain the region properties of each grain's top view area. Boundary grains that are not fully included within the analyzed frame are blanked out. Thus a misleading statistical analysis of the characteristic diameter is avoided. Finally, confirmed grain areas are replaced with ellipses of the same normalized second central moments.

Image based screenings belong to surface sampling methods that are typically applied without contact to the bed – and without destroying it. Therefore the accuracy of the estimates a' and b' of the 'real' *a*- and *b*-axes suffers if grains are partly buried, if the photographic plane is not parallel to the area spanned by the grains' main axes ("foreshortening"), or if overlapping or even imbricated bed-surface structures are

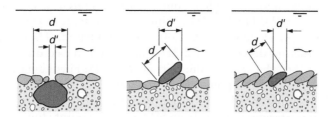

Figure 4.6.6 Three potential causes for underestimation of grain sizes object detected on top view photographs (redrawn from Graham *et al.*, 2010). From left to right: burial, foreshortening, and overlapping (imbrication).

present (Graham *et al.*, 2010). Figure 4.6.6 is a depiction of possible errors. Graham *et al.* (2010) summarize related biases (including further error sources associated with the methods used to collect the data) that have been reported since the 1970's. The relation is b'/b = 0.80–0.99 (= 0.01–0.33 ϕ-units), indicating that the errors that underestimate the b-axis are of the same order of magnitude as errors associated with determining grain-sizes using square hole sieves (Church *et al.*, 1987).

Several investigations have helped to define a representative area for photographic areal sampling. Fripp and Diplas (1993) proposed a criterion that a sampling area should comprise at least 100 times the area of the largest particle. Depending on the sampling method and the required precision, Graham *et al.* (2010) recommend analyzing image areas of 50–200(400) times the area of the largest particle.

The minimum detectable grain size depends mainly on the resolution of the photograph. Effects like texture, illumination and wetting are of minor significance. Graham *et al.* (2005b) found that for their object detection procedure, the measurement error increases substantially for grains with an apparent b-axis smaller than 23 px (approximately 16 mm for the scale of their imagery). However, the software BASEGRAIN indicates that this limit for the detectable b-axis may be halved to at least 10–15 px (typically 4–8 mm for a customary hand held camera).

4.A APPENDIX

4.A.1 Structure from Motion and MultiView Stereo

A promising photogrammetric approach to get georeferenced high resolution digital elevation models is the Structure from Motion (SfM) technique combined with MultiView Stereo (MVS) algorithms. SfM-MVS is well established in structure analysis of cultural heritage (*e.g.*, Guidi *et al.*, 2004; Pereira Uzal, 2016), but is increasingly used in geoscience (*e.g.*, Carrivick *et al.*, 2016) and to survey river environments (*e.g.*, Javernick *et al.*, 2014; Smith & Vericat, 2015) as well. A detailed description of SfM-MVS methods is given in Smith *et al.* (2016).

SfM uses a series of input images with overlapping view perspectives to estimate 3D camera positions and reconstructs 3D scene geometry points. It applies feature detectors such as SIFT (Lowe, 2004) or SURF (Herbert *et al.*, 2008), tracks them by the KLT feature tracker (Lucas & Kanade, 1981; Tomasi & Kanade, 1991) or similar matching algorithms, and then filters out incorrect matches, *e.g.*, by RANSAC (Fischler & Bolles, 1981). Finally, 3D positions of the feature trajectories and the camera positions (*e.g.*, Dellaert *et al.*, 2000) are estimated. They can be used to recover a scene structure described by a 3D point cloud. Based on the these results, MVS methods (*e.g.*, Strecha *et al.*, 2006; Seitz *et al.*, 2006; Furukawa & Ponce, 2007) are then used to improve the point cloud accuracy. As MVS increases the point density typically by several orders of magnitude, the final output is called a 'dense point cloud' and the intermediate one from SfM a 'sparse point cloud'. MVS methods construct rigid structures only, and moving objects like vessels, fauna or tracer particles are automatically ignored. The output of MVS is a set of oriented points with the 3D coordinate color information and the surface normal vector at each oriented point. Both the sparse and the dense point clouds can easily be georeferenced from a small number of ground control points collected in the field or from measurements of camera positions directly saved on the Exif-data of each photo. The 3D point clouds, if referenced in a global or local system, can then be used to create a variety of digital elevation products.

Morgan *et al.* (2017) summarized image acquisition and processing routines for SfM-MVS applications in laboratory flumes that are also valid for outdoor conditions. Their findings indicate that SfM-MVS can provide topographic data at higher resolution and of similar accuracy to TLS but at lower cost. Woodget *et al.* (2015) used a simplified refraction correction to show that airborne through-water measuring is possible by SfM-MVS for water depths smaller than 0.70 m. Recently, Dietrich (2017) improved their approach by additionally considering camera positions. For the case of ideal conditions (*e.g.*, clear water, horizontal water surface) he found accuracies of ~0.02% of the flying height and precisions of ~0.1% of the flying height. Moreover, Detert *et al.* (2017) present an initial proof of concept for a solely image-based method which showed that even a crude computation of flow rates is possible without direct contact to water. Flow discharge is determined by PIV surface velocity fields based on airborne orthophotos obtained via SfM-MVS and by bathymetry estimates based on both through-water SfM-MVS and turbulence metrics.

Two illustrations related to SfM-MVS are given below. Figure 4.A.1.1 gives an example reported in Detert *et al.* (2016a) where SfM-MVS was applied in the laboratory to determine the evolution of a movable bed. In comparison with reference values based on a laboratory laser scanner, the elevation models created using a hand-held high resolution single reflex camera showed to be of similar millimeter accuracy. An example of a field application is shown in Figure 4.A.1.2. Here, Detert *et al.* (2016a) used SfM-MVS to get a measure of the roughness of gravel patches by the standard deviation of the vertical coordinate of the gravel surface. The measure is directly related to characteristic grain sizes like d_{50} or d_{84} (*e.g.*, Aberle & Nikora, 2006; Heritage & Milan, 2009, see also Section 4.6.4.2).

4.A.2 Sediment size classification

Characteristic particle sizes are typically used as a proxy to describe or classify sediment mixtures (Table 4.A.2.1). To determine characteristic sizes the frequency distribution of different grain-sizes in a sampling volume has to be known. In many cases it follows

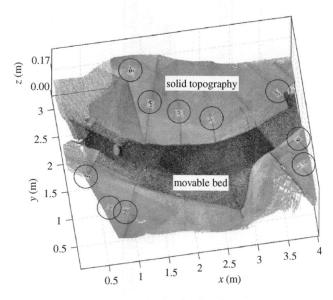

Figure 4.A.1.1 3D-view of a typical dense point cloud of a laboratory model created with SfM software PhotoScan (Agisoft, 2016) based on 26 photographs taken with a Nikon D7100. Ten coded ground control points are also visible (highlighted by circles).

Figure 4.A.1.2 3D-view of a dense point cloud covering a 0.5×0.5 m² cut-out of a gravel patch, generated with PhotoScan (Agisoft, 2016) based on 25 photographs taken with a Nikon D7100 resulting in a point cloud density of 1 million points/m².

approximately a lognormal distribution. The ϕ-scale (Krumbein, 1938), a modification of the Wentworth scale, is inspired by these findings. It expresses particle size d as

$$\phi = -\log_2(d) \tag{4.A.2.1}$$

with d in (mm). Numerous authors recast Equation (4.A.2.1) as $d = 2^{-\phi}$ to classify soils by integer values of ϕ to define the different class limits. For instance, ISO 14688-1

Table 4.A.2.1 Grain size classification by ISO 14688-1 (2002), supplemented with ϕ-and ψ-scalevalues.

Particle Sizes (mm)	0.002	0.0063	0.02	0.063	0.2	0.63	2.0	6.3	20	37.5	63	200	63	>630
ϕ (-)	+9.0			+4.0			-1.0				-5.98			
ψ (-)	-9.0			-4.0			+1.0				+5.98			
Soil fraction	Fine soil						Coarse soil					Very coarse soil		
Sub-fraction	Clay	Silt			Sand			Gravel			Cobble	Boulder	Large Boulder	
		Fine	Medium	Coarse	Fine	Medium	Coarse	Fine	Medium	Coarse				

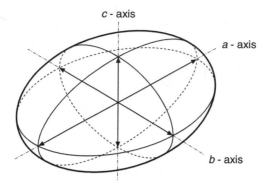

Figure 4.A.2.1 Three main axes of a grain.

(2002), see Table 4.A.2.1, gives a basic soil classification most commonly used for general engineering purposes. Depending on the authors, the chosen class limits differ considerably (*e.g.*, Williams *et al.*, 2006). Note that $\phi = 0$ represents a diameter of 1 mm and that with increasing ϕ the diameter d becomes smaller. The ϕ-transformation is convenient for studies that focus on sand and finer sediment. However, for studies in gravel-bed rivers, the ψ-scale was developed, with $\psi = -\phi$, to get increasingly larger values as particle sizes increase from sand to boulders.

The shape of a single grain can be approximated by an ellipsoid as shown in Figure 4.A.2.1. The shape is described by ratios of the length of the three main body axes a, b, c representing the longest, intermediate and shortest axis. Two detailed and slightly different definitions can be found in Yuzyk and Winkler (1991) and Gordon *et al.* (1992) regarding the determination of a and b. Typically, the b-axis (or, depending on the sieving method, a derivative thereof) is used as a standard proxy to a characteristic diameter in river hydraulics and morphodynamics, as the b-axis is decisive during the sieving process. However, larger boulders or blocks used for river engineering structures are handled differently. Typically they are not described by a characteristic length scale but by the weight of a sphere of equivalent volume (from which a characteristic diameter may be derived), as the mass of a block with a b-axis of, *e.g.*, 1.8 m may vary between 4 and 10 tons. More details are given in Whittaker and Jäggi (1986).

4.A.3 Spatial variability

At the surface of a gravel bed river, the flow causes sorting of the bed material. Finer material is more likely to be washed out or transported – whereas coarser material is more likely to be deposited and, in turn, prevents fines from being washed out. Thus, a coarser layer forms at the bed surface. Detailed, partly different explanations for the formation and presence of a coarse-grained surface layer are given by Carling and Reader (1982), Parker *et al.* (1982), Dunkerley (1994) and Dietrich *et al.* (1989). If the surface material shields the subsurface material from the flow it is often called pavement or an armoring layer (Sutherland, 1987). However, finer material can be transported and deposited on top on the coarsened surface layer – when such finer material is

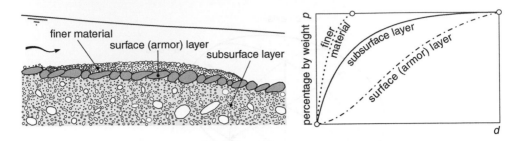

Figure 4.A.3.1 Armor and subsurface layer with mobilized, finer material on top. Left: illustration of an idealized stratigraphy. Right: schematic plot of related grain size distributions.

supplied from upstream. Thus, an armor layer in turn can be locally covered by finer material. Figure 4.A.3.1 illustrates this phenomenon.

Mean particle sizes of bed-material may vary longitudinally, transversally, and vertically. This variability occurs at various spatial scales, *i.e.*, the local scale, a stream section, and the stream reach. Figure 4.A.3.2 shows a typical variation of grain-diameter d_m along a river reach of 40 km. The spatial variability has to be considered when choosing a 'characteristic' site for sampling and selecting appropriate numbers and sizes of samples. Consequently, sampling and the subsequent sieve-analysis must always be performed at several sites of a river bed and ideally with several methods to get an accurate estimate of the grain-size distribution.

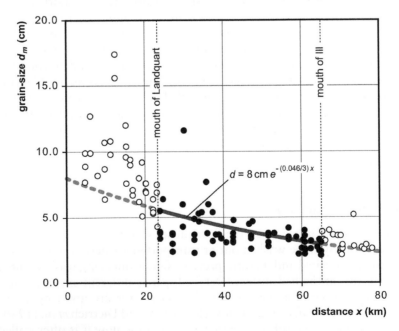

Figure 4.A.3.2 Variations of the mean grain size in the Alpine Rhine River between the mouth of rivers Landquart and Ill due to grain sorting and abrasion.

REFERENCES

Aberle, J. (2007) Measurements of armour layer roughness geometry function and porosity. *Acta Geophys*, doi:10.2478/s11600-11006-10036-11605.

Aberle, J. & Nikora, V. (2006) Statistical properties of armored gravel bed surfaces. *Water Resour Res*, 42, W11414, doi:11410.11029/12005WR004674.

Agisoft (2016) *Agisoft PhotoScan Professional Edition* (version 1.2.5). www.agisoft.com.

Ainslie, M.A. (2010) *Principles of Sonar Performance Modelling*. Berlin, Springer.

Alho, P., Vaaja, M., Kukko, A., Kasvi, E., Kurkela, M., Hyyppä, J., & Kaartinen, H. (2011) Mobile laser scanning in fluvial geomorphology: Mapping and change detection of point bars. *Z Geomorphol*, 55, 31–50.

American National Standards Institute (2007) ANSI Z136.1. *Safe use of lasers*. ANSI.

American Society for Photogrammetry and Remote Sensing (ASPRS) (2013) *Manual of Photogrammetry, 6th edition*. McGlone, C. (ed.) Bethesda, Maryland, American Society for Photogrammetry and Remote Sensing.

Antonarakis, A.S., Richards, K.S., Brasington, J., & Bithell, M. (2009) Leafless roughness of complex tree morphology using terrestrial lidar. *Water Resour Res*, 45, W10401, doi:10.1029/2008WR007666.

Antonarakis, A.S., Richards, K.S., Brasington, J., & Muller, E. (2010) Determining leaf area index and leafy tree roughness using terrestrial laser scanning. *Water Resour Res*, 46, W06510 doi:10.1029/2009WR008318.

Ashmore, P.E., & Church, M. (1998) Sediment transport and river morphology: A paradigm for study. In: Klingeman, P.C., Beschta, R.L., Komar, P.D., & Bradley, J.B. (eds.) *Gravel-Bed Rivers in the Environment*. Water Resources Publications LLC. Highlands Ranch, Colorado. pp. 115–148.

ASTM E1382-97 (2010) *Standard Test Methods for Determining Average Grain Size Using Semiautomatic and Automatic Image Analysis*. [Online] ASTM International. Available from: www.astm.org/Standards/E1382.htm [Accessed 15th February 2017].

Atkinson, K.B. (1996) *Close Range Photogrammetry and Machine Vision*. Latheronwheel, Scotland. Whittles Publishing.

Bangen, S.G., Wheaton, J.M., Bouwes, N., Bouwes, B., & Jordan, C. (2014) A methodological intercomparison of topographic survey techniques for characterizing wadeable streams and rivers. *Geomorphology*, 206, 343–361.

Bates, P.D., Marks, K. J., & Horritt, M. S. (2003) Optimal use of high resolution topographic data in flood inundation models. *Hydrol Proc*, 17(3), 537–557.

Beraldin, J., Blais, F., Boulanger, P., Cournoyer, L., Domey, J., El-Hakim, S. F., & Taylor, J. (2000) Real world modelling through high resolution digital 3D imaging of objects and structures. *ISPRS J. Photogramm Remote Sens*, 55(4), 230–250.

Bertin, S. & Friedrich, H. (2014) Measurement of gravel-bed topography: Evaluation study applying statistical roughness analysis. *J Hydraul Eng*, 140(3), 269–279.

Besl, P.J. & McKay, N.D. (1992) Method for registration of 3-D shapes. *IEEE Trans Pattern Anal Mach Intell*, 14(2), 239–256.

Billi, P. & Paris, E. (1992) Bed sediment characterization in river engineering problems. *Erosion and Sediment Transport Monitoring in River Basins*. IAHS Report Number 210. 11–20.

Bouratsis, P., Diplas, P., Dancey, C.L., & Apsilidis, N. (2013) High-resolution 3-D monitoring of evolving sediment beds. *Water Resour Res*, 49(2), 977–992.

Bradley, S. (2007). *Atmospheric Acoustic Remote Sensing: Principles and Applications*. Boca Raton, CRC press.

Brasington, J., Langham, J., & Rumsby, B. (2003) Methodological sensitivity of morphometric estimates of coarse fluvial sediment transport. *Geomorphology*, 53(3), 299–316.

Brasington, J., Vericat, D., & Rychkov, I. (2012) Modeling river bed morphology, roughness, and surface sedimentology using high resolution terrestrial laser scanning. *Water Resour Res, 48*, W11519, doi:10.1029/2012WR012223.

Brinker, R.C. & Minnick, R. (2012) *The Surveying Handbook*. 2nd edition. Berlin, Springer.

Bunte, K. & Abt, S.R. (2001) *Sampling surface and subsurface particle-size distributions in wadable gravel- and cobble-bed streams for analyses in sediment transport, hydraulics, and streambed monitoring*. Gen. Tech. Rep. RMRS-GTR-74. Fort Collins, CO. U.S. Department of Agriculture, Forest Service, Forest Service, Rocky Mountain Research Station. [Online] Available from: http://www.fs.fed.us/rm/pubs/rmrs_gtr074.pdf [Accessed 30th January 2017].

Buscombe, D., Rubin, D.M., & Warrick, J.A. (2010) A universal approximation of grain size from images of non-cohesive sediment. *J Geophys Res, 115*, F02015, doi:10.1029/2009JF001477.

Butler, J.B., Lane, S.N., & Chandler, J.H. (1998) Assessment of DEM quality for characterizing surface roughness using close range digital photogrammetry. *Photogramm Rec, 16*(92), 271–291.

Cameron, S.M., Nikora, V.I., Albayrak, I., Miler, O., Stewart, M., & Siniscalchi, F. (2013) Interactions between aquatic plants and turbulent flow: A field study using stereoscopic PIV. *J Fluid Mech, 732*, 345–372.

Carbonneau, P.E., Lane, S.N., & Bergeron, N.E. (2004) Catchment-scale mapping of surface grain size in gravel bed rivers using airborne digital imagery. *Water Resour Res, 40*, W07202, doi:10.1029/2003WR002759.

Carling, P.A. & Reader, N.A. (1982) Structure, composition and bulk properties of upland stream gravels. *Earth Surf Proc Land, 7*(4), 349–365.

Carrivick, J.L., Smith, M.W., & Quincey, D.J. (2016) *Structure from motion in the Geosciences*. Chichester, Wiley-Blackwell. ISBN 978-1-118-89584-9.

Chandler, J., Ashmore, P., Paola, C., Gooch, M., & Varkaris, F. (2002) Monitoring river-channel change using terrestrial oblique digital imagery and automated digital photogrammetry. *Ann Assoc Am Geogr, 92*(4), 631–644.

Christ, R.D., & Wernli, R.L. (2014) *The ROV Manual – a User Guide for Remotely Operated Vehicles*. 2nd edition. Oxford, Butterworth-Heinemann.

Church, M.A., McLean, D.G., & Wolcott, J.F. (1987) River bed gravels: Sampling and analysis. In: Thorne, C.R., Bathurst, J.C., & Hey, R.D. (eds.). *Sediment transport in gravel-bed rivers*. Chichester, John Wiley & Sons. pp. 43–88.

Coleman, S.E., Nikora, V.I., & Aberle, J. (2011) Interpretation of alluvial beds through bed-elevation distribution moments. *Water Resour Res, 47*(W11505), doi:10.1029/2011WR010672.

Dai, F., Feng, Y., & Hough, R. (2014) Photogrammetric error sources and impacts on modeling and surveying in construction engineering applications. *Visualization in Engineering 2*:2.

de Jong, C. (1995) Temporal and Spatial interactions between river bed roughness, geometry, bed load transport and flow hydraulics in mountain streams—Examples from Squaw Creek (Montana, USA) and Lainbach/Schmiedlaine (Upper Bavaria, Germany). *Berlin Geogr Abh, 59*, 1–229.

Dellaert, F., Seitz, S., Thorpe, S., & Thrun, S. (2000) *Structure from Motion without Correspondence*. Technical Report CMU-CS-91-132. Pittsburgh, Carnegie Mellon University.

DeVries, P. & Goold, D.J. (1999) Levelling rod base required for surveying gravel river bed surface ele-vations. *Water Resour Res, 35*(9), 2877–2879.

Detert, M. & Weitbrecht, V. (2012) Automatic object detection to analyze the geometry of gravel grains – stand-alone tool. In: Muños, R.M. (ed.) *Proceedings of the River Flow 2012*. London, Taylor & Francis. ISBN 978-0-415-62129-8. pp. 595–600.

Detert, M., Kadinski, L., & Weitbrecht, V. (2016a) Feasibility tests to airborne gravelometry for prealpine rivers. In: Wieprecht, S. & Noack, M. (eds.) *Proceedings of the International Symposium on River Sedimentation, Stuttgart, Germany*. London, Taylor & Francis. ISBN 978-1-138-02945-3, pp. 1258–1263.

Detert, M., Huber, F., & Weitbrecht, V. (2016b) Unmanned Aerial Vehicle-based surface PIV experiments at Surb Creek. In: Constantinescu, G., Garcia, M., & Hanes, D. (eds.) *Proceedings of the River Flow 2016*, pp. 563–568.

Detert, M., Johnson, E.D., & Weitbrecht, V. (2017) Proof of concept for low-cost and non-contact synoptic airborne river flow measurements. *Int J Remote Sens*, 38(8–10), 2780–2807, doi: 10.1080/01431161.2017.1294782

Dietrich, J.T. (2017) Bathymetric Structure-from-Motion: Extracting shallow stream bathymetry from multi-view stereo photogrammetry. *Earth Surf Proc Land*, 42, 355–364.

Dietrich, W.E., Kirchner, J.W., Ikeda, H., & Iseya, F. (1989) Sediment supply and the development of the coarse surface layer in gravel-bedload rivers. *Nature*, 340, 215–217.

Diplas, P. (1992) Surface and subsurface granulometry. In: *Grain Sorting Seminar*. Laboratory of Hydraulics, Hydrology and Glaciology (VAW), ETH Zurich, Zurich, Switzerland. Report number: 117, pp. 157–162. [Online] Available from: http://www.vaw.ethz.ch/das-institut/vaw-mitteilungen/1990-1999.html [Accessed 7th February 2017].

Dix, M., Abd-Elrahman, A., Dewitt, B., & Nash, L., Jr. (2012) Accuracy evaluation of terrestrial LIDAR and multibeam sonar systems mounted on a survey vessel. *J Surv Eng*, 138(4), 203–213.

Dowling, D.R. & Sabra, K.G. (2015) Acoustic Remote Sensing. *Annu Rev Fluid Mech*, 47(1), 221–243.

Dunkerley, D.L. (1994) Discussion: Bulk sampling of coarse clastic sediments for particle-size analysis. *Earth Surf Proc Land*, 19(3), 255–261.

Eilertsen, R.S., Olsen, N.R.B., Rüther, N., & Zinke, P. (2013) Channel-bed changes in distributaries of the lake Øyeren delta, southern Norway, revealed by interferometric sidescan sonar. *Norw J Geol*, 93, 25–35.

Fehr, R. (1986) A method for sampling very coarse sediments in order to reduce scale effects in movable bed models. In: *Proceeding of the IAHR Symposium 86 on Scale Effects in Modelling Sediment Transport Phenomena*, Toronto Ontario, August 1986. pp. 383–397.

Fehr, R. (1987a) *Geschiebeanalysen in Gebirgsflüssen (Grain size analysis in torrents)*. Laboratory of Hydraulics, Hydrology and Glaciology (VAW), ETH Zurich, Zurich, Switzerland. Report number: 92. [Online] Available from: http://www.vaw.ethz.ch/das-institut/vaw-mitteilungen/1980-1989.html [Accessed 7th February 2017].

Fehr, R. (1987b) Einfache Bestimmung der Korngrössenverteilung von Geschiebematerial mit Hilfe der Linienzahlanalyse (Simple detection of grain size distribution of sediment material using line-by-number analysis). (in German) *Schweizer Ingenieur und Architekt*, 105(38), 1104–1109. [Online] Available from: http://retro.seals.ch/ [Accessed 31th January 2017].

Fischler, M.A. & Bolles, R.C. (1981) Random sample consensus: A paradigm for model fitting with applications to image analysis and automated cartography. *Commun ACM*, 24(6), 381–395.

Franceschi, M., Teza, G., Preto, N., Pesci, A., Galgaro, A., & Girardi, S. (2009) Discrimination between marls and limestones using intensity data from terrestrial laser scanner. *ISPRS J Photogramm Remote Sens*, 64(6), 522–528.

Fripp, J.B. & Diplas, P. (1993) Surface sampling in gravel streams. *J Hydraul Eng*, 119(4), 473–490.

Fuller, W.B. & Thomson, S.E. (1907) The laws of proportioning concrete. *Proc Am Soc Civ Eng*, 33(3), 222–298.

Furukawa, Y. & Ponce, J. (2010) Accurate, dense, and robust multiview stereopsis. *IEEE Trans Pattern Anal Mach Intell*, 32(8), 1362–1376.

Gislao, M. (2009) *Performing a grains analysis in 5 easy steps*. [Online] Available from: http://www.olympusamerica.com/seg_industrial/files/Grains123_092509.pdf [Accessed 31st January 2017].

Godding, R., Hentschel, B., & Kauppert, K. (2003) Videometrie im wasserbaulichen Versuchswesen. *Wasserwirtschaft, HY*(4), 36–40.

Gordon, N.D., McMahon, T.A., & Finlayson, B.L. (1992) *Stream Hydrology. An Introduction for Ecologists.* Chichester, John Wiley & Sons.

Graham, D.J., Reid, I., & Rice, S.P. (2005a) Automated sizing of coarse-grained sediments: Image-processing procedures. *Int Ass Math Geol, 37*(1), 1–21.

Graham, D.J., Rice, S.P., & Reid, I. (2005b) A transferable method for the automated grain sizing of river gravels. *Water Resour Res, 41*, W07020, doi:10.1029/2004WR003868.

Graham, D.J., Rollet, A.-J., Piégay, H., & Rice, S.P. (2010) Maximizing the accuracy of image-based surface sediment sampling techniques. *Water Resour Res, 46*, W02508, doi:10.1029/2008WR006940.

Graham, D.J., Rollet, A.-J., Rice, S.P., & Piégay, H. (2012) Conversions of surface grain-size samples collected and recorded using different procedures, *J Hydraul Eng, 138*(10), 839–849.

Guidi, G., Beraldin, J.A., & Atzeni, C. (2004) High accuracy 3D modelling of cultural heritage: The digitizing of Donatello. *IEEE Trans Image Process, 20*(13), 370–380.

Heays, K.G., Friedrich, H., & Melville, B.W. (2014) Laboratory study of gravel-bed cluster formation and disintegration. *Water Resour Res, 50*(3), 2227–2241.

Henning, M. (2013). *Mehrdimensionale statistische Analyse räumlich und zeitlich hoch aufgelöster Oberflächen von Dünenfeldern.* Mitt. Leichtweiß-Institut für Wasserbau No. 160. Braunschweig, Technische Universität Braunschweig.

Henning, M., Hentschel, B., & Hüsener, T. (2008) Determination of channel morphology and flow features in laboratory models using 3D-photogrammetry. In: *Proceedings of the River Flow 2008, Kubaba,* pp. 2383–2390.

Herbert, B., Ess, A., Tuytelaars, T., & Van Gool, L. (2008) SURF: Speeded Up Robust Features. *Comput Vis Image Und, 110*(3), 346–359.

Heritage, G.L. & Milan, D.J. (2009) Terrestrial laser scanning of grain roughness in a gravel-bed river. *Geomorphology, 113*(1–2), 4–11.

Hodge, R., Brasington, J., & Richards, K. (2009a) In situ characterization of grain-scale fluvial morphology using Terrestrial Laser Scanning. *Earth Surf Proc Land, 34*(7), 954–968.

Hodge, R., Brasington, J., & Richards, K. (2009b) Analysing laser-scanned digital terrain models of gravel bed surfaces: Linking morphology to sediment transport processes and hydraulics. *Sedimentology, 56*(7), 2024–2043.

Horn, B.K.P. (1987) Closed-form solution of absolute orientation using unit quaternions. *J Opt Soc Am A, 4*(4), 629–642.

Huber, A. (1966) *Interner Bericht über die Durchführung von Geschiebeanalysen im Felde (Internal report to conduct field-analysis of bedload).* (Unpublished Report, in German). Zurich. Laboratory of Hydraulics, Hydrology and Glaciology (VAW), ETH Zurich, Switzerland.

ISO 14688-1 (2002) *Geotechnical Investigation and Testing – Identification and Classification of Soil* [Online]. International Organization for Standardization. Available from: http://www.iso.org/iso/iso_catalogue/catalogue_tc/catalogue_detail.htm?csnumber=25260 [Accessed 30th July 2015].

James, M.R. & Robson, S. (2012) Straightforward reconstruction of 3D surfaces and topography with a camera: Accuracy and geoscience application. *J Geophys Res., 117* (F03017, doi:10.1029/2011JF002289).

Javernick L, Brasington J., & Caruso B. (2014) Modelling the topography of shallow braided rivers using Structure-from-Motion photogrammetry. *Geomorphology, 213*, 166–182.

Johnson, H.P. & Helferty, M. (1990) The geological interpretation of side-scan sonar. *Rev Geophys, 28*(4), 357–380.

Kellerhals, R. & Bray, D.I. (1971) Sampling procedures for coarse fluvial sediments. *J Hydraul Div, 97*(HY8), 1165–1180.

Kozakiewicz, J. (2013) Automated image analysis for measuring size and shape of martian sands. *44th Lunar and Planetry Science Conference*. Extended abstract no. 2906. [Online] Available from: http://www.lpi.usra.edu/meetings/lpsc2013/programAbstracts/view/ [Accessed 31st January 2017].

Kraus, K. (2011) *Photogrammetry – Geometry from Images and Laser Scans*. 2nd edition. De Gruyter.

Krumbein, W.C. (1938) Size frequency distribution of sediments and the normal phi curve. *J Sed Petrol, 8*, 84–90.

Kuperman, W. & Roux, P. (2007) Underwater Acoustics. In: Rossing, T.D. (ed.) *Springer Handbook of Acoustics*. Berlin, Springer. pp. 149–204.

L-3 Communications SeaBeam Instruments (2000). Multibeam Sonar: Theory of operation. [Online] https://www.ldeo.columbia.edu/res/pi/MB-System/sonarfunction/SeaBeamMultibeam TheoryOperation.pdf. [Accessed 15th February 2017).

Lane, S.N. (2000) The measurement of river channel morphology using digital photogrammetry. *Photogramm Rec, 16*(96), 937–961.

Lane, S.N., James, T.D., & Crowell, M.D. (2000) Application of digital photogrammetry to complex topography for geomorphological research. *Photogramm Rec, 16*(95), 793–821.

Lane, S.N., Chandler, J.H., & Porfiri, K (2001) Monitoring river channel and flume surfaces with digital photogrammetry. *J Hydraul Eng, 127*(10), 871–877.

Leon, J.X. Roelfsema, C.M., Saunders, M.I., & Phinn, S.R. (2015) Measuring coral reef terrain roughness using 'Structure-from-Motion' close-range photogrammetry. *Geomorphology, 242*(1), 21–28.

Lichti, D.D. (2007) Error modelling, calibration and analysis of an AM–CW terrestrial laser scanner system. *ISPRS J Photogramm Remote Sens, 61*(5), 307–324.

Lichti, D. D., & Jamtsho, S. (2006) Angular resolution of terrestrial laser scanners. *The Photogramm Rec, 21*(114), 141–160.

Lichti, D.D. & Skaloud, J. (2010) Registration and Calibration. In: Vosselman, G. & Maas, H-G. (eds.) *Airborne and Terrestrial Laser Scanning*. Dunbeath, Whittles Publishing. pp. 83–133.

Little, W.C. & Mayer, P.G. (1976) Stability of channel beds by armoring. *J Hydraul Div, 102* (HY11), 1647–1661.

Lowe, D.G. (2004) Distinctive image features from scale-invariant keypoints. *Int J Comput Vis, 60*, 91–110.

Lucas, B.D. & Kanade, T. (1981) An iterative image registration technique with an application to stereo vision. In: *Proceedings of the 7th International Joint Conference on Artificial Intelligence*. San Francisco, Morgan Kaufmann Publishers Inc. 2, pp. 674–679.

Luhmann, T., Robson, S., Kyle, S., & Boehm, J. (2014) *Close-Range Photogrammetry and 3D Imaging*. 2nd edition. Berlin, de Gruyter.

Lurton, X. (2010) *An Introduction to Underwater Acoustics: Principles and Applications*. 2nd edition. London, Springer Praxis Publishing.

Lurton X. (2016) Modelling of the sound field radiated by multibeam echosounders for acoustical impact assessment. *Appl Acoust, 101*, 201–221.

Maerz, N.H., Palangio, T.C., & Franklin, J.A. (1996) WipFrag image based granulometry System. In: *Proceedings of the FRAGBLAST 5 Workshop on Measurement of Blast Fragmentation*, Canada, Montreal. pp. 91–99.

Mann, R. (1998) *Field Calibration Procedures for Multibeam Sonar Systems*. Report TEC-0103. Alexandria, VA. US Army Coprs of Engineers Topographic Engineering Center.

Milan, D.J., Heritage, G.L., & Hetherington, D. (2007) Application of a 3D laser scanner in the assessment of erosion and deposition volumes and channel change in a proglacial river. *Earth Surf Proc Land, 32*(11), 1657–1674.

Morgan, J.A., Brogan, D.J., & Nelson, P.A. (2017) Application of Structure-from-Motion photogrammetry in laboratory flumes. *Geomorphology, 276*, 125–143.

Nield, J.M., Wiggs, G.F.S., & Squirrell, R.S. (2011) Aeolian sand strip mobility and protodune development on a drying beach: Examining surface moisture and surface roughness patterns measured by terrestrial laser scanning. *Earth Surf Proc Land*, *36*(4), 513–522.

Nield, J. M., King, J., & Jacobs, B. (2014) Detecting surface moisture in aeolian environments using terrestrial laser scanning. *Aeolian Res*, *12*(March 2014), 9–17.

Parker, G., Klingeman, P.C., & McLean, D.G. (1982) Bedload and the size distribution of paved gravel-bed streams. *J Hydraul Div*, *108*(HY4), 544–571.

Pereira Uzal, J.M. (2016) 3D modelling in cultural heritage using Structure from Motion Techniques. *PH investigación*, *6*, 49–59.

Petrie, G. & Toth, C. K. (2009a) Introduction to laser ranging, profiling, and scanning. In: Shan, J. & Toth, C.K. (eds.) *Topographic Laser Ranging and Scanning Principles and Processing.* Boca Raton, CRC Press. pp. 1–27.

Petrie, G. & Toth, C. K. (2009b) Terrestrial laser scanners. In: Shan, J. & Toth, C.K. (eds.) *Topographic Laser Ranging and Scanning Principles and Processing.* Boca Raton, CRC Press. pp. 87–128.

Sagnes, P. (2010) Using multiple scales to estimate the projected frontal surface area of complex three-dimensional shapes such as flexible freshwater macrophytes at different flow conditions. *Limnol Oceanogr-Meth*, *8*, 474–483.

Saleh, M. & Rabah, M. (2016) Seabed sub-bottom sediment classification using parametric sub-bottom profiler. *NRIAG J Astron Geophy*, doi:10.1016/j.nrjag.2016.01.004.

Sampson, C.C., Fewtrell, T.J., Duncan, A., Shaad, K., Horritt, M.S., & Bates, P.D. (2012) Use of terrestrial laser scanning data to drive decimetric resolution urban inundation models. *Adv Water Resour*, *41*, 1–17.

Schenk T (2005) *Introduction to Photogrammetry*. Columbus, The Ohio State University.

Seitz, S.M., Curless, B., Diebel, J., Scharstein, D., & Szeliski, R. (2006) A comparison and evaluation of multi-view stereo reconstruction algorithms. In: *Proceedings of the 2006 IEEE Computer Society Conference on Computer Vision and Pattern Recognition. 1*, pp. 519–528.

Singal. S.P. (ed.) (1997) *Acoustic Remote Sensing Applications*. Lecture Notes in Earth Sciences, *69*, Springer, Berlin.

Smart, G.M., Aberle, J., Duncan, M., & Walsh, J. (2004) Measurement and analysis of alluvial bed roughness. *J Hydraul Res*, *42*(3), 227–237.

Smith, M.W. & Vericat, D. (2014) Evaluating shallow-water bathymetry from through-water terrestrial laser scanning under a range of hydraulic and physical water conditions. *River Res Appl*, *30*(7), 905–924.

Smith, M.W. & Vericat, D. (2015) From experimental plots to experimental landscapes: Topography, erosion and deposition in sub-humid badlands from Structure-from-Motion photogrammetry. *Earth Surf Proc Land*, *40*, 1656–1671.

Smith, M.W., Carrivick, J.K., & Quincey, D.J. (2016) Structure from motion photogrammetry in physical geography. *Prog Phys Geog*, *40*(2), 247–275.

Storz-Peretz, Y. & Laronne, J.B. (2015) Morphotextural characterization of dryland braided channels. *Geol Soc Am Bull*, *125*(9–10), 1599–1617.

Strecha, C., Fransens, R., & Van Gool, L. (2006) Combined depth and outlier estimation in Multi-View Stereo. In: *Proceedings of the 2006 IEEE Computer Society Conference on Computer Vision and Pattern Recognition.* doi:10.1109/CVPR.2006.78.

Strom, K.B., Kuhns, R.D., & Lucas, H.L. (2010) Comparison of automated photo grain sizing to standard pebble count methods. *J Hydraul Eng*, *136*(8), 461–473.

Sutherland, A.J. (1987) Static armor layers by selective erosion. In: Thorne, C.R., Bathurst, J.C., & Hey, R.D. (eds.) *Sediment Transport in Gravel-Bed Rivers*. Chichester, John Wiley. pp. 243–267.

Tomasi, C. & Kanade, T. (1991) *Detection and Tracking of Point Features*. Technical Report CMU-CS-91-132. Pittsburgh, Carnegie Mellon University.

Urick, R.J. (1983) *Principles of Underwater Sound*. New York, McGraw-Hill Inc.

Vaaja, M., Hyyppä, J., Kukko, A., Kaartinen, H., Hyyppä, H., & Alho, P. (2011) Mapping topography changes and elevation accuracies using a mobile laser scanner. *Remote Sens, 3*(3), 587–600.

Wackrow, R. & Chandler, J.H. (2008) A convergent image configuration for DEM extraction that minimises the systematic effects caused by an inaccurate lens model. *Photogramm Rec, 23*(121), 6–18.

Wackrow, R. & Chandler, J.H. (2011) Minimising systematic error surfaces in digital elevation models using oblique convergent imagery. *Photogramm Rec, 26*(133), 16–31.

Wehr, A. & Lohr, U. (1999) Airborne laser scanning – an introduction and overview. *ISPRS J Photogramm Remote Sens, 54*(2), 68–82.

Weichert, R., Wickenhäuser, M., Bezzola, G.R., & Minor, H.-E. (2004) Grain size analysis for coarse river beds using digital imagery processing. In: Greco, M., Carravetta, A., & Della Morte, R. (eds.) *Proceedingsof the River Flow 2004*, Naples, Italy. pp. 753–760.

Westoby, M.J., Brasington, J., Glasser, N.F., Hambrey, M.J., & Reynolds, J.M. (2012) 'Structure-from-Motion' photogrammetry: A low-cost, effective tool for geoscience applications. *Geomorphology, 179*, 300–314.

Wheaton, J.M., Brasington, J., Darby, S.E., & Sear, D.A. (2010) Accounting for uncertainty in DEMs from repeat topographic surveys: Improved sediment budgets. *Earth Surf Proc Land., 35*(2), 136–156.

Whittaker, J.G. & Jäggi, M.N.R. (1986) *Blockschwellen (Block sills)*. Laboratory of Hydraulics, Hydrology and Glaciology (VAW), ETH Zurich, Zurich, Switzerland. Report no. 91. [Online] Available from: http://www.vaw.ethz.ch/das-institut/vaw-mitteilungen/1980–1989.html [Accessed 7th February 2017].

Williams, R.D., Brasington, J., Vericat, D., Hicks, D. M., Labrosse, F., & Neal, M. (2011) Monitoring braided river change using terrestrial laser scanning and optical bathymetric mapping. In: Smith, M., Paron, P., & Griffiths, J.S. (eds.) *Geomorphological Mapping Methods and Applications*. Elsevier. pp. 507–532.

Williams, R.D., Brasington, J., Hicks, M., Measures, R., Rennie, C.D., & Vericat, D. (2013) Hydraulic validation of two-dimensional simulations of braided river flow with spatially continuous aDcp data. *Water Resour Res, 49*, 5183–5205, doi:10.1002/wrcr.20391.

Williams, R.D., Brasington, J., Vericat, D., & Hicks, D.M. (2014) Hyperscale terrain modelling of braided rivers: Fusing mobile terrestrial laser scanning and optical bathymetric mapping. *Earth Surf Proc Land, 39*(2), 167–183.

Williams, S.J., Arsenault, M.A., Buczkowski, B.J., Reid, J.A., Flocks, J.G., Kulp, M.A., Penland, S., & Jenkins, C.J. (2006) *Surficial sediment character of the Louisiana offshore continental shelf region: A GIS Compilation*. U.S. Geological Survey Open-File Report 2006-1195. [Online] Available from: http://pubs.usgs.gov/of/2006/1195/index.htm [Accessed 15th February 2017].

Wolman, M.G. (1954) A method of sampling coarse river-bed material. *Trans Am Geoph Union., 35*, 951–956.

Woodget, A., Carbonneau, P.E., Visser, F., & Maddock, I.P. (2015) Quantifying submerged fluvial topography using hyperspatial resolution UAS imagery and structure from motion photogrammetry. *Earth Surf Process Land, 40*, 47–64.

Wu, Y., Chen, Y., Yang, K., Ding, J., & Tang, Q. (2010) On methods of multi-beam data error correction. In: *Proceeding of the OCEANS 2010*, 24–27 May 2010, *Sydney*. IEEE. doi:10.1109/OCEANSSYD.2010.5603632.

Yuzyk, T.R. & Winkler, T. (1991) *Procedures for Bed-Material Sampling. Lesson Package No. 28*. Ottawa, Environment Canada, Water Resources Branch, Sediment Survey Section.

Chapter 5

Sediment Transport

5.1 INTRODUCTION

Accurate measurement of sediment transport is required for many applications of water resources management, such as prediction of reservoir filling rates, quantification of contaminant transport, and assessment of morphodynamic stability. Depending upon the problem at hand, measurements of suspended load and/or bedload will be required. Suspended load is measured within the water column whereas bedload is measured immediately above the stable bed. In general, bed material transport is measured as bedload, but saltating sand bed material can be captured as suspended load. Suspended load also includes fine silt and clay particles transported through the system as washload. The spatiotemporal distribution of the measurements may also be important to consider; for example, suspended sediments are usually measured as a particle concentration in a time-integrated water sample at a discrete location, while estimation of total suspended sediment flux requires integration of these concentrations with velocity distributed across the channel section. This section describes measurement techniques for both bedload and suspended load, including conventional physical sampling methods and recent advances in surrogate measurement techniques. Table 5.1.1 summarizes the main instruments described.

5.2 BEDLOAD

5.2.1 Physical traps and samplers

5.2.1.1 Principles of measurement

Physical traps and samplers are designed to obtain a sample of bedload particles transported into the measuring device during a known period. Typically particles are deposited in the device due to a reduced flow and particle velocity or because they are retained behind a sampler screen. Some methods enable collecting the entire grain size distribution of the bedload particles transported over the entire stream width, other methods are limited to sampling only some of the transported grain sizes. The latter methods are more restricted regarding spatial and temporal coverage. The bedload samples are subsequently analyzed to determine total mass as well as mass by grain-size classes ranging from sand to cobbles. Detailed discussions of physical traps and samplers can be found in Gray *et al.* (2010), Ryan *et al.* (2005) and Megahan (1999).

Table 5.1.1 Instruments and methods for measuring bedload and suspended load transport

Method	Type	Method	Lab/field	Scale of measurement	Temporal resolution
Retention basin	Bedload	direct	field	all bedload particles	very low
Trough sampler	Bedload	direct	field/lab	across stream width	low
Pit sampler	Bedload	direct	field/lab	parts of streambed	low
Bedload sampler	Bedload	direct	field	point	low
Hydrophones	Bedload	acoustic	field/lab	parts of streambed	high
Impact pipes	Bedload	acoustic	field	point/small line	high
Plate geophones	Bedload	seismic	field/lab	point/small area	high
Bed form tracking	Bedload	optical/ acoustic	field/lab	depends on resolution of bathymetry measurement	high
Tracer particle	Bedload	–	field/lab	used to study transport length; not suitable for bedload measurements	very low to high;
Particle tracking	Bedload	optical	lab	small area	high
ADCP bottom tracking	Bedload	acoustic	field	point along a line	high
Point sampler	Suspended	direct	field/lab	local	low
Depth integrating sampler	Suspended	direct	field	local over water depth	low
Optical backscatter	Suspended	optical	field/lab	local; calibration required	high
Laser diffraction	Suspended	optical	field/lab	point	high
Focused beam reflectance	Suspended	optical	field/lab	point	high
ADCP	Suspended	acoustic	field	profile	high
Acoustic particle flux-profiler	Suspended	acoustic	field/lab	profile	high
Multi-beam echo sounders	Suspended	acoustic	field	profile	high

5.2.1.2 Samplers and deployment methods

5.2.1.2.1 Sediment retention basins

In a closed and impounded sediment retention basin all bedload particles are deposited. Regular surveys of the deposits in the basin allow for determination of the cumulative sediment transport in the interval between two surveys. Bedload samples from the basin may be analyzed for grain size composition. A useful example of a sediment retention basin built for bedload transport measurements is given by Rickenmann (1997) (Figure 5.2.1). Sediment retention basins are often built for hazard prevention or mitigation purposes in mountain areas, but they are typically confined by an open type check dam structure (*e.g.* a slit-type check dam with horizontal beams), and thus the sampling efficiency is limited. If water flow is separated upstream of the retention basin, all coarser particles will deposit, and sediment accumulation in the basin may be measured automatically at shorter time intervals during a flood event, such as in the case of the Rio Cordon experimental site in Italy (Lenzi *et al.*, 1999). Other types of retention basins may include hydropower reservoirs, or settling basins built at water intakes for hydroelectric power production. Reservoirs will trap all particles transported

Figure 5.2.1 Example for a sediment retention basin (Erlenbach stream, Switzerland, after flood event of 20 June 2007; Photo D. Rickenmann).

as bedload (like natural lakes). Water intake settling basins, on the other hand, typically include a Tyrolean weir that shunts coarse particles over a rack past the settling basin entrance, which limits the maximum particles to be captured (Rickenmann & McArdell, 2008).

5.2.1.2.2 Trough or pit samplers

Similar to retention basins, trough or pit samplers are permanent installations in a stream, typically including a container fixed in the streambed (Figure 5.2.2). Troughs usually cover the entire stream width and are known for example as vortex and conveyor belt samplers (*e.g.* Milhous, 1973; Hayward & Sutherland, 1974; Tacconi & Billi, 1987). Pits typically span only a part of the streambed and have no *in situ* weighing apparatus. Being of the same type, Reid-type (Birkbeck) slot samplers automatically and continuously weigh the accumulated sediment (*e.g.* Reid *et al.*, 1980; Laronne *et al.*, 2003). According to Gray *et al.* (2010), troughs and pits tend to produce the most reliable bedload data, provided that they are not full, have slots that span the channel, are capable of capturing the largest bedload particles, and possess a slot length that exceeds the maximum saltation length. Therefore, Reid-type (Birkbeck) slot samplers or vortex and conveyor belt samplers are probably the most suitable methods to calibrate other physical samplers or surrogate samplers.

Figure 5.2.2 Example for a trough or pit sampler (Nahal Eshtemoa, Israel; photo D. Rickenmann).

5.2.1.2.3 Bedload samplers

Bedload samplers retain particles using a screen and include netframe samplers, such as the Helley-Smith (HS) sampler, and bedload traps. Helley-Smith samplers are an example of pressure-difference bedload samplers, which are designed so that the entrance velocity at the sampler is similar to the ambient stream velocity (Helley & Smith, 1971; Childers, 1999; Figure 5.2.3). These samplers collect particles small enough to enter the nozzle but larger than the mesh size of the collection bag. Helley-Smith samplers

Figure 5.2.3 Examples of bedload samplers. From left to right: US BL-84 (3 inch opening), Helley-Smith sampler (3 inch opening), Elwha sampler (8 by 4 inch opening), Toutle River 2 sampler (12 by 6 inch opening). Photo courtesy of Kristin Bunte, Colorado State University.

undersample particles with dimensions greater than about half the nozzle opening, which limits their use to smaller sediment sizes. The original HS sampler and the more recent BL-84 sampler have an opening of 7.6 by 7.6 cm and are designed for collection primarily of sand to medium gravel. Such samplers are typically operated manually by a single person, and thus operation at high flows may be difficult, although heavier versions are also available that may be deployed from a boat. There are also HS samplers with larger opening sizes, allowing larger grain sizes to be sampled: the 15.2 by 15.2 cm HS, the 20.3 by 10.2 cm Elwha sampler, the 30.4 by 15.2 cm Toutle River II sampler, and the 30.4 by 30.4 cm BTMA-2 sampler (Bunte et al., 2010a; Figure 5.2.3). The sampling efficiency of bedload samplers varies with grain size, and has been the subject of extensive research (e.g., Childers, 1999). Bunte-type samplers are an example of portable bedload traps (Bunte et al., 2004, 2010b) (BLT). The traps are mounted on ground plates anchored to the bed. They can be deployed in wadeable, coarse gravel and cobble bed streams up to moderately strong flows. At the Erlenbach stream, a small torrent, movable metal baskets of $1m^3$ size are installed at the downstream side of a large check dam; they can be automatically positioned at the centerline of the flow to collect bedload, and they are used to calibrate the Swiss plate geophone system (Rickenmann et al., 2012).

5.2.1.3 Spatial and temporal resolution

According to their design and deployment characteristics, the various sampler methods have different typical ranges of spatial and temporal resolution. For example, sediment retention basins provide temporally integrated samples which are representative of bedload transport at a given location along a stream channel. Trough or pit samplers can typically sample smaller volumes of accumulated bedload particles, while the samples represent bedload transport characteristics at a given stream cross-section, either covering the entire width or parts thereof. In principle, vortex and conveyor belt samplers can sample over longer time periods, if the samples can be either stored away or be continuously analyzed during the measurement. Bedload samplers are more flexible with regard to the selection of sampling locations both along and across a stream channel. Sampling duration and thus time-averaging depends on the volume of the sampler and on the transport rate. Typically, sampling duration is smaller compared to the other main sampler types, i.e. retention basins and trough or pit samplers. Still, it is generally difficult to obtain high spatial or temporal resolution from the spatially discrete and time-integrated samples collected with samplers.

5.2.1.4 Error sources

Typically, the main objective of these measurements is to derive bedload transport rates, either as a bulk quantity for a range of grain sizes or as fractional transport rates for given grain-size classes. There may be an interest in bedload transport rates with a high spatial and temporal resolution, or it may be more important to know a mean bedload transport rate over the channel width and averaged over some time. Additionally, the main focus of the measurements may be on the total sediment yield integrated over longer time durations or for example due to a sporadic bedload-transporting flow event in a smaller stream or catchment. These objectives influence the choice of a suitable measuring technique and the need to assess possible error sources.

The main error sources may be broadly classified into deficiencies of the measuring apparatus or deficiencies due to inadequate sampling strategies for a given measurement objective. The typically large temporal and spatial variability of bedload transport rates in natural streams is an additional important challenge if the measurement objective is to determine a mean value for a given time duration and channel reach (Bunte & Abt, 2005; Singh *et al.*, 2009).

Many bedload samplers need to be calibrated. According to Gray *et al.* (2010), most calibration studies were performed in flume experiments. However, flume conditions may not be fully representative of the natural variability of streambed and bedload transport characteristics. Therefore a given bedload sampling method may be tested against another bedload sampling technique that is thought to be more reliable under natural field conditions. For example, Bunte *et al.* (2008) compared coarse bedload transport measured with bedload traps (BT) and HS samplers. They found that bedload transport rates measured with HS samplers in the same conditions yielded higher transport rates during low flow conditions and increased less steeply with discharge than those measured with BT samplers. They considered BT samplers to be more reliable than HS samplers, and they attributed the oversampling of HS at low flows to short sampling times, the flared HS sampler design, and to involuntary particle pickup during sampler placement (Bunte & Abt, 2005). A correction function was proposed by Bunte *et al.* (2010b).

Possible measurement errors and some examples of determining relative sampling characteristics of different techniques are discussed in more detail for example in Gray *et al.* (2010) and in Megahan (1999). When using HS type samplers, horizontal movements upon sampler emplacement or retrieval can cause the sampler to dredge bed material (Childers, 1999). Gaeuman and Jacobson (2006) reported that the presence of a sampler on a sand bed frequently produces scour that can result in either oversampling or undersampling, or even cause the sampler to literally sink into the bed. Based on field and lab calibration of Bunte-type bedload traps, it was found that they matched true transport rates to within 56–117% in field measurements, and to within 50–200% for most of the flume samples, even though flume conditions were unfavorable for bedload traps (Bunte *et al.*, 2010a). Using detailed measurements from eight gravel-bed streams, Bunte *et al.* (2008) reported that during low flow conditions bedload transport rates measured with HS samplers yielded higher transport rates than those measured with BT; the differences can be up to 2 or 3 orders of magnitude. Near conditions of bank full flow, measurements with both devices yielded similar results, and at higher flows, the small HS opening and the short sampling time caused undersampling, yielding smaller bedload transport rates than measured with BT (Bunte *et al.*, 2008).

5.2.2 Passive acoustic measurements

5.2.2.1 Principles of measurement

Many passive acoustic measurements of bedload transport essentially record naturally generated signals, *i.e.* the sound or vibration induced by moving bedload particles.

Inter-particle collisions among moving particles or between moving particles and the streambed are the main source of sediment-related noise detected by hydrophones, *i.e.* underwater microphones. Other measuring devices record the impacts of sediment particles onto an impact structure, such as a pipe, plate, or column, and use different types of sensors, such as microphones, geophones, accelerometers, and piezoelectric sensors to measure the resulting vibrations of the structure (Rickenmann, 2017). Being an indirect or surrogate measuring technique, passive acoustic measurements require direct measurements of bedload transport for calibration, preferably at the same field site and synchronized with the surrogate measuring system.

5.2.2.2 Available samplers and deployment methods

Hydrophones (underwater microphones) are placed in a river environment with a supporting structure (*e.g.* close to one bank) and sense acoustic waves or the under-water sound generated by inter-particle collisions during bedload movement (Barton *et al.*, 2010; Geay *et al.*, 2017). These sensors are also sensitive to any other noise present in the river, *e.g.* generated due to turbulence (hydrodynamic noise) or due to vessel traffic (Bassett *et al.*, 2013).

Impact pipes (also called Japanese pipe microphones / hydrophones) measure the sound within an air-filled steel pipe generated by impacting bedload particles, and they are typically partly embedded in the streambed and aligned transverse to the stream flow direction (Mizuyama *et al.*, 2010a, 2010b). Steel plates equipped with a geophone sensor underneath and mounted flush with the streambed (also called Swiss plate geophones) measure vibrations of the steel plate generated by moving bedload particles; specifically, velocity of the displacement of the geophone sensor due to the vibration is recorded (Rickenmann *et al.*, 2012) (Figure 5.2.4). An example of an impact column is the gravel-transport sensor (GTS) developed by Downing (2010), which is installed vertically at the streambed. When a particle strikes the column, an electric charge is generated, the magnitude of which depends on the force of impact and

Figure 5.2.4 Example of the Swiss plate geophone (Photo Swiss Federal Research Institute WSL).

the momentum of the particle. The design of the system has been optimized through laboratory testing (Papanicolaou & Knapp, 2010).

Hydrophone (underwater microphone) measurements represent integrated recordings of sediment-related noise covering an (unknown) area of the riverbed. An advantage of the integrative measurement is that it is not local and provides information on average bedload fluxes, while a disadvantage is that the signal response is dependent on the location of the acoustic sources (Geay *et al.*, 2017). Using multiple hydrophones at different places may help to define better the major source areas of sediment-related noise (Bassett *et al.*, 2013).

In contrast to hydrophones, impact sensors record bedload transport activity at the location of the streambed where they are deployed. Using arrays of impact plates or impact pipes allows monitoring the spatial variability; for example, across the entire width of the stream. The same is true in principle for impact columns; however, this technique has the disadvantage of introducing an obstacle near the streambed, influencing the local flow hydraulics.

In principle, both hydrophones and impact sensors permit continuous monitoring of bedload transport activity in time. However, if the raw signal is to be recorded continuously there are practical limits with regard to storing the signal from multiple sensors over longer time periods. Appropriate procedures for online processing of the raw signal and retrieving summary values may partly solve this technical problem (*e.g.*, Geay *et al.*, 2017; Rickenmann *et al.*, 2014).

5.2.2.3 Calibrations and error sources

Most studies using indirect bedload monitoring techniques have focused on quantifying bedload flux. Regarding hydrophone measurements, several flume experiments have shown that the energy of the resulting acoustic signal, expressed as the root mean square value of the measured acoustic signal, is related to bedload flux with power laws (*e.g.* Thorne, 1986; Voulgaris *et al.*, 1995). In addition, these studies also demonstrated that the centroid frequency of the sediment related noise is a power function of the grain diameter for uniform particles; however, it is more difficult to infer grain size distribution for a mixture of gravel particles (Thorne, 1986). The frequency spectrum is mainly altered by sediment characteristics including size, shape and mineralogy (Thorne & Foden, 1988). A major challenge of hydrophone measurements in natural gravel-bed rivers with large roughness and elevated turbulence (notably, mountain rivers and streams) is the distinction between sediment-related noise and noise due to the turbulence of water flow (P. Belleudy, pers. comm. 2012).

Field calibration efforts have been made particularly for total bedload mass or flux with impact sensors; *e.g.*, the Japanese pipe microphones and the Swiss plate geophones. For both of these techniques it was found that an approximately linear relationship exists between number of impulses and bedload mass (Mizuyama *et al.*, 2010b; Rickenmann *et al.*, 2012), in which the impulses represent a summary value of the raw signal counting the number of exceedances of the amplitude above a threshold. While the lower limit of grain-size detection is 4 to 8 mm for the Japanese pipe microphones (Mizuyama *et al.*, 2010a), it is 20 to 30 mm for the Swiss plate geophones (Rickenmann *et al.*, 2012, 2014). For the Swiss plate geophones, it was found that the root mean square value of the measured acoustic signal is a power-law function of bedload flux. Complementary flume experiments showed that the geophone signal response appears to be mainly influenced by particle size, mean flow or particle velocity and bed roughness, which in turn influence the type of particle motion over the steel plate (Rickenmann *et al.*, 2014; Wyss *et al.*, 2016a). The calibration measurements with the Swiss plate geophones indicate an uncertainty (standard error) between estimated and measured geophone summary values of 18% in a Swiss mountain stream, and between 45% and 65% in three Austrian glacial streams (Rickenmann *et al.*, 2014). A similar system with piezoelectric bedload impact sensors was used earlier in the Erlenbach, and the error in estimating the measured total sediment loads deposited in the retention basin was assessed to be approximately a factor of two (Rickenmann & McArdell, 2007). Only a few acoustic measuring techniques have been shown to allow for a determination of bedload transport by grain size classes (Mao *et al.*, 2016; Wyss *et al.*, 2016b).

5.2.3 Active acoustic measurements

5.2.3.1 Principles of measurement

Acoustic Doppler current profilers (ADCPs) were introduced in Chapter 3 of Volume II for measurement of water velocity. An ADCP employs bottom tracking (Doppler sonar) to measure the velocity of the instrument (or the boat to which it is attached) relative to the river bed. If the bed is mobile, the bottom-track velocity is biased by the movement of the sediment along the bed. This bias can be extracted if a second measure of boat velocity is available, such as from a Differential Global Positioning System (DGPS). The difference between the biased bottom track velocity (v_{bt}) and the DGPS velocity (v_{DGPS}) is the apparent bedload velocity (v_b):

$$v_b = v_{DGPS} - v_{bt} \tag{5.2.1}$$

The apparent bedload velocity can be correlated to the bed-load transport rate (Rennie *et al.*, 2002; Rennie & Villard 2004; Rennie *et al.*, 2017).

5.2.3.2 Available samplers and deployment methods

Several ADCPs are commercially available. Any ADCP can be used to measure v_b, as long as the ADCP has bottom tracking functionality and is integrated with DGPS. It should be recognized, however, that the acoustic response of a given sediment particle depends on instrument frequency. Furthermore, various ADCPs can employ different bottom track algorithms. Thus, measured v_b for a given transport condition will be instrument specific (Ramooz & Rennie, 2010).

The ADCP must be deployed facing downwards for measurements of v_b. The instrument can be fixed in a stationary position on an in-stream structure (in which case DGPS is not necessary), but generally is deployed on a mobile platform. The ADCP may be permanently mounted or fixed directly to a boat. Alternatively, small floating platforms (tethered boats) are available for mounting most ADCPs. Remote controlled boats are also available, but have not yet been utilized for v_b measurements.

5.2.3.3 Spatial and temporal resolution

The bottom track sampling rate of most ADCPs is on the order of 1 Hz, which sets the temporal resolution of measurement. However, measurements of v_b have high variance due to both instrument noise and temporal variability of bedload transport (Rennie et al., 2002; Rennie & Millar, 2007), thus averaging is required. For example, for stationary measurements of relatively weak partial transport in a gravel-bed river, Rennie et al. (2002) determined that 25 minutes of sampling was required to obtain stable estimates of the mean and standard deviation of v_b. Fortunately, shorter sampling durations are sufficient when homogenous transport is present; Rennie and Villard (2004) found that 10 to 15 minutes was adequate in sand-bed reaches of a large river. For calibration exercises, five minute sample durations are typically employed for both v_b and the concurrent physical bedload sample.

Moving boat v_b measurements have also been conducted to map spatial distributions of relative bedload transport (Rennie & Millar, 2004; Gaeuman & Jacobson, 2006; Rennie & Church, 2010; Jamieson et al., 2011; Williams et al., 2015). In this case, spatial averaging is employed to reduce observed variance. The achievable spatial resolution of the resulting maps depends on both the variance of v_b and the spatial density of the survey (Rennie & Church, 2010). Rennie and Church (2010) produced v_b maps with 25 m grid resolution from surveyed transects spaced 110 m apart in a large 500 m wide wandering gravel-bed river, while Williams et al. (2015) achieved 2 m grid resolution by spacing transects < 2 m apart in a medium sized braided gravel-bed river anabranch with width on the order of 40 m.

5.2.3.4 Calibration and error sources

The first calibration exercise was performed by Rennie et al. (2002). Physical sampler measurements were compared to stationary v_b measured in the gravel-bed reach of Fraser River, Canada. This step was followed by sand-bed calibrations (Rennie & Villard, 2004; Gaeuman & Jacobson, 2007), and laboratory calibrations for both gravel-bed and sand-bed experiments (Ramooz & Rennie, 2010). For individual river reaches it is possible to generate a linear calibration between mean v_b and bedload transport rate. However, the calibration is site and instrument specific, as v_b depends on the instrument frequency, bottom track algorithm and pulse length, and the size distribution of bed material transport. Outstanding uncertainties include the location of measurement within the near-bed mobile layer and the relative contribution of mobile particles of various grain sizes and velocities. Rennie et al. (2017) reviewed all previous calibrations as well as new data from the Rees River, New Zealand, and examined in detail the influence of bedload grain size on v_b calibration.

Based on spatially distributed measurements in the Fraser River gravel-bed reach, Rennie and Church (2010) developed a v_b uncertainty model based on quantified error sources. The error sources included bottom track errors due to both acoustic noise and beam heterogeneity as quantified by the bottom track error velocity (mean ~ 0.025 m/s), compass errors (up to 0.03 m/s), and DGPS errors (~0.026 m/s). The uncertainty model also included an estimate of the true temporal variability of bedload velocity (mean ~0.04 m/s), which was required because a single ping measurement of v_b does not necessarily reflect the local mean bedload velocity. Overall, the true variability of bedload velocity was the dominant source of v_b uncertainty, which had a mean value of 0.07 m/s. This uncertainty was comparable to the mean v_b signal in the reach.

5.2.4 Monitoring of bed form movement

5.2.4.1 Principles of measurement

The basic principle for the determination of bedload transport rates through the monitoring of bed form movement, also known as bed form tracking, is related to the analysis of bed profiles or digital elevation models measured at successive time intervals. These data can be used to extract information on the sediment volume in motion and its migration rate, and the product of these parameters is proportional to the bedload rate, given that a negligible amount of bed sediments will be transported in suspension and that the transport rate below the bed form trough is negligible (*e.g.*, Simons *et al.*, 1965).

The idea of determining sediment transport rates by such an approach emerged already in the 19th century (*e.g.*, Hubbell & Sayre, 1964) and the technique was significantly advanced in the 1960s through the development of acoustic methods to monitor bed elevations (*e.g.*, Richardson *et al.*, 1961; Simons *et al.*, 1965; Crickmore, 1967). An initial assumption in this approach was that bed forms move as two-dimensional (2D) steadily propagating bodies of unchanging shape so that the bedload rate can be related to the product of the thickness of the sediment layer in movement and its migration rate, *i.e.* through the application of the Exner equation of sediment continuity. The layer thickness depends on the bed form shape and has often been expressed as the product of the average dune height and a form factor. Both bed form geometry and migration rate have been estimated using cross-correlation techniques (*e.g.*, Nikora *et al.*, 1997; Duffy & Hughes-Clarke, 2005; Aberle *et al.*, 2012; Muste *et al.*, 2016), h-level crossing analyses (*e.g.*, Shen & Cheong, 1977; van der Mark *et al.* 2008), or cross-spectral density functions (*e.g.*, Willis & Kennedy, 1977).

Today, the attractiveness of this method is expanding, as novel acoustic and optical measurement techniques and instrumentation allow for increasingly extensive bed-surface measurements at finer temporal and spatial resolutions, both in the laboratory and in the field (*e.g.*, Dinehart, 2002; Hoekstra *et al.*, 2004; Nittrouer *et al.*, 2008; Henning *et al.*, 2009; Aberle *et al.*, 2010; Henning *et al.*, 2010; Abraham *et al.*, 2011; Aberle *et al.*, 2012; Muste *et al.*, 2016). Such data can be organized into digital elevation models and analyzed, allowing not only for the determination of averaged 2D-parameters but also for the analysis of instantaneous multidirectional migration and bedload transport rates at smaller spatial scales in the bed parallel plane, and hence also bed form dispersion (see Figure 5.2.5).

For illustration purposes, we describe the data acquisition and processing steps involved in the determination of bedform movement in a laboratory experiment (Muste *et al.*, 2015). The raw bedform maps were obtained by scanning the channel bottom with a linear arrangement of acoustic sensors while the flow in the channel was running. If bedform scanning is conducted with a sufficient density of data points, the discrete depth values acquired by each sensor can be interpolated to obtain a map of the channel bottom, as illustrated in Figure 5.2.5. Subsequently, acoustic maps are created as a continuous depth-data layer covering the target area of the channel bottom by interpolating the measured depths. This step is the basis for all sounding techniques and

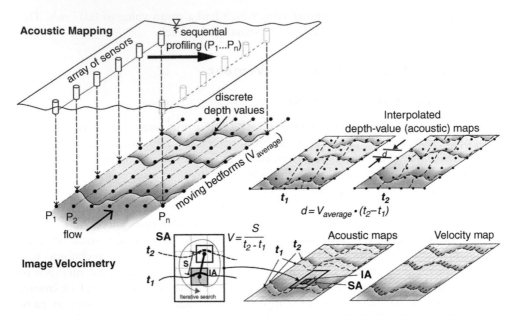

Figure 5.2.5 Example of time sequenced bathymetry of a dune field obtained by acoustic mapping velocimetry (adapted from Muste *et al.*, 2016).

produces the conventional bathymetric maps. The acoustic maps provide a full 3-D description of the geometry of the bed forms. In the second step, the acoustic maps are converted to "image-equivalent" maps by resampling the raw information in pixel coordinates (Muste *et al.*, 2015). Images obtained at different time steps are subsequently processed using an image velocimetry technique (Muste *et al.*, 2008). Bedload transport rates are obtained by combining the outcomes of the acoustic mapping carried out in the first step along with the velocity fields obtained in the second step using analytical relationships for predicting sediment transport rates for the bedload (Vanoni, 2006). Alternatively, the velocity fields in step 2 can be determined in DEM-subsections through cross-correlation in Cartesian coordinates instead of pixel coordinates (Henning *et al.*, 2010).

5.2.4.2 Available technologies

The geometry and migration of bed forms can be best captured using optical and acoustic measurement systems. Optical methods are based on the principles of photogrammetry (see Section 4.4, Volume II) and require a clear view of the moving sediment bed and hence clear water conditions with a negligible amount of suspended sediments. Such conditions are rare in the natural environment and therefore photogrammetric methods have mainly been applied in the laboratory (*e.g.*, Henning *et al.*, 2010; Henning, 2013). Acoustic methods are based on the principles of echo-sounding (see Section 4.3, Volume II) and can be applied in both laboratory and field environments with and without clear visibility (see also the next section).

5.2.4.3 Spatial and temporal resolution

The spatial and temporal resolution of photogrammetric and acoustic measurement systems differ to some extent. Photogrammetric methods allow for the instantaneous determination of bed elevation fields at a high temporal rate. The spatial resolution depends on the camera resolution (small scale) and field of view (large scale), while the maximum achievable temporal resolution is generally determined by the storage capacity and data transfer rate of the photogrammetric system. If the photogrammetric system is attached to a moving carriage, a wider spatial field can be covered but the temporal resolution will decrease, as moving the system to a new position (and possibly recalibrating the system) requires time.

The spatial and temporal resolution of the acoustic method depends strongly on the measurement setup and if the measurements are carried out in the laboratory or the field. In the laboratory, a single echo-sounder or multiple echo-sounders can be mounted to a carriage which can be pushed along the flume to obtain a system specific spatial and temporal resolution (*e.g.*, Coleman *et al.*, 2005) which depends on the measurement frequency and carriage speed. Alternatively, transducer arrays can be attached to a rigid frame to get a higher spatial resolution. Such frames can be used for sequential profiling at one and the same position or at different positions (*e.g.*, Jetté & Hanes, 1997; Muste *et al.*, 2016) to enhance the spatial scale covered by such "rigid" setups. The achievable spatial resolution depends on the footprint size of the acoustic sensors as the individual acoustic beams should not interfere. For measurements at a single position, high temporal resolution is possible. For sequential profiling the temporal resolution depends on how fast the frame can be moved into a new position, and if data are collected continuously while the frame is repositioned, the acoustic ping rate will influence the spatial resolution.

In the field, acoustic surveys are typically vessel based using multiple single beam echo-sounders attached to booms, side-scan or multi-beam echo-sounders, or the bottom tracking function of ADCPs. The achievable spatial resolution depends, as for the laboratory measurements, on the footprint size of the sensors. However, as such measurements are typically carried out with a moving vessel it also depends on vessel speed in combination with the measurement frequency. The temporal resolution depends on the time required by the vessel to cover the spatial field of interest.

If optical or acoustic instruments are moved during the measurements, it should be considered that the measured bed elevation field will not be instantaneous. This means that the measurements include a time-lag which is introduced through moving the measurement system. However, the migration rate of bed forms is typically much slower than the time required to move the measurement system so that "quasi-instantaneous" bed elevation fields can be measured.

5.2.4.4 Error sources

The technique of bed-form tracking is prone to several uncertainties including the depth of the bedload layer below the dune trough, the identification and definition of both 2D- and 3D-bed form shapes, the occurrence of multi-scale bed forms and consequently bed-form superposition, as well as multidirectional bed-form movement (*e.g.*, Abraham *et al.*, 2011; Aberle *et al.*, 2012; Muste *et al.*, 2016). The depth of the bed layer below the dune trough is difficult to determine by the described methods and remains therefore largely unknown. Moreover, bed forms are hardly characterized by a constant shape. The occurrence of 3D-bed forms, multidirectional bed form-movement, and shape changes caused by superposition of bed forms of different spatial scales may hamper not only the identification of the bed form dimensions but also the applicability of cross-correlation techniques which are used to derive the bed form migration rate from elevation fields (as compared to seeding particles used in image-based velocimetry, see Section 3.7, Volume II). The aforementioned issues related to spatial (*e.g.*, overlapping acoustic footprints and sensor interference) and temporal resolution (time-lag during measurements of larger spatial areas) should also be taken into account during data analysis. Nonetheless, both optical and acoustic methods have been proven to provide reliable data on bedload transport rates and velocity fields of dunes (*e.g.*, Abraham *et al.*, 2011; Henning, 2013; Muste *et al.*, 2016).

5.2.5 Bedload particle tracers

5.2.5.1 Principles of measurement

Use of tracers has a long tradition in fluvial geomorphology (Hassan & Ergenzinger, 2003). The tagging of bed particles allows study of their movement along the channel bed. The passive tracer techniques allow identifying start and end positions before and after a transporting flow event or measuring (automatically) the passage of particles at one or several locations along the channel. In principle, active tracer techniques allow continuous monitoring of the particle trajectory over time. While measurements of bedload transport at a point or at a cross-section (with physical traps or passive acoustics) are equivalent to an Eulerian approach, the tracing of bedload particles corresponds to a Lagrangian approach of monitoring bedload (Habersack *et al.*, 2010).

5.2.5.2 Available particle tracing methods

Particles may be marked by painting (Laronne & Carson, 1976), inserting magnets (Hassan *et al.*, 1984), or using naturally magnetic (Ergenzinger & Custer, 1983; Hassan *et al.*, 1984) or radioactive (Hubbel & Sayre, 1964) properties. More recently, particle tracer studies using passive integrated transponder (PIT) tags combined with a radio frequency identification (RFID) technique have become more widespread in particle tracer studies (Nichols, 2004; Lamarre *et al.*, 2005; Liébault *et al.*, 2012; Gronz *et al.*, 2016; Figure 5.2.6). Another technique is based on active transponders which have radio transmitters (Ergenzinger *et al.*, 1989; Chacho *et al.*, 1989, 1994; Habersack, 2001; McNamara & Borden, 2004).

Typically, the tagged particles are artificially inserted at selected locations in the streambed. After initial movement, particle tracing with the more modern methods allows identification of the particle position without the need to remove it or disturb the streambed. This is important when investigating the critical flow conditions for initiation of particle motion. For techniques requiring a manual instream survey, due to safety and accessibility reasons, it should be noted that particle positions in the streambed can be recovered only once discharge has receded below a critical level.

Figure 5.2.6 RFID transponder inserted into tracer stone. The transponder of Texas Instruments is 4 mm in diameter and 23 mm long. (Photo Swiss Federal Research Institute WSL).

5.2.5.3 Spatial and temporal resolution

Passive tagging of bedload particles allows their position to be identified at different times, enabling the determination of particle displacement as a function of the integrated flow history between two successive surveys. Automatic detection of the passage of passive tracers at an instrumented cross-section is possible for magnetic tracers (Ergenzinger & Custer, 1983; Rempel *et al.*, 2010), for RFID tracers (Schneider *et al.*, 2010, 2014; Johnston *et al.*, 2009), and for active transponders (Habersack, 2001; McNamara & Borden, 2004). The main advantage of using active transponders is that continuous monitoring of tracer particle position is possible.

5.2.5.4 Error sources

For a given tracer particle, the main task is to determine its position in the streambed. Typically this is found by surveying techniques, with corresponding accuracies related to measurements of relative or absolute positions. To include a statistically significant number of tracer particles, the recovery rate is an important parameter to be considered. Recovery and/or detection rates depend on the degree of burial of particles and on the fraction of particles exported from the study reach, which may be due to rapid dispersion of the tracers with frontrunners (the fastest moving tracer particles) traveling very far. Frontrunners may not be easily detectable if they reach a different stream environment, such as a lower order stream with high discharge or a sediment retention basin. For example, having inserted 451 RFID tracers in a gravel-bed river in France in spring 2008 some 2 km upstream of the confluence with a larger river, Liébault *et al.* (2012) reported detection rates of 78% in 2008, 45% in 2009, and 25% in 2010, the re-surveys having been made in July each year. Having inserted more than 733 RFID tracers in a steep torrent in Switzerland in June 2009 and May 2010, Schneider *et al.* (2014) noted detection rates below 20% at the end of July 2010 (after a large flood event), with a substantial fraction of tracer particles having been buried too deeply for identification. McNamara and Borden (2004) had equipped five cobbles with motion-sensing radio transmitters using different frequencies as active tracers, and they reported on some technical limitations regarding data acquisition such as radio interference.

5.3 SUSPENDED LOAD

5.3.1 Physical sampling for suspended sediment

The material in this section is based on several sources and is intended to give enough information to choose an appropriate method for collecting suspended-sediment samples for the purpose of determining sediment concentration. It is recommended that more detailed sources be consulted before embarking on an experimental program. Detailed practice manuals as well as other useful compilations of information on collecting physical samples of suspended sediment include Davis (2005), Edwards and Glysson (1999), Garcia (2008), Gray and Gartner (2009), Kuhnle (2013), as well as the United States Geological Survey (USGS) Field Manual (http://water.usgs.gov/owq/FieldManual; accessed 28th February 2017) and several publications on the U.S. Federal Interagency Sedimentation Project (FISP) website (http://water.usgs.gov/fisp/publications.html; accessed 28th February 2017).

Physical samples of the water-sediment mixture are necessary, in spite of many recent advances in instrumentation for sediment measurement. There are several important reasons that the need for physical sampling persists. Physical sampling is the only accepted reference technique; each new automated or instrumentation-based method must be compared to physical samples. When large-scale projects, such as dam construction or removal, river modifications, the design of water intake structures, etc., depend on sediment data, the risk and expense of the work justify the time and expense of physical sampling. Physical samples can also yield a grain size distribution, which can be invaluable for characterizing the sediment load at a site.

5.3.1.1 Collecting a representative sample

To obtain accurate samples, it is important to understand the concept of a representative sample. A sample is representative if it correctly reflects the characteristics of the sampled area. Concentration gradients in the stream must be correctly represented. For instance, if a point-integrating sampler is lowered to within 20 cm of the bed in a 3m deep stream, the sample that is collected will not properly represent the coarse fraction of suspended-sediment concentration in that cross-section, since there will be a much higher concentration of sand in the unsampled zone near the bed.

Isokinetic sampling is critical to the collection of a representative sample of the water/sediment mixture. The intake nozzle of an isokinetic sampler collects a water-sediment sample from a stream at a velocity equal to the stream velocity at the nozzle entrance.

This allows the water and sediment to flow into the nozzle in the same proportions as they were in the stream, resulting in a sample that is representative of the sediment load at that point. If the intake nozzle velocity is higher than the local stream velocity, the sample will have too much water, since sand particles, with their increased density relative to water, will have more momentum than water particles and will not follow the same path as water particles near the nozzle. If the nozzle velocity is too low, the sample will have too much sand, since some fraction of the water will be excluded from the sample, and the momentum of particles will cause them to enter the nozzle in greater amounts than the ambient concentration. Isokinetic conditions become more important for larger particle sizes. For example, sampling at a rate of 30% of the stream velocity results in negligible error for 0.01 mm particles, while 0.45 mm particles will be oversampled by 70% (see Gray and Gartner (2009) for a more comprehensive treatment of particle-size induced error in physical samples). Samplers that have been tested and approved by the U.S. FISP collect an isokinetic sample at minimum stream velocities of 0.5 to 0.6 m/s. An exception to this rule is samplers that use rigid containers, such as the DH-81. For these, at velocities less than 0.5 m/s, there is a slight static head between the air exhaust and the nozzle so that the sampler will still collect a sample. The user must understand that the sample collected under these conditions is not an isokinetic sample.

In designing an experiment or monitoring program, distinguishing between fines or wash-load sediment ($D_{50} < 62$ µm) and sand-sized particles (62 µm $< D_{50} < 2000$ µm) will affect the sampling methodology. A simple dip sample can be representative if the concentration of fine sediments is required, since fine particles are typically distributed through a channel cross-section in a homogeneous manner. The location from which the sample is taken will have little effect on the measured concentration.

5.3.1.2 Depth integrating vs. point integrating samplers

When the total load of a river or stream is needed, depth integrating samplers are typically used (Figure 5.3.1). These samplers collect the water/sediment mixture while being lowered and then raised at a constant rate. The resulting sample is an integrated representation of the suspended-sediment concentration throughout the water column.

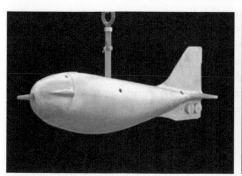

Figure 5.3.1 US DH-74 Depth integrating sampler (left) and US P-61-A1 point integrating sampler (right) (The pictures are sourced from the United States Federal Interagency Sedimentation Project (FISP) (http://water.usgs.gov/fisp/catalog_index.html; accessed 28th February 2017), U.S. Geological Survey).

This process can be repeated through a cross-section to assess the spatially averaged transport rate. The number of verticals needed to characterize load at a location and the methodology for combining the data can be found in Edwards and Glysson (1999) and is discussed briefly below.

There is an unsampled zone near the channel bed that must be accounted for when using depth-integrated samples. This zone is caused by the design of samplers, and the height of the unsampled zone varies depending on the sampler. This situation is unavoidable, since a sampler with the nozzle near the bottom may sample directly from the bed, contaminating the sample with bedload and adding a high bias to the measured concentration. The selection of a depth-integrating sampler is based on the depth and velocity of the stream. Larger, heavier samplers are used in deeper, swifter rivers.

Point integrating samplers (Figure 5.3.1) have an electric valve that can be opened remotely when the sampler is in the desired position. After the valve is opened, the sample chamber is filled isokinetically over time, resulting in an integration of the concentration at that point. Samples from a single vertical can be composited to arrive at a mean concentration through a vertical. There are specific situations that may call for the use of a point-integrating sampler. Comparison of the concentration yielded by a point-integrating sampler to another measurement technique is one example.

In both laboratory and field settings, low water depths may limit the type of sampler that can be used; however, sediment transporting flows at low depths are more common in laboratory experiments due to typical low sidewall heights in flumes. For physical samples in such cases, an isokinetic vacuum sampler can be used. An L-shaped nozzle attached to a vacuum line can be positioned using a point gauge or other apparatus to maintain a known elevation relative to the sediment bed. The strength of the vacuum can be varied to yield a withdrawal rate that will produce isokinetic conditions at the nozzle.

5.3.1.3 Choosing a sampler

It is important, if at all possible, to use standardized samplers so that results may be compared to previous measurements or to measurements from other rivers. The United States Federal Interagency Sedimentation Project (FISP) is widely recognized as a source for suspended sediment samplers with designs that have been thoroughly tested. The FISP website is at http://water.usgs.gov/fisp/ [accessed 1st November 2016] and includes a catalog with descriptions of suspended sediment samplers. The catalog is a reference that is used worldwide. Figure 5.3.2 provides a flowchart to guide selection of the appropriate sediment sampler for various measurement scopes and conditions (also available at http://water.usgs.gov/fisp/catalog_index.html; accessed 1st November 2016).

5.3.1.4 Sampling throughout a channel cross-section

The concept of the representative sample also drives the need for sampling throughout a cross-section if total suspended sediment load is the goal of the measurement. Edwards and Glysson (1999) give detailed descriptions of how to sample a stream cross-section in a manner that yields a correctly depth and discharge weighted suspended-sediment load. Only a brief summary will be included here. The Equal-Discharge Increment (EDI)

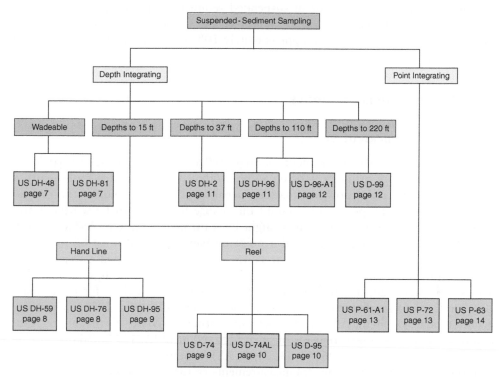

Figure 5.3.2 Guidance for choosing the correct FISP sampler. The figure is sourced from the United States Federal Interagency Sedimentation Project (FISP) (http://water.usgs.gov/fisp/catalog_index. html; accessed 28th February 2017), U.S. Geological Survey.

requires the division of the stream cross-section into four to five panel sections of equal discharge. Samples are collected at the centroid of each of the sections. The vertical transit rate of the samplers can vary among the sampled locations. The EDI method requires detailed measurements of stream velocity and bathymetry in order to identify the discharge increments. In the Equal-Width Increment (EWI) method, the stream is sampled at 10–20 equally spaced locations. Care must be taken to use the same vertical transit rate for all locations, and the suspended-sediment discharge can be calculated from the mean concentration multiplied by the mean flow discharge during the sample collection period.

5.3.1.5 Error sources

Typical error sources in physical suspended sediment sampling include:

- Lack of isokinetic sampling and correction
- Errors in elevation and depth measurement
- Scooping from the bed for depth-integrating samplers
- Unequal transit time error for depth-integrating samplers
- Errors introduced in sample analysis

Additional discussion of error in suspended sediment sampling can be found in Topping *et al.* (2011). For their data from the Grand Canyon of the Colorado River, concentration errors ranged from approximately 10% to 100%, with increased error for larger grain sizes.

5.3.2 Optical measurements

5.3.2.1 Optical backscatter

Optical Backscatter (OBS) instruments use light reflected from suspended particles to measure the turbidity of water, which can be produced by suspended and dissolved matter, including clay, silt, organic matter, microbes, or dyes. Turbidity is an expression of the optical properties of a liquid that cause light rays to be scattered and absorbed rather than transmitted in straight lines through a sample (ASTM, 2003). Optical Backscatter instruments yield a point measurement and must be moved or deployed in an array if coverage of a larger area is needed. Suspended material in the sand-size range can also produce turbidity and will register readings on turbidity probes; however, light attenuation is less for suspended sand than for equal mass concentrations of smaller particle sizes (Davies-Colley & Smith, 2001). There are several standards from the American Society for Testing and Materials that are relevant to optical backscatter measurements of suspended sediment (ASTM Standard D7512-09, 2009; ASTM Standard D6855-03, 2003; ASTM Standard D6698-07, 2007). In addition, a key resource for optical measurements of turbidity is the aforementioned USGS National Field Manual, in particular, the turbidity chapter of Anderson (2005) at http://pubs.water.usgs.gov/twri9A6/ [accessed 1st November 2016].

The relatively low cost and ability to monitor turbidity over time with good temporal resolution (on the order of 1 Hz) make these instruments attractive. Optical devices can operate based on either backscatter or transmission of light. Based on reports in the literature, OBS devices (*e.g.*, Figure 5.3.3) appear to be in more widespread use, having the advantage of being less flow intrusive than transmissometers, due to the source and receiver electronics being located near one another. OBS instruments are typically cylinders with a diameter of 2–3 cm, a length of 12–16 cm and a mass of 200–300 grams. The relatively small size and weight, along with low cost, make this type of instrument attractive for use on remote field sites. Transmissometers are more sensitive at low particle concentrations, while optical backscatter sensors are linear over a broader range of turbidity (Downing, 1996). Like other optical devices, both OBS instruments

Figure 5.3.3 The RBR solo Tu as an example of an OBS turbidity meter; sensor head to the right of image (Photo: Rennie).

and transmissometers can be susceptible to biological fouling. When the source of the light is occluded, the output can be reduced, or the light may be reflected back to the sensor. In either case, a spurious reading that does not properly represent the sediment concentration in the sampling volume is produced.

A consequence of the availability of multiple instrument designs is that turbidity measured from instruments with different optical designs can differ by factors of two or more for a given sample, even with identically calibrated instruments. Data from instruments with different designs or manufacturers are not directly interchangeable, and the data are not directly comparable without additional work to establish relations between the instruments. It is beyond the scope of this section to provide detailed guidance for calibrating with standards; however, Anderson (2005) offers guidelines on calibrating with standards and properly recording the results.

If turbidity data are to be converted to suspended sediment concentration, particle size dependence must be accounted for. The relationship between attenuated or back-scattered light and concentration varies with particle size, so instruments must be calibrated with particle sizes similar to those to be measured in the field before accurate measurements of suspended sediment concentration can be made. For a given particle concentration, the response of an OBS device increases with decreasing particle size (Conner & DeVisser, 1992; Downing, 1996; Sutherland et al., 2000). OBS signal response is greatly affected by particles that are smaller than about 44 μm (Conner & De Visser, 1992), while there is a smaller effect for particles in the range of 200–400 μm (Conner & De Visser, 1992). Bunt et al. (1999) documented a change in grain size from medium sand to fine silts that led to a 100 fold increase in instrument response. OBS instruments may not be well suited to locations where large changes in particle size may occur or where particle sizes are predominantly less than 100 μm (Conner & DeVisser, 1992).

The maximum measureable concentration for OBS devices is dependent on the size of the suspended particles, as well as the electronic configuration of the instrument (Gray & Gartner, 2009). For example, an OBS device was shown by Ludwig and Hanes (1990) to be linear for concentrations less than about 2 g/L for clay and silt and 10 g/L for sand. Kineke and Sternberg (1992) were able to measure concentrations up to 320 g/L by using the nonlinear region of the response curve. The specification sheet for the OBS-3+ manufactured by Campbell Scientific lists a concentration range up to 500 g/L. The upper suspended sediment concentration limit for transmissometers depends on optical path length, but can be as low as 0.05 g/L (D&A Instrument Company, 1991). In general for optical instruments, the wider the measurement range the less precise the within-range turbidity data, and vice versa.

The accuracy of OBS sensors is dependent on many variables, as discussed above. Achieving the manufacturer's specified accuracy is dependent on careful deployment and calibration. An example of specifications follows so that the reader will have a general idea of the concentration ranges that can be measured with an OBS instrument. The manufacturer of the OBS-3+, Campbell Scientific, states a concentration accuracy of 2% of reading or 1 mg/l (whichever is larger) for mud and 4% of reading or 10 mg/l (whichever is larger) for sand. The OBS 3+ range of concentration for mud is 5,000–10,000 mg/l and for sand it is 50,000–100,000 mg/l. It should be reiterated here that OBS accuracy is dependent on calibration data and steady particle size distribution.

5.3.2.2 Laser diffraction

Laser diffraction particle-sizing instruments are common in laboratory settings, and there are also commercially available instruments that can be used in short and long-term field deployments. Here, the focus will be on instruments that can be used in the hydraulics laboratory or in the field to give *in situ* measurements of particle size and concentration. These instruments yield a point measurement and must be moved or deployed in an array if coverage of a larger area is needed. There is a good description of laser diffraction measurement principles in Agrawal and Pottsmith (2000). A laser beam is directed into the sample volume where particles in suspension scatter, absorb, and reflect the beam. The scattered laser light is received by a series of ring shaped detectors of progressive diameters that allow measurement of the scattering angle of the beam. Particle size can be calculated from knowledge of this angle, using the Fraunhofer approximation or the exact Lorenz-Mie solution. By basing concentration measurements on these measured particle sizes, particle size dependency is eliminated, thus both particle size and concentration can be measured simultaneously (Swithenbank *et al.*, 1976; Riley & Agrawal, 1991; Agrawal & Pottsmith, 1994, 2000). In the absence of additional information, particle density must be assumed.

The only commercially available device of this type that is known to the authors is the Laser In Situ Sediment Transmissometer (LISST) series from Sequoia Scientific. The specifications for these instruments can be found on the company website, and Figure 5.3.4 shows the commonly used LISST-100X and the more recent LISST-200X particle size analyzers. Successful field deployments of LISST instruments are described in Mikkelsen and Pejrup (2001) and Melis *et al.* (2003). Melis *et al.* (2003) point out the need to maintain the optics carefully. If biological matter occludes the laser, the results obtained by the system will be affected. It is also necessary to be aware of expected particle size distributions, since laser diffraction instruments are typically designed to operate over a limited size range (Mikkelsen & Pejrup, 2001).

The manufacturer of the LISST-100X states that the size range is either 1.25–250 µm or 2.5–500 µm and the sediment concentration range is typically 1–800 mg/l for the standard 50 mm optical path. The actual range of concentration is dependent on grain size, with a lower concentration range for finer particles. A higher concentration range is possible with a shorter path length. There are several sizes of LISST instruments, but the LISST-100X will be used as a representative for size. It is 13.3 cm in diameter, 87 cm long, and weighs 11 kg. The size of the instrument can make deployment challenging, particularly in flowing water.

Figure 5.3.4 The LISST-100X (above) and LISST-200X (below) particle size analyzers (Photo: Sequoia Scientific).

Proper beam alignment is very important for the operation of laser diffraction instruments, so deployment design must take protection of the instrument into account. The cost of the instrument is on the order of $30,000 (in 2016), which also suggests that careful deployment is necessary.

5.3.2.3 Focused beam reflectance

In focused beam reflectance measurement, a laser beam focused to a spot (< 2 μm^2) in the sample volume is rotated very quickly (many times per second). As it rotates, the beam hits particles that reflect the beam back to the receiver. The time of this reflection event is used to determine the chord length of the particles in the path of the laser. This information is used to calculate the volume of a sphere representing the particle (Phillips & Walling, 1995; Law *et al.*, 1997). There is no particle size dependence, since concentration measurements from focused beam reflectance are based on measured particle sizes. The instrument has a wide particle diameter measuring range of roughly 1–1000 μm with measurements of over 1000 μm possible. The concentration range is 10–50 mg/l. Good results from field studies have been reported (Phillips & Walling, 1995, Law *et al.*, 1997).

The FBRM-G400 and FBRM-G600 from Mettler Toledo are commercially available instruments that utilize this technology for laboratory application. These instruments do not appear to be optimized for sediment measurement, but they are specified to measure particle sizes in the 0.5–2000 μm diameter size range. The wetted part of the instruments is relatively small. For example, the PBRM-G400 has dimensions of 8.9x23.7x4.92 cm for the electronics and the wetted probe is 9.5 mm in diameter with options of 91/206/400 mm length. Several authors (*e.g.* Law *et al.*, 1997; Bale *et al.*, 2002; Greaves *et al.*, 2008) have used FBRM for sediment-related work. Law *et al.* (1997) describe a laboratory testing process that resulted in a rescaling method for improving accuracy for a range of particle types. Greaves *et al.* (2008) tested FBRM against known particle sizes and found that it could detect changes in particle size but could be misleading in its estimate of actual particle size. FBRM typically overestimates particle size for diameters less than 150 μm (De Clercq *et al.*, 2004).

5.3.3 Acoustic methods

5.3.3.1 Acoustic backscattering

A sound wave interacting with particles along the sound propagation path is the underlying principle behind Acoustic Backscattering Instruments (ABIs) which permits measuring velocity, as described in Section 3.2 of Volume II. The measurement of velocity is associated with the estimation of the Doppler frequency shift of the emitted sound produced by the moving particles (Figure 3.2.1b). For the measurement of suspended material concentrations with ABIs, the intensity of the return signal and its travel time are used to determine the number, type and location of particles along the sound propagation path. Already in the 1980s, ABIs were being used to measure sediment transport in marine environments. They are powerful instruments for such studies, since they can measure co-located and simultaneous velocity and intensity profiles in a non-intrusive manner, and this with high spatial resolution. In addition, downward facing instruments can provide information on bed forms. Since the 1990s, acoustic Doppler current profilers (ADCPs; see Section 3.3, Volume II) have become a widely used tool for streamflow monitoring, and they have been increasingly used to monitor fluvial suspended sediments. The commercial development of fixed side-looking ADCPs has also led to the development of new interpretation techniques.

A significant number of review papers on the use of acoustic data for measuring suspended particle size and concentration have been published in recent years, including those of Thorne and Hanes (2002), Landers (2010) and Thorne and Hurther (2014), to name a few. The theory presented in these and other papers is given in more detail in the appendix of this section. In this section, the word beam is synonymous with both acoustic transducer and the sound it emits, backscatter refers to the intensity of the backscattered or echoed signal, and cell, bin or gate refer to measurement (sampling) volumes (see Sections 3.2 and 3.3, Volume II).

5.3.3.2 Principles of measurement

The intensity or power per unit area of the sound backscattered from a suspension of particles depends on the density and elasticity of the particles, their concentration and the ratio of the size of the suspended particles to the wavelength of the acoustic waves. Suspended sediment concentration can therefore be determined using the backscattered intensity. In some cases it can also be obtained from the attenuation that is calculated from the intensity recorded at two or more ranges. The most appropriate method for determining suspended sediment concentration (either intensity or attenuation) is a function of the frequency of the instrument and the size and concentration of suspended sediment. The background principles for such measurements and details on how this can be achieved can be found in the Appendix 5.A, Volume II.

5.3.3.3 Instruments

5.3.3.3.1 Acoustic Doppler Current Profiler (ADCP)

Whenever velocity data are recorded with an ADCP, backscattered intensity data are also recorded. These data can provide both qualitative and quantitative information on the spatial distribution of suspended sediment throughout the cross section. As for limitations of the instruments, for downward facing deployments, for beams that are angled at 20 degrees from the vertical, data collected within the bottom 6% of the water column must be discarded because scattering of the side-lobes of the acoustic transducers from the river bed interferes with scattering from the suspended particles near the bed (e.g. Morlock, 1996). These data are automatically discarded by most data collection and post processing software for acoustic discharge measurements. For horizontal deployments, interception of the beams with the surface must be avoided since scattering from capillary waves on the river surface can be misinterpreted as high levels of suspended sediment (Moore, 2012; Moore et al., 2012).

While ADCPs are the most commonly used acoustic surrogate for monitoring suspended sediment, other acoustic instruments such as multi-beam echo sounders (see Chapter 4.3, Volume II) have been used to monitor suspended sediment dynamics in fluvial environments (Simmons et al., 2010). These instruments typically operate in the 100 kHz to 1 MHz range, measuring data in a wide swath perpendicular to the track of the survey vessel. The swath width or cross-track subtended angle is typically 120° so that the swath width is about 5 times the water depth. The spatial resolution of multi-beam echo sounders can be as small as 1 cm, and data are typically collected at a repetition rate of 1 Hz.

5.3.3.3.2 Acoustic Doppler Velocity Profiler (ADVP)

In addition to taking velocity measurements, Acoustic Doppler Velocity Profilers (ADVPs) (see Section 3.2.3.4, Volume II) can be used to measure sediment transport, since the backscatter intensity obtained with ADVPs provides sediment concentration profile information (details in Section 5.A.1.1). Most important, however, 3D velocity and suspended particle concentration profile information is collected simultaneously and co-located in the same sampling volumes within the profile. Blanckaert *et al.* (2017) demonstrated that an ADVP can resolve the bedload layer and measure bedload particle velocities as well. This instrument system, also called an Acoustic Doppler Particle Flux Profiler (ADFP), is suitable for application in laboratory and field studies in rivers, estuaries, oceans and lakes. Different from optical systems, it is not limited by the particle concentration. Furthermore, it resolves fluxes at turbulence scales in time and space. Although the velocity determined from the Doppler phase does not require any calibration, an initial calibration of the backscattering intensity for a given particle size distribution has to be established. This can be done by suction sampling or from Optical Backscattering Systems (OBSs) in the same flow (Thorne & Hanes, 2002).

Several ADFP systems have been proposed. For example, by integrating the backward/forward scattering approach into an existing ADVP, Shen and Lemmin (1997) developed an ADFP that resolved the 2D velocity field and the corrected suspended particle concentration field co-located in the same sampling volumes of the profile. This configuration allowed calculation of instantaneous particle fluxes in all profile gates quasi-instantaneously. The use of this acoustic Doppler flux profiler provided new insight into the dynamics of suspended particle transport and, in particular, demonstrated the importance of coherent structures in sediment transport (Shen & Lemmin, 1999; Hurther & Lemmin, 2001; Cellino & Lemmin, 2004). Zedel and Hay (1999) and Smyth *et al.* (2002) also investigated the possibility of co-located velocity and concentration profiling using ABIs. Hurther and Thorne (2011) established that in particle-laden flows, an ADFP can measure the velocity profile and distinguish between suspended transport, bedload transport and the position of the non-moving bed. This confirms that the ADFP is well suited for studies in particle-laden flows in the laboratory and in the field, since it can simultaneously provide all three components of the measurement triad, as defined by Thorne and Hanes (2002), *i.e.* hydrodynamic quantities, suspension quantities and bed parameters.

Multi-frequency backscattering intensity profiling has been integrated into ADVPs. For example, a two-frequency system was developed by Hurther *et al.* (2011) and Naqshband *et al.* (2014a, 2014b), and Hay *et al.* (2012a, 2012b, 2012c) presented a broadband multi-frequency ADFP, called MFDop, capable of 0.0009 m vertical resolution at 85 Hz. This system allows for estimation of both particle concentration and grain size (Crawford & Hay 1993; Thorne & Hardcastle 1997; Wilson & Hay 2015a, 2015b). Bagherimiyab and Lemmin (2012) showed that a two-frequency ADFP using 1 MHz and 1.66 MHz carrier frequencies can distinguish between hydrogen bubbles and suspended sediment. In an experiment where sediment was suspended only in the lower 40% of the profile, hydrogen bubbles indicated the full depth profile of velocity at 1 MHz. At 1.66 MHz, however, the hydrogen bubbles disappeared as tracers, and sediment particles were able to trace the velocity profile due to the particle motion, but only in the lower 40% of the water column.

5.3.3.4 Error sources

Due to the statistical nature of the backscattered intensity, the normalized standard error of the average intensity is approximately $1/\eta^{0.5}$, where η is the number of independent samples (Thorne & Hurther, 2014). Since concentration is proportional to the backscattered intensity when given in linear units, the normalized standard error of the concentration also scales with $1/\eta^{0.5}$. This implies that if 100 profiles were averaged and the instrument and sediment constants were properly accounted for, the estimated mean concentration would have an accuracy of ± 0.1 times the sample standard deviation. Error will be greater if the samples are not independent, which occurs when the sampling window is smaller than the emitted pulse. Based on Monte Carlo simulations of 100 samples, Simmons et al. (2010) found that when there was 50% overlap of the sampled bins there was a 12.5% error in the concentration estimate compared with 10.4% when there was no overlap. To ensure that the backscatter values recorded in all bins of a given profile are independent, most commercial acoustic backscatter systems are designed such that the sampled volume for the backscattered intensity is equal to or less than the ensonified volume (Tessier, 2006, pp. 14–15), even though the velocity in each depth cell is often a triangular weighted average of two depth cells (RD Instruments, 1996, p. 17).

As for the accuracy of multi-frequency acoustic sizing, using a three-frequency system Hay and Sheng (1992) obtained mean size estimates with 10–20% accuracy and concentrations with 10% accuracy for their field measurements. Through their simulations of a range of acoustic inversion techniques, Thorne and Hurther (2014) assessed how various factors impact acoustic estimates of suspended particle size and concentration. Other sources of error that are more difficult to quantify include scattering from non-sedimentary particles such as bubbles and organic matter (algae, zooplankton).

5.A APPENDIX

5.A.1 Concentration measurements with Acoustic Backscattering Instruments (ABIs)

Appendix 5.A.1 is also linked to Sections 3.2 and 3.3 of Volume II, where details on the notation used can be found.

5.A.1.1 Principles of concentration measurements with Acoustic Backscattering Instruments (ABIs)

Acoustic Backscattering Instruments (ABIs) can profile the vertical distribution of sediment concentration under both laboratory and field conditions (Thorne et al., 1991, 1993; Admiraal & Garcia, 2000; Thorne et al., 2009). Using these instruments, profiles of the intensity of the backscattered sound from suspended sediment are collected and this intensity is then related to the suspended sediment concentration at each point of the profile through an initial calibration. The intensity is proportional to the pressure scattered by the cloud of particles. The mean voltage V received by the transducer is proportional to the square root of the intensity and can be expressed as

$$\left\langle V^2(r) \right\rangle = S_N S_{res} F_f(r,a) C_M \, exp \left(-4 \int_0^r \langle a_t(r) \rangle dr \right) \qquad (5.A.1.1)$$

S_N is the overall system sensitivity constant. S_{res} is the system response factor depending on the propagation loss factor $1/r^2$, where r is the sound path range, and the medium attenuation loss factor a_0, i.e., $S_{res}(r) = r^{-2} \exp(-4ra_0)$. S_{res} does not depend on particle characteristics and can easily be calculated. $F_f(r,a)$ is the size-averaged form function,

$F_f(r,a) = \langle |f_\infty(\pi,a)|^2 \rangle / \langle a \rangle$, where a is particle radius. The term $exp\left(-4 \int_0^r \langle a_t(r) \rangle dr \right)$

represents the attenuation due to the particles suspended along the sound path r. This is a critical parameter for correctly determining the sediment concentration along the acoustic beam. The term C_M is the concentration by mass and can be given as

$$C_M = \rho_1 C_v = (4\pi \rho_1 N/3) \int_0^\infty a^3 n(a) da \qquad (5.A.1.2)$$

where $C_v = (4\pi(N)/3)$ is the volume concentration for uniform size particles with N being the number of particles per unit volume and ρ_1 is the density of the particles. For non-uniform size particles, a size distribution $n(a)$ results in $\langle a \rangle = \int_0^\infty an(a) da$ and

$C_v = (4\pi N/3) \int_0^\infty a^3 n(a) da$, which is used in the above Equation (5.A.1.2) for C_M. It should be noted that all terms in the above equations are functions of distance r from the transducer.

The attenuation due to the presence of particles along the sound path should be the average over the whole particle size distribution $n(a)$,

$$\langle a_t \rangle = \int_0^\infty a_t n(a) da \qquad (5.A.1.3)$$

Rearranging the above equations, the local sediment mass concentration, $C_M(r,a)$, can be formulated as

$$C_M(r,a) = B_1 B_2(r, C_M) \langle V^2(r) \rangle \qquad (5.A.1.4)$$

$$B_1 = S_N^{-1} F_f^{-1}(r,a) \qquad (5.A.1.5)$$

$$B_2(r, C_M) = S_{res}^{-1}(r) exp \left(4 \int_0^r \langle a_t(r) \rangle dr \right) = r^2 exp \left(4a_0 r + B_3 \int_0^r C_M(r) dr \right)$$

$$(5.A.1.6)$$

where B_3 becomes a constant for incoherent scattering and when multiple scattering effects are ignored (valid when $C_v < 1\%$). B_1 depends on the system sensitivity and form functions. $B_2(r, C_M)$ is an attenuation loss term, comprising the spreading loss and attenuation loss due to the ambient fluid and the suspended sediment particles. Calibration or semi-empirical relations can be used to determine B_1 and $B_2(r, C_M)$ (Thorne & Hanes, 2002). In low particle concentrations ($\ll 1$ g/L), the acoustic attenuation is negligible and the exponential term in $B_2(r, C_M)$ can be dropped. For the case of higher concentrations, profiles over longer distance, or higher acoustic frequencies, attenuation effects become more pronounced and attenuation compensation has to be implemented. However, no explicit expression can be obtained from the above equation and this equation has to be inverted in order to determine $C_M(r, a)$. Estimation of $C_M(r, a)$ is affected by the non-linear sediment attenuation due to the presence of suspended particles along the whole travel path of the acoustic pulse. Erroneous estimations of the attenuation terms may result, and an accumulation of the gate-to-gate propagation of these errors along the profile will bias the time-averaged sediment concentration profile estimates.

In order to obtain profiles of concentration from a single transducer operating at a fixed frequency, knowledge of the sediment grain size distribution is needed. As a first order approximation, it can be assumed that $n(a)$ is constant with height above the bed. This supports weak grain sorting effects along the profile, implying that the Rouse number exceeds unity at all locations in the profile, even for the largest size fraction in the grain size distribution.

The two most commonly used methods to determine $C_M(r, a)$ are the iterative and the explicit inversion methods (Thorne & Hanes, 2002), as summarized below.

5.A.1.2 Acoustic inversion methods

5.A.1.2.1 Iterative inversion method

The iterative inversion method was proposed by Thorne et al. (1993). It relies on the assumption of negligible sediment attenuation in the first profile gate. An iterative gate-to-gate solution can then be applied between consecutive gates from the transducer face to the last gate in the profile. The iterative inversion method is sensitive to the numerical discretization error of the attenuation integral between consecutive gates. Errors linked to the negligible attenuation assumption in the first gate and to numerical approximation of the attenuation integral are cumulative when the concentration profile is monotonic. Furthermore, the integral estimate error propagates with the iteration steps so that even the smallest error at a given gate generates errors in all the following gates in the profile. More details are given in Hurther et al. (2011).

5.A.1.2.2 Explicit inversion method

The explicit inversion method was proposed by Lee and Hanes (1995) who showed that the derivative of $C_M(r, a)$ with range r can be integrated. As with the iterative inversion method, it is assumed that intensity attenuation due to sediments between the transducer and the first point in the profile is negligible. This method is computationally more efficient than the iterative inversion method. The error propagates from gate-to-gate as it does with the iterative inversion method. Thus, an error at any gate of the profile will affect the concentration estimates at all subsequent gates. This is discussed in Thorne and Hanes (2002), Bricault (2006) and Hurther *et al.* (2011). Pedocchi and Garcia (2012) used this method for concentration profiling in oscillatory flow with an Ultrasonic Velocity Profiler (UVP).

The results obtained with the application of these two methods are particularly unsatisfactory when an error is made at the first gate in the profile where attenuation is assumed to be negligible, since this will cause an accumulation of errors over the entire profile. Furthermore, the above inversion methods do not perform well for attenuating media found in highly turbulent benthic suspension flows. Therefore, the methods presented in Sections 5.A.1.3.1 and 5.A.1.3.2 were developed.

5.A.1.3 Attenuation compensation using hardware

5.A.1.3.1 Backward and forward scattering

The problems encountered with the above inversion methods can be overcome by using backscattered and forward scattered profile signals over the same water column, thus providing attenuation compensation even in high particle concentrations, as long as multiple scattering is avoided (Shen & Lemmin, 1996). The system gives excellent results in terms of concentration range profiling up to 200 kg/m^3 over a distance of more than a decimeter above a channel bed. The spatial resolution is 6 mm with a temporal resolution in the 100 ms range (Shen & Lemmin, 1996). The system configuration requires the alignment of two emitting / receiving transducers mounted face-to-face. This configuration may be limited in its use for profiling above strongly mobile beds, since one of the transducers must be mounted at bed level. Shen and Lemmin (1998) used a Least Mean Square (LMS) compensation algorithm in conjunction with this hardware solution in order to improve the quality of sediment concentration profiling. It was shown that this algorithm is more robust than the above inversion algorithms, and errors do not accumulate along the profile path.

5.A.1.3.2 Two-frequency method

The inconvenience of the backward/forward scattering solution for field applications of the system was overcome by an approach based on exploiting backscattering intensity at two (or more) emitted frequencies. Hay (1991), Hay and Sheng (1992), Crawford and Hay (1993), Thosteson and Hanes, (1998) used widely different frequencies. Bricault (2006) and Hurther *et al.* (2011) showed that two relatively close frequencies (*e.g.* 1.5 MHz and 2 MHz) completely resolve the concentration field of fine particles typically found in benthic boundary layer applications. Two advantages of the two-frequency approach are the following: the boundary condition that the intensity attenuation due to sediments between the transducer and the first point in the profile is negligible is no longer necessary in order to perform the inversion, and the absence of integrals eliminates errors caused by their numerical approximation. As a result, the propagation of the local errors along the profile via the integrals is removed and this ensures gate independent errors over the entire concentration profile.

In order for this bi-frequency method to work properly, the acoustic backscattering parameter ka_s for the lower of the two acoustic frequencies must be smaller than 10. k and a_s are the acoustic wavenumber and the mean radius of the suspended sediments, respectively. This condition can easily be respected for acoustic frequencies varying between 1 MHz and 5 MHz when using sand of typical mean radius between 50 μm and 250 μm. Furthermore, due to the high sensitivity to frequency of the attenuation coefficient in the range $ka_s < 1$ which roughly corresponds to the Rayleigh scattering regime, differences between acoustic frequencies in the range of 100 kHz induce ratio values of attenuation coefficients at the two frequencies sufficiently different from unity. This frequency range can be handled by a single emitter transducer without specific large band characteristics, greatly simplifying the hardware arrangement over previous multi-frequency, multi-transducer systems.

5.A.2 Concentration measurements with ADCPs

Appendix 5.A.2 is linked to Section 5.3.3 of Volume II, where details on the notation used can be found. This appendix deals specifically with the determination of sus-pended sediment concentrations from ADCP measurements. This Appendix provides similar information as the preceding Appendix 5.A.1 focusing on ABIs, but with the specific focus on ADCPs.

5.A.2.1 Scattering formulations

The intensity of the sound scattered from an object is often given in logarithmic units relative to a reference intensity. Since acoustic intensity is proportional to the square of the pressure (P), the backscattered intensity can be expressed in terms of the sound pressure level, SPL, which is given in logarithmic units relative to the reference pressure (P_r) as (Clay & Medwin, 1977):

$$SPL = 20log_{10}\frac{P}{P_r} \text{ dB re } P_r \tag{5.A.2.1}$$

where "re" is short for "relative to". The sound pressure level depends on the difference between the source level, SL, and the loss of intensity during signal transmission (i.e. the transmission loss, TL):

$$SPL = SL - TL \tag{5.A.2.2}$$

Therefore in order to estimate the amount of sound scattered by suspended particles in a given range bin, the backscattered sound received by the instrument must be corrected for beam spreading and sound absorption in the water column. Depending on the instrument, the backscattered intensity is either recorded as raw intensity (e.g. by Teledyne RD Instruments' ADCPs) or as the signal to noise ratio (e.g. by Sontek's M9). For ADCPs that record the raw intensity, the volume backscattering level of the ensonified particles in decibels, S_v, can be calculated from the backscattered intensity using the following expression (Gostiaux & van Haren, 2010):

$$S_v = 10log_{10}\left(10^{k_cE/10} - 10^{k_cE_{noise}/10}\right) + 20log_{10}(R) + 2\alpha R + A \qquad (5.A.2.3)$$

where k_c is a beam specific echo intensity scale factor [dB/count], E is the echo intensity [counts], E_{noise} [counts] is the intensity reference level of each beam in the ambient noise, R is the distance along each beam (i.e., the slant distance), α is the attenuation of the acoustic signal by the suspension along the beam path (see below) and A is a constant which depends on the instrument, its settings and the water temperature. To determine the volume backscattering level for instruments that record the signal to noise ratio, SNR [dB], Equation (5.A.2.3) becomes:

$$S_v = SNR + 20log_{10}(R) + 2\alpha R + A \qquad (5.A.2.4)$$

For ADCPs constructed by Teledyne RD Instruments[1], the constant A can be calculated as follows (Deines, 1999):

$$A = C + 10log_{10}(T_x + 273.16) - 10log_{10}(\text{slant bin size, m})$$
$$- 10log_{10}(\text{transmit power, Watts}) + 10log_{10}(4\pi) \qquad (5.A.2.5)$$

where C is a constant which is specific to the ADCP model and T_x is the temperature of the transducer in degrees Celsius. Values of C are given in Table 1 of Deines (1999); they are accurate to within +/– 3dB.

The volume backscattering level is related to the concentration, size, elasticity and density of particles from which the sound is scattered. The theoretical volume backscattering strength for a suspension of particles can be calculated using the following expression:

$$S_v = 10log_{10}\left(\frac{3C_MR_1k_s^2}{16\pi}\right) \qquad (5.A.2.6)$$

where R_1 is the reference distance (1 m) and C_M is the mass concentration. The term k_s is the sediment backscattering function which depends on the size and composition of the sediment through its dependence on the far field form factor, f_∞:

$$k_s^2 = \frac{\int_0^\infty |f_\infty(x)|^2 a^2 n(a)da}{\rho_s \int_0^\infty a^3 n(a)da} \qquad (5.A.2.7)$$

where n is the probability density function of the number distribution of the sediment (cf. Seinfeld & Pandis, 1998), ρ_s is the density of the sediment, a is the particle radius and $x = k_s a$ is the non-dimensional wavenumber. Equation (5.A.2.6) applies to incoherent scattering from suspensions with concentrations less than 1% by volume (mass concentrations < 30 g L^{-1}), which is typically the case in fluvial applications where the particles are randomly distributed.

Expressions for f_∞ for sediment suspensions have been developed by adapting the theory for scattering from rigid spheres to results from laboratory experiments of sound scattering from non-cohesive sediments. At the time of writing, the most recently published expression is that of Moate and Thorne (2012):

$$\frac{|f_\infty(x)|^2}{\rho_s} = \left(\frac{x^2 \left(1 - 0.25e^{-((x-1.5)/0.35)^2} \right) \left(1 + 0.6e^{-((x-2.9)/1.15)^2} \right)}{42 + 28x^2} \right)^2 \qquad (5.A.2.8)$$

It is similar to previously published expressions (*e.g.* Thorne & Meral, 2008), but was fit to a data set that included scattering measurements from sediments of different densities including a non-cohesive plate-shaped sediment, while previous studies only examined scattering from more rounded quartz-like sediments.

From Equations (5.A.2.6) – (5.A.2.8) it can be seen that when $x \ll 1$, the far field form factor is proportional to x^4, meaning that backscatter increases strongly with frequency. This is known as the Rayleigh scattering regime. When $x \gg 1$ the far field form factor is a constant and no longer changes with particle size. This is known as the geometric scattering regime. The far field form factor is plotted as a black solid curve in Figure 5.A.2.1 alongside the backscattering function at 3 MHz.

The attenuation term in Equations (5.A.2.3) and (5.A.2.4) is the sum of attenuation due to water, α_w and attenuation due to sediment α_s. The attenuation due to water is a function of water density and of the instrument's acoustic frequency. For fluvial environments it can be calculated using Fisher and Simmons (1977)'s expression, accounting for temperature, but neglecting the salinity and pressure terms

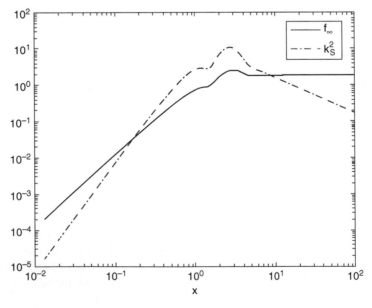

Figure 5.A.2.1 Far field form factor (solid line) and the square of the backscattering function (k_s^2) at 3 MHz (-.-) as a function of non-dimensional wave number x.

since the former is zero for fresh water and the latter is zero for depths less than 100 m:

$$\alpha_w \left[\frac{dB}{m}\right] = 8.686 \times (55.9 - 2.37T + 4.77 \times 10^{-2}T^2 - 3.48 \times 10^{-4}T^3) \times 10^{-15}f^2$$

$$(5.A.2.9)$$

where T is the water temperature in degrees Celsius and f is the acoustic frequency in Hz.

Attenuation due to suspended sediments is a function of acoustic frequency as well as sediment concentration and grain size. When suspended sediment concentrations are less than about 0.1 kg m^{-3} (0.1g L^{-1}) and acoustic frequencies are on the order of 1MHz or less, the attenuation due to sediment is minimal for ranges less than a few meters, but as frequency, range or concentration increase, the attenuation due to the suspended sediment also increases. The sediment attenuation can be due to the viscous drag between the fluid and the suspended particles or due to scattering of the sound in directions other than that of the instrument. For typical ADCP frequencies (0.3 to 3 MHz), viscous absorption occurs in the presence of silt and clay sized sediments, while scattering attenuation is caused by sand sized particles. The expression for the total sediment attenuation is (e.g. Hay, 1983; Thorne & Meral, 2008):

$$\alpha_s \left[\frac{dB}{m}\right] = \frac{8.686C_M}{\rho_s} \left(\frac{\int_0^\infty \xi(a)\, a^3 n(a) da}{\int_0^\infty a^3 n(a) da} + \frac{3\int_0^\infty \chi(a) a^2 n(a) da}{4\int_0^\infty a^3 n(a) da} \right) \qquad (5.A.2.10)$$

The first term in brackets is the ensemble averaged viscous attenuation while the second term is the ensemble averaged scattering attenuation.

The viscous attenuation parameter ξ can be calculated using the theoretical expression for viscous absorption by a suspension of spherical particles with radius a (Urick, 1948):

$$\xi = \frac{k(\gamma - 1)^2}{2} \left(\frac{s}{s^2 + (\gamma + \tau)^2} \right),$$

$$s = \frac{9}{4\beta a}\left[1 + \frac{1}{\beta a}\right], \qquad (5.A.2.11)$$

$$\gamma = \frac{\rho_s}{\rho}, \quad \tau = \frac{1}{2}\left[1 + \frac{9}{2\beta a}\right], \quad \beta = \sqrt{\frac{\omega}{2v}}$$

where ρ and ρ_s are respectively the densities of the water and the sediment, v is the kinematic viscosity of water and ω is the angular acoustic frequency. Excellent agreement has been found between Equation (5.A.2.11) and observations of viscous absorption from both kaolin and fine sand (see Figure 6 of Urick, 1948).

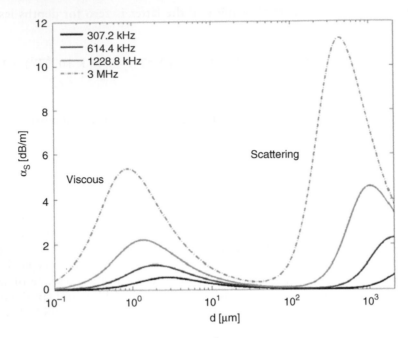

Figure 5.A.2.2 Theoretical attenuation from a 1 kg m^{-3} suspension of mono-sized particles of diameter d.

There are a number of empirical expressions for the total scattering cross section (χ) of suspensions of non-cohesive sediments, but the most recent is that of Moate and Thorne (2012):

$$\frac{\chi}{\rho_s} = \frac{0.09x^4}{1380 + 560x^2 + 150x^4} \qquad (5.A.2.12)$$

This empirical relationship was obtained for measurements of attenuation from suspensions of various non-cohesive sediments, including non-cohesive plate-shaped sediments.

Making use of Equations (5.A.2.10) – (5.A.2.12), the total attenuation from a 1 kg m^{-3} (1 g L^{-1}) suspension of mono-sized particles is plotted as a function of particle diameter in Figure 5.A.2.2 for common ADCP frequencies. From this figure it can be seen that viscous attenuation is caused by silts and clays while scattering attenuation is caused by sands. It can also be seen that for a given particle size, attenuation increases with frequency.

5.A.2.2 Obtaining concentration from backscattered intensity

It can be seen that in order to determine suspended sediment concentrations using the equations presented in the preceding section, the instrument must be properly calibrated and the size distribution and density of the suspended sediment must be known. Since acoustic devices are generally used when the suspension characteristics are unknown, an assumption must be made about the size of the suspended sediment. An estimate can be obtained using measurements of the bed sediments. This is the common practice in coastal studies. Bed grab samples are much harder to obtain in fast flowing rivers than they are in coastal zones, particularly when gravel and cobbles impede collection of the finer interstitial sediment.

Since the recorded backscatter depends on S_v, which is a linear function of $10log_{10}(C_M)$ through Equation (5.A.2.6), one can obtain an empirical relationship between the recorded backscatter and the concentration using concurrent and collocated water samples and ADCP measurements. This approach was introduced by Thevenot et al. (1992) and popularized by Gartner (2004), Landers (2010), and others. Using the fluid corrected backscatter FCB, which is the backscattered signal corrected for spreading and attenuation due to the water:

$$FCB = 10log_{10}\left(10^{k_c E/10} - 10^{k_c E_{noise}/10}\right) + 20log_{10}(R) + 2\alpha_w R \qquad (5.A.2.13)$$

the method consists of using the data to find the constants K_1 and K_2 in the following equation

$$10log_{10}(C_M) = K_1 + K_2 FCB \qquad (5.A.2.14)$$

Once K_1 and K_2 are determined for a given site and instrument, Equation (5.A.2.14) can be used to convert time series of backscatter to concentration. This method has the advantage that the instrument constant is not required (compare Equation (5.A.2.14) to Equation (5.A.2.3)), but it relies on the assumption that suspended sediment size is invariant in space and time. Figure 37 of Thevenot et al. (1992) is reproduced in Figure 5.A.2.3 to demonstrate how the relationship between backscatter and concentration

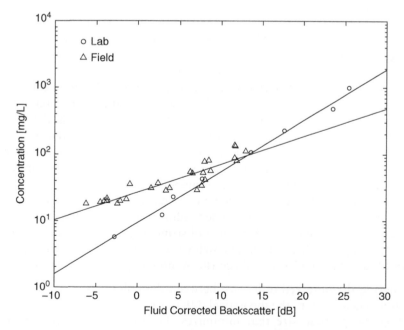

Figure 5.A.2.3 Data presented in figure 37 of Thevenot et al. (1992) showing the relation between sediment concentration and fluid corrected backscatter for a 2.4 MHz ADCP.

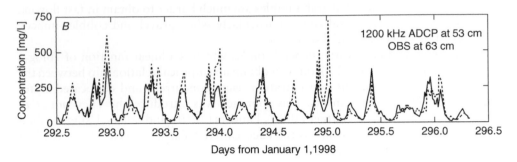

Figure 5.A.2.4 Modified version of figure 4b of Gartner (2004). Time series of concentration estimated from optical (OBS) and acoustic (ADCP) sensors at similar heights above the bed. The ADCP was calibrated with the optical instrument.

can vary depending on site conditions. The two relationships were obtained for the same instrument. The laboratory backscatter measurements were made using suspensions of particles collected in bottom grab samples from the project site.

As an example of the application of Thevenot *et al.* (1992)'s method, we use a figure from Gartner (2004). Gartner calibrated a 1200 kHz ADCP using optical back-scatter data collected at 1 m from the bed. Although this is different than using concentrations measured in physical samples, the procedure is the same. The resulting concentration estimated from the ADCP data collected 0.53 m above the bed is plotted in Figure 5.A.2.4; it is compared to estimations from optical backscatter data collected at 0.63m. It can be seen that there is good agreement between the acoustical and optical estimates of concentration.

From Equations (5.A.2.6) – (5.A.2.8) it can be seen that in the Rayleigh regime, back-scatter strength is more influenced by changes in particle size than by changes in concentration since S_v depends on $10\log_{10}(a^3)$ while it depends on only $10\log_{10}(C_M)$. This means that if concentration remains constant but particle diameter doubles, S_v will increase by 9 dB. Therefore, if sediment diameter were to double after the calibration was performed, the resulting increase in backscatter could be misinterpreted as a multiplication of the concentration by a factor of 8. This problem is commonly referred to as the size-concentration ambiguity. It can be overcome somewhat by using an additional instrument, such as an optical turbidity meter. Optical instruments are typically more affordable than acoustic instruments, however they are not suitable for long term deployments (weeks to months) because their performance is compromised by biofouling.

A second way to facilitate determination of concentration from backscatter data is to obtain a representative size of the suspended sediment using simultaneous backscatter data at two or more frequencies. A number of sizing methods exist, including those of Sheng (1991), Hay and Sheng (1992), Crawford and Hay (1993), Thorne and Hurther (2014) and Jourdin *et al.* (2014). All acoustic grain sizing methods involve two steps:

1. Comparing the observed difference between the volume backscatter strength at two frequencies (*i, j*) to the theoretical difference, and
2. Finding the particle size that minimizes the difference between theory and observations.

5.A.2.3 Obtaining concentration from attenuation

Much like the acoustic backscatter, the sediment attenuation is also a linear function of concentration (recall Equation (5.A.2.10)). Therefore, field calibrations can be done to establish linear relationships between concentration and acoustic attenuation provided the attenuation can be calculated from the backscatter profiles. In order to calculate attenuation, particle concentration and grain size must be uniform over two or more measurement bins, therefore this method is most relevant to side-looking ADCP deployments when grain size and concentration are typically homogeneous across all measurement bins. The method was developed by Topping *et al.* (2007) and it has since been implemented by Wright *et al.* (2010), Moore *et al.* (2012) and Moore *et al.* (2013), among others.

The first step is to compute the fluid corrected backscatter at all ranges using Equation (5.A.2.13). The second step is to compute the slope of the profiles of *FCB* with range in order to calculate the sediment attenuation as follows:

$$\alpha_s = -\frac{1}{2}\frac{\Delta FCB}{\Delta R} \tag{5.A.2.15}$$

where ΔR is the range over which *FCB* changes by ΔFCB.

It has been shown that the acoustic attenuation can be a more accurate suspended sediment surrogate than the backscatter when the suspended matter is fine (*e.g.* Moore *et al.*, 2012). This is because the viscous attenuation is proportional to the particle diameter, while the backscatter and scattering attenuation are proportional to particle volume when $x \ll 1$, so that if particle size changes post-calibration more uncertainty is introduced when using the backscatter than when using the attenuation method.

An example of application of the attenuation inversion method is plotted in Figure 5.A.2.5. This figure is taken from Moore *et al.* (2012); it corresponds to

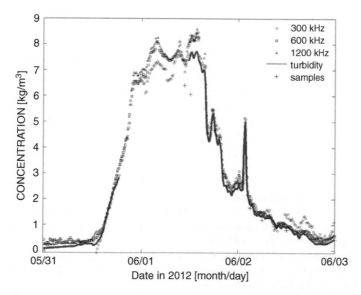

Figure 5.A.2.5 Estimation of suspended sediment concentration during a flood using an optical turbidity meter, physical samples and acoustic attenuation at 300, 600 and 1200 kHz (Moore *et al.*, 2012).

measurements made using side-looking ADCPs during a spring flood. For each ADCP, a relationship was fit between the attenuation and the optical turbidity data, this relationship was then used to estimate concentration when the turbidity meter failed from May 31 to June 1, 2012. The acoustic and optical estimates are also compared to concentrations measured via filtration of physical samples.

NOTE

1   The focus on Teledyne RD Instruments profilers in this subchapter is due to the fact that they are the most commonly used ADCP, it does not indicate endorsement by the author.

REFERENCES

Aberle, J., Coleman, S.E., & Nikora, V.I. (2012) Bed load transport by bed form migration. *Acta Geophys*, 60(6), 1720–1743.

Aberle, J., Nikora, V., Henning, M., Ettmer, B., & Hentschel, B. (2010) Statistical characterization of bed roughness due to bed forms: A field study in the Elbe River at Aken, Germany. *Water Resour Res*, 46, W03521, doi:10.1029/2008WR007406.

Abraham, D., Kuhnle, R.A., & Odgaard, A.J. (2011) Validation of bed-load transport measurements with time-sequenced bathymetric data. *J Hydraul Eng*, 137(7), 723–728, doi:10.1061/(ASCE)HY.1943-7900.0000357.

Admiraal, D. & Garcia, M. (2000) Laboratory measurement of suspended sediment concentration using an acoustic concentration profiler (ACP). *Exp Fluids*, 28(2), 116–127.

Agrawal, Y.C. & Pottsmith, H.C. (1994) Laser diffraction particle sizing in STRESS. *Cont Shelf Res*, 14, 1101–1121, doi:10.1016/0278- 4343(94)90030-2.

Agrawal, Y.C. & Pottsmith, H.C. (2000) Instruments for particle size and settling velocity observations in sediment transport. *Mar Geol* 168(1), 89–114.

Anderson, C.W. (2005) Turbidity (version 2.1): U.S. Geological Survey Techniques of Water-Resources Investigations, book 9, chap. A6, section 6.7, 55 p. [Online] Available from http://water.usgs.gov/owq/FieldManual/Chapter6/Section6.7_v2.1.pdf, [accessed 28th February 2017].

ASTM Standard D7512–09 (2009) *Standard Guide for Monitoring of Suspended-Sediment Concentration in Open Channel Flow Using Optical Instrumentation*. West Conshohocken, Pennsylvania, USA, ASTM International.

ASTM Standard D6855–03 (2003) *Standard Test Method for Determination of Turbidity Below 5 NTU in Static Mode*. West Conshohocken, Pennsylvania, USA, ASTM International.

ASTM Standard D6698–07 (2007) *Standard Test Method for On-Line Measurement of Turbidity Below 5 NTU in Water*. West Conshohocken, Pennsylvania, USA, ASTM International.

Bagherimiyab, F. & Lemmin, U. (2012) Fine sediment dynamics in unsteady open-channel flow studied with acoustical and optical systems. *Cont Shelf Res*, 46, 2–15.

Bale, A.J., Uncles, R.J., Widdows, J., Brinsley, M.D., & Barrett, C.D. (2002) Direct observation of the formation and break-up of aggregates in an annular flume using laser reflectance particle sizing. *Proc Mar Sci*, 5, 189–201.

Barton, J.S., Slingerland, R.L., Pittman, S., & Gabrielson, T.B. (2010) Monitoring coarse bedload transport with passive acoustic instrumentation: A field study. In: Gray, J.R., Laronne, J.B., & Marr, J.D.G. (eds.) *Bedload surrogate Monitoring Technologies*. U.S. Geological Survey Scientific Investigations Report 2010–5091. US Geological Survey: Reston, VA. pp 38–51. [Online] Available from: http://pubs.usgs.gov/sir/2010/5091/papers/ [Accessed 28th February 2017].

Bassett, C., Thomson, J., & Polagye, B. (2013) Sediment-generated noise and bed stress in a tidal channel. *J Geoph Res Oceans*, *118*, doi:10.1002/jgrc.20169.

Blanckaert, K., Heyman, J., & Rennie, C.D. (2017) Measurements of bedload sediment transport with an Acoustic Doppler Velocity Profiler (ADVP). *J Hydraul Eng*, doi:http://dx.doi.org/10.1061/(ASCE)HY.1943-7900.0001293

Bricault, M. (2006) Rétrodiffusion acoustique par une suspension en milieu turbulent: Application à la mesure de concentration pour l'étude de processus hydro-sédimentaires. *PhD thesis*, France, Grenoble Institut National Polytechnique.

Bunt, J.A., Larcombe, P., & Jago, C.F. (1999) Quantifying the response of optical backscatter devices and transmissometers to variations in suspended particulate matter. *Cont Shelf Res*, *19*(9). 1199–1220.

Bunte, K., Abt, S.R., Potyondy, J.P., & Ryan, S.E. (2004) Measurement of coarse gravel and cobble transport using portable bedload traps. *J Hydraul Eng*, *130*(9), 879–893.

Bunte, K. & Abt, S.R. (2005) Effect of sampling time on measured gravel bed load transport rates in a coarse-bedded stream. *Water Resour Res*, *41*(11), W11405, doi:10.1029/2004WR003880.

Bunte, K., Abt, S.R., Potyondy, J.P., & Swingle, K.W. (2008) A comparison of coarse bedload transport measured with bedload traps and Helley-Smith samplers. *Geodinamica Acta*, *21*(1–2), 53–66.

Bunte, K., Swingle, K.W., & Abt, S.R. (2010a) Necessity and difficulties of field calibrating signals from surrogate techniques in gravel-bed streams: Possibilities for bedload trap samplers. In: Gray, J.R., Laronne, J.B., & Marr, J.D.G. (eds.) *Bedload surrogate Monitoring Technologies*, U.S. Geological Survey Scientific Investigations Report 2010–5091, US Geological Survey: Reston, VA. pp 107–129. [Online] Available from: http://pubs.usgs.gov/sir/2010/5091/papers/ [Accessed 28th February 2017].

Bunte, K., Abt, S.R., Swingle, K.W., & Potyondy, J.P. (2010b) Functions to adjust transport rates from a Helley-Smith sampler to bedload traps in coarse gravel-bed streams (rating curve approach). In: *Session on Fluvial Geomorphology, paper presented at Proceedings of the 4th Federal Interagency Hydrologic Modeling Conference and the 9th Federal Interagency Sedimentation Conference* Las Vegas, 27 June–1 July 2010.

Cellino, M. & Lemmin, U. (2004) Influence of coherent flow structures on the dynamics of suspended sediment transport in open-channel flow. *J Hyd Eng*, *130*(11), 1077–1088.

Chacho, E.F., Burrows, R.L., & Emmett, W.W. (1989) Detection of coarse sediment movement using radio transmitters. In: *Proceedings of the XXIII Congress of the International Association for Hydraulic Research, August 1989, Ottawa, Ontario, Canada*. The National Research Council of Canada. B367–B373.

Chacho, E.F., Burrows, R.L., & Emmett, W.W. (1994) Monitoring gravel movement using radio transmitters. In: Cotroneo, G.V. & Rumer, R.R. (eds.) *Hydraulic Engineering '94, Proceedings of the 1994 ASCE National Conference on Hydraulic Engineering, August 1994 Buffalo, NY*. pp. 785–789.

Childers, D. (1999) Field comparison of six-pressure-difference bedload samplers in high energy flow. *U.S. Geological Survey Water-Resource Investigations Report 92–4068*, 59. [Online] Available from: http://pubs.er.usgs.gov/pubs/wri/wri924068/[Accessed 28th February 2017].

Clay, C.S. & Medwin, H. (1977) *Acoustical Oceanography: Principles and Applications*. Toronto, John Wiley & Sons.

Coleman, S.E., Zhang, M.H., & Clunie, T.M. (2005) Sediment-wave development in subcritical water flow. *J Hydraul Eng*, *131*(2), 106–111.

Conner, C.S. & De Visser, A.M. (1992) A laboratory investigation of particle size effects on optical backscatterance sensor, *Mar Geol*, *108*, 151–159, doi:10.1016/0025-3227(92)90169-I.

Crawford, A.M. & Hay, A.E. (1993) Determining suspended sand size and concentration from multifrequency acoustic backscatter. *J Acoust Soc Am*, *94*, 3312–3324.

Crickmore, M.J. (1967) Measurement of sand transport in rivers with special reference to tracer methods. *Sedimentology 8*(3), 175–228, doi: 10.1111/j.1365-3091.1967.tb01321.x.

D&A Instrument Company (1991). *OBS-1 & 3 Instruction Manual. 41,* Port Townsend, Washington.

Davies-Colley, R.J. & Smith, D.G. (2001) Turbidity, suspended sediment, and water clarity: A review. *J Am Wat Resour Ass, 37*(5), 1085–1101.

Davis, B.E. (2005) Federal interagency sedimentation project Report QQ: A guide to the proper selection and use of federally approved sediment and water-quality samplers. Vicksburg, Mississippi. 35.

De Clercq, B., Lant, P.A., & Vanrolleghem, P.A. (2004) Focused beam reflectance technique for in situ particle sizing in wastewater treatment settling tanks. *J Chem Techn Biotechn, 79*(6), 610–618.

Deines, K.L. (1999) Backscatter estimation using broadband acoustic Doppler current profilers. In: *Proceedings of the IEEE 6th Working Conference on Current Measurement, March 11–13, 1999, San Diego, CA, USA.* pp. 249–253.

Dinehart, R.L. (2002) Bedform movement recorded by sequential single-beam surveys in tidal rivers. *J Hydrol, 258*(1–4), 25–39, doi:/10.1016/S0022-1694(01)00558-3.

Downing, J.P. (1996) Suspended sediment and turbidity measurements in streams: What they do and do not mean. *Paper presented at Automatic Water Quality Monitoring Workshop, B. C. Water Quality Monitor. Agree. Coord. Comm.,* Richmond, British Columbia, Canada.

Downing, J. (2010) Acoustic gravel-momemtum sensor. In: Gray, J.R., Laronne, J.B., & Marr, J. D.G. (eds.) *Bedload surrogate Monitoring Technologies.* U.S. Geological Survey Scientific Investigations Report 2010–5091, US Geological Survey: Reston, VA. pp. 143–158. [Online] Available from: http://pubs.usgs.gov/sir/2010/5091/papers/ [Accessed 28th February 2017].

Duffy, G.P. & Hughes-Clarke, J.E. (2005) Application of spatial cross correlation to detection of migration of submarine sand dunes. *J Geophys Res, 110,* F04S12, doi: 10.1029/2004JF000192.

Edwards, T.K. & Glysson, G.D. (1999) *Field Methods for Measurement of Fluvial Sediment.* U.S. Geological Survey Techniques of Water Resources Investigations, Book 3, Chapter 2, 89.

Ergenzinger, P. & Custer, S.G. (1983) Determination of bedload transport using naturally magnetic tracers: First experiences at Squaw Creek, Gallatin County, Montana. *Water Resour Res, 19*(1), 187–193, doi:10.1029/WR019i001p00187.

Ergenzinger, P., Schmidt, K.D., & Busskamp, R. (1989) The pebble transmitter system (PETS): First results of a technique for studying coarse material erosion, transport and deposition. *Zeitschrift für Geomorphologie, 33,* 503–508.

Fisher, F.H. & Simmons, V.P. (1977) Sound absorption in sea water. *J Acoust Soc Am, 62,* 558–564.

Gaeuman, D. & Jacobson, R.B. (2006) Acoustic bed velocity and bedload dynamics in a large sand-bed river. *J Geophys Res, 111,* F02005, doi:10.1029/2005JF000411.

Gaeuman, D. & Jacobson, R.B. (2007) Field assessment of alternative bed-load transport estimators. *J Hydraul Eng, 133*(12), 1319–1328.

Garcia, M.H. (ed.) (2008) *Sedimentation Engineering, Processes, Measurements, Modeling, and Practice.* ASCE Manuals and Reports on Engineering Practice No. 110. Reston, VA, USA, American Society of Civil Engineers.

Gartner, J.W. (2004) Estimating suspended solids concentrations from backscatter intensity measured by acoustic Doppler current profiler in San Francisco Bay, California. *Mar Geol, 211,* 169–187.

Geay, T., Belleudy, P., Gervaise, C., Habersack, H., Aigner, J., Kreisler, A., Seitz, H. & Laronne, J.B. (2017) Passive acoustic monitoring of bedload discharge in a large gravel bed river. *J Geophys Res – Earth Surf, 122*(2), 528–545, doi:10.1002/2016JF004112.

Gostiaux, L. & van Haren, H. (2010) Extracting meaningful information from uncalibrated backscattered echo intensity data. *J Atm Ocean Techn, 27*(5), 943–949.

Gray, J.R. & Gartner, J.W. (2009) Technological advances in suspended-sediment surrogate monitoring. *Water Resour Res, 45*, W00D29, doi:10.1029/2008WR007063.

Gray, J.R., Laronne, J.B., & Marr, J.D.G. (2010) Bedload-surrogate Monitoring Technologies. US Geological Survey Scientific Investigations Report 2010–5091. Reston, VA. US Geological Survey. [Online] Available from: http://pubs.usgs.gov/sir/2010/5091/papers/ [Accessed 28th February 2017].

Greaves, D., Boxall, J., Mulligan, J., Montesi, A., Creek, J., Dendy Sloan, E., & Koh, C.A. (2008) Measuring the particle size of a known distribution using the focused beam reflectance measurement technique. *Chem Eng Sci, 63*(22), 5410–5419.

Gronz, O., Hiller, P.H., Wirtz, S., Becker, K., Iserloh, T., Seeger, M., Aberle, J., & Casper, M.C. (2016) Smartstones: A small 9-axis sensor implanted in stones to track their movements. *Catena, 142*, 245–251.

Habersack, H.M (2001) Radio-tracking gravel particles in a large braided river in New Zealand: A field test of the stochastic theory of bed load transport proposed by Einstein. *J Hydrol Proc, 15* (3), 377–391.

Habersack, H., Seitz, H., & Liedermann, M. (2010) Integrated automatic bedload transport monitoring. In: Gray, J.R., Laronne, J.B., & Marr, J.D.G. (eds.) *Bedload surrogate Monitoring Technologies*. Reston, VA. U.S. Geological Survey Scientific Investigations Report 2010–5091. US Geological Survey. pp 218–235. [Online] Available from: http://pubs.usgs.gov/sir/2010/5091/papers/ [Accessed 28th February 2017].

Hassan, M.A. & Ergenzinger, P. (2003) Use of tracers in fluvial geomorphology. In: Kondolf, G. M. & Piégay, H. (eds.) *Tools in Fluvial Geomorphology*. Chichester, John Wiley & Sons. pp. 397–423.

Hassan, M.A., Schick, A.P., & Laronne, J.B. (1984) The recoveroy of flood-dispersed coarse sediment particles, a three-dimensional magnetic tracing method. *Catena Supplement, 5*, 153–162.

Hay, A.E. (1983) On the remote acoustic detection of suspended sediment at long wavelengths. *J Geophys Res, 88*(C12), 7525–7542.

Hay, A.E. (1991) Sound scattering from a particle laden turbulent jet. *J Acoust Soc Am, 90(4)*, 2055–2074.

Hay, A.E. & Sheng, J. (1992) Vertical profiles of suspended sand concentration and size from multifrequency acoustic backscatter. *J Geophys Res, 97*(C10), 15661–15677.

Hay, A.E., Zedel, L., Cheel, R., & Dillon, J. (2012a) Observations of the vertical structure of turbulent oscillatory boundary layers above fixed roughness beds using a prototype wideband coherent Doppler profiler: 1. The oscillatory component of the flow, *J Geophys Res Oceans, 117*(3), C03005, doi:10.1029/2011JC007113.

Hay, A.E., Zedel, L., Cheel, R., & Dillon, J. (2012b) Observations of the vertical structure of turbulent oscillatory boundary layers above fixed roughness using a prototype wideband coherent Doppler profiler: 2. Turbulence and stress, *J Geophys Res Oceans, 117*(3), C03006, doi:10.1029/2011JC007114.

Hay, A.E., Zedel, L., Cheel, R., & Dillon, J. (2012c) On the vertical and temporal structure of flow and stress within the turbulent oscillatory boundary layer above evolving sand ripples: Observations using a prototype wide-band coherent Doppler profiler. *Cont Shelf Res, 46*, 31–49.

Hayward, J.A. & Sutherland, A.J. (1974) The Torlesse stream vortex-tube sediment trap. *J Hydrol (NZ), 13*(1), 41–53.

Helley, E.J. & Smith, W. (1971) *Development and Calibration of a Pressure-Difference Bed Load Sampler*. U.S. Geological Survey Open File Report, Washington, USA.

Henning, M., Hentschel, B., & Hüsener, T. (2009) Photogrammetric system for measurement of dune movement. In: *Conference Proceedings of the 33rd IAHR Congress "Water Engineering for a Sustainable Environment"*, Vancouver, Canada, *August 9–14, 2009*, 4966–4972 (CD-ROM).

Henning, M., Aberle, J., & Coleman, S. (2010) Analysis of 3D-bed form migration rates. In: Dittrich, A., Koll, K., Aberle, J., & Geisenhainer, P. (eds.) *Proceedings of the International Conference on Fluvial Hydraulics River Flow 2010, 8–10 September 2010, Braunschweig, Germany*. Karlsruhe, Bundesanstalt für Wasserbau. pp. 879–885.

Henning, M. (2013) *Mehrdimensionale statistische Analyse räumlich und zeitlich hoch aufgelöster Oberflächen von Dünenfeldern*. Mitt. Leichtweiß-Institut für Wasserbau No. 160, Technische Universität Braunschweig, Braunschweig (in German).

Hoekstra, P., Bell, P., van Santen, P., Roode, N., Levoy, F., & Whitehouse, R. (2004) Bedform migration and bedload transport on an intertidal shoal, *Cont Shelf Res*, 24(11), 1249–1269, doi:10.1016/j.csr.2004.03.006.

Hubbell, D.W. & Sayre, W.W. (1964) Sand transport studies with radioactive tracers. *J Hydraul Div*, 90(3), 39–68.

Hurther, D. & Lemmin, U. (2001) Discussion of "Equilibrium near-bed concentration of suspended sediment" by Z. Cao. *J Hydraul Eng*, 127(5), 430–434.

Hurther, D., Thorne, P.D., Bricault, M., Lemmin, U., & Barnoud, J-M. (2011) A multi-frequency acoustic concentration and velocity profiler (ACVP) for boundary layer measurements of fine-scale flow and sediment transport processes. *Coast Eng*, 58(7), 594–605.

Hurther, D. & Thorne, P.D. (2011) Suspension and near-bed load sediment transport processes above a migrating, sand-rippled bed under shoaling waves. *J Geophys Res*, 116, doi:10.1029/2010JC006774.

Jamieson, E.C., Rennie, C.D., Jacobson, R.B., & Townsend, R.D. (2011) 3-D flow and scour near a submerged wing dike: ADCP measurements on the Missouri River. *Water Resour Res*, 47, W07544, doi:10.1029/2010WR010043.

Jetté, C.D. & Hanes, D.M. (1997) High-resolution sea-bed imaging: An acoustic multiple transducer array. *Meas Sci Technol*, 8, 787–792.

Johnston, P., Bérubé, F., & Bergeron, N.E. (2009) Development of a flatbed passive integrated transponder antenna grid for continuous monitoring of fishes in natural streams. *J Fish Biol*, 74(7), 1651–1661.

Jourdin, F., Tessier, C., Le Hir, P., Verney, R., Lunven, M., Loyer, S., Lusven, A., Filipot, J.-F., & Lepesqueur, J. (2014) Dual-frequency ADCPs measuring turbidity, *Geo-Marine Letters*, 34 (4), 381–397.

Kineke, G.C. & Sternberg, R.W. (1992) Measurements of high concentration suspended sediments using the optical backscatterance sensor. *Mar Geol*, 108, 253–258. doi:10.1016/0025-3227(92)90199-R.

Kuhnle R.A. (2013) Suspended Load. In: Shroder, J.F. (ed.) *Treatise on Geomorphology*. 9. San Diego, Academic Press. pp. 124–136.

Lamarre, H., MacVicar, B., & Roy, A.G. (2005) Using passive integrated transponder (PIT) tags to investigate sediment transport in gravelbed rivers. *J Sed Res*, 75, 736–741.

Landers, M.N. (2010) Review of methods to estimate fluvial suspended sediment characteristics from acoustic surrogate metrics, In: *Second Joint Federal Interagency Conference, Las Vegas, June 27–July 1, 2010*. [Online] Available from: http://acwi.gov/sos/pubs/2ndJFIC/ [Accessed 28th February 2017].

Laronne, J.B. & Carson, M.A. (1976) Interrelationships between bed morphology and bed material transport for small, gravel-bed channel. *Sedimentology*, 23, 67–85.

Laronne, J.B., Alexandrov, Y., Bergman, N., Cohen, H., Garcia, C., Habersack, H., Powell, D.M., & Reid, I. (2003) The continuous monitoring of bedload flux in various fluvial systems. In: Bogen, J., Fergus, T., & Walling, D. (eds.) *Erosion and Sediment Transport Measurement in Rivers – Technological and Methodological Advances*. IAHS Publication no. 283. pp. 134–145.

Law, D.J., Bale, A.J., & Jones, S.E. (1997) Adaptation of focused beam reflectance measurement to in-situ particle sizing in estuaries and coastal waters. *Mar Geol*, *140*(1), 47–59.

Lee, T.H. & Hanes, D.M. (1995) Direct inversion method to measure the concentration profile of suspended particles using backscattered sound. *J Geophys Res*, *100*(C2), 2649–2657, doi:10.1029/94JC03068.

Lenzi, M.A., D'Agostino, V., & Billi, P. (1999) Bedload transport in the instrumented catchment of the Rio Cordon. Part I: Analysis of bedload records, conditions and threshold of bedload entrainment. *Catena*, *36*, 171–190.

Liébault, F., Bellot, H., Chapuis, M., Klotz, S., & Deschâtres, M. (2012) Bedload tracing in a high-sediment-load mountain stream. *Earth Surf Proc Land*, *37*(4), 385–399.

Ludwig, K.A. & Hanes, D.M. (1990) A laboratory evaluation of optical backscatterance suspended solids sensors exposed to sand-mud mixtures. *Mar Geol*, *94*, 173–179. doi:10.1016/0025-3227(90)90111-V.</REF>

Mao, L., Carrillo, R., Escauriaza, C., & Iroume, A. (2016) Flume and field-based calibration of surrogate sensors for monitoring bedload transport. *Geomorphology*, *253*, 10–21.

McNamara, J. & Borden, C. (2004) Observations on the movement of coarse gravel using implanted motion-sensing radio transmitters. *Hydrol Proc*, *18*, 1871–1884.

Megahan, W.F. (1999) Scale considerations and the detectability of sedimentary cumulative watershed effects. Technical Bulletin No. 776. National Council of the Paper Industry for Air and Stream Improvement.

Melis, T.S., Topping, D.J., & Rubin, D.M. (2003) Testing laser-based sensors for continuous in situ monitoring of suspended sediment in the Colorado River, Arizona. In: Bogen, J., Fergus, T., & Walling, D. (eds.) *Erosion and Sediment Transport Measurement in Rivers—Technological and Methodological Advances*. IAHS Publication no. 283, pp. 21–27.

Mikkelsen, O. & Pejrup, M. (2001) The use of a LISST-100 laser particle sizer for in-situ estimates of floc size, density and settling velocity. *Geo-Marine Letters*, *20*(4), 187–195.

Milhous, R.T. (1973) Sediment transport in a gravel-bottomed stream. *Ph.D. Thesis*. Corvallis, Oregon State University.

Mizuyama, T., Oda, A., Laronne, J.B., Nonaka, M., & Matsuoka, M. (2010a) Laboratory tests of a Japanese pipe hydrophone for continuous monitoring of coarse bedload. In: Gray, J.R., Laronne, J.B., & Marr, J.D.G. (eds.) *Bedload surrogate Monitoring Technologies*. Reston, VA. U.S. Geological Survey Scientific Investigations Report 2010–5091. US Geological Survey. pp 319–335. [Online] Available from: http://pubs.usgs.gov/sir/2010/5091/papers/ [Accessed 28th February 2017].

Mizuyama, T., Laronne, J.B., Nonaka, M., Sawada, T., Satofuka, Y., Matsuoka, M., Yamashita, S., Sako., Y., Tamaki, S., Watari, M., Yamaguchi, S., & Tsuruta, K. (2010b) Calibration of a passive acoustic bedload monitoring system in Japanese mountain rivers. In: Gray, J.R., Laronne, J.B., & Marr, J.D.G. (eds.) *Bedload Surrogate Monitoring Technologies*. Reston, VA. U.S. Geological Survey Scientific Investigations Report 2010–5091. US Geological Survey. pp 296–318. [Online] Available from: http://pubs.usgs.gov/sir/2010/5091/papers/ [Accessed 28th February 2017].

Moate, B.D. & Thorne, P.D. (2012) Interpreting acoustic backscatter from suspended sediments of different and mixed mineralogical composition. *Cont Shelf Res*, *46*, 67–82.

Moore, S.A. (2012) Monitoring flow and fluxes of suspended sediment in rivers using side-looking acoustic Doppler current profilers. *Ph.D. Thesis*. University of Grenoble.

Moore, S.A., Le Coz, J., Hurther, D., & Paquier, A. (2012) On the application of horizontal ADCPs to suspended sediment transport surveys in rivers. *Cont Shelf Res*, *46*(1), 50–63.

Moore, S.A., Le Coz, J., Hurther, D., & Paquier, A. (2013) Using multi-frequency acoustic attenuation to monitor grain size and concentration of suspended sediment in rivers. *J Acoust Soc Am*, *133*(4), 1959–1970.

Morlock, S.E. (1996) *Evaluation of Acoustic Doppler Current Profiler measurements of river discharge*. U.S. Geological Survey Water-Resources Investigations Report 95–4218.

Muste, M., Fujita, I., & Hauet, A. (2008) Large-scale Particle Image Velocimetry for measurements in riverine environments. *Water Resour Res*, 44, W00D19, doi:10.1029/2008WR006950.

Muste, M., Baranya, S., Tsubaki, R., Kim, D., Ho, H.-C., Tsai, H.-W., & Law, D. (2015) Acoustic mapping velocimetry proof-of-concept experiment. In: *Proceedings of the 36th IAHR World Congress, 28 June–3 July, The Hague, The Netherlands*.

Muste, M., Baranya, S., Tsubaki, R., Kim, D., Ho, H., Tsai, H., & Law, D. (2016) Acoustic mapping velocimetry. *Water Resour Res*, 52(5), 4132–4150.

Naqshband, S., Ribberink, J.S., Hurther, D., Barraud P.A., & Hulscher, S.J.M.H. (2014a) Experimental evidence for turbulent sediment flux constituting a large portion of the total sediment flux along migrating sand dunes. *Geophys Res Lett*, 41(24), 8870–8878, doi:10.1002/2014GL062322.

Naqshband, S., Ribberink, J.S., Hurther, D., & Hulscher, S.J.M.H. (2014b) Bed load and suspended load contributions to migrating sand dunes in equilibrium. *J Geophys Res Earth Surf*, 119(5), 1043–1063, doi:10.1002/2013JF003043.

Nichols, M.H. (2004) A radio frequency identification system for monitoring coarse sediment particle displacement. *Appl Eng Agricult*, 20(6), 783–787.

Nikora, V.I., Sukhodolov, A., & Rowinski, P.M. (1997) Statistical sand wave dynamics in one-directional water flows. *J Fluid Mech*, 351, 17–39, doi:10.1017/S0022112097006708.

Nittrouer, J.A., Allison, M.A., & Campanella, R. (2008) Bedform transport rates for the low-ermost Mississippi River. *J Geophys Res*, 113, F03004, doi:10.1029/2007JF000795.

Papanicolaou, A.N. & Knapp, D. (2010) A particle tracking technique for bedload motion. In: Gray, J.R., Laronne, J.B., & Marr, J.D.G. (eds.) *Bedload surrogate Monitoring Technologies*. Reston, VA. U.S. Geological Survey Scientific Investigations Report 2010–5091. US Geological Survey. pp. 352–366. [Online] Available from: http://pubs.usgs.gov/sir/2010/5091/papers/ [Accessed 28th February 2017].

Pedocchi, F. & Garcia, M.H. (2012) Acoustic measurement of suspended sediment concentration profiles in an oscillatory boundary layer. *Cont Shelf Res*, 46, 87–95.

Phillips, J.M. & Walling, D.E. (1995) Measurement in situ of the effective particle-size characteristics of fluvial suspended sediment by means of a field-portable laser backscatter probe: Some preliminary results. *Mar Freshwat Res*, 46, 349–357.

Ramooz, R. & Rennie, C.D. (2010) Laboratory measurement of bedload with an ADCP. In: Gray, J.R., Laronne, J.B., & Marr, J.D.G. (eds.) *Bedload surrogate Monitoring Technologies*. Reston, VA. U.S. Geological Survey Scientific Investigations Report 2010–5091. US Geological Survey. pp. 367–386. [Online] Available from: http://pubs.usgs.gov/sir/2010/5091/papers/ [Accessed 28th February 2017].

RD Instruments (1996) *Acoustic Doppler Current Profiler Principles of Operation a Practical Primer*. P/N 951-6069-00.

Reid, I., Layman, J.T., & Frostick, L.E. (1980) The continuous measurements of bedload discharge. *J Hydraul Res*, 18(3), 243–249.

Rempel, J., Hassan, M.A., & Enkin, R. (2010) Laboratory calibration of a magnetic bed load movement detector. In: Gray, J.R., Laronne, J.B., & Marr, J.D.G. (eds.) *Bedload surrogate Monitoring Technologies*. Reston, VA. U.S. Geological Survey Scientific Investigations Report 2010–5091. US Geological Survey. pp. 400–406. [Online] Available from: http://pubs.usgs.gov/sir/2010/5091/papers/ [Accessed 28th February 2017].

Rennie, C.D. & Millar, R.G. (2004) Measurement of the spatial distribution of fluvial bedload transport velocity in both sand and gravel. *Earth Surf Proc Land*, 29(10), 1173–1193.

Rennie, C.D. & Villard, P.V. (2004) Site specificity of bed load measurement using an acoustic Doppler current profiler. *J Geophys Res*, 109, F03003, doi:10.1029/2003JF000106.

Rennie, C.D. & Millar, R.G. (2007) Deconvolution technique to separate signal from noise in gravel bedload velocity data. *J Hydraul Eng*, *133*(8), 845–856.

Rennie, C.D. & Church, M. (2010) Mapping spatial distributions and uncertainty of water and sediment flux in a large gravel bed river reach using an acoustic Doppler current profiler. *J Geophys Res*, *115*, F03035, doi:10.1029/2009JF001556.

Rennie, C.D., Millar, R.G., & Church, M.A. (2002) Measurement of bedload velocity using an acoustic Doppler current profiler. *J Hydraul Eng*, *128*(5), 473–483.

Rennie, C.D., Vericat, D., Williams, R.D., Brasington, J., & Hicks, M. (2017) Calibration of aDcp apparent bedload velocity to bedload transport rate. In: Tsutsumi, D. & Laronne, J. (eds.) *Gravel-Bed Rivers: Processes and Disasters*. Wiley-Blackwell.

Richardson, E.V., Simons, D.B., & Posakony, G.J. (1961) Sonic depth sounder for laboratory and field use. *Geological Survey Circular*. Washington, USA, US Department of the Interior.

Rickenmann, D. (1997) Sediment transport in Swiss torrents. *Earth Surf Proc Land*, *22*(10), 937–951.

Rickenmann, D. (2017) Bedload transport measurements with geophones and other passive acoustic methods. Journal of Hydraulic Engineering, 60th Anniversary State-of-the-Art Reviews, doi:10.1061/(ASCE)HY.1943-7900.0001300.

Rickenmann, D. & McArdell, B.W. (2007) Continuous measurement of sediment transport in the Erlenbach stream using piezoelectric bedload impact sensors. *Earth Surf Proc Land*, *32* (9), 1362–1378.

Rickenmann, D. & McArdell, B.W. (2008) Calibration of piezoelectric bedload impact sensors in the Pitzbach mountain stream. *Geodinamica Acta*, *21*(1/2), 35–52.

Rickenmann, D., Turowski, J.M., Fritschi, B., Klaiber, A., & Ludwig, A. (2012) Bedload transport measurements at the Erlenbach stream with geophones and automated basket samplers. *Earth Surf Proc Land*, *37*(9), 1000–1011.

Rickenmann, D., Turowski, J.M., Fritschi, B., Wyss, C., Laronne, J., Barzilai, R., Reid, I., Kreisler, A., Aigner, J., Seitz, H., & Habersack, H. (2014) Bedload transport measurements with impact plate geophones: Comparison of sensor calibration at different gravel-bed streams. *Earth Surf Proc Land*, *39*, 928–942.

Riley, J.B. & Agrawal, Y.C. (1991) Sampling and inversion of data in diffraction particle sizing. *Appl Optics*, *30*(33), 4800–4817.

Ryan, S.E., Bunte, K. & Potyondy, J.P. (2005) Breakout Session II, Bedload-Transport Measurement: Data Needs, Uncertainty, and New Technologies. In: Gray, J.R. (ed.) *Proceedings of the Federal Interagency Sediment Monitoring Instrument and Analysis Research Workshop, September 9–11, 2003, Flagstaff, Arizona*. [Online] Available from: http://pubs.usgs.gov/circ/2005/1276 [Accessed 28th February 2017].

Schneider, J., Hegglin, R., Meier, S., Turowski, J.M., Nitsche, M., & Rickenmann, D. (2010) Studying sediment transport in mountain rivers by mobile and stationary RFID antennas. In: Dittrich, A., Koll, K., Aberle, J., & Geisenhainer, P. (eds.) *Proceedings of the International Conference on Fluvial Hydraulics River Flow 2010, 8–10 September 2010, Braunschweig, Germany*. Karlsruhe, Bundesanstalt für Wasserbau. pp. 1723–1730.

Schneider, J.M., Turowski, J.M., Rickenmann, D., Hegglin, R., Arrigo, S., Mao, L., & Kirchner, J.W. (2014) Scaling relationships between bed load volumes, transport distances, and stream power in steep mountain channels. *J Geophys Res Earth Surf*, *119*, 533–549, doi:10.1002/2013JF002874.

Seinfeld, J.H. & Pandis, S.N. (1998) *Atmospheric Chemistry and Physics: From Air Pollution to Climate Change*, Chapter 7: Properties of the Atmospheric Aerosol. Wiley.

Shen, C. & Lemmin, U. (1996) Ultrasonic measurements of suspended sediments: A concentration profiling system with attenuation compensation. *Meas Sci Technol*, *7*(9), 1191–1194.

Shen, C. & Lemmin, U. (1997) A two-dimensional acoustic sediment flux profiler. *Meas Sci Technol*, *8*(8), 880–884.

Shen, C. & Lemmin, U. (1998) Improvements in acoustic sediment concentration profiling using an LMS compensation algorithm. *J Oceanic Eng (IEEE)*, 23(2), 96–104.

Shen, C. & Lemmin, U., (1999) Application of an acoustic particle flux profiler in particle-laden open-channel flow. *J Hydraul Res*, 37(3), 407–419.

Shen, H.W. & Cheong H.-F. (1977) Statistical properties of sediment bed profiles. *J Hydraul Div*, 103(HY11), 1303–1321.

Sheng, J. (1991) Remote determination of suspended sediment size and concentration by multi-frequency acoustic backscatter. *Ph.D. thesis*. Dept. of Physics, Memorial University of Nfld.

Simmons, S., Parsons, D.R., Best, J.L., Orfeo, O., Lane, S.N., Kostaschuk, R., Hardy, R.J., West, G., Malzone, C., Marcus, J., & Pocwiardowski, P. (2010) Monitoring suspended sediment dynamics using MBES. *J Hydraul Eng*,136(1), 45–49.

Simons, D.B., Richardson, E.V., & Nordin, C.F. (1965) Bedload equation for ripples and dunes. *Geological Survey Professional Paper 462-H*, Washington, USGS.

Singh, A., Fienberg, K., Jerolmack, D.J., Marr, J., & Foufoula-Georgiou, E. (2009) Experimental evidence for statistical scaling and intermittency in sediment transport rates. *J Geophys Res Earth Surf*, 114, F01025, doi:10.1029/2007JF000963.

Smyth, C.E., Zedel, L., & Hay, A.E. (2002) Coherent Doppler profiler measurements of nearbed suspended sediment fluxes and the influence of bedforms. *J Geophys Res*, 107(C8), doi:10.1029/2000JC000760, 2002.

Sutherland, T.F., Lane, P.M., Amos, C.L., & Downing, J. (2000) The calibration of optical backscatter sensors for suspended sediment of varying darkness levels. *Mar Geol*, 162(2–4), 587–597.

Swithenbank, J., Beer, J.M., Taylor, D.S., Abbot, D., & McCreath, G.C. (1976) A laser diagnostic technique for the measurement of droplet and particle size distribution. New York, AIAA. 76–79.

Tacconi, P. & Billi, P. (1987) Bed load transport measurement by a vortex-tube trap on Virginio Creek, Italy. In: Thorne, C.R., Bathurst, J.C. & Hey, R.D. (eds.) *Sediment Transport in Gravel-Bed Rivers*. Chichester, Wiley. pp. 583–615.

Tessier, C. (2006) Caractérisation et dynamique des turbidités en zone côtière : l'exemple de la région marine Bretagne Sud (Characterization of water turbidity and its dynamics: The South Brittany coastal zone (France) as a case study). *Ph.D. Thesis*, Bordeaux University.

Thevenot, M.M., Prickett, T.L., & Kraus, N.C. (1992) *Tylers Beach, Virginia, dredged material plume monitoring project, 27 September to 4 October 1991*. Technical Report: DRP-92-7, US Army Corps of Engineers.

Thorne, P.D. (1986) Laboratory and marine measurements on the acoustic detection of sediment transport. *J Acoust Soc Am*, 80, 899–910.

Thorne, P.D & Foden, D.J. (1988) Generation of underwater sound by colliding spheres. *J Acoust Soc Am*, 84, 2144–2152.

Thorne, P.D. & Hanes, D.M. (2002) A review of acoustic measurement of small-scale sediment processes. *Cont Shelf Res*, 22(4), 603–632.

Thorne, P.D. & Hardcastle, P.J. (1997) Acoustic measurements of suspended sediments in turbulent currents and comparison with in-situ samples, *J Acoust Soc Am*, 101(5), 2603–2614.

Thorne, P.D. & Meral, R. (2008) Formulations for the scattering properties of sandy sediments for use in the application of acoustics to sediment transport. *J Cont Shelf Res*, 28, 309–317.

Thorne, P.D. & Hurther, D. (2014) An overview on the use of backscattered sound for measuring suspended particle size and concentration profiles in non-cohesive inorganic sediment transport studies. *Cont Shelf Res*, 73, 97–118.

Thorne, P.D., Davies, J.S., & Bell, P.S. (2009) Observations and analysis of sediment diffusivity profiles over sandy rippled beds under waves. *J Geophys Res*, 114, C02023, doi:10.1029/2008JC004944.

Thorne, P.D., Hardcastle, P.J., & Soulsby, R.L. (1993) Analysis of acoustic measurements of suspended sediments. *J Geophys Res*, *98*(C1), 899–910, doi:10.1029/92JC01855.

Thorne, P.D., Vincent, E.C., Hardcastle, P.J., Rehman, S., & Pearson, N. (1991) Measuring suspended sediment concentrations using acoustic backscattering devices. *Mar Geol*, *98*(1), 7–16.

Thosteson, E.D. & Hanes, D.M. (1998) A simplified method for determining sediment size and concentration from multiple frequency acoustic backscatter measurements. *J Acoust Soc Am*, *104*, 820–830.

Topping, D.J., Wright, S.A., Melis, T.S., & Rubin, D.M. (2007) High-resolution measurements of suspended-sediment concentration and grain size in the Colorado River in Grand Canyon using a multi-frequency acoustic system, In: *Proceedings of the 10th International Symposium on River Sedimentation. August 1–4, Moscow, Russia.*

Topping, D.J., Rubin, D.M., Wright, S.A., & Melis, T.S. (2011) Field evaluation of the error arising from inadequate time averaging in the standard use of depth-integrating suspended-sediment samplers. *U.S. Geological Survey Professional Paper*, *1774*. 95.

Urick, R.J. (1948) The absorption of sound in suspensions of irregular particles. *J Acoust Soc Am*, *20*, 283–289.

van der Mark, C.F., Blom, A., & Hulscher, S.J.M.H. (2008) Quantification of variability in bedform geometry. *J Geophys Res*, *113*, F03020, doi:10.1029/2007JF000940.

Vanoni, V.A. (2006) *Sedimentation Engineering: Theory, Measurements, Modeling, and Practice*. Manuals and Reports on Engineering Practice No. 54. Reston, Va., Am. Soc. of Civ. Eng. Publ.

Voulgaris, G., Wilkin, M.P. & Collins, M.B. (1995) The in situ passive acoustic measurement of shingle movement under waves and currents: Instrument (TOSCA) development and preliminary results. *Cont Shelf Res*, *15*(10), 1195–1211.

Williams, R.D., Rennie, C.D., Brasington, J., Hicks, D.M. & Vericat, D. (2015) Linking the spatial distribution of bed load transport to morphological change during high-flow events in a shallow braided river. *J Geophys Res Earth Surf*, *120*, 604–622. doi:10.1002/2014JF003346.

Willis, J.C. & Kennedy, J.F. (1977) *Sediment Discharge of Alluvial Streams Calculated from Bed Form Statistics*. IIHR Report No. 202. Iowa City, Iowa Institute of Hydraulic Research.

Wilson, G.W. & Hay. A.E. (2015a) Acoustic backscatter inversion for suspended sediment concentration and size: A new approach using statistical inverse theory. *Cont Shelf Res*, *106*, 130–149.

Wilson, G.W. & Hay, A.E. (2015b) Measuring two-phase particle flux with a multi-frequency acoustic Doppler profiler. *J Acoust Soc Am*, *138*(6), 3811–3819.

Wright, S.A., Topping, D.J., & Williams, C.A. (2010) Discriminating silt-and-clay from suspended-sand in rivers using side-looking acoustic profilers. In: *Proceedings of the 2nd Joint Federal Interagency Sedimentation Conference, June 2010, Las Vegas, USA.* [Online] Available from: http://acwi.gov/sos/pubs/2ndJFIC [Accessed 28th February 2017].

Wyss, C.R., Rickenmann, D., Fritschi, B., Turowski, J.M., Weitbrecht, V., & Boes, R.M. (2016a) Laboratory flume experiments with the Swiss plate geophone bedload monitoring system. Part I: Impulse counts and particle size identification. *Water Resour. Res.*, *52*(10), 7744–7759.

Wyss, C.R., Rickenmann, D., Fritschi, B., Turowski, J.M., Weitbrecht, V., & Boes, R.M. (2016b) Measuring bedload transport rates by grain-size fraction using the Swiss plate geophone signal at the Erlenbach. *J. Hydraul. Eng.*, 10.1061/(ASCE)HY.1943-7900.0001090, 04016003.

Zedel, L. & Hay, A.E. (1999) A coherent Doppler profiler for high resolution particle velocimetry in the ocean: Laboratory measurements of turbulence and particle flux. *J Atm Ocean Technol*, *16*, 1102–1117.

Chapter 6

Auxiliary Hydraulic Variables

6.1 INTRODUCTION

The preceding chapters presented detailed information on instrumentation and experimental methods with regard to the measurement of velocities, discharge, bathymetry, roughness, and sediment transport. Experimental studies also require information on auxiliary hydraulic variables such as water depth, water surface slopes, and bed slopes in order to obtain a complete data set. Depending on the scope of the investigation, the measurement of further auxiliary hydraulic variables, such as pressure, bed shear stress, and drag forces may also be required.

The measurement of auxiliary hydraulic variables can be challenging and may require sophisticated experimental strategies and the use of advanced technologies that have been developed for other disciplines, such as aeronautics and robotics. There exist many different instruments and methods so that a detailed description of each method and instrument is beyond the scope of this chapter. Consequently, the objective of this chapter is to provide a summary of routine instruments and experimental methods for the measurement of water depth, slopes, pressure, bed shear stress, drag forces and some fluid properties. Table 6.1.1 provides a general overview of measurement methods often used in the laboratory and the field to measure auxiliary variables. Ensuing sub-sections contain more detail about the instruments identified in the table as well as references to additional information regarding the instruments and their application.

Measurement of fluid properties such as density, dynamic viscosity and surface tension is beyond the scope of this text. For hydraulics experiments, it is often adequate to use tables based on temperature and salinity to obtain necessary fluid properties. If needed, detailed information on instrumentation to measure fluid properties can be found in Wakeham *et al.* (2007).

6.2 WATER DEPTH

6.2.1 Principles of measurement

Water depth h is defined as the distance from the water surface level h_{ws} to the bed level h_b, i.e., $h = h_{ws} - h_b$. Determination of h thus requires the measurement of both h_{ws} and h_b or their difference. Water depth measurements need to account for spatial and temporal scales, which in turn are governed by hydrodynamic boundary conditions as well as bed bathymetry and roughness. Examples of temporal variations of the water

Table 6.1.1 Methods and instruments for measurements of auxiliary hydraulic variables

Variable	Section	Principles of measurement	Temporal resolution
Water depth	6.2	mechanical, optical, acoustic, electrical	time averaged to instantaneous
Pressure	6.4	metering; deformation of reference bodies; force measurements; piezoelectric; optical;	time averaged to instantaneous
Bed shear stress	6.5	velocity and turbulence measurements; force measurements; incipient motion of reference bodies; water depth and slope measurements; pressure measurements; heat- and mass transfer; magnetic; optical;	time averaged to instantaneous
Drag forces	6.6	strain gauges; springs; digital load cells; piezoelectric;	time averaged to instantaneous
Density, Conductivity, Temperature	6.7	mechanical, electrical	time averaged to instantaneous

depth include unsteady flow conditions and flows over mobile beds (*e.g.*, bed form movement and erosion and sedimentation processes). For these examples both h_{ws} and/ or h_b vary with time.

Spatial variations of the water depth are apparent for non-uniform flow conditions, flows at high Froude numbers, flows in channels with a non-uniform cross-sectional geometry, and rough bed flows. In non-uniform flow conditions, such as gradually or rapidly varied flow, the water depth varies per definition in the direction of the flow, and in flows characterized by high Froude numbers surface waves are present, *i.e.*, the water surface undulates. In channels with non-uniform cross-sectional geometry (*e.g.*, compound channels) local water depths vary due to changes in the cross-sectional geometry and hence differ from cross-sectional averaged values. These, in turn, can be different from reach averaged values. In rough bed flows, the bed elevation is non-uniformly distributed and hence local values of $h_{ws} - h_b$ can vary significantly, depending on the size of roughness elements.

The latter examples show that it is important to clearly define the vertical location of the bed reference level (origin of the vertical axis), *i.e.*, if it refers for example to the geodetic height of the mean bed level (most common) or to any other height of the bed elevations derived from the statistical distribution of bed elevations (*e.g.*, minimum, maximum). In fact, depending on the bed reference level definition, estimates of important hydraulic parameters such as water depth and bed shear stress can vary significantly within the same dataset (*e.g.*, Aberle *et al.*, 2008). For highly variable flow conditions, such as flows in step-pool systems, a reach averaged water depth may be determined from the continuity equation by measuring the discharge and mean flow velocity by means of tracer-measurements (*e.g.*, the salt-dilution method; see Section 3.10, Volume II).

Water depth measurements can generally be achieved using a variety of instruments and methods based on hydromechanical, acoustic, and optical principles. The following sub-sections provide details on various instruments and experimental methods for measuring water depth (summarized in Table 6.2.1). These details are complemented by considerations related to spatial and temporal scales and potential error sources. The focus of this sub-chapter is mainly on the determination of the water surface

Table 6.2.1 Summary table of instrument types for measurement of water depth and water surface elevation

Instrument	Principles	Spatial resolution	Temporal resolution
Lead lines, rulers, sounding poles, water gauges, float gauges, water level transmitters	mechanical	local	low
Point gauge	mechanical	local	low
Capacitance gauge, resistance gauge	electrical	local	high
Piezometer	pressure	local	low
Ultrasonic sensors	acoustic	local	high
Laser displacement	Optical	local	high
Photogrammetric	Optical	areal	low to high
Remote sensing	acoustic/ optical	large scale	low

elevation as instrumentation and methods for bed level measurements are summarized in Chapter 4 (Volume II).

6.2.2 Available instrumentation

Instrumentation for the measurement of the water surface and bed level, and hence water depth, can be applied in both the laboratory and field. However, laboratory measurements are carried out under controlled conditions so that higher accuracy can and should be achieved than in field measurements. Moreover, the suitability of the different methods and instruments depends also on the experimental boundary conditions and the objective of the experiments.

6.2.2.1 Mechanical instruments and methods

Mechanical systems such as lead lines, sounding poles, ordinary water gauges or rulers have been the earliest tools used for water depth measurements. Lead lines and sounding poles provide information on the distance between the local bed and the water surface level and have mainly been used in the field. Such measurements are generally of low accuracy when applied in shallow flows. Ordinary water gauges (or rulers) can be used both in the laboratory and the field to provide water surface level measurements (or depth) within a mm to cm accuracy range.

The most common instrument for water surface measurements is the point gauge (Figure 6.2.1). Point gauges have an accuracy of up to 0.1 mm. They are predominantly used in laboratory investigations, and are either equipped with needles that are lowered toward the water surface or with hooks so that the water surface is approached from below. For both types the gauge tip is brought into contact with the water surface which may be observed visually or indicated electronically. The corresponding height information is subsequently read from an attached scale or displayed electronically. The use of point gauges for field measurements is not common due the fact that the gauges must be fixed at a steady position and must be accessible for a wide range of flow conditions in order to take readings.

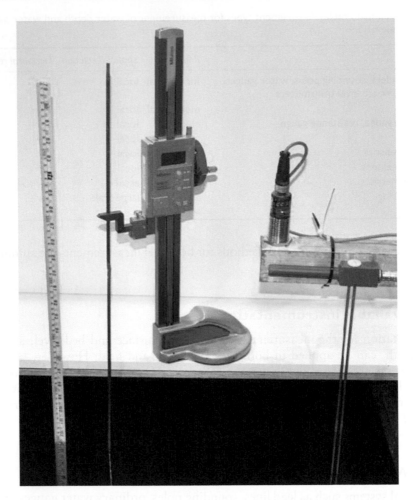

Figure 6.2.1 Example of water level measurements with a ruler, point gauge, acoustic sensor, and a resistance probe (from left to right).

Float gauges and water level transmitters can also measure the water surface elevation with high accuracy. Float gauges determine the height of a floating object, while water level transmitters are submersed and are based on measurements of the hydrostatic pressure. The latter principle is also applied by means of piezometers, *i.e.*, by connecting small diameter holes in the wall or bed of a flume (referred to as pressure taps) to a glass or Perspex container in which the water surface elevation can be directly measured. The basic working principles of methods based on the hydrostatic pressure are described below in Section 6.4.

A technique mainly used in coastal and maritime engineering studies to measure the water surface elevation with a high temporal resolution is to use capacitance or resistance wave probes. Capacitance probes measure the length of the submerged part of the probe, based on change in capacitance of a circuit including the probe wire, which can be related to the instantaneous water level. Resistance probes (Figure

6.2.1) represent twin-probes through which water level changes are measured by monitoring the conductivity between the wires. Further details on the measurement principles of wave gauges can be found in *e.g.*, Hughes (1993) and Arndt *et al.* (2007).

6.2.2.2 Acoustic and optical instruments and methods

Acoustic sensors measure the distance from the sensor head to the surface, which reflects the acoustic signal (see Figure 6.2.1). Such sensors, which are also described in Chapters 3 and 4 (Volume II), are manufactured by many different suppliers and can be installed above (or below) the water surface for the determination of the water (or bed) level. Their accuracy depends on the instrument type, sounding range, and footprint size due to the acoustic cone, and is usually provided by the manufacturer in the instrument specifications.

In submersed conditions acoustic sensors are generally placed at the bed in order to measure the vertical distance to the water surface. Such measurements are often supplemented by additional pressure measurements so that the pressure and acoustic sensor work in tandem to determine the water level. If the acoustic sensor is installed above the water surface, the water surface is used as the acoustic target. In this case, a bed reference level is needed to determine the water depth.

Within navigable rivers and streams, the water depth may be measured using acoustic instruments installed on vessels. The vessel represents a float from which the instruments are deployed. Using the acoustic sensor to measure the distance from the sensor head of the deployed instrument (*e.g.*, ADCPs, sonars, ADVs) to the bed, and knowing the submergence of the sensor head below the surface, allows for the determination of the water depth. Such measurements are affected by vessel motion and need to be corrected for yaw, pitch and roll.

Optical methods used for the measurement of bathymetry, such as photogrammetric and laser based methods, may also be used to determine information on water surface levels in both the laboratory and the field by using floating objects on the water surface as optical targets. The working principles of these instruments have been described in Chapter 4 (Volume II) and are not repeated here.

Modern remote sensing techniques allow for both local and large scale measurements of water levels. In field studies, local surveying can be used to determine the water level at or close to the banks and to relate the measured heights in a global co-ordinate system. Air-borne measurements (*e.g.*, Radar, Airborne light detection and ranging (LiDAR); Hopkinson *et al.*, 2011) or satellite-based measurements (*e.g.*, Pan & Nichols, 2013) allow for large scale measurements of the water level and water depth. The application of these methods can be cost-restrictive.

6.2.2.3 Spatial and temporal resolution

The instruments and methods for the determination of water depth are characterized by different spatial and temporal resolutions. Point based measurements, *i.e.*, water depth measurements at a single location, can only be up-scaled to a larger spatial scale if steady uniform flow conditions prevail, the water surface is flat, and if the bed reference level is clearly defined. However, even for such conditions it is advisable to measure the water depth at multiple locations so that an average value can be determined to minimize potential errors. For more complex flows, the measurement of water surface levels (and hence) water depth at multiple locations is required. In flume studies, most of the aforementioned sensors can be mounted to (automated) carriages, enabling the measurement of water surface elevations at many positions. Spatial information from point based measurements can also be obtained by (simultaneous) measurements with arrays of two or more sensors.

Point gauge measurements are mainly carried out manually and therefore it is not possible to use them for applications that require high temporal sampling frequency. This limits their applicability to steady or slowly varying water levels. In the latter case, the contact point needs to be constantly observed or methods need to be applied to adjust the gauges automatically (Novak *et al.*, 2010). When using pressure based methods for the determination of the water surface level in unsteady flows, the response time of the instrument is important. Slowly fluctuating water levels can be measured with higher temporal resolution than manual techniques using float gauges, optical, and acoustic methods. These methods may also be used to measure rapidly changing water levels, although the latter are mainly measured using capacitance and resistance wave probes which can be operated at very high sampling frequencies.

The spatial and temporal resolutions of remote sensing techniques for field measurements of water depth depend on each specific remote sensing method and associated instrumentation. In general, remote sensing techniques provide measurements of water surface elevations over large spatial scales but with limited spatial resolution. The temporal resolution of many remote sensing techniques is limited by slow repetition rates (*e.g.*, satellite-based techniques usually have very low temporal resolutions).

6.2.2.4 *Error sources*

Although water depth is a basic parameter in hydraulic studies, there exist many potential error sources which are related to the individual instruments and methods, so that a detailed description of error sources is beyond the scope of this section. Thus, the section contains only a rather broad description of potential error sources.

All of the aforementioned sensors measure the water and bed surface levels within an "instrument-related" and hence arbitrary coordinate system. Thus it is important to establish a common datum and coordinate system for the sensors, the flume or study region, and the other instruments used for the measurements.

As indicated above, point based measurements should be collected at multiple locations within a measurement section to get a robust measurement of the water depth based on an average value of the water surface elevation and with a clearly defined bed reference level. Moreover, it is preferable that measurements in flumes be acquired along the flume centerline to reduce the influence of side walls. Water surface level measurements should be determined along the flow axis at a spatial interval that also enables the determination of water surface slopes (see Section 6.3). It is advisable that manual point gauge readings of water depth all be taken by the same person in order to reduce cumulative biases.

If water surface or bed levels are determined using submersed measurement instruments, care should be taken to ensure that the instruments do not affect the geometry of mobile beds, as the submersed instruments may induce flow that causes scouring. Therefore, bed level measurements are often carried out in studies with mobile beds after draining the flume/model or by reducing the flow rate. In these cases, special care must be taken that the bed geometry is not affected by a sudden change in the hydraulic boundary conditions. Taking readings with acoustic/optical instruments, the measurement range of the instruments needs to be taken into account in the planning stage of the experiments and the calibration of the instruments should be checked before the experiments.

6.3 WATER SURFACE AND BED SLOPE

6.3.1 Principles of measurement

The water surface slope and bed slope are gradients, and their determination requires the measurement of the vertical change of the water surface and bed level, respectively, over a known horizontal distance along the main flow direction. The required data for the determination of the slope can therefore be obtained using the instruments and methods described within the sections about bathymetry (Chapter 3, Volume II) and water depth measurements (Section 6.2 above).

In general, a gradient can be determined by the measurement of the vertical change between two points over a given horizontal distance. However, the determination of the slope by using only two measurement points is only adequate for situations where the slope is constant, such as steady uniform flow conditions and flat beds, or for situations where a global estimate of the slope is sufficient. In the presence of surface waves, for example flows with high Froude numbers, it is of major importance to determine the water surface slope from more than two points in order to determine the mean slope, as significant errors can be introduced due to the non-uniformity of the water surface (*e.g.*, local measurements of the water surface elevation in the wave trough or at the wave crest in case of standing waves). The same applies for the determination of the bed slope when bedforms or bed features are present (*e.g.*, step-pool or riffle pool systems, dunes, rough armor layers etc.). In this case the mean bed slope should be determined by fitting a (linear) trend-line to the measured bed surface elevation data thereby ensuring that enough data points are available for a reliable estimate of the bed slope. In case spatial data are available, it may be preferable to estimate the slopes through spatial fitting procedures, for example by fitting a plane to the data to determine the slope in the transverse and longitudinal direction.

Gradually varied and/or rapidly varied flows are characterized by a change of the water depth with position. Thus, the determination of the corresponding water surface slopes requires water surface profiling, *i.e.*, the measurement of the water surface elevation at given horizontal increments along the flow axis so that changes in the gradient can be adequately determined.

6.4 PRESSURE

6.4.1 Principles of measurement

A fluid in contact with a boundary produces a force perpendicular to the loaded area of the boundary. The boundary may be either a solid boundary in contact with the fluid or, for purposes of analysis, an imaginary plane within the fluid. The resulting force is called a pressure force, and when divided by the unit area over which it acts is called the pressure. The SI unit for pressure is Pascal (Pa), equal to one Newton per square meter

Table 6.4.1 Pressure conversion of 1 kPa in other units

Unit	kPa	bar	Psi	at	Atm	Torr	mmH$_2$O[*)]
value	1.0	0.01	$14.50377 \cdot 10^{-2}$	$1.0197 \cdot 10^{-2}$	$0.98692 \cdot 10^{-2}$	7.5006	101.972

*) at 4°C

(N/m^2). Nonstandard units of pressure include bars (bar), pounds per square inch (psi), technical atmospheres (at) and standard atmospheres (atm), torrs (1 Torr = 1 mmHg at 0 °C), and millimeters of water (mm H$_2$O, also mm WC). Their relation to 1,000 Pa = 1 kPa is listed in Table 6.4.1.

Pressure, a scalar quantity, can be measured based on three different principles related to a pressure reference value: (i) absolute pressure, (ii) gauge or relative pressure, and (iii) differential pressure. Absolute pressure is measured against a perfect vacuum so that the resulting measurement value reflects both the fluid and local atmospheric pressure p_{atm}. Gauge, or relative pressure, means that the measured pressure value is zero-referenced against the local ambient air pressure, while differential pressure measurements determine the pressure difference between two points.

In principle, pressure can be sensed based on three basic methods. The simplest and classical method using a manometer involves metering a liquid column of known density that is balanced against the unknown pressure. This method is most suitable for measurements of time averaged and steady pressure, as liquid manometers respond slowly to changes in input pressures. The second method relies on pressure transducers which measure the deflection, deformation, or strain of a flexible membrane or cylinder (*e.g.*, diaphragm, bellow, etc.) on which the unknown pressure acts. The sensed deflection or deformation is converted into a mechanical or electrical signal which is related to the input pressure that causes the deflection or deformation. The third method makes use of a pressure sensing medium that undergoes very little deformation when subjected to the input pressure. Instead, a measureable property of the sensing medium (*e.g.*, resistivity, color, etc.) changes in response to changes in the input pressure. Both the second and third method allow for the measurement of temporal pressure fluctuations with a high sampling frequency. The instruments using these three basic principles can be configured for laboratory or field conditions. Some of them use the same instrument configuration for both measurement situations.

Pressure measurements are carried out by mechanical devices or electronic pressure transducers. The main difference between mechanical pressure sensors and electronic pressure transducers is that the latter require energy from an external source. Electronic pressure transducers are often grouped into categories of inductive, capacitive and resistive sensors. Optical (*e.g.*, fiber Bragg gratings), resonant or thermal (*e.g.*, Pirani gauge) techniques belong to electronic transducers as well, but do not really fit into the aforementioned three main groups.

6.4.2 Available instrumentation

6.4.2.1 Steady pressure measurements

Numerous types of pressure sensors are available that vary in technology, design, performance, application suitability, and cost. Figure 6.4.1 presents a broad overview

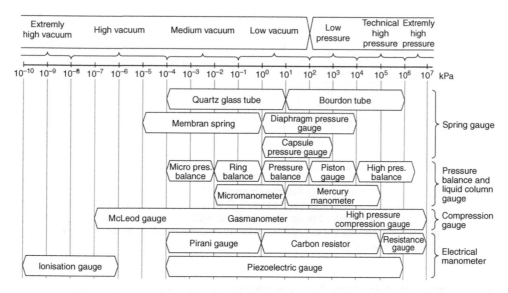

Figure 6.4.1 Applicable pressure ranges for pressure-sensing instruments (after Bild 7.1-1 from Wuest, 1994, with permission of de Gruyter).

of different sensors and their applicable pressure range for various engineering applications (see also Holman, 1989). A more refined but not exhaustive overview with regard to instruments that can be used for measurements of both fluctuating and steady pressure in hydraulic engineering applications is provided in Table 6.4.2. Many of the instruments for steady pressure measurements, such as Piezometers, U-shaped

Table 6.4.2 Selected instruments to measure pressure in hydraulic engineering studies (the list is not exhaustive).

Instrument	Principle	Method	Range
Piezometer	differential pressure	liquid column	$1 - 10^6$ Pa
U-shaped manometer	differential pressure	liquid column	$1 - 10^6$ Pa
Pitot tube[1]	differential pressure	liquid column, flexible membrane, resulting forces or electrical resistance	$1 - 10^5$ Pa
Bourdon pressure gauge	gauge or relative pressure	flexible membrane	$10^3 - 10^9$ Pa
Piezoelectric sensor	differential pressure	electrical field generation	$0.1 - 10^8$ Pa
Piezo-resistive sensor	differential pressure	electrical resistance	$0.1 - 10^6$ Pa
Capacitive sensor	gauge or absolute pressure	electrical strain	$10^3 - 10^5$ Pa
Pressure Sensitive Paint	gauge or relative pressure	fluorescence depending on the external air pressure / oxygen	$10^3 - 10^5$ Pa

1 Dependent on type, this instrument can be used to determine static pressure, dynamic pressure, or total pressure. This instrument is also used for measurements of flow velocity.

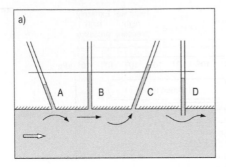
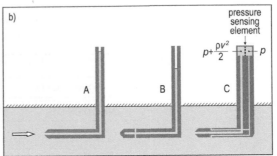

Figure 6.4.2 a) Piezometer installation: Unless the piezometer is vertical as in B, the water elevation will be drawn down as in A or increased as in C. D shows the tube pushed into the flow, causing the flow to curve under the tip which pulls the water level down (USBR, 2001; used with permission from the Bureau of Reclamation). b) Different Pitot tubes A) simple, B) static and C) pitot-static tube.

manometers, Pitot-tubes, and Bourdon pressure gauges, and their principles of measurement have been extensively described in textbooks (*e.g.*, McKeon & Engler, 2007) so that they will not be described in detail in the following. Note also that for low pressure ranges (below 10 mm H$_2$O), special instrumentation is required for accurate pressure measurements (see Figure 6.4.1). Piezometers and Pitot-tubes are shown in Figure 6.4.2.

6.4.2.2 Fluctuating pressure measurements

Typical instruments for the measurement of fluctuating pressure contain either flexible membranes, capacitive sensors or piezotechnical elements. The working principle of sensors that are predominantly used in hydraulic engineering applications will be described in the following.

6.4.2.2.1 Strain gauge transducers

The most common flexible membrane transducer type for measurements in liquids is the strain-gauge transducer. In general, acoustic transducers could also be used for these purposes (*e.g.*, condenser microphones) but they are generally not used in liquids due to their dependency on humidity. Strain-gauge transducers like capacitive ceramic sensors utilize a strain gauge to translate membrane displacement into an electric signal. The resistance of the strain-gauge, mounted to the pressure sensor membrane, changes in response to a change in pressure when the membrane is deflected. The output of the strain gauge is connected to one arm of a Wheatstone bridge (cf. Hufnagel & Schewe, 2007) and amplified to measureable levels. Strain-gauge transducers are useful for applications where clearances are small so that other transducer types cannot be used. In general, such sensors respond to pressure fluctuations of up to 5 kHz. Strain gauge transducers are less sensitive to vibration than piezoelectric transducers (see next section). Although their main drawbacks are drift of the zero frequency and changes in sensitivity, in general, they are versatile and reliable.

6.4.2.2.2 Piezoelectric transducers

A piezoelectric transducer is a device that uses the piezoelectric effect to convert pressure to an electrical signal. Piezoelectric materials are crystalline substances that, when deformed, produce an electric field in which the generated voltage is proportional to the applied force. Conversely, if an electric field is imposed along one of the piezo-electric axes, the crystal will be deformed in proportion to the applied voltage.

Piezoelectric sensors are excellent for measuring rapid changes in pressure, but unfortunately, they cannot be used for true static measurements. A static pressure applied to a piezoelectric sensor generates a fixed amount of charge on the piezoelectric material. Since conventional materials are not perfect insulators, the generated charge decays rapidly, leading to a constant shift of the signal. Moreover, piezoelectric units are sensitive to vibration so that the desired pressure-induced signals are often masked by noise induced by the vibrations.

6.4.2.2.3 Piezo-resistive transducers

The piezo-resistive effect causes, in contrast to the piezoelectric effect, only a change in resistance, *i.e.*, it does not produce electrical charges. The core of piezo-resistive pressure sensors is a micromechanical silicon wafer with implanted piezo-resistors on its bending panel (see Figure 6.4.3 for a sketch of the principle layout). A pressure load and the resulting mechanical bending stress causes a change of these piezo-resistors, suitably combined in a bridge circuit by anodic bonding. The output of the powered bridge is a voltage signal in the range of [mV] which is proportional to pressure. In most cases, the silicon device is connected with a glass base to ensure restraint at the edges. In order to be able to apply the reference pressure, the glass base typically has a hole (in Figure 6.4.3 the differential pressure is measured with reference to atmospheric pressure p_{atm}). Temperature errors can be compensated electronically. Dependent on the configuration, piezo-resistive transducers (or other sensors) can be used to measure both static and dynamic pressure. When the transducer pinhole opening is parallel to the flow, the sensor measures static pressure while, when it faces the flow it measures stagnation pressure, it indicates the intensity of pressure drag. Thus, the sensed pressure reflects the effective force per pinhole area in direction of the pressure tube.

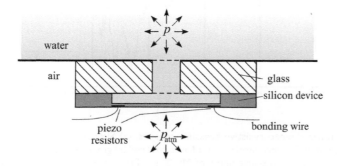

Figure 6.4.3 Schematic cross section of a piezo-resistive sensor to measure the differential pressure (Detert, 2008).

Detert (2008) and Detert *et al.* (2010b) used miniaturized piezo-resistive sensors (MPPS) to experimentally investigate hydrodynamic processes at the water-sediment interface in streambeds by the simultaneous measurement of pressure and velocities. Figure 6.4.4a gives a dimensional sketch of a miniaturized piezo-resistive pressure sensor (MPPS) used by Detert (2008) and Detert *et al.* (2010b), while Figure 6.4.4b gives an example of an array of MPPS to enhance the spatial resolution of the measurements. For this particular setup, flexible PVC tubes were used to provide atmospheric pressure in the pick-up section of the transducers (the section where the piezo-resistors were installed). The pick-up sections were encapsulated with slowly hardening epoxy resin and sealed up with clear varnish to make them water resistant. The ready-built sensors were finally point-calibrated so that the 1–9 V output represented a pressure range of 0–4 kPa with an accuracy of better than 1.0% full scale. The response time guaranteed by the manufacturer was <10 ms, limited due to signal conditioning by amplifying blankets. It is interesting to note that the tests carried out by Detert (2008)

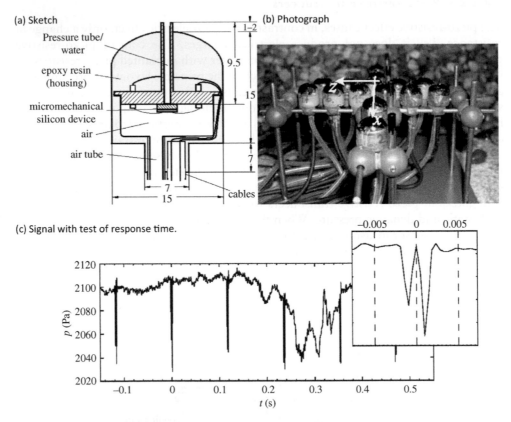

Figure 6.4.4 Miniaturized piezo-resistive pressure sensors. (a) Sketch of a pressure pickup (mm). (b) Array of MPPSs mounted on a grid, not covered by gravel yet. This arrangement was used in synchronous measurements of MPPS and PIV. (c) Measurement to test the response time of the MPPS, where the sensor reacts on the double pulse of a PIV laser sheet of $\Delta t = 2$ ms (Detert, 2008).

revealed that the MPPS were able to react within 2 ms. This became apparent from simultaneous Particle-Image Velocimetry (PIV) and pressure measurements in which four sensors were aligned within the laser sheet. The energy of the laser double pulses was clearly identifiable in the signal, as shown in Figure 6.4.4c, and this unforeseen effect could hence be used to validate the synchronization of the PIV and the MPPS. This example shows that unforeseen effects can also be used in positive ways.

6.4.2.3 Pressure Sensitive Paint

Pressure Sensitive Paint (PSP) is used for optical measurements of surface-pressure distributions. The working principles of PSP are based on the phenomenon of oxygen quenching of luminescence, also known as the Kautsky-effect. The PSP technique enables a non-contact investigation of changes in surface-pressure without disturbing the flow. It has the capability of providing areal measurements over the entire surface of a body. The higher the pressure, the higher the partial pressure of the oxygen and the more the intensity emitted by the coating is attenuated. The PSP technique has many advantages over traditional pressure measurement techniques. Therefore it has been extensively used in fluid dynamics, mainly in wind tunnel experiments in the field of aviation.

Conceptually a PSP system is composed of a model coated with PSP, an illumination source, and a camera. The PSP is distributed over the model surface. During the measurement the surface is illuminated at a wave length that causes the PSP to luminesce, and the luminescent intensity is recorded by the camera. The conversion of the video data to a pressure signal requires previous calibration in a pressure chamber at known pressure and ambient velocity conditions. A major source of error is the temperature sensitivity of PSP. Fundamentals of the use of PSP can be found in McLachlan and Bell (1995), Liu *et al.* (1997), McKeon and Engler (2007), and Gregory *et al.* (2008).

6.4.3 SPATIAL AND TEMPORAL RESOLUTION

The spatial and temporal resolution of pressure measurements depends significantly on the instrumentation used. All the instruments described above, except for the PSP-technique, represent point-based measurements, as the measurement is related to the opening of the corresponding sensor (*e.g.*, wall tappings, pressure tubes, etc.). Thus, using such techniques, the only possible way to obtain higher spatial resolution is to use arrays of pressure sensors. An example of the use of pressure-transducer arrays and the subsequent reconstruction of the pressure field can be found in Detert (2008) and Detert *et al.* (2010a) (see Figure 6.4.4b).

Pressure measurements based on metering a liquid column of known density have a slow response time and can therefore only be used to determine time averaged pressures. However, novel instrumentation as described in Section 6.4.2.2, such as piezo-resistive pressure transducers, enable measurements of pressure fluctuations with a time resolution up to the inertial subrange (kHz range), allowing pressure measurements over the complete turbulence spectrum. Care should be taken when choosing the sampling frequency in order to avoid aliasing effects due to high frequency noise.

6.4.4 ERROR SOURCES

Error sources related to pressure measurements are varied, ranging from instrument related errors through errors with their installation to further diffuse causes. Instrument related errors are associated with incorrect calibration, and calibrations should be checked before data collection. Care also needs to be taken when installing the instruments with regard to the correct setup (see also Figure 6.4.2a). For example, if tubes are used to meter the pressure against a liquid column it must be ensured that air-bubbles have been evacuated from the tubing. Moreover, the alignment of transducers, probes, and wall openings needs to be carefully checked to avoid, for example, dynamic pressure effects on measurements designed for static pressure conditions.

Spatial resolution errors, aerodynamic interference, and acoustic reflection are the three leading physical causes of errors, especially for dynamic pressure measurements. Spatial resolution errors are related to the size of the transducer opening used to sense the pressure. A small opening is associated with good spatial resolution and high-frequency response, while a larger size is associated with good sensitivity and low-power response. Thus, the design of the experiments and transducers is generally a trade-off between the size of the opening and the signal strength (see the design of the MPPS by Detert, 2008). This subject is of importance for all types of measurements in convected pressure fields and is described in detail in Goldstein (1996). In general, this issue can be addressed by connecting tubes or cavities to the transducers (*e.g.*, the cavity shown Figure 6.4.3) which act as transmission lines that have characteristic masses, resistances, and capacitances. Such setups behave as linear harmonic oscillators with a series of resonance frequencies. At each of these frequencies the pressure in the tube/cavity resonates so that its value at the transducer is larger than at the opening.

Dynamic pressure measurements can be affected by aerodynamic interference caused by turbulence in the form of gustiness of turbulent eddies. Eddies produce velocity fluctuations so that a measured signal contains not only the required pressure fluctuations, but also turbulence induced fluctuations. This problem is analogous to the sensing of mean static pressure flow for which the measured static pressure p_m exceeds the true static pressure p_s, dependent on the distribution of turbulent energy between the normal components given that the probe is correctly aligned (McKeon & Engler, 2007). In case of acoustic measurements, "pseudosound" due to the flow as well as acoustic reflection may affect the measurements. These error sources are not discussed further here as such sensors are not commonly used in liquids.

6.5 BED SHEAR STRESS

6.5.1 Principles of measurement

A solid boundary is subject to normal and tangential stresses when a fluid flows past it. The instrumentation for the measurement of normal stress, *i.e.*, pressure, has been described in the preceding section. This section addresses instrumentation for the measurement of tangential stress at the flow boundary, or in other words bed or wall shear stress τ_0. Bed shear stress and a related parameter called the shear velocity $u_* = (\tau_0/\rho)^{0.5}$, where ρ denotes the fluid density, are key-parameters in hydraulics. The bed shear stress varies across spatial and temporal scales (local/instantaneous and/or spatially/temporal averaged bed shear stress) and depends on the nature of the investigated problem (*e.g.*, smooth or rough bed flows).

Bed shear stress expresses the shear force per unit area and is rather difficult to measure directly. Consequently, the literature is rich in the descriptions of methods for the evaluation of bed shear stress or shear velocity based on surrogate hydraulic measurements such as near bed flow velocity profiles, near bed turbulence characteristics, water depth, and slope. (*e.g.*, Kim *et al.*, 2000; Ackerman & Hoover, 2001; Biron *et al.*, 2004; Rowiński *et al.*, 2005; Bagherimiyab & Lemmin, 2013). Table 6.5.1 provides an overview of indirect methods for the evaluation of the time averaged bed shear stress (or the shear velocity) relying on measurements with instruments that are described in other sections of this book and which will not be repeated here. We note that a description of the corresponding methods and hydraulic principles is beyond the scope of this chapter as it focuses on instrumentation, and the interested reader is therefore referred to Volume I as well as to the scientific literature (*e.g.*, Rowiński *et al.*, 2005; Bagherimiyab & Lemmin, 2013) for more details on the principles and methods. Table 6.5.1 is not exhaustive as further methods may be used to derive bed shear stress from data and analytical considerations such as the application of the Saint-Venant equation for unsteady flows (*e.g.*, Rowiński *et al.*, 2000; Ghimire & Deng, 2011). It must be emphasized that the applicability of the summarized methods depends on the temporal resolution of the instruments, hydraulic boundary conditions, as well as spatial and temporal flow variability, especially in unsteady and/or rough bed flows. For the latter, considerations based on the double averaging methodology (DAM) are required for the upscaling of local bed shear stress estimates to larger spatial scales (see Section 2.3.2 in Volume I).

In addition to the methods summarized in Table 6.5.1, there exist specific instruments and methods for the direct measurement of the tangential force per unit area and

Table 6.5.1 Methods for determination of time averaged bed shear stress and shear velocity based on surrogate hydraulic measurements as described in Rowiński et al. (2005) and Bagherimiyab and Lemmin (2013).

Measurement of	Parameter	Determination/parameterization of
Velocity profile	τ_0	near bed velocity gradient in laminar boundary layers
Velocity profile	u_*	near bed velocity profile in the viscous sublayer of turbulent boundary layers
Velocity profile	u_*	velocity gradient or profile in the logarithmic sublayer
Velocity profile	u_*	power law velocity distribution
Reynolds-stress distribution	u_*^2	near bed value of Reynolds-stress in 2D-flows
Turbulence	u_*^2	near bed value of turbulent kinetic energy (TKE)
Turbulence	u_*	turbulent energy production and dissipation
Turbulence	u_*	spectra of turbulent fluctuations
Water depth and slope	τ_0	momentum equation (depth-slope product)
Incipient motion	τ_0	incipient motion of particles of known density and diameter using the Shields-diagram
Energy loss in pipe flows	τ_0	friction loss based on Darcy-Weisbach friction factor in conjunction with the Moody-diagram

the indirect determination of bed shear stress from pressure, heat- and mass-transfer measurements in the near bed region from which the bed shear stress can be derived. Many of these instruments and methods have been extensively discussed in the scientific literature and books (*e.g.*, Hanratty & Campbell, 1996; Ackerman & Hoover, 2001; Tavoularis, 2005; Klewicki *et al.*, 2007), so that the following section will provide only a brief overview on the corresponding instrumentation and methods.

6.5.2 Available instrumentation

The available instrumentation for the measurement of bed shear stress can be classified into direct and indirect techniques. Direct techniques require no assumptions with regard to the flow conditions while indirect techniques rely on the measurements of surrogate parameters from which the bed shear stress can be estimated. Most of the indirect techniques described below rely on the measurement of flow characteristics in the viscous sublayer, thus restricting their applicability range quite significantly to flows where the probes can be placed in the viscous sublayer. Table 6.5.2 summarizes the instruments described in the following. It is worth mentioning that many sensors (*e.g.*, shear plates, thermal sensors, laser based sensors) can be manufactured at small scales using microelectromechanical systems (MEMS) by extending silicon-based integrated circuit (IC) micromachining in order to assemble miniature engineering systems (Naughton & Sheplak, 2002; Gad-el-Hak, 2006).

6.5.2.1 Direct techniques

6.5.2.1.1 Shear plates

Shear plates, or floating-element balances, represent a direct method for the determination of bed shear stress with a high temporal resolution by measuring the local tangential force per unit area. This is achieved by directly allowing some portion of the surface to be movable against a restoring force and by either measuring the displacement of the element or the force required to keep it in a fixed position. The

Table 6.5.2 Summary of instruments for direct and indirect measurements of bed shear stress

Instrumentation	Method and principle	Requirements/restrictions
Shear plate	direct; tangential force per unit area	force measurements
Oil-film interferometry	direct; thinning of oil film	only possible over smooth surface
Ferrofluids	direct; deformation of ferrofluid	magnetic field
Liquid crystal coating	direct; color change of crystals	optical access
Preston tubes	obstacle; pressure difference	Calibration
Stanton gauge	obstacle; pressure measurement	viscous sublayer
Sublayer fence	obstacle; pressure difference	viscous sublayer
FST-hemispheres	obstacle; movement of hemispheres	can affect near bed flow conditions
Thermal probes	heat transfer; constant temperature	calibration; viscous sublayer
Mass transfer probes	mass transfer; electrochemical reaction	electricity; chemicals
Pulsed wall probes	indirect; heat tracer	viscous sublayer
Micro-optical sensors	indirect; velocity measurement	viscous sublayer

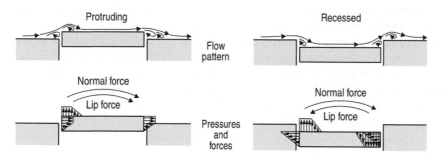

Figure 6.5.1 Hydraulic effects affecting shear stress measurements with a shear plate (Figure 12.8 of Klewicki *et al.*, 2008, redrawn with permission of Springer, permission conveyed through Copyright Clearance Center, Inc.).

use of shear plates requires the installation of a false floor in flumes so that they can be flush mounted with the bed in case the flume cannot be modified from below. At the same time, a gap between the shear plate and bed is required to guarantee free movement of the plate. This gap can cause undesired hydraulic effects which can affect the shear stress measurements (see Figure 6.5.1). Moreover, in the case of movable beds, fine particles may deposit in the gap and can therefore affect the movability of the plate.

The size of the plate is basically a trade-off between the need to have measurable forces acting on the surface and the desire to have local bed-shear stress values. In rough bed flows, surface friction and form drag of the bed elements will be measured with such devices, *i.e.*, the total shear stress. Examples for the design and successful use of shear plates, mainly in coastal engineering applications, are found in *e.g.*, Rankin and Hires (2000), Barnes and Baldock (2007), Barnes *et al.* (2009), Seelam *et al.* (2011) and Gmeiner *et al.* (2012).

6.5.2.1.2 Surface coating methods

Surface coating methods utilize visualization techniques to measure changes in surface properties of materials that are applied to the bed (Ackerman & Hoover, 2001). The use of these methods is not common in environmental hydraulics as their applicability is mainly restricted to smooth bed flows and because they may result in surface fouling.

Oil-film interferometry is based on the determination of the thinning of an oil film on a smooth surface which can be related to bed shear stress. The technique makes use of the fact that the light reflected from the top and bottom of the very thin film (typically of the order of several wavelengths of visible light) forms interference fringes, which can be interpreted as contour lines of the oil-film surface. Assuming that the film thickness is only affected by bed shear stress, the thinning of the oil-film can then be related to bed shear stress by solving the thin oil-film equation. Further details on this technique can be found in Fernholz *et al.* (1996) and Naughton and Hind (2013).

Recent developments showed that it is also possible to determine bed shear stress using magnetic fluids, *i.e.*, ferrofluids, by creating a ferrofluid spike at the bed through the use of permanent magnets. The bed shear stress can subsequently be determined by measuring the deformation of the spike due to the flow attack. Details of this novel methodology may be found in Musumeci *et al.* (2014, 2015).

Liquid crystal coating can be used to determine the surface shear magnitude and direction at every point on a surface by determining the color change of the coated surface. The color-change is due to the alignment of the molecules within the coating in response to the frictional forces acting on it. More details on this method may be found in Reda et al. (1997) and Buttsworth et al. (2000). The applicability of this method is, however, rather restricted due to the complicated calibration processes, required optical access, and most importantly the fact that the liquid crystal coatings are rather expensive and must be frequently reapplied.

6.5.2.2 Indirect techniques

Besides the indirect methods based on the measurements of velocities, turbulence, water depth, and incipient motion summarized in Table 6.5.1, there exists a variety of instruments that can be used for the determination of bed shear stress by measuring surrogate flow parameters near the bed. These instruments will be briefly described in the following subsections.

6.5.2.2.1 Obstacle flow techniques

Obstacle flow techniques rely on instrumentation which is placed on the bed. Although the instruments, strictly speaking, interfere with the near bed flow field, they have been shown to provide reliable data on bed shear stress.

Preston tubes (Preston, 1954) are a common instrument for the measurement of local bed shear stress values (Figure 6.5.2). A Preston tube is a wall mounted pitot tube with which the stagnation pressure is measured at the probe opening. The simultaneous determination of the static pressure by a nearby wall tap allows determination of point based bed shear stress based on the pressure differences in the boundary layer. The Preston tube must be small enough to be within the wall layer, and pressure differences must be large for the detection by a manometer. For most practical applications calibration procedures need to be applied as the tube opening will primarily be located above the viscous sublayer in the logarithmic portion of the flow. Note that the use of Preston tubes is not restricted to the laboratory environment, as such probes can also be used in field studies (Molinas et al., 1998; Ackerman & Hoover, 2001). Further details on this instrument and its application can be found in the scientific literature and textbooks (e.g., Ackerman & Hoover, 2001; Klewicki et al., 2007; Thornton et al., 2008).

A Stanton gauge is a small apparatus that measures the pressure within the viscous sub-layer, restricting its application to hydraulically smooth flows. Stanton gauges are usually small and flattened impact tubes, having the wall as one of their sides. A widely used variant has a thin blade attached to the wall partly blocking the opening of a static pressure tap (Tavoularis, 2005). Since the instrument is applied in the viscous sublayer, the blade thickness must be smaller than the thickness of the viscous sublayer. The bed shear stress is determined from the pressure difference between the blocked tab and the local static pressure. Stanton gauges require calibration and are generally characterized by uncertainties of $\pm 3\%$. The uncertainty increases in flows with large pressure gradients (Tavoularis, 2005).

A sublayer fence, also called a surface fence, consists of a wall tap in which a thin blade is mounted for the measurement of magnitude and direction of the skin friction in

Figure 6.5.2 The Preston tube used by Thornton *et al.* (2008) on the surface of a gravel bed (Photo from
Thornton *et al.*, 2008)

the viscous sublayer. The blade partitions the tap into an upstream and downstream
half and the blade tip extends slightly into the flow but must, at the same time, remain
immersed in the viscous sublayer (Tavoularis, 2005). The flow over the blade consists
of a recirculation zone both in front and in the back of the blade (or fence). The pressure
difference between the two halves is linearly related to the wall shear stress. This
instrument is, as is the Stanton gauge, independent of the validity of the logarithmic
law of the wall (Fernholz *et al.*, 1996) but its use is restricted to smooth bed flows as it
relies on measurements in the viscous sublayer. The advantages of sublayer fences over
Stanton gauges are that they almost double the pressure reading, eliminate the necessity
of a static-pressure tap, and can be used to determine the direction of the bed shear
stress. However, in flows with strong pressure gradients the sublayer fence is of
unknown accuracy (Hanratty & Campbell, 1996).

The FST-hemisphere method (FST: *Fließwasserstammtisch*) is a technique for the
determination of the magnitude of local bed shear stress values in rivers and streams
(Statzner & Müller, 1989; see Figure 6.5.3). A FST-hemisphere set consists of 24
(or 20) hemispheres of identical size and surface properties but varying density.
Starting with the lightest hemisphere and using a standardized procedure, the hemispheres are subsequently placed on a horizontal plate (perspex or steel) and exposed to
the flow. The heaviest hemisphere moved by the flow is then used to give an indication
of the local shear stress. In general, FST-hemisphere results are strongly influenced by

flow characteristics associated with irregularities of the bottom topography (roughness type) and submergence (Dittrich & Schmedtje, 1995). Additional errors may arise if the bottom plate is not in a horizontal position, and it must be considered that the plate on which the hemispheres are resting modifies the natural flow conditions. Further details on this method and its accuracy can be found in Statzner and Müller (1989), Dittrich and Schmedtje (1995), and Wang *et al.* (2007).

6.5.2.2.2 *Heat and mass transfer techniques*

Heat- and mass-transfer probes consist of very small films or wires embedded in the wall that maintain a constant temperature (heat transfer) or, equivalently, a constant concentration (mass transfer), enabling the determination of the rate of heat or mass transfer between the sensor and the adjacent fluid. Such sensors must be small in order to measure the heat or mass transfer within the thermal or concentration boundary layer, which in turn must lie within the viscous sublayer (Gust, 1988). Heat transfer, mass transfer, and momentum transfer are physically dependent on the same transport properties of fluids, allowing the development of a theoretical relation between them. The resulting relation between heat or mass transfer and the bed shear stress (momentum transfer) generally requires *in situ* calibration of the sensor. Heat- and mass-transfer techniques are less intrusive than obstacle flow techniques and can provide information on bed shear stress with a high temporal resolution. Theoretical details with regard to these techniques will not be discussed here and can be found in *e.g.,* Hanratty & Campbell (1996) and Tropea *et al.* (2007).

Figure 6.5.3 A FST-hemisphere placed in a rough bed (Photo from Koch, 1994).

Thermal probes such as hot wire or hot film probes (hot film probes are predominantly used in liquid flows) are heated electrically by a constant temperature anemometer circuit (CTA). This setup enables the measurement of the electric current needed to keep the probe at a constant temperature which is higher than the ambient temperature, and hence leads to heat transfer. Hot film probes have the advantage that they are small and flush mounted to the surface so that they do not interfere with the flow and that velocity profiling is not required. However, they need to be calibrated in a flow where the bed shear stress is known, they are sensitive to substrate temperature, and they must be inside the viscous sublayer (Thompson *et al.*, 2003a).

Mass-transfer probes are based on similar principles as heat-transfer probes. Measuring with such probes, an electrochemical reaction is carried out on the surface of a flush mounted electrode and the voltage of the electrode is kept large enough so that the reaction rate is mass-transfer controlled but yet small enough to avoid side-reactions. The current produced by the chemical reaction can be related to the velocity gradient at the electrode-surface. Mass transfer probes have the advantage that problems are avoided with regard to substrate heat loss and that they, in principle, can be calibrated analytically. However, they have a lower frequency response and their use is restricted to fluids and equipment that is compatible with the chemicals used. Further details on this method may be found in Hanratty & Campbell (1996).

6.5.2.2.3 Pulsed wall probes

Pulsed wall probes are an adaptation of the pulsed-wire anemometer, consisting of three parallel wires mounted in a plane parallel to the wall and at a height within the viscous sublayer (Fernholz *et al.*, 1996). The central wire is pulsed by a very short electrical pulse (about 5 µs) generating a heat tracer which is subsequently detected by the up- or downstream sensor wire depending on the direction of flow. The travel time of the heat tracer can be used to determine the instantaneous velocity and bed shear stress (Fernholz *et al.*, 1996). These probes require calibration, and the instantaneous shear stress should not create a time of flight of the heated fluid parcel shorter than the minimum time which the pulsed wire anemometer is able to measure or severe errors may be generated. Therefore, care must be taken to adjust the probe geometry (wire spacing and wall distance) to the flow conditions under investigation. More details on this probe type can be found in Fernholz *et al.* (1996) and references therein.

6.5.2.2.4 Micro-optical shear stress sensor

Micro-optical shear stress sensors are designed to determine the shear stress a few hundred microns from the bed within the viscous sublayer based on Laser Doppler Anemometry. The micro-sensor is flush mounted on the bottom and emits laser light through two rectangular slits, thereby generating a linearly diverging fringe pattern in the fluid. Particles in the flow passing through the fringes scatter light to a detector with a frequency which is proportional to the velocity and inversely proportional to the fringe-spacing. Within the viscous sublayer, the measured frequency is hence directly proportional to the bed shear stress (Modarress *et al.*, 2006). The method is basically restricted to laboratory environments. It is worth mentioning that the method can also be used in reversing flows by shifting the frequency of one of the slits.

6.5.3 Spatial and temporal resolution

6.5.3.1 Spatial resolution

The measurement principles of the instruments described in the preceding sections differ significantly and so do their spatial and temporal resolutions. The techniques based on velocity and turbulence measurements listed in Table 6.5.1 provide, in general, local estimates of time averaged bed shear stress (or the shear velocity) at the position where the corresponding measurements are taken. Local measurements above rough beds will be biased due to the spatial flow heterogeneity. Consequently, bed shear stress estimates from velocity and turbulence measurements should be done at multiple locations and incorporate spatial averaging (double averaging methodology). The depth-slope product (as well as the pressure loss method in pipe flows) provides spatially averaged bed shear stress values, although side-wall corrections may be necessary depending on the aspect ratio (*e.g.*, Vanoni & Brooks, 1957).

Shear plates (or floating balances) provide an integrated value of bed shear stress over a fixed area and their spatial resolution is consequently directly related to the size of the shear plate (or balance) as described above. Depending on the wall boundary conditions (hydraulically smooth or rough), contributions from viscous and turbulent drag will both be measured. Surface coating methods provide information on the magnitude and direction of the bed shear stress over the coated area and the corresponding spatial resolution can be very high. In fact, liquid crystal coatings provide shear stress information at every point on the coated surface with a resolution down to molecule-size. On the other hand, the resolution of the ferrofluid technique is governed by the size of the ferrofluid spike.

Indirect methods (obstacle flow techniques, heat- and mass-transfer techniques, pulsed wall probe, micro-optical sensors) provide local bed shear stress values at the scale of the sensor, which in turn may affect the accuracy of the measurement. Moreover, the applicability of many of these methods is restricted to measurements in the viscous sublayer, preventing their applicability in hydraulically rough flows and imposing a significant restriction with regard to the vertical scale of the instruments. For some of the methods, the spatial resolution may be increased if the sensors can be constructed as MEMS. An exception is the FST-technique, as the size of the hemispheres is much larger than the viscous sublayer.

6.5.3.2 Temporal resolution

The number of techniques that can be used to measure instantaneous (time-resolved) shear stress values is considerably lower than the number of techniques for the determination of time averaged stress. This is due to the fact that the theoretical foundations of most indirect techniques are based on time-averaged considerations so that the techniques cannot be unambiguously applied to instantaneous measurements.

Furthermore, the accurate quantification of instantaneous shear stress values requires small probes with a high frequency response (Klewicki *et al.*, 2007).

Consequently, the methods listed in Table 6.5.1 and the described obstacle based techniques provide only time-averaged bed shear stress estimates as they rely on principles of the near bed flow field derived from time-averaged equations. Oil film interferometry also provides time-averaged estimates as it has been shown that shear stress fluctuations do not affect the movement of the film (Fernholz *et al.*, 1996). The response time of liquid crystal coatings may not be sufficient to measure instantaneous values in turbulent flows. Ferrofluids have been successfully used for bed shear stress measurements under wave action (Musumeci *et al.*, 2014).

For bed shear stress measurements on hydraulically smooth beds, mass and heat transfer methods as well as pulsed wall probes can be used to measure the instantaneous velocity gradient and hence bed shear stress fluctuations (Klewicki *et al.*, 2007), with the temporal resolution dependent on the frequency response of the sensor. However, most of these instruments require calibrations which are based on correlation-based relationships so that they cannot be interpreted as an *independent means for measuring the time averaged shear stress* (Klewicki *et al.*, 2007). Shear plates (or floating-element balances) require force measurements which can be carried out with a high temporal frequency, so that they can provide instantaneous values of bed shear stress. However, the spatial scale of these instruments must be considered in the interpretation of such data.

Finally, near-bed velocity measurements in the viscous sublayer may also be used for the determination of the instantaneous velocity gradient and thus the instantaneous shear stress. However, such measurements are sophisticated and their feasibility depends on experimental boundary conditions.

6.5.4 Limitations and error sources

The preceding sections highlighted an important error source and limitation of many of the instruments used for bed shear stress measurements – the requirement that measurements are within the viscous sublayer. For such instruments, significant error is introduced if the probe extends out of the viscous sublayer. This limits such instruments to experiments with hydraulically smooth flow conditions. Moreover, nearly all of the described instrument based methods are impracticable above movable beds (Arndt *et al.*, 2007). Therefore it is not surprising that many different indirect methods for the determination of bed shear stress, based on velocity, turbulence measurements or a momentum balance have been developed and are used in both practical and scientific studies, especially for rough bed flow conditions and flows over movable beds. An exception may be the use of ferrofluids which have recently be used to measure bed shear stress in the presence of suspended impurities and moderate sediment transport (Musumeci *et al.*, 2014).

A detailed consideration of the error sources related to the methods summarized in Table 6.5.1 and Table 6.5.2 is beyond the scope of the present section. Instead, the most important error sources as well as further limitations beyond the one described above will be briefly presented, mainly for the instruments that can also be used in rough bed conditions.

Shear plates (floating balances) are "simple devices" for bed shear stress measurements but are prone to various errors. The most important ones are associated with spatial scale (already discussed above), misalignment of the plate, the effect of the gaps around the plate, and pressure-gradient forces. The misalignment of the plate (protruding or recesses, see Figure 6.5.1) and the presence of gaps can induce additional forces acting on the plate which were not present during calibration. If shear plates are used in flows with pressure gradients, the measurements of total shear stress should be corrected by subtracting the contribution of the pressure gradient arising through the pressure differences in the up- and downstream gap. Finally, care has to be taken that the shear plate can move freely during the experiments (see also Klewicki *et al.*, 2007) and it should be considered that ambient vibrations (caused *e.g.*, by flume vibrations) may affect the readings and introduce errors.

The accuracy of surface coating methods can be affected by impurities of the surface and depends also on the visualization technique used. Oil-film interferometry may be significantly affected by surface imperfections and cannot be used in rough bed flows.

When using Preston tubes, care needs to be taken to carefully align the probe and to ensure that the reading from the static-pressure tap is not affected by blockage caused by the probe. Preston-tubes should not be used in highly accelerating flows, detached boundary layers or in flows where the logarithmic portion of the flow is significantly altered, and smooth wall calibrations should not be applied to the measurements carried out in rough bed flows (Ackerman & Hoover, 2001; Klewicki *et al.*, 2007). It should also be noted that the Preston tube is an intrusive method that disturbs the flow.

Stanton gauges give inaccurate readings in flows with large pressure gradients. The FST-method is characterized by a range of uncertainties which are associated with the deployment platform and deployment procedure (see Dittrich & Schmedtje, 1995; Wang *et al.*, 2007).

Heat-transfer methods rely also on measurements in the viscous sublayer (as discussed before), and require careful calibration. Static calibrations may be different from dynamic calibrations as the temperature field may change for these situations. The accuracy of the measurements can be affected by heat transfer within the substrate that holds the constant temperature film, and the spatial scale of the film should be taken into account when interpreting the measured values. Similarly, data from mass-transfer probes should be interpreted with regard to the spatial scale and the response time of the concentration boundary layer.

6.6 DRAG FORCES

6.6.1 Principles of measurement

A body exposed to flow is subject to a drag force F, which is generally parameterized by the classical form drag equation $F = 0.5\rho C_D A_C U_c^2$, where ρ is the fluid density, C_D is the drag coefficient, A_C is the characteristic area (usually the projected frontal area) and U_c is the approach velocity. The drag coefficient C_D includes both components arising due to viscous and pressure drag, and depends on the flow conditions which are usually characterized by the object Reynolds-number (*e.g.*, Nakayama, 1999). However, theoretical computation of the viscous and pressure drag components is generally impractical, except for bodies of simple shape (*e.g.*, cylinders, spheres) for which functional relationships between the drag coefficient and object Reynolds number can be found in textbooks. For more complex and/or flexible bodies, one must rely on measurements in order to determine drag forces and hence to parameterize C_D. Reporting such measurements, it is important that both A_c and U_c are clearly defined so that the results can be generalized and/or used by other researchers (*e.g.*, Statzner *et al.*, 2006).

The measurement of both steady and fluctuating drag forces acting on a body has a long history in wind tunnel experiments, and there exist excellent overviews on measurement principles for such experimental investigations (*e.g.*, Hufnagel & Schewe, 2007). However, not all of the principles and instruments which are common in wind tunnel experiments can be easily applied in hydraulic engineering applications as the medium water imposes problems. Waterproof sensors or special experimental setups such as those described below are required.

The most common drag force measurement systems consist of mechanical or digital force gauges (or transducers). Optical and electromagnetic methods are also sometimes used, but are generally restricted to small scale applications (nanosensing). A classical mechanical force gauge is the spring balance, whose measurement principle is based on Hooke's Law, which states that the force needed to compress or extend a spring by a certain distance is proportional to that distance. Digital force transducers (or load cells), on the other hand, create an electric signal with a magnitude that is directly proportional to the measured force. Digital load cells consist of basic mechanical components that convert strain into an electrical signal (Cutkosky *et al.*, 2008); this is typically achieved using strain gauges or piezoelectric crystals. Depending on the setup, such transducers enable the measurement of forces and torques in all three Cartesian coordinates (or in an arbitrary coordinate system), *i.e.*, of six different components (three forces along coordinate axes and the moments about these axes). A further method to measure drag forces is based on inverse dynamics, *i.e.*, on the determination of the acceleration of a body of known mass, thus enabling the calculation of drag using Newton's 2nd law. The acceleration may be measured directly using accelerometers or derived from measurements of the movement of a body with high temporal resolution (*e.g.*, Diplas et *al.*, 2008). However, such methods are not common in hydraulic engineering applications.

Drag force measurements in water flows are challenging and can broadly be subdivided into two categories related to sensor implementation: above the water surface (or the flume) to keep the sensor dry; and submersed, *i.e.*, bottom mounted (see Figure 6.6.1a). Locating the transducer above the water surface means that a part of the test-object will be in air (as in towing tank experiments, Figure 6.6.1b) and that the object cannot be bottom mounted. Alternatively, the test object can be connected to the transducer via a special support setup (*e.g.*, a stiff rod or a pole) so that the total drag measured by the sensor consists of both the drag of the object and the support installation. Bottom mounted transducers typically require the installation of a false bottom in the flume so that they can be flush mounted (Figure 6.6.1a). Examples of the different setups and their application in experiments may be found in the studies summarized in Table 6.6.1 and the references therein.

 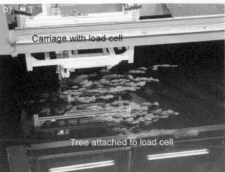

Figure 6.6.1 a) The submersed drag force measurement system used by Schoneboom *et al.* (2008) and
b) photograph of the towing tank test by Jalonen & Järvelä (2014) using a load cell located
above the water surface to measure drag forces exerted by a full scale tree (Photos: a)
J. Aberle; b) J. Järvelä).

Table 6.6.1 Studies using custom-made drag force transducer setups in the laboratory and the field (the
list is not exhaustive). SG denotes strain gauge and WS water surface.

Study	Sensor arrangement	Details
Albayrak *et al.* (2011)	above WS	1D SG load cell; cantilever in bending
Armanini *et al.* (2005)	submersed	3D SG load cell; cantilever in bending
Azinfar and Kells (2009)	above WS	1D SG load cell; cantilever in bending
Callaghan *et al.* (2007)	submersed	Compression load-cell; trolley moving horizontally
Carrington (1990)	submersed	1D SG load cell; cantilever in bending
Dwivedi *et al.* (2010)	submersed	3D SG load cell; cantilever in bending
Fathi-Maghadam and Kouwen (1997)	submersed	1D SG force-balance system beneath flume
Hygelund and Manga (2003)	above WS	torque wrench
Jalonen and Järvelä (2014)	above WS	1D SG load cell; cantilever in bending
Kothyari *et al.* (2009)	above WS	1D SG load cell; cantilever in bending
Oplatka (1998)	submersed	2D-SG load cell
Sand-Jensen (2003)	above WS	spring balance; trolley moving horizontally
Schoneboom *et al.* (2008)	submersed	1D SG load cell; cantilever in bending
Shields and Alonso (2012)	above WS	2D-SG load cell; dual circuitry of differential SG bridges
Statzner *et al.* (2006)	submersed	compression load-cell; trolley moving horizontally
Thompson *et al.* (2003b)	submersed	tensile load cell
Tinoco and Cowen (2013)	submersed	drag plate; strain-beam load cell
Whittaker *et al.* (2013)	above WS	3D SG dynamometer
Wilson *et al.* (2008)	submersed	1D SG load cell; cantilever in bending

6.6.2 Available instrumentation

In general, force transducers can be bought off the shelf from various manufacturers as they are heavily used in industrial applications. However, in most hydraulic studies, force transducers have been custom-built to suit the experimental setup. Various studies using different setups for the measurements are summarized in Table 6.6.1. The majority of these studies used strain-gauge measurements, with the strain-gauges typically being configured in a Wheatstone bridge circuit. A review of the corresponding measurement technique is beyond the scope of the present chapter and the interested reader will find much more detailed information in Hufnagel and Schewe (2007) with regard to strain-gauges and the wiring of Wheatstone bridges.

Strain-gauge based load cells can be built with different sensitivities, as many different strain-gauges are commercially available. Load cells may be designed as "bending-type load cells", "shear-type load cells", and "tension and compression load cells", *i.e.*, by measuring the strain of a deforming body caused by bending, shear, or tensile stress, respectively (see Hufnagel & Schewe, 2007 for details). Each of these load cell types can be constructed as a single load cell, enabling the determination of single force components. A combination of these individual components enables the measurement of forces and torques in multiple directions. Sensors should be selected such that the hydraulic load causes a sufficiently large deformation of the load cell. This requirement can be problematic for the measurement of small forces.

Piezoelectric force transducers are less common in hydraulic engineering studies (see also Section 6.4.2.2.2). Typical piezoelectric transducers resemble a washer, as a quartz ring is mounted between two steel plates in the transducer case. A pair of shear and pressure sensitive rings which measure the orthogonal components of a shear force and a compression force on the transducer, respectively, act in combination as a multi-component transducer (see Hufnagel & Schewe, 2007).

6.6.3 Spatial and temporal resolution

The spatial resolution of drag force sensors can be very high, especially if they are constructed as microelectromechanical systems (MEMS). However, most of the strain-gauge based sensors have a somewhat larger spatial resolution as the sensors need to be larger in order to guarantee sufficient deformation, so that the resulting strain can be adequately measured. The temporal resolution of drag force sensors can be very high up to the kHz range (for examples see studies summarized in Table 6.6.1).

6.6.4 Error sources

Drag force measurements rely on the functionality of the load cells which should be carefully calibrated before the experiments. The transducers also need to be carefully monitored with regard to drift of the signal during the measurements, although temperature drift is usually compensated through Wheatstone bridge configurations. Probably the most important error source for drag force measurements is to ensure the correct alignment of the sensor, especially if sensors are used that measure the force in only one direction. Similarly, the sensors need to be chosen with regard to the expected flow forces and hence the expected strain in case of strain-gauge based load cells. For example, considering a cantilever in bending; it can be shown on the basis of strain gauge specifications that the beam must be rather long (and thin) in order to measure low forces. This can limit the applicability of such sensors, especially in conditions with limited space. Using trolley-based load cells, the weight of the trolley may have some effect due the friction associated with the trolley movement. If the sensors are mounted above the water surface, calibration measurements are required to evaluate the drag forces associated with the structure holding the test-element in place. Finally, drag force sensors may be sensitive and prone to vibrations within the experimental facility which can be flume related or triggered externally (*e.g.*, pumps).

6.7 CONDUCTIVITY-TEMPERATURE-DEPTH PROBES FOR FLUID DENSITY

This section provides some additional information on measurements of density, conductivity, and temperature as these quantities are also often recorded in hydraulic experiments. In fact, many flows are driven or modulated by density gradients. For example, the difference in density between an effluent and a receiving water body can result in a positively or negatively buoyant plume, in which buoyancy determines the nature and rate of mixing of the effluent with the receiving body (*e.g.*, Malacic, 2005). A river entering a water body of different density can also be analyzed as a buoyant plume (*e.g.*, Horner-Devine, 2009; Nekouee *et al.*, 2013). Mixing in oceans and lakes is often driven by density gradients. Differences in density between adjacent water bodies can also drive exchange flow between the water bodies, with lower density fluid moving on top in one direction and higher density fluid moving in the opposite direction below. An example is the exchange flow through the Strait of Gibraltar (Armi & Farmer, 1988; Millot, 2009). A shear layer occurs between the two layers, with associated flow instabilities (Lawrence, 1993; Zhu & Lawrence, 2000; Tedford *et al.*, 2009). In order to understand such flows, it is necessary to determine fluid density accurately. The density of a fluid varies with composition, temperature, and pressure, all of which must be measured to obtain reliable estimates of density. Gill (1982) provided an equation of state to calculate density as a non-linear function of salinity, temperature, and pressure. More recently, the TEOS-10 standard suggested the use of thermodynamic principles based on absolute salinity for accurate calculation of density (see http://www.teos-10. org/; accessed 9th December 2016).

Conductivity-Temperature-Depth (CTD) probes are designed specifically to determine water density based on measurements of conductivity, temperature and pressure (see Figure 6.7.1 for an example). CTDs have been used since the 1950s by oceanographers and limnologists (Stewart, 2008). CTDs have also been used to assess mixing processes in rivers, particularly in estuaries (*e.g.*, Kawanisi, 2004). Determination of vertical profiles of density can be achieved by lowering a continuously recording CTD throughout the water column. CTDs can also be towed to obtain longitudinal data. The translation speed and the sampling frequency determine the measurement resolution. As an example, the Seabird SBE 911 plus CTD has a sampling frequency of 24 Hz.

Conductivity is measured in units of siemens/m (S/m) as a surrogate for chemical composition, because the electrical conductivity of water increases with salinity. The measured conductivity is calibrated using a standard solution of known salinity.

Figure 6.7.1 A CTD-probe (Seabird SBE19plusV2; Photo: A. Pilechi).

Conductance is the inverse of electrical resistance, and thus can be obtained by dividing a measured electrical current by the voltage of the electric field. The conductance depends on both the conductivity of the fluid and the length and cross-sectional area of the electrical current flow. A conductivity probe basically consists of a flow cell with either electrodes or transformers that apply voltage to induce current. The specific design of the sensor determines accuracy, which can be degraded by uncertainty in the electrical current cross-sectional area and resistivity of the sensor components. It is worth noting that both of these sensor properties can be altered by biofouling and application of anti-biofouling agents, which is particularly problematic if a portion of the electrical current is external to the flow cell. The SBE 4 conductivity sensor used in Seabird profiling CTDs minimizes these uncertainties by employing three low-resistance platinum electrodes placed within the flow cell such that the electrical current is entirely contained within the cell (http://www.seabird.com/document/conductivity-sensors-moored-and-autonomous-operation; accessed 19th October 2016). The rated precision and accuracy of the SBE 4 are 0.00004 S/m and ±0.0003 S/m, which corresponds to a practical salinity accuracy of ±0.01 [dimensionless, but approximately units of g/kg] using the Practical Salinity Scale 1978 (PSS-78; *IEEE Journal of Oceanic Engineering*, Vol. OE-5, No. 1, January 1980) at standard temperature and pressure. The absolute salinity of seawater of a certain composition from a given location can be calculated from measurements of practical salinity using the TEOS-10 standard.

Temperature can be measured using a thermistor, *i.e.*, a resistor in which resistance varies with temperature. An electrical circuit is employed to generate a current, and resistance of the thermistor is calibrated to ambient temperature. Measurement accuracy can be on the order of $\pm 0.001°C$ with drift in the calibration of $0.002°C/$yr. High sampling rates can be achieved (*e.g.*, 20 Hz), which allows for micro-structure profiling of temperature fluctuations associated with flow turbulence (*e.g.*, Ward *et al.*, 2014).

Depth (h) is estimated by measuring pressure (p) and assuming hydrostatic conditions (*i.e.*, $h = p/\gamma$, where $\gamma = \rho g$ is fluid specific weight, ρ is fluid density, and g is gravitational acceleration). Pressure is measured using a transducer consisting of either a digital strain gauge to measure pressure directly (typical accuracy 0.3% to 2.5% depending upon temperature) or a piezoelectric crystal whose frequency of oscillation varies with ambient pressure (typical accuracy 0.01%); see also Section 6.4.

REFERENCES

Aberle, J., Koll, K., & Dittrich, A. (2008) Form induced stresses over rough gravel-beds. *Acta Geophys*, 56(3), 584–600.

Ackerman, J.D. & Hoover, T.M. (2001) Measurement of local bed shear stress in streams using a Preston tube. *Limnol Oceanogr*, 46(8), 2080–2087.

Albayrak, I., Nikora, V., Miler, O., & O'Hare, M. (2011) Flow-plant interactions at a leaf scale: effects of leaf shape, serration, roughness and flexural rigidity. *Aquat Sci*, doi:10.1007/s00027-00011-00220-00029.

Armi, L. & Farmer, D.M. (1988) The flow of Mediterranean water through the Strait of Gibraltar. *Prog Oceanogr*, 21, 1–105.

Armanini, A., Righetti, M., & Grisenti, P. (2005) Direct measurement of vegetation resistance in prototype scale. *J Hydraul Res*, 43(5), 481–487.

Arndt, R.E.A., Kawakami, D., Wosnik, M., Perlin, M., Duncan, J.H., Admiraal, D.M., & Garcia, M.H. (2007) Hydraulics. In: Tropea, C., Yarin, A.L., & Foss, J.F. (eds.) *Springer Handbook of Experimental Fluid Mechanics*. Berlin, Springer. pp. 959–1042.

Azinfar, H. & Kells, J.A. (2009) Flow resistance due to a single spur dike in an open channel. *J Hydraul Res*, 47(6), 755–763.

Bagherimiyab, F. & Lemmin, U. (2013) Shear velocity estimates in rough-bed open-channel flow. *Earth Surf Process Land*, 38(14), 1714–1724.

Barnes, M.P. & Baldock, T.E. (2007) Direct bed shear stress measurements in laboratory swash. *J Coast Res*, SI 50, 641–645.

Barnes, M.P., O'Donoghue, T., Alsina, J.M., & Baldock, T.E. (2009) Direct bed shear stress measurements in bore-driven swash. *Coast Eng*, 56(8), 853–867.

Biron, P.M., Robson, C., Lapointe, M.F., & Gaskin, S.J. (2004) Comparing different methods of bed shear stress estimates in simple and complex flow fields. *Earth Surf Process Land*, 29(11), 1403–1415.

Buttsworth, D.R., Elston, S.J., & Jones, T.V. (2000) Skin friction measurements on reflective surfaces using nematic liquid crystal. *Exp Fluids*, 28(1), 64–73.

Callaghan, F.M., Cooper, G.G., Nikora, V.I., Lamoroux, N., Statzner, B., Sagnes, P., Radford, J., Malet, E., & Biggs, B.J.F. (2007) A submersible device for measuring drag forces on aquatic plants and other organisms. *New Zeal J Mar Fresh Res*, 41, 119–127.

Carrington, E. (1990) Drag and dislodgment of an intertidal macroalga: consequences of morphological variation in Mastocarpus papillatus Kützing. *J Exp Mar Biol Ecol*, 139, 185–200.

Cutkosky, M., Howe, R., & Provancher, W. (2008) *Force and tactile sensors*. In: Siciliano, B & Khatib, O. (eds.) *Springer handbook of robotics*. Berlin, Springer. pp. 455–476

Detert, M. (2008) *Hydrodynamic processes at the water-sediment interface of streambeds*. Doctoral thesis, University of Karlsruhe, [Online] Available from: http://dx.doi.org/10.5445/KSP/1000008267 [Accessed 10th February 2017].

Detert, M., Nikora, V., & Jirka, G.H. (2010a) Synoptic velocity and pressure fields at the water sediment interface of streambeds. *J Fluid Mech*, 660, 55–86.

Detert, M., Weitbrecht, V., & Jirka, G.H. (2010b) Laboratory measurements on turbulent pressure fluctuations in and above gravel beds. *J Hydraul Eng*, 136(10), 779–789.

Diplas, P., Dancey, C.L., Celik, A.O., Valyrakis, M., Greer, K., & Akar, T. (2008) The role of impulse on the initiation of particle movement under turbulent flow conditions. *Science*, 322 (5902), 717–720.

Dittrich A. & Schmedtje, U. (1995) Indicating shear stress with FST-hemispheres – effects of stream-bottom topography and water depth. *Freshwat Biol*, 34, 107–121.

Dwivedi, A., Melville, B.W., Shamseldin, A.Y., & Guha, T.K. (2010) Drag force on a sediment particle from point velocity measurements: A spectral approach. *Water Resour Res*, 46 (W10529), doi:10.1029/2009WR008643.

Fathi-Maghadam, M. & Kouwen, N. (1997) Nonrigid, nonsubmerged, vegetative roughness on floodplains. *J Hydraul Eng*, 123(1), 51–57.

Fernholz, H.H., Janke, G., Schober, M., Wagner, P.M., & Warnack, D. (1996) New developments and applications of skin-friction measuring techniques. *Meas Sci Technol*, 7, 1396–1409.

Gad-el-Hak, M. (ed.) (2006). *The MEMS – Handbook: MEMS applications*. Boca Raton, CRC-Press.

Ghimire, B. & Deng, Z.Q. (2011) Event flow hydrograph-based method for shear velocity estimation. *J Hydraul Res*, 49(2), 272–275.

Gill, A.E. (1982) *Atmosphere-Ocean* Dynamics, Academic Press, New York.

Gmeiner, P., Liedermann, M., Tritthart, M., & Habersack, H. (2012) Development and testing of a device for direct bed shear stress measurement. In: *Water – Infinitely Deformable but Still Limited – Proceedings of the 2nd IAHR Europe Congress, 27–29 June 2012, Technische Universität München, München, Germany*.

Goldstein, R.J. (1996) *Fluid Mechanics Measurement*. 2nd edition. London, Taylor & Francis.

Gregory, J.W., Asai, K., Kameda, M., Liu T., & Sullivan, J.P. (2008) A review of pressure-sensitive paint for high-speed and unsteady aerodynamics. *Proc Inst Mech Eng, Part G: J Aero Eng*, 222(2), 249–290.

Gust, G. (1988) Skin friction probes for field applications. *J Geophys Res*, 93(C11), 14,121–114,132.

Hanratty, T.J. & Campbell, J.A. (1996) Measurement of wall shear stress. In: Goldstein, R.J. (ed.) *Fluid Mechanics Measurements*. 2nd edition. London, Taylor and Francis. pp. 575–648.

Holman, J.P. (1989) *Experimental Methods for Engineers.* 5th Edition. New York, McGraw-Hill Book Company. 199–226.

Hopkinson, C., Crasto, N., Marsh, P., Forbes, D., & Lesack, L. (2011) Investigating the spatial distribution of water levels in the Mackenzie Delta using airborne LiDAR. *Hydrolog Process,* 25(19), 2995–3011.

Horner-Devine A. (2009) The bulge circulation in the Columbia River plume. *Cont Shelf Res, 29* (1), 234–251.

Hufnagel, K. & Schewe, G. (2007) *Force and moment measurement.* In: Tropea, C., Yarin, A.L., & Foss, J.F. (eds.) *Springer Handbook of Experimental Fluid Mechanics.* Berlin, Springer. pp. 563–618.

Hughes, S.A. (1993) *Physical models and laboratory techniques in coastal engineering.* Advanced series on Ocean Engineering, Vol. 7. Singapore and River Edge, NJ, World Scientific.

Hygelund, B. & Manga, M. (2003) Field measurements of drag coefficients for model large woody debris. *Geomorphology, 51,* 175–185.

Jalonen, J. & Järvelä, J. (2014) Estimation of drag forces caused by natural woody vegetation of different scales. *J Hydrodyn Ser B, 26(4),* 608–623.

Kawanisi, K. (2004) Structure of turbulent flow in a shallow tidal estuary. *J Hydraul Eng, 130(4),* 360–370.

Kim, S.C., Friedrichs, C.T., Maa, J.P.Y., & Wright, L.D (2000) Estimating bottom stress in a tidal boundary layer from acoustic Doppler velocimeter data. *J Hydraul Eng, 126(6),* 399–406.

Klewicki, J.C., Saric, W.S., Marusic, I., & Eaton, J.K. (2007) *Wall-bounded flows.* In: Tropea, C., Yarin, A.L., & Foss, J.F. (eds.) *Springer Handbook of Experimental Fluid Mechanics.* Berlin, Springer. pp. 871–908.

Koch, W. (1994) Ermittlung lokaler Sohlschubspannungen in Fließgewässern mittels FST-Halbkugeln. Diplomarbeit, Institut für Wasserbau und Kulturtechnik, Universität Karlsruhe (TH), Januar 1994.

Kothyari, U.C., Hashimoto, H., & Hayashi, K. (2009) Drag coefficient of unsubmerged rigid vegetation stems in open channel flows. *J Hydraul Res, 47(6),* 691–699.

Lawrence, G.A. (1993) The hydraulics of steady 2-layer flow over a fixed obstacle. *J Fluid Mech,* 254, 605–633.

Liu, T., Campbell, B.T., Burns, S.P., & Sullivan, J.P. (1997) Temperature- and pressure-sensitive lu-minescent paints in aerodynamics. *Appl Mech Rev, 50,* 227–246.

Malacic, V. (2005) Initial spread of an effluent and the overturning length scale near an underwater source in the northern Adriatic. *J Mar Systems, 55*(1–2), 47–66.

McKeon, B.J. & Engler, R.H. (2007) Pressure measurement systems. In: Tropea, C., Yarin, A.L., & Foss, J.F. (eds.) *Springer Handbook of Experimental Fluid Mechanics.* Berlin, Springer. pp. 179–214.

McLachlan, B.G. & Bell, J.H. (1995) Pressure-sensitive paint in aerodynamic testing, *Exp Therm Fluid Sci, 10*(4), 470–485.

Millot, C. (2009) Another description of the Mediterranean Sea outflow. *Prog Oceanogr, 82*(2), 101–124.

Modarress, D., Svitek, P., Modarress, K., & Wislon, D. (2006) Micro-optical sensors for boundary layer flow studies. *Proceedings of the FEDSM2006, July 17-20, Miami, USA.* pp. 1037–1044.

Molinas, A., Kheireldin, K., & Wu, B. (1998) Shear stress around vertical wall abutments. *J Hydraul Eng, 124*(8), 822–830.

Musumeci, R.E., Marletta, V., Ando, B., Baglio, S., & Foti, E. (2014) Measurement of wave near-bed velocity and bottom shear stress by Ferrofluids. *IEEE Trans Instrum Meas, 99,* doi:10.1109/TIM.2014.2359521.

Musumeci, R.E., Marletta, V., Ando, B., Baglio, S., & Foti, E. (2015) Ferrofluid measurements of bottom velocities and shear stresses. *J Hydrodyn Ser B*, *27*(1), 150–158.

Nakayama, A. (1999) *Introduction to Fluid Mechanics*. Oxford, Butterworth-Heinemann.

Naughton, J.W. & Hind, M.D. (2013) Multi-image oil-film interferometry skin friction measurements. *Meas Sci Technol*, *24*, doi:10.1088/0957-0233/24/12/124003.

Naughton, J.W. & Sheplak, M. (2002) Modern developments in shear-stress measurement. *Prog Aerosp Sci*, *38*(6–7), 515–570.

Nekouee, N., Roberts, P.J.W., Schwab, D.J., & McCormick, M.J. (2013) Classification of buoyant river plumes from large aspect ratio channels. *J Hydraul Eng*, *139*(3), 269–309.

Novak, P., Guinot, V., Jeffrey, A., Reeve, D.E (2010) *Hydraulic modelling – an introduction*. London, Spon Press.

Oplatka, M. (1998) Stabilität von Weidenverbauungen an Flussufern (Stability of bank fixations by willow trees). *VAW-report 166*, Versuchsanstalt für Wasserbau, Hydrologie und Glaziologie (VAW), ETH Zürich.

Pan, F. & Nichols, J. (2013) Remote sensing of river stage using the cross-sectional inundation area-river stage relationship (IARSR) constructed from digital elevation model data. *Hydrolog Process*, *27*(25), 3596–3606.

Preston, J.H. (1954) The determination of turbulent skin friction by means of Pitot tubes. *J Roy Aeronaut Soc*, *58*(2), 109–121.

Rankin, K.L. & Hires, R.I. (2000) Laboratory measurement of bottom shear stress on a movable bed. *J Geophys Res Oceans*, *105*(C7), 17011–17019.

Reda, D.C., Wilder, M.C., Farina, D.J., & Zilliac, G. (1997) New methodology for the measurement of surface shear stress vector distributions. *AIAA Journal*, *35*(4), 608–614.

Rowiński, P.M., Czernuszenko, W., & Pretre, J.M. (2000) Time dependent shear velocities in channel routing. *Hydrolog Sci J*, *45*(6), 881–895.

Rowiński, P.M., Aberle, J., & Mazurczyk, A. (2005) Shear velocity estimation in hydraulic research. *Acta Geophys Pol*, *53*(4), 567–583.

Sand-Jensen, K. (2003) Drag and reconfiguration of freshwater macrophytes. *Freshwat Biol*, *48*, 271–283.

Schoneboom, T., Aberle, J., Wilson, C.A.M.E., & Dittrich, A. (2008) Drag force measurements of vegetation elements. In: *Proceedings of the ICHE 2008, Nagoya, Japan, Papers on CDROM*.

Seelam, J.K., Guard, P.A., & Baldock, T.E. (2011) Measurement and modeling of bed shear stress under solitary waves. *Coast Eng*, *58*(9), 937–947.

Shields, F.D. & Alonso, C.V. (2012) Assessment of flow forces on large wood in rivers. *Water Resour Res*, *48*(W04516), doi:10.1029/2011WR011547.

Statzner B. & Müller R. (1989) Standard hemispheres as indicators of flow characteristics in lotic benthic research. *Freshwat Biol*, *21*, 445–446.

Statzner, B., Lamoroux, N., Nikora, V., & Sagnes, P. (2006) The debate about drag and reconfiguration of freshwater macrophytes: comparing results obtained by three recently discussed approaches. *Freshwat Biol*, *51*(11), 2173–2183.

Stewart, R.H. (2008) Introduction to Physical Oceanography, self-published, http://www.colorado.edu/oclab/sites/default/files/attached-files/stewart_textbook.pdf [accessed 10th February 2017].

Tavoularis, S. (2005) *Measurement in Fluid Mechanics*. Cambdrige, Cambdrige University Press.

Tedford, T., Pieters, R., & Lawrence, G.A. (2009) Symmetric Holmboe instabilities in a laboratory exchange flow. *J Fluid Mech*, *636*, 137–153.

Thompson, C.E.L., Amos, C.L., Jones, T.E.R., & Chaplin, J. (2003a) The manifestation of fluid-transmitted bed shear stress in a smooth Annular flumes – A comparison of methods. *J Coast. Res*, *19*(4), 1094–1103.

Thompson, A.M., Wilson, B.N., & Hustrulid, T. (2003b) Instrumentation to measure drag on idealized vegetal elements in overland flow. *Trans ASAE*, *46*(2), 295–302.

Thornton, C.I, Cox, A.L., & Sclafani, P. (2008) Preston tube calibration. Colorado State University, Engineering Research Center, Fort Collins, Colorado, July 2008. [Online] Available from https://www.usbr.gov/research/projects/download_product.cfm?id=1184 [Accessed 9th December 2016].

Tinoco, R.O. & Cowen, E.A. (2013) The direct and indirect measurement of boundary stress and drag on individual and complex arrays of elements. *Exp Fluids*, *54*(4), 1–16.

Tropea, C., Yarin, A.L., & Foss, J. (eds.) (2007) *Springer Handbook of Experimental Fluid Mechanics*. Berlin, Springer. ISBN 978-3-540-25141-5.

United States Department of the Interior, Bureau of Reclamation (USBR) (2001) Water measurement manual. [Online] Available from http://www.usbr.gov/tsc/techreferences/mands/wmm/ [Accessed 13th October 2016].

Vanoni, V.A. & Brooks, N.H. (1957) Laboratory studies of the roughness and suspended load of alluvial streams. Pasadena, Sedimentation Laboratory, California Institute of Technology.

Wakeham, W.A., Assael, M.J., Marmur, A., De Coninck, J., Blake, T.D., Theron, S.A., & Zussman, E. (2007) *Material properties: Measurement and Data*. In: Tropea, C., Yarin, A. L., & Foss, J.F. (eds.) *Springer Handbook of Experimental Fluid Mechanics*. Berlin, Springer. pp. 85–179.

Wang, X., Bockelmann-Evans, B., & Liang, D. (2007) Examination of FST-hemispheres for evaluating boundary shear stress in streams. *Int J River Basin Manag*, *5*(2), 155–163.

Ward, B., Fristedt, T., Callaghan, A.H., Sutherland, G., Sanchez, X., Vialard, J., & Doeschate, A. (2014) The Air-Sea Interaction Profiler (ASIP): An autonomous upwardly rising profiler for microstructure measurements in the upper ocean. *J Atmosph Ocean Tech*, *31*(10), 2246–2267.

Whittaker, P., Wilson, C., Aberle, J., Rauch, H.P., & Xavier, P. (2013) A drag force model to incorporate the reconfiguration of full-scale riparian trees under hydrodynamic loading. *J Hydraul Res*, *51*(5), 569–580.

Wilson, C.A.M.E., Hoyt, J., & Schnauder, I. (2008) Impact of foliage on the drag force of vegetation in aquatic flows. *J Hydraul Eng*, *134*(7), 885–891.

Wuest, W. (1994) Messung hydrostatischer und hydrodynamischer Größen. In: Profos, P. & Pfeifer, T. (Hrsg) *Handbuch der industriellen Meßtechnik*. 6. Auflage. München, R. Oldenbourg Verlag. pp. 771–794.

Zhu, D. & Lawrence, G. (2000) Hydraulics of exchange flows. *J Hydraul Eng*, *126*(12), 921–928.

Chapter 7

Discharge

7.1 INTRODUCTION

Discharge, Q, is the volume of fluid passing through a flow cross-section per unit time. Discharge measurement methods and instruments have a long tradition of development, and continue to advance through adoption of new configurations and technologies aimed at increasing measurement efficiency and accuracy.

Laboratory and field hydraulics concern open-channel (also free-surface) and closed-conduit flows. The measurement of discharge in natural streams (a.k.a. streamflow) is of close interest also for hydrology and water-resource management. While the operation and the configuration principles for discharge-measurement instruments for open channels differ widely for laboratory and field conditions, those for pipe flows are quite similar. Detailed overviews of conventional techniques for measuring discharges in closed-conduit and free-surface flow situations are provided by Tavoularis (2005), Baker (2000), and Goldstein (1996), among many references. Therefore, this chapter only briefly reviews the techniques. More attention is given to new instruments and methods for measuring discharges in natural streams, as this area has experienced significant recent progress and continues to develop at a fast pace.

Practical approaches for discharge estimation through measurement are based on a wide variety of principles and methods. The most accurate method uses the volumetric principle whereby water passing through a flow cross section is collected over a period of time. The method is suitable for laboratory settings and it is used as a primary standard for calibration of discharge instrumentation (Section 5.2.2 in Volume I). The vast majority of instruments and methods for measuring discharge in open channels use the mass-continuity equation, $Q = UA$; where U is the cross-sectional average velocity (a.k.a. bulk velocity) and A is the flow cross section area (the area of the flow perpendicular to its mean flow direction). The average velocity, U, can be obtained by integrating velocities measured over the flow cross section. The cross-section area, A, is typically obtained by surveying/measuring the channel boundaries and relating the area to the flow depth (i.e., area rating).

Discharge measurements can be discrete (a single measurement acquired at a given instant) or continuous (measurements taken over time, typically at a pre-established sampling frequency). Discrete discharge measurements characterize the status of the flow in a natural channel or check the flow rate during an experiment. Continuous discharge measurements are used to capture short-term (unsteady or periodic flows) or long-term (of the order of hours, days, seasons or years) variations of flow rates.

Continuous measurements are often conducted with instruments equipped to automatically collect and transmit the data to the user.

Measurements in laboratory conditions use a wide variety of instruments (*e.g.*, Tavoularis, 2005). None of these instruments measure discharge directly. Rather, they measure one or more fluid flow properties (pressure, conductivity, or temperature) or flow characteristics (*e.g.*, the water head on a weir) at the measurement location. These measurements are subsequently related to discharge values measured simultaneously with an alternative method of higher accuracy (notably the volumetric method). The process of obtaining the relationship between the probe response and the discharge values is defined as calibration. Calibration can be conducted by the instrument producer (factory calibration) or by the user. Most discharge measurements in laboratories are determined using a factory-set instrument and its associated calibration relationship (NIST, 2006). These instruments are used where flow conditions are similar to the calibration conditions. If the flow conditions considerably differ, the measured flow needs to be conditioned prior to measurement (Section 4.3.3.2 in Volume I).

Field measurements use a wide range of instruments and methods for measuring discharge. However, there are no widely-recognized reference instruments or methods for verifying the accuracy of the discharges acquired in natural-scale open-channel flows. Discrete values of discharge can be obtained by measuring one or more flow variables and subsequently combining them through analytical methods to estimate discharge. Contemporary instruments often use the same housing for probes to measure different flow variables. This arrangement enables discharge measurement from one deployment in the stream (*e.g.*, the Acoustic Doppler Current Profiler described in Section 3.3, Volume II). Alternatively, continuous discharge can be obtained indirectly from measurements of water level, free-surface slope, or velocities in conjunction with governing laws for open channels (conservation of mass, energy and momentum) or with empirical relationships that link discharge to flow variables measured at one or more locations in the stream. These relationships (termed ratings) are developed through calibration measurements acquired over a wide range of flows with alternative measurements. The established relationships are valid only for the site where the measurements were acquired. As conditions at a site can change due to natural causes (*e.g.*, changes in the stream morphology or vegetation), periodic field surveys may be needed to verify the reliability of the discharge estimates over time.

This chapter groups differing types of instruments and methods for discharge estimation. Table 7.1.1 lists instruments commonly used for discrete or continuous monitoring of discharge in laboratory conditions. They are typically referred to as flowmeters. Some of these instruments are also used in field conditions (*e.g.*, weirs, acoustic travel-time discharge meters). Table 7.1.2 lists instruments and methods for discrete measurement of discharge in field conditions. This latter category of instruments is advancing rapidly, taking advantage of the capabilities of new acoustic-, laser-, and image-based techniques to measure non-intrusively. Of particular relevance are the acoustic methods. They have revolutionized measurements through superior efficiency, safety and accuracy compared with the conventional methods (Muste *et al.*, 2007; Mueller *et al.*, 2013).

Table 7.1.3 lists methods frequently used for continuous estimation of stream discharges. For small streams, custom-designed methods are set at the location of the monitoring site (*e.g.*, Boiten, 2000; Herschy, 2009). With the advancements in instrumentation technology, continuous monitoring of discharge is conveniently attained by

Table 7.1.1 Selected discharge measurement instruments used in laboratories (adapted from Tavoularis, 2005). Sections mentioned in column one refer to Experimental Hydraulics, Volume II.

Instrument or method		Flow type	Intrusive	Measurement type
Volumetric				
Weighers		Pipe/Channel	No	Discrete
Nutating disks		Pipe/Channel	Yes	Continuous
Gear and lobed impellers		Pipe	Yes	Continuous
Differential energy/momentum				
Orifices (Section 7.2.3)		Pipe	Yes	Continuous
Venturi (Section 7.2.3)		Pipe/Channel	Yes	Continuous
Nozzles (Section 7.2.3)		Pipe	Yes	Continuous
Weirs (Section 7.2.1)		Channel	Yes	Continuous
Sluice gates (Section 7.2.2)		Channel	Yes	Continuous

Table 7.1.1 (Cont.)

Instrument or method		Flow type	Intrusive	Measurement type
Variable area		Pipe	Yes	Continuous

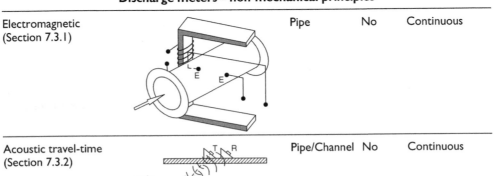

Discharge meters – non-mechanical principles				
Electromagnetic (Section 7.3.1)		Pipe	No	Continuous
Acoustic travel-time (Section 7.3.2)		Pipe/Channel	No	Continuous

Table 7.1.2 Selected discharge instruments and methods for discrete measurements used in fieldwork (sections mentioned in the second column refer to Experimental Hydraulics, Volume II)

Instrument or method		Flow type	Intrusive	Measurement type
Dilution methods	Section 3.10	Channel	No	Discrete
Floats	Section 3.9	Channel	No	Discrete
Mid section	Section 7.4.1	Channel	Yes	Discrete
Mean section	Section 7.4.1	Channel	Yes	Discrete
Velocity depth	Section 2.3, Herschy (2009)	Channel	Yes	Discrete
Velocity contour	Section 2.3, Herschy (2009)	Channel	Yes	Discrete
ADCP transects	Section 7.4.2	Channel	No	Discrete
Acoustic tomography	Section 3.4	Channel	No	Discrete/Continuous
LSPIV	Section 3.7.4	Channel	No	Discrete

simultaneous measurements of several variables (*e.g.*, stage and velocity). These advances extend our capabilities to monitor unsteady and non-uniform flows; these flows have been difficult to quantify with conventional methods (Sections 7.5.1 and 7.6 below). Challenges still remain in measuring velocities and their spatial variation for complex flow conditions; *e.g.*, the presence of secondary flows, large turbulence structures, temporal variations in the flow during the acquisition of the data, and the presence of disturbances such as those caused by high concentrations of sediment. These flow complexities are site specific and typically associated with larger flows events as discussed below.

Table 7.1.3 Selected discharge instruments and methods for continuous measurements used in fieldwork (sections mentioned in the second column refer to Experimental Hydraulics, Volume II)

Instrument or method		Flow type	Intrusive	Measurement type
	Rating methods based on water level measurements			
Stage discharge	Section 7.5.1	Channel	No	Continuous
Slope-area	Section 7.5.1	Channel	No	Discrete/ Continuous
Stage-fall	Section 7.5.2	Channel	No	Continuous
	Rating methods based on water level and velocity measurements			
Index velocity	Section 7.6.1	Channel	No	Continuous
	Rating methods based on hydrodynamic models			
Velocity distribution	Section 7.6.2	Channel	No	Continuous
Hybrid	Section 7.6.3	Channel	No	Continuous
	Flow measuring structures			
Broad-crested weir	Section 7.2.1	Channel	Yes	Continuous
Flumes	Chapter 7, WMO (2010a)	Channel	Yes	Continuous

Establishing reliable relationships for continuous discharge measurement in field conditions is a challenging task for both normal as well as large flows. Challenges arise because estimation of discharge depends not only on the instruments and methods used for the measurement but also on the site characteristics including accessibility. The most challenging measurement situations occur in natural channels where the flow is affected by changes in boundary geometry or variable backwater (*e.g.*, due to the presence of structures in the stream or of a confluence). Challenges also occur where a substantial portion of the discharge is conveyed by floodplains and/or the flow becomes unsteady (*e.g.*, propagation of a natural flood wave). These special situations require the development of customized empirical or analytical methods for building the relationships for continuous monitoring. Streams affected by variable backwater are sometimes approached by empirical stage-fall-discharge rating curves (Rantz *et al.*, 1982b; Herschy, 2009) whereas unsteady flow situations are corrected using analytical methods (Rantz *et al.*, 1982b; Schmidt, 2002). A flow situation that potentially combines variable backwater and unsteady effects occurs in tidal channels. An additional complexity in such channels is flow stratification (freshwater over seawater) that is troublesome for acoustic-based instruments. These complex measurement situations are beyond the scope of the present handbook and are only briefly mentioned in the present chapter.

This chapter provides only brief reviews of the well-documented conventional methods for discrete and continuous measurements of discharge used primarily in laboratories (Sections 7.2, 7.3.1, and 7.3.2) and fieldwork (Sections 7.3.3, 7.4.1, 7.5.1). Particular attention is devoted to contemporary methods for discharge monitoring in fieldwork (Sections 7.4.2 and 7.6). The contemporary methods often entail a rejuvenation of earlier measurement methods. The rejuvenation is facilitated by new sensing technology (acoustic-, image-based) and combination of the field measurements with numerical models for estimating discharge. This chapter ends with practical suggestions regarding discharge measurements at field sites.

7.2 INTRUSIVE FLOWMETERS

Table 7.1.1 lists instruments for measuring water (and for some of the instruments air) discharge in hydraulic laboratories. Some of the instruments are used also for field measurements. This section describes the operation principle and configuration of selected instruments listed in the table. Most of the instruments involve the principle of conservation of energy in order to estimate discharge; e.g., devices using a flow contraction. Other devices, such as broad-crested weirs, are designed to induce critical flow depth from which discharge can be calculated directly. In general, a calibrated discharge coefficient is required to account for energy loss as flow passes through the measurement device. Therefore these instruments require calibration after their construction.

7.2.1 Weirs

Many discharge weir and flume designs exist, and they can be classified as broad-crested weirs and sharp-crested weirs.

Sharp-crested weirs utilize conservation of energy to estimate water discharge as a function of the upstream flow depth above the weir crest (Δh) (see Table 7.1.1). For a rectangular sharp-crested weir, the following equation applies:

$$Q_{ideal} = \frac{2}{3} L \sqrt{2g} \left[\left(\Delta h + \frac{U_1^2}{2g} \right)^{3/2} - \left(\frac{U_1^2}{2g} \right)^{3/2} \right] \qquad (7.2.1)$$

where L is length of the weir crest, U_1 is the upstream flow velocity, and g is gravitational acceleration. In practice, a discharge coefficient (C_d) is required to account for energy loss, and it is often assumed that upstream flow velocity is negligible. Accordingly,

$$Q = C_d \frac{2}{3} L \sqrt{2g} \, (\Delta h)^{1.5} = C_w L \, (\Delta h)^{1.5} \qquad (7.2.2)$$

The weir coefficient (C_w) is typically on the order of 1.84 in SI units (Daugherty et al., 1985). It is necessary to measure Δh sufficiently upstream so that it is unaffected by the drawdown of the water surface near the weir, but not so far upstream that energy loss between the measurement point and the weir becomes significant.

To measure low magnitudes of water discharge, it can be advantageous to use a triangular (notched) sharp-crested weir. Compared to a rectangular weir, a triangular weir produces greater head differential for smaller flow increments and ensures that the flow fully separates from the weir crest, albeit at the expense of greater head loss. The corresponding discharge formula is:

$$Q = C_d \left(\frac{8}{15} \right) \sqrt{2g} \, tan \left(\frac{\theta}{2} \right) (\Delta h)^{2.5} \qquad (7.2.3)$$

where θ is the angle of the weir notch. The discharge coefficient C_d depends on θ and Δh, thus calibration is recommended, but for $\theta = 90°$, C_d is approximately 0.585 (Daugherty et al., 1985).

Broad-crested weirs are common. A special type of broad-crested weir is obtained by constricting the flow with a hump located on the channel bed and, or, narrowing the channel to induce critical flow conditions. Examples of such weirs include Parshall flumes and, at larger scale, calibrated Ogee spillways (Herschy, 2009). Given that critical depth occurs over the weir, specific-energy theory can be used to derive C_d analytically (Henderson, 1966). For example, C_w in Equation (7.2.2) for critical flow over a broad-crested weir is 1.70, assuming negligible approach flow velocity and no energy loss over the weir.

7.2.2 Sluice gates

Energy principles can also be employed to calculate water discharge through sluice gates and are well-documented in multiple resources (Boiten, 2000; Herschy, 2009). The discharge equation through a sluice gate can be written as (see Table 7.1.1):

$$Q = C_L C_C h_g h_1 L \sqrt{\frac{2g}{h_1 + C_c h_g}} \qquad (7.2.4)$$

where h_1 is the flow depth upstream of the gate, h_g is the gate opening height, L is the width of the gate, C_c is the contraction coefficient for the flow past the gate (approximately 0.61), and C_L is a loss coefficient (often assumed equal to 1.0).

7.2.3 Differential pressure flow meters: orifice-plate and Venturi meters

Energy principles can also be applied in pipe flows, wherein measurement of the change in pressure through a constriction is related to the velocity of flow through the constriction (Tavoularis, 2005). Orifice plate and Venturi meters respectively employ sharp and gradual reductions in pipe cross-sectional area. An orifice plate results in more energy loss, but is relatively easy to install. The discharge equation for an orifice plate is (Daugherty *et al.*, 1985):

$$Q = A_2 U_2 = A_2 \frac{C_v}{\sqrt{1 - \left(A_2/A_1\right)^2}} \sqrt{2g(p_1 - p_2)} = A_2 C_d \sqrt{2g(p_1 - p_2)} \quad (7.2.5)$$

Positions 1 and 2 are upstream and at the constriction, respectively, as illustrated in Table 7.1.1. A is cross sectional area, p is the pressure expressed as piezometric head (length dimensions), C_v is a coefficient to account for frictional losses, and C_d is a discharge coefficient. For an orifice, $A_2 = C_c A_o$, where A_o is the orifice cross-sectional area and C_c is the contraction coefficient. The loss coefficient C_v depends on pipe size, constriction ratio, and Reynolds number, but is about 0.98 in a Venturi meter. The energy-loss coefficient associated with an orifice plate should be calibrated.

7.3 NON-INTRUSIVE FLOWMETERS

The instruments described in this section estimate discharge non-intrusively (Table 7.1.1). They were originally designed for laboratory measurements. Subsequently, the instruments have been applied to field conditions. These instruments indirectly measure flow velocity over the flow cross section (bulk flow velocity), include measurement of the flow cross-section area, and involve the flow-continuity principle to estimate discharges. Given that the inferred flow velocity is only sampled from a portion of the cross section, they need some form of velocity indexing to estimate section-averaged velocity from the measured velocity (Section 7.6). Manufacturers typically calibrate instruments for laboratory use, and users may have to additionally calibrate instruments when using them at field sites.

7.3.1 Electromagnetic discharge meters

Electromagnetic, water-discharge meters measure velocity using Faraday's principle. In open channel flows, electromagnetic current meters provide velocity measurements at a point within the cross section (Section 3.5.2.2, Volume II). In pipes, however, electromagnetic discharge meters encompass the entire pipe and thereby induce a magnetic field over the flow cross section, and thus the measured velocity is the section-averaged velocity (Tavoularis, 2005). Typical discharge errors are about 1%.

7.3.2 Acoustic travel-time discharge meters

The operating principle of these water-discharge instruments is based on the time-of-flight (travel time) taken by a high-frequency (acoustic) pressure wave between an emitter and a receptor of the wave pulse, as illustrated in Table 7.1.1. The receptor is typically a form of piezometer. The wave pulse moving downstream or upstream in a current will propagate faster or slower, respectively. Using two transducers fixed upstream and downstream and on opposite sides of the channel, acoustic pulses are sent against and with the current. With the speed of sound, distance between the transmitter known, and the detection of the travel time for the pulse pairs, the velocity of the current can be obtained (Tavoularis, 2005). Using the so-obtained velocity component as an index velocity, a calibration curve relating the measured and independently estimated bulk channel velocity is constructed for a range of flows.

Commercial systems for this instrument are available for pipe flows, in which transducers are typically attached to the outer pipe walls. Knowledge of the pipe wall thickness and materials is required, as this influences the acoustic transmission. It is possible to employ such measurements in partially filled pipes, but better results are obtained in full pipes. An extension of the method for use in open channels in the field is acoustic tomography (Section 3.4, Volume II). Measurement of discharge using acoustic tomography in open channels is still being developed, but the first commercial systems may soon be available.

7.4 DISCRETE STREAMFLOW MEASUREMENTS BASED ON VELOCITY INTEGRATION

With the exception of dilution, tracer methods, and acoustic tomography, which yield reach-averaged velocity (Chapter 3, Volume II), most discharge measurements in open channels are based on point velocities. For discharge estimation, the point velocity measurements are integrated across the section. Point measurements densely distributed across natural river cross sections are useful not only for accurate discharge estimation, but also for capturing the non-uniform distribution of the velocities in the vertical and transverse directions.

The first velocity point measurements were acquired with paddle wheels as early as the first century B.C. (Frazier, 1974). Conventional point-velocimeters, such as propeller meters (Section 3.5.2.1 in Volume II), have been extensively used to estimate discharges in rivers for centuries. More recently, the advent of acoustic velocimetry using Acoustic Doppler Current Profilers (ADCPs, Sections 7.4.2 and 3.3 of Volume II) has considerably changed the conventional measurement approaches by yielding spatially intensive data collected in a fraction of the time needed by propeller meters.

Streamflow estimation based on point-velocities has been used extensively in natural open channels of all sizes. The methods of integration of the point-velocity datasets have evolved in parallel with the instrumentation development, such that currently there are several alternatives for discharge estimation. These alternatives are configured to accommodate the measurement approach as well as the type of discharge estimation; *i.e.*, discrete measurements (one measurement at a time at any location) or continuous (repeated measurements at fixed points). Given the large variety of instruments and methods, the terminology of the domain is also diverse. Figure 7.4.1a introduces the terminology for the definitions of various depth-related parameters used in this chapter.

7.4.1 Mid-section and mean-section methods

Mid-section and mean-section methods have been extensively used in conjunction with instruments acquiring velocity at local points over a flow cross section. These methods determine discharge by acquiring point velocities and depths at discrete verticals distributed across the section (Herschy, 2009). Verticals may contain multiple point velocities. Pairs of adjacent verticals create panels in which an elementary discharge is calculated, as illustrated in Figures 7.4.1b and 7.4.1c.

In the mid-section method the discharge passing a panel centered on each vertical is calculated using the depth-averaged velocity multiplied by the depth at the measurement location and the panel width (defined as the sum of half the width of the adjacent verticals to the measurement location (Figure 7.4.1b)). The depth-averaged velocity measured in the vertical i is assumed to represent the average velocity through the panel.

In the mean-section method, velocities and depths collected in the verticals defining the panel, *i.e.*, between i and $i+1$, are averaged and multiplied by the distance between adjacent verticals (Figure 7.4.1c). The procedure for discharge estimation is identical for each panel and the total discharge is obtained as the sum of the panel discharges covering the cross section. The discharges at the edges of the cross sections are obtained using the same relationships as for the central panels assuming that the velocities decrease gradually to zero from the last measured point to the bank (Herschy, 2009).

a)

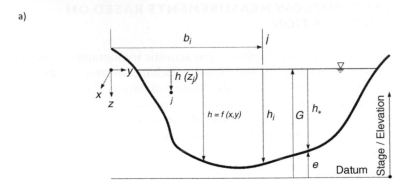

b_i = horizontal distance from reference (initial) point

h = local flow depth; $h = f(x, y)$

h_* = adjusted stage

G = water surface elevation (stage)

e = stage for zero flow in the channel

h_i = depth at vertical (profile) i

$h(z_j)$ = depth at point j in the vertical

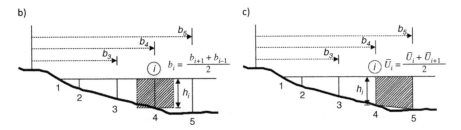

Figure 7.4.1 Schematic of discharge estimation methods: a) definition of terms for natural open channels; b) mid-section; c) mean section. $\overline{U}_i$ = the depth-averaged velocity at vertical i.

Alternative assumptions can be made regarding edge discharge estimation; *e.g.*, Fulford and Sauer (1986). While the differences between the two methods are less than 1%, the mid-section method is more often used than the mean-section method.

Depending upon the degree of non-uniformity of the flow and cross-section bathymetry, velocity measurement at a larger number of verticals and at more points in each vertical yields more robust discharge estimates. The guidelines regarding the density of the point measurements are continuously updated, as they account for a variety of factors. For example, ISO 748 (2007) recommends that the discharge through individual panels should not be more than 5%–10% of the section discharge. Turnipseed and Sauer (2010) recommend 25 to 30 verticals, none of which passing more than 10% of the section discharge. Measurements distributed throughout the vertical profile yield the best estimate of local depth-averaged velocity, $\overline{U}_i$. If only a single point velocity measurement is obtained to estimate $\overline{U}_i$, then usually it is assumed that the log law of the wall holds throughout the vertical profile, in which case the point velocity, $\bar{u}_{i,j}$, (the time-averaged velocity at point j at a given depth in the vertical i) at a relative depth of 0.4 of flow depth above the bed is taken to equal $\overline{U}_i$. Standard procedures for conducting such measurements, computing discharge, and calculating uncertainty of discharge estimates are available in Turnipseed and Sauer (2010) and WMO (2010a, 2010b).

7.4.2 ADCP transects

The Acoustic Doppler Current Profiler (ADCP) described in Section 3.3, Volume II was first used for measuring riverine discharge in the late 1980s. ADCP use for river-flow measurements has profoundly changed the way that velocities and discharge are collected in streams and man-made channels. One method for determining discharge with an ADCP is by individually collecting velocity profiles at fixed points across a section (the "section-by-section" method). The mid-section or mean section integration methods presented in Section 7.4.1 can be subsequently employed to estimate discharge. Most often, however, an ADCP is deployed from a moving boat or platform (the "moving-boat" method). In this approach, the ADCP collects velocities with high-spatial density that are subsequently converted to flow fluxes and integrated across the stream to estimate the discharge. Mueller *et al.* (2013) provide detailed guidance on procedures for conducting moving-boat ADCP measurements.

Section 3.3, Volume II describes the successive transformations applied by the instrument software to the ADCP measured velocities to obtain the vertical profile of velocity. In the first phase of these transformations, the water velocity in individual bins, $\vec{V}_w$, and the velocity at which the instrument is moving with respect to the river bed, $\vec{V}_b$, are expressed in instrument coordinates. Note that the water velocity vector, $\vec{V}_w$, in this section is obtained through summation of the three velocity vectors defined in Section 3.3.2.3.3, Volume II. Velocities then are transformed to Earth coordinates using a calibrated internal compass and an estimate of local magnetic declination (Section 3.3.3.5, Volume II). If boat velocity is measured using ADCP bottom tracking, then both water and boat velocities are measured in the same coordinate system. However, near-bed sediment transport can bias bottom velocities (Rennie *et al.*, 2002), necessitating either "moving bottom" bias correction procedures or use of Global Positioning System (GPS) data to measure boat velocities. GPS boat velocities are measured directly in Earth coordinates; accurate compass calibration and local magnetic declination are essential to obtain unbiased velocity and discharge measurements (Section 3.3.3.5 in Volume II).

Calculation of discharge using the moving-boat approach is based on the estimation of the elemental flow passing through an individual bin, Q_{bin}, that is the portion of the cross-sectional area defined as ds in the vertical plane along the ADCP's path between two consecutive pings, as illustrated in Figure 7.4.2a. Following Christensen and Herrick (1982) (similar equations, integrated across the section, were developed in Gordon, 1989), the discharge through an individual bin illustrated in Figure 7.4.2 is:

$$Q_{bin} = \left(\vec{V}_w \cdot \vec{n} \right) \left| \vec{V}_b \right| dz\, dt \qquad (7.4.1)$$

where $\vec{V}_w$ is the water velocity vector measured within a depth cell (bin), $\vec{V}_b$ is the boat velocity vector, $\vec{n}$ is a unit vector normal outward to the differential area ds, dz is the

a)

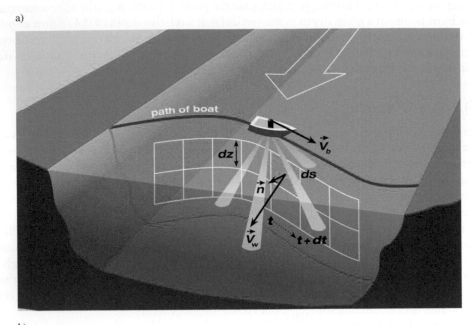

b)

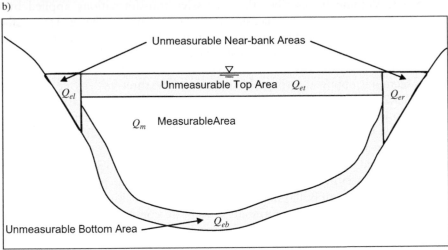

Figure 7.4.2 Estimation of the cross-sectional discharge from ADCP transects: a) variables involved in the calculation of the elemental (Q_{bin}) discharge; b) mapping of the ADCP-measured and unmeasured areas over the cross section.

height of the ADCP bin (*i.e.*, the vertical resolution of the instrument), and dt is the time between two consecutive pings (profiles). An equivalent form of Equation (7.4.1) can be written as:

$$Q_{bin} = (V_{wx}V_{by} - V_{wy}V_{bx})dz\,dt \tag{7.4.2}$$

As Equation (7.4.2) accounts for the directionality of the water flux, the moving ADCP measurement method gives the net discharge measured in a bin. Consequently, for the moving-boat method it is not necessary for the transect line to be perfectly straight (*i.e.*, orthogonal) to the river banks.

Software provided with the ADCP computes the total discharge as soon as the measurement is finalized. The integration of all discharges in elemental bins acquired using the ADCP provides the total discharge sensed directly by the instrument over the cross section. This quantity, called *measured discharge* and labeled Q_m in Figure 7.4.2b, is obtained by summing all the bins between the lower and upper limits of the measured water column across the ADCP transect.

The total discharge in ADCP transect measurements is estimated as the summation of the following sub-components (Figure 7.4.2b): (i) the discharge computed from quantities directly measured by the ADCP and external devices, Q_m; (ii) the discharge estimated by the ADCP post-processing algorithms when part of the directly measured data are missing, Q_{em}; (iii) the discharge estimated in the unmeasurable area near the free surface, Q_{et}; (iv) the discharge estimated in the unmeasurable area near the channel bed, Q_{eb}; and, (v) the discharge estimated in unmeasured areas near the left and right channel edges, Q_{el} and Q_{er}, respectively,

$$Q_t = Q_m + Q_{em} + Q_{et} + Q_{eb} + Q_{el} + Q_{er} \qquad (7.4.3)$$

Poor data quality (cf. Simpson, 2001) can result in missing velocity values in individual cells (bins) or entire verticals (ensembles) that must subsequently be interpolated. Furthermore, velocity in unmeasured portions of transects (too close to the water surface, channel edges, and channel bed) must be estimated (Section 3.3.2.3.2 in Volume II). Velocities at the top and bottom of each vertical profile are obtained by extrapolating an assumed or fitted velocity profile for the streamwise velocity, most often assumed to follow a 1/6-power law, although the power exponent should be calibrated. Discharge at channel edges, Q_{el} and Q_{er}, is obtained from the depth-averaged velocity, $\overline{U}_e$, measured in the ensembles nearest the edge, the channel depth at those ensembles, h_e, the edge area (based on distance to the edge, b_e), and an edge coefficient (C_e) that accounts for the cross-sectional shape of the channel edge:

$$Q_e = C_e \overline{U}_e b_e h_e \qquad (7.4.4)$$

ADCP discharge measurements acquired from moving boats are typically repeated several times by crossing the stream in both directions (reciprocal transects). Subsequent averaging of the repeated measurements can improve the quality of the measurement and minimize directional bias. While not uniformly accepted, some measurement protocols suggest four to six reciprocal transects with observed variance of less than 5% (current United States Geological Survey guidelines recommend at least two reciprocal transects and a total measurement exposure time of at least 720 s according to Mueller *et al.*, 2013). Moving-boat ADCP discharge measurements depend upon many factors (environmental, instrument, operator), and the refinement of the measurement protocols in conjunction with uncertainty estimation remains an area of active research (Chapter 7, Volume I, as well as Gonzalez-Castro & Muste, 2007; Le Coz *et al.*, 2016; Moore *et al.*, 2016; Mueller *et al.*, 2016).

7.5 CONTINUOUS STREAMFLOW MONITORING USING STAGE MEASUREMENTS

Traditional hydrometric guidelines group the methods for continuous monitoring of discharge in terms of the independent flow variables directly measured for the estimation: free-surface water level (or stage), free-surface slope, and velocity (*e.g.*, Rantz, 1982a, 1982b; Herschy, 2009; WMO, 2010b). Given that the first two groups of methods use the continuous and direct measurement of stage, this section discusses both methods. Velocity-based methods for continuous discharge estimation are subsequently discussed in Section 7.6.

Stage-based methods were historically developed first and are still common in continuous stream monitoring. Water surface level (or stage) is measured continuously at one location for the stage-discharge method (SDM) or at multiple (two or more) locations for the continuous slope-area method (CSAM). Stage-based discharge estimation methods relate the water-surface level above a local datum with the corresponding discharge and/or slope of the channel through an empirical relationship (rating). Ratings are established from a series of calibration measurements collected for a wide range of flows using standardized methods (*e.g.*, WMO, 2010b). The stage is determined with a staff gauge, pressure or radar-based sensor (see Section 6.2 in Volume II). Options for direct discharge measurements in natural channels are discussed in Section 7.4.

The most common approach for stage-based, discharge estimation is the simple stage-discharge method (SDM). Additional methods from this category have been developed to complement the simple SDM capabilities for situations where direct measurements are difficult or impossible to acquire (flood events) or when the accuracy of the SDM is limited due to complexities in the flow (*e.g.*, non-uniform or unsteady flows). The methods tackling the first concern aim to extrapolate SDM in the area of high flows combining discrete measurements with analytical relationships for open-channel flows (*e.g.*, the slope-area and conveyance methods). For complex flow situations, the less accurate SDM estimates are adjusted using additional measured variables that account for non-uniformity (*e.g.*, slope in the stage-fall-discharge method) or corrected using analytical relationships for unsteady flows (review of the multiple methods is provided by Schmidt, 2002). Given that most SDM-related methods are extensively documented in the hydrometric literature (*e.g.*, Rantz, 1982b; WMO, 2010b), they are only briefly mentioned here. This section provides a short introduction of the extensively-used SDM method (Section 7.5.1) along with emerging methods used to measure discharge in complex flow situations (Section 7.5.2).

7.5.1 Simple stage-discharge ratings

Most practical problems related to open-channel flows describe flow characteristics using one-dimensional flow equations applied to channel segments (Section 2.4.3.4 in Volume I). These governing equations relate changes in depth and velocity along the stream with the type of energy losses in the channel, i.e., distributed or local. Distributed losses include variations in boundary roughness and channel geometry in the streamwise direction, while local losses are associated with features such as sudden change in cross-section geometry. The interplay between the two types of energy losses along with the stage decides the flow regime and location of flow controls (*e.g.*,

Henderson, 1966). Accordingly, rating-method terminology distinguishes between channel and section controls. This distinction, introduced by Kennedy (1984), is useful for selecting a gauging site and for developing rating curves. The ratings are simple and reliable when the gauging location is in a channel reach where flow is dominated only by one control. However this is rarely the case, as flow controls might change with flow regime (depth), even if there are no changes in channel morphology. Therefore, it is necessary to understand the nature of the flow controls for all flow ranges passing through a gauging station. For a rating curve to remain accurate over time, it is essential that a gauging station is in a location with minimal erosion and deposition.

Stage-discharge ratings can be categorized as simple or complex (Holmes, 2017). Simple ratings assume a unique relationship between stage and discharge, whereas complex ratings assume that a stage-discharge relationship depends on additional independent flow variables. The type of rating for a given site is not known *a priori*, so preliminary analyses are needed to decide what type of curve and dependencies to choose. Subsequent to the above considerations, a simple stage-discharge rating can be "controlled" by a section of channel (local control located upstream or downstream from the station) or by a channel reach on which the station is installed (channel control). Channel control is defined as the combined action of all channel features in the vicinity of the station (*i.e.*, longitudinal and cross-section geometry, slope, and boundary roughness). For a station with section control, a specific section of the channel (*e.g.*, natural or man-made weirs across the stream) can uniquely define the relation between stage and discharge depending on the flow regime, with the most common being the subcritical one (Holmes, 2017).

Section control is common for the lower discharge portion of rating curves (Figure 7.5.1). Flow over a section control for any stage can be described with an exponential function of the form:

$$Q = \alpha h_*^{\beta} \qquad (7.5.1)$$

where $h_* = G - e$ is an adjusted stage, with G being the water surface stage in a local datum and e being the stage for zero flow in the control section (Figure 7.4.1). This stage definition takes into account only the "active" flow depth over the local control, as the channel can have zero flow with a non-zero stage in the channel; *e.g.*, flow ponding produced by a local control (Holmes, 2017). The coefficients α and β are obtained from regression analysis applied to calibration points plotted typically in log-log coordinates. The coefficients are variable and depend on the geometry and configuration of a control section. When a section control is artificial (*e.g.*, a man-made weir installed in the stream), analytical hydraulic methods for estimation of discharge can be

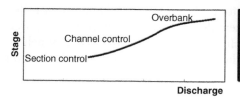
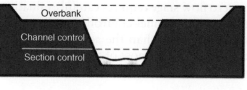

Figure 7.5.1 Simple SDM rating curve indicating various controls

developed using the appropriate equations for the installed hydraulic structure. A better fit for Equation (7.5.1) can be obtained using a polynomial function:

$$Q = \alpha + \beta h_* + \gamma h_*^2 + \dots \qquad (7.5.2)$$

Choosing a higher order for the polynomial improves the goodness of fit, but because Equation (7.5.2) does not have a physical origin, the uncertainty of the regression coefficients can increase using this equation. The uncertainty can be reduced by increasing the number of points used for the relationship construction. Equations (7.5.1) and (7.5.2) are mainly used for guidance as the approaches used for construction of stage-discharge ratings may differ across hydrometric agencies.

As flow increases to the medium to high range, channel control typically replaces local control. Consequently, for a relatively straight channel and a quasi-uniform and steady flow in the vicinity of the gauging station, head losses due to friction with the stream bed and banks (S_f) balance the energy created by the bed slope (S_0), such that the longitudinal water surface level is parallel to the channel bed (*i.e.*, $S_0 - S_f = 0$). The flow-resistance equation for this case can be expressed using Manning's equation[1]:

$$Q_0 = \frac{k S_0^{1/2} R^{2/3} A}{n} \qquad (7.5.3)$$

where Q_0 indicates discharge in equilibrium conditions, R is hydraulic radius, A is cross-sectional area, n is the Manning's coefficient representing channel roughness, and k is a conversion factor between SI and English units ($k=1$ for SI units, and $k=1.49$ for Imperial units). Provided Q_0, S_0 and n are fixed, and A and R are functions of stage only, Manning's equation represents an unambiguous relation between stage and discharge that can be taken as a guide in shaping the rating. Alluvial rivers and channels tend to adjust their bed slope toward conditions of uniform flow, because non-uniform flow implies gradients in sediment transport capacity that tend to normalize the flow through erosion and deposition. Bed adjustments continue until an equilibrium is reached in which flow is uniform, or quasi uniform. As a consequence of this general geomorphologic evolution toward equilibrium, the uniform flow assumption often holds well, particularly in straighter reaches with uniform section geometry, and is frequently taken as the natural state for discharge estimation.

Channel geometry, including bed slope, can be quantified with surveying methods of variable degrees of accuracy. The methods range from inferences based on historical geodetic information to high-precision surveys using a multi-beam echo-sounder integrated with a Real Time Kinetic (RTK) GPS unit. A longitudinal channel survey is utilized to establish the zero flow stage (see explanation for Equation 7.5.1). Due to the irregularity of the natural river bed, the selection of the location and measurement of the zero flow stage requires careful evaluation for both section- and channel- controlled stations. For channel-controlled stations, the zero flow stage is an "effective" stage that is usually higher than the actual bed stage (Holmes, 2017). For flat-bed channels of regular geometry (most of them man-made), the zero flow stage is the actual elevation of the bed.

---

[1] An equivalent formalism is offered by the Chézy equation, which is obtained from the Manning equation after substituting $C = R^{1/6}/n$.

Roughness coefficients can be estimated from a rapid assessment, based on tables summarizing empirical values of Manning's *n* for different types of coverage of the channel bed and banks (Hicks & Mason, 1998) or be inferred more accurately from *in situ* measurements of the surface level slope (*e.g.*, Kean & Smith, 2005; Eekhout & Hoitink, 2015). A limited number ($\approx$ 10) of *in situ* discharge measurements yields (via Equation 7.5.3) a relatively reliable estimate roughness coefficient. For compound channels, as it is the case when the streamflow through the floodplains is included in the rating, or when the banks are covered by dense vegetation, values of Manning's *n* may be stage dependent.

The typical appearance of a simple stage-discharge rating is illustrated in Figure 7.5.1. Often times, the ratings have a section control in the lower discharge ranges and a channel control in the higher range. An abrupt change in the gradient of the stage-discharge relation occurs at the stage where the floodplains become inundated (above the bankfull stage) as a result of the drastic changes in the geometry and roughness. As a result, the stage-discharge rating has multiple segments with different controls (Holmes, 2017). The calibration measurements acquired over the range of possible flows along with Equation (7.5.3) or with the slope-conveyance and step-backwater hydraulic analysis are useful in shaping the rating curve for the vast majority of gauging stations (Holmes, 2017). These methods require surveying of one or more stream cross sections downstream from the control section.

The simple rating curves are developed starting with fitting regression lines to the initial calibration measurements. Subsequently, segmentation of the curve for different types of flow through the station (*i.e.*, low, medium, high) are tested and analyzed to attain the rating curve for a channel site. Conventional means for statistical assessment of the rating quality may be inappropriate, because of the changing physical circumstances of flow (such statistics include goodness-of-fit measures, such as the sum of squares due to error, R-square, and root mean squared error). The selection of the zero-flow stage, the shape for the rating (which is a function of the cross section geometry), and decision on re-adjustment of the ratings over time (due to temporary or permanent changes in section control geometry or flow regime) require that rating curve developers be experienced, know the gauging site, and possess good hydraulic-analysis skills. Guidance on the protocols for developing ratings is available in many resources (*e.g.*, WMO, 2010b, Holmes, 2017).

The range over which *in situ* flow measurements (notably for calibration data) are taken limits the domain for which the discharge can be estimated. The high range of the rating curve is typically difficult to construct as there are few (if any) *in situ* discharge measurements for peak flows. Extrapolation of rating curves beyond the range of direct measurements is risky, in particular for the polynomial-based regression models, which lack a physical basis. Overcoming this limitation is typically circumvented by the use of the slope-area method when high mark stages visible after passing of the flood events are available. A survey of the high marks conducted at two or more locations enables estimation of energy slope and application of the slope-area method (Rantz *et al.*, 1982b; Dalrymple & Benson, 1984). Implementation of the slope-area approach requires information on the cross-sectional area of the stream at the location of stage measurements, the distance between stage sensors, the velocity in the cross section for determining the energy gradient, and knowledge of the roughness of the stream where the method is applied (Herschy, 2009).

Of paramount importance for obtaining a reliable and enduring simple rating curve for a gauging station is careful selection of the gauging location (Rantz *et al.*, 1982a). Deviations from guidelines for location selection might encounter local flow disturbances not accounted for in assumptions of rating curve construction. The disturbances affect calibration measurements and they become uncertainty sources. Even with all precautions in place, the quality of rating curves needs to be verified periodically by direct observations, as the stage-discharge ratings vary in time because of changes in the physical features at the gauging station site. Changes in roughness due to seasonal variation in aquatic vegetation within the channel can also complicate ratings. The changes or shifts in ratings can be gradual or abrupt, temporary or permanent. A shift of the rating curve is recommended when new calibration points differ from the established rating with more than a pre-established amount (*e.g.*, ±5%, 8% or 10% according to the rules-of-thumb applied in the U.S., New Zealand, and France, respectively). Guidance on shift detection, establishment of periods for shifting controls, and application of corrections to the rating are described in WMO (2010b) and Rantz *et al.* (1982b).

7.5.2 Complex stage-discharge ratings

Given the assumptions on which rating relationships are built, simple SDMs are limited to flow situations with propensity for uniform and steady flow conditions. When these conditions are not fulfilled, the ratings become more involved both in terms of number of calibration points needed for constructing the curves and the need for additional analysis. Moreover, recourse to numerical or physical modeling may be needed for capturing complex flow features. Details on practical approaches to adjust simple rating curves for non-uniform and unsteady flow conditions are extensively treated in Rantz *et al.* (1982b) and WMO (2010b). Basic considerations and practical aspects of these approaches are briefly presented below.

Non-uniform flow conditions might occur due to the presence of local hydraulic controls (*e.g.*, confluences and naturally-occurring features such as riffles) likely to cause backwater or drawdown affecting in turn the simple stage-discharge relationships at the station. For backwater effects, the discharge–stage relationship is a function of both stage and the stream energy slope (Rantz *et al.*, 1982b). Consequently, observation sites experiencing backwater should be equipped with stage-fall-discharge ratings (slope ratings). These ratings are determined empirically through the observation of discharge and of stages at a base gauge and an auxiliary gauge (typically located downstream from the base). The water surface slope is determined using stage readings at two gauge locations, taken with a sufficient frequency to capture the change in the slope with discharge. By analyzing the relationship between the slope and discharge established from direct measurements, it can be observed if the fall recorded in each measurement is affected by backwater at all stages, or only above a given value of the discharge. Accordingly, alternative stage-fall-discharge ratings can be used, based on the type of relationship developed between stage and fall (*i.e.*, constant, linear or curvilinear functions). Construction of each of these special cases of dependencies is detailed in Rantz *et al.* (1982b) and WMO (2010b).

Unsteady flow conditions add another complexity in the use of the simple rating curves for estimating discharges. Dealing with this type of complexity is usually done through the use of adjustment factors determined analytically or numerically, based on the cause of the temporal disturbance (*i.e.*, channel storage or flow unsteadiness). It is worth mentioning that this complex flow is quite ubiquitous, as each storm triggers the propagation of a flood wave through the stream network. Due to flow acceleration and deceleration during the passage of a flood wave, the relation between stage and discharge is not unique (*e.g.*, Henderson, 1966). The flow unsteadiness produces a loop in the stage-discharge relationship with discharges being larger for the same stage on the rising limb compared to the falling limb of the storm hydrograph (a.k.a. hysteresis). This non-unique stage relationship can confuse the construction and use of stage-discharge curves (Rantz *et al.*, 1982b).

Methods to correct for unsteadiness effects on simple stage-discharge ratings use the one-dimensional continuity and momentum equations for unsteady, open-channel flows:

$$\frac{\partial A}{\partial t} + \frac{\partial Q}{\partial x} = 0 \tag{7.5.4}$$

$$\frac{\partial Q}{\partial t} + \frac{\partial}{\partial x}(\frac{Q^2}{A}) + gA\frac{\partial h}{\partial x} - gA(S_0 + S_f) = 0 \tag{7.5.5}$$

$$\underbrace{\underbrace{\underbrace{\phantom{\frac{\partial Q}{\partial t} + \frac{\partial}{\partial x}(\frac{Q^2}{A})}}_{dynamic\ wave}}_{diffusion\ wave}}_{kinematic\ wave}$$

where h is the flow depth (varies only in the streamwise direction for the one-dimensional model), S_o is the channel bottom slope, and S_f is the energy slope. Equations (7.5.4) and (7.5.5) are referred to as the Saint Venant equations and can be solved only numerically. To substantiate the departure of the loop from the steady hydrograph for which $S_0 = S_f$, according to Ponce and Simons (1977), the terms in Equations (7.5.5) can be grouped to delineate types of flood waves. Each type of wave is associated with a flow type as illustrated in Equation (2.4.73) in Volume I. Overall, the following wave-flow types can be distinguished: the kinematic wave for steady uniform flows, diffusion wave for steady non-uniform flows, and dynamic wave for unsteady, non-uniform flows. Equation (7.5.5) also indicates the contribution of each slope term to the loop thickness. It can be inferred that the dynamic waves produce the largest loop while the kinematic waves do not lead to loops at all. Furthermore, the loop thickness is more prominent during rapid streamflow changes (typical of intense storms) through stations located on low-gradient slope streams.

As flow unsteadiness occurs with each storm, flow unsteadiness impacts all storm hydrographs to varying extent. Implementation of a complex rating is, however, not required in all cases, as for some cases the loop thickness might be within the uncertainty of the discharge measurements used in the hysteresis detection. The practical issue of assessing when hysteresis is significant enough to take action in operations is still an open research question (Muste *et al.*, 2015; Holmes, 2017). A reliable method for detection of hysteresis is to compare discharges estimated with simple ratings with *in situ* discharge measurements collected continuously during the rising and falling limbs of several flood wave hydrographs. If differences are found, one of the available corrective methods should be implemented (Rantz *et al.*, 1982b).

Correction methods amend the simple SDM approach through the use of the one-dimensional momentum equation for unsteady, gradually varied flow; Equations (7.5.4) and (7.5.5). These methods require additional measurements acquired during the propagation of the unsteady flows or use of analytical models to account for unsteadiness (Jones, 1916; Rantz *et al.*, 1982b; Petersen-Øverleir, 2006). All the correction methods are associated with simplifying assumptions. Therefore, engineering judgment is needed to assess the hydrodynamic situation at the station and to choose the appropriate method for each site. Schmidt (2002) and Dottori *et al.* (2009) offer extensive reviews of the methods for correcting the simple SDM for the effect of unsteady flows. Practical implementation protocols for the corrections are methodically documented in Rantz *et al.* (1982b), Kennedy (1984), and WMO (2010b).

There is no question that the most reliable approach for monitoring flow during unsteady flows is the direct acquisition of discrete discharge measurements during the whole duration of flow variation. For this purpose, an event-based monitoring approach needs to be devised whereby flow is tracked with direct measurements using sampling rates commensurate with the time scale of the flood wave propagation. With the advent of the new generation of instruments and communication technologies these measurements are increasingly feasible. This is well-illustrated by the measurements acquired with ADCPs by various agencies for strengthening the data samples used for building and validating the SDM ratings (Figures 7.5.2a and 7.5.2b). The SDM hysteretic behavior illustrated in Figures 7.5.2a and 7.5.2b is associated with flow unsteadiness propagating in large and medium rivers, respectively. Figure 7.5.2b also illustrates the discharge evolution during the storm as estimated with the Fread (1975) method. Among the available correction methods, Fread's approach is one of the most economical as it requires only one surveyed cross section, in addition to the stage readings at the gauging station.

An emerging approach for capturing unsteady flows with stage-discharge measurements is the continuous slope-area method (CSAM). This method builds on the conventional slope-area method (Dalrymple & Benson, 1984) fitted with continuous slope measurements (Smith *et al.*, 2010). The free-surface slope can currently be detected with cost-efficient sensing technology (predominantly acoustic). Results of direct measurements with a simplified CSAM version are shown in Figure 7.5.2c (Muste *et al.*, 2016; Lee & Muste, 2017). They reveal that the loop characterizing the hysteresis can be captured even for small streams.

While the impact of hysteresis on the simple SDM is well-known theoretically, efforts to detect and correct for this impact in routine gauging practice are lagging, as there is a perception that hysteresis effects are small and cannot be discerned from the uncertainty of the instruments and methods used to build the SDM ratings. Consequently, most of the unsteady events on small and medium inland rivers remain undocumented. Hysteresis has only received attention in flood-prone large rivers (*e.g.*, the Mississippi or the Rhine Rivers) with the purpose to better understand the process and provide more accurate data for the streamflow forecasting models. Emerging research aims at systematically evaluating the impact of hysteresis by analyzing existing data for sites prone to this phenomenon (Holmes, 2017). The acquisition of such experimental evidence and supplementary analyses can support the sound evaluation of the effects of hysteresis on current monitoring practices and will eventually lead to optimized algorithms for dealing with this rating complexity in practical situations.

The most complex situation in rating construction and usage occurs when flow non-uniformity and unsteadiness occur simultaneously. This might be the case for most storms, as the rate of change in discharge is most likely associated with flow non-uniformity due to inherent variabilities of natural channels. For this situation, WMO

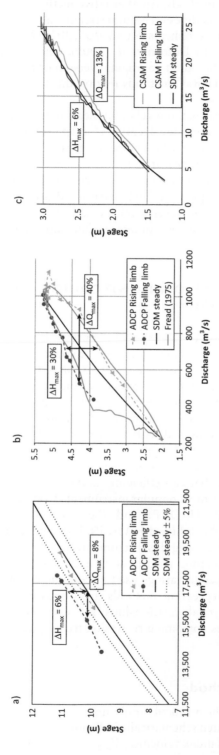

Figure 7.5.2 Illustration of hysteresis in the SDM for various rivers and events: a) Mississippi River, USA – 21 days, ADCP measurements (adapted from Holmes, 2017); b) Ebro River, Spain – 1 day, ADCP measurements (adapted from Muste & Lee, 2013); and c) Clear Creek, USA – 2 days, CSAM (adapted from Muste *et al.*, 2016).

(2010b) recommends fall-rating methods. An alternative method to account for both non-uniform flow effects and unsteadiness in natural stream reaches uses the slope-area method with continuous slope measurements (Smith *et al.*, 2010; Muste *et al.*, 2016). A more complex method is offered by Dottori *et al.* (2009), who introduce a dynamic rating curve (DyRaC) approach based on an original version of the full dynamic flow equations. The CSA-based and DyRaC methods rely on the measurement of surface level at two or more locations.

The practical limitation of these methods relies on measurement accuracy for determining free-surface slope. Use of only a few locations for determining the slope may be insufficient. Local gradients in water surface slope are typically small (possibly even smaller than the uncertainty of the measurement instrument) and affected by other measurement obstacles (transverse surface gradients, wind and turbulence waviness). Moreover, lowland areas are also prone to backwater and drawdown effects that are not accounted for by the methods. However, implementation of these methods is more efficient for both normal and extreme flow conditions than adjusting simple ratings for complex flow situations.

7.6 CONTINUOUS STREAMFLOW MONITORING USING VELOCITY MEASUREMENTS

The limitations of the rating methods discussed in the previous section (especially the presence of backwater and flow unsteadiness) have led to development of monitoring techniques that include continuous measurement of an index-velocity in addition to stage. There are many instruments that can continuously measure *in situ* velocities at a point (Section 3.5, Volume II), along a line (Section 3.3.5, Volume II), or over a surface (Section 3.7.4, Volume II). These measurements can be related (via an index) to the bulk-flow velocity and further combined with continuous measurement of the cross sectional area to readily determine discharges.

Recently, the velocity-index method has become widely used due to the adoption of acoustic and radar velocimetry that can efficiently and continuously measure velocities across natural streams. The most popular instruments from this category are the horizontally-positioned ADCP (HADCP) and the vertically-positioned ADCP (VADCP). Given the wider applicability of HADCPs, implementation of the index-velocity method is best illustrated using this instrument. Rantz *et al.* (1982b) provide extensive descriptions of this method used in conjunction with other instruments. This section discusses three types of ratings used in conjunction with the measurement of the index velocity. Regression methods are discussed first, followed by methodologies that use analytical relationships for establishing flow profiles. Subsequently, hybrid approaches combining regression techniques and hydrodynamic models are presented.

7.6.1 Index-velocity method

Horizontal ADCP (HADCP). The protocol for obtaining discharge estimates using the index-velocity method is illustrated schematically in Figure 7.6.1. Repeated calibration measurements of stages (G) and index velocities, V_{index}, are used to develop associated

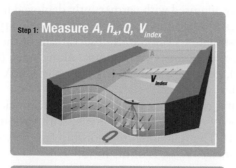

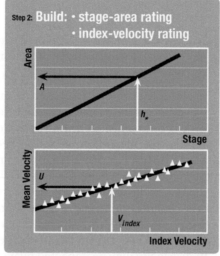

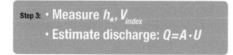

Figure 7.6.1 Schematic diagram of the estimation of discharges using the index-velocity method.

ratings for the channel cross section (A) and mean channel velocity U, respectively. V_{index} is the average value of the individual in-bin velocities measured by the ADCP across the channel width. The channel cross-section is determined from a detailed survey of the river bathymetry at the gauge location. The mean channel velocity is determined using the direct discharge measurement divided by the cross-section area at the time of the calibration measurement. Discharges can be measured with any of the conventional methods (*e.g.*, a propeller-type current meter) but recently, the most often used measurement method relies on ADCPs deployed on moving boats (Mueller *et al.*, 2013).

A "stage-area" rating is obtained by relating the stage to the corresponding cross-section area. Stages are measured with methods used by the stage-discharge method described in Section 7.5.1 in this chapter. Multiple mean-channel velocities and simultaneously acquired index velocities are paired to build the "index-velocity rating", typically by using least-square regression. Direct and continuous measurements of depth and index-velocity acquired with permanently deployed instruments in conjunction with the two ratings provide area and mean channel velocity that, multiplied, give continuous discharge estimates. Continuous observations of the index velocity can be obtained by averaging measurements at a point, along lines (*e.g.*, with a HADCP, shown in Figure 7.6.1), or over a surface (*e.g.*, for the LSPIV method described in Section 3.7, Volume II).

The index-velocity ratings are conveniently built with regression models that find the best-fit for pairing velocity measurements obtained with various instruments and the cross-section velocity, U. The functional relationship (g_o) between the mean channel and measured index velocity, V_{index} is of the form:

$$U = g_o(V_{index}) \qquad (7.6.1)$$

The index velocity, V_{index}, can be the average of a velocity time series acquired at a point ($\bar{u}_{i,j}$), or an average taken over the measurement volume of the flow velocity measurement device ($V_{index} = \sum_{i=1}^{n} \bar{u}_{i,j}$, for j = constant, *i.e.*, a horizontal line) as shown in Figure 7.6.1. The relation between V_{index} and U may display a degree of ambiguity when the flow varies in magnitude or over time for several reasons. For example, for gauging stations close to bends, the variation of flow velocity close to the channel banks generally precedes the mean flow variation because of inertia that is most pronounced in that region (Muste *et al.*, 2015). In compound channels, flow velocity variation in shallow regions lags behind mean flow variation. For simple channel geometries where the law of the wall applies, the location that best represents the mean velocity is close to the channel center, at a relative depth of about 0.4, where it is assumed that the local velocity equals the depth-averaged velocity. Closer to the bed, the velocity magnitude decreases and hence the measurement error in V_{index} is amplified. Closer to the surface, flow velocity may be impacted by wind shear or by side wall effects, causing velocity to peak below the surface.

When residuals of the linear regression model show systematic patterns in the index-velocity rating, several steps can be considered to expand the model complexity. A multi-linear or nonlinear function can be employed to achieve a better fit to the pairs of V_{index} versus U points. These approaches are appropriate when the relation between these variables is unambiguous. Theoretically, any form of systematic behavior of the residuals can be accounted for by increasing model complexity. A multiple linear regression approach may be required to eliminate heteroscedasticity, the systematic behavior of the variance of the residuals. Typically, the water surface stage, G, adjusted by the zero-flow elevation, is the first candidate to be added to the regression model (Levesque & Oberg, 2012):

$$U = aV_{index} + bh_* + ch_*V_{index} + I \qquad (7.6.2)$$

where a, b and c are coefficients and I is an intercept. Analysis of numerous data sets has shown that the term representing the stage variable alone, bh_*, is typically insignificant for practical situations (Levesque & Oberg, 2012). However, there is no statistical basis to guarantee that a model accurate in one or more situations can directly be applied to other circumstances. A more rigorous approach would be to refrain from applying any form of operation to the velocity data prior to the regression analysis, and to consider finding the optimal function for the expression $U = f(h_*, V_{index})$. If the data set available for model development is large, it may be worth considering advanced regression model development techniques based on multivariate statistics (e.g., Johnson & Wichern, 1992; Härdle et al., 2012). It is worth noting that an extremely good model fit for historical data does not guarantee accuracy of future predictions. For example, dredging activity may change the index velocity relation in a way that can never be predicted from historical data.

The accuracy of the index-velocity method for discharge estimation is intrinsically linked to the quality of the regression model adopted to translate index velocity to cross-section averaged velocity. For channels with a simple geometry, this relationship is often linear. Compound channels require more complex relations, which is an ongoing field of research (e.g., Uchida & Shoji, 2014). For example, if the rating is developed as a continuum for the main channel and its floodplain, a bimodal rating, as described in Ruhl and Simpson (2005), should be constructed. The index velocity method is assumed to be relatively accurate for measurements of non-uniform flows. For unsteady flows, it is expected to improve the accuracy of the estimation as the method is based on two direct and simultaneous measurements collected continuously at a high sampling rate.

For flood situations, the relationship between V_{index} and U may differ for the rising and falling limbs of the flood hydrograph because of inertia in the process of floodplain inundation and emptying. To account for such circumstances, Levesque and Oberg (2012) propose a more complex relationship (i.e., Equation 7.6.2). The remaining question is if the index-velocity itself is completely free of hysteresis. The novelty of the index-velocity method compared to the more-often and extensively used SDM, as well as the cost-intensive measurements for acquiring continuous data during unsteady flows, hamper sufficient experimental and analytical data to definitively answer the above question, or to anticipate potential complexities associated with the use of unique regression coefficients for unsteady events.

Another version of the index-velocity method uses bottom-mounted acoustic Doppler current profilers termed vertical ADCP (VADCP). These instruments are extensively used in smaller, man-made channels. VADCPs are fixed on the bed, typically in the channel center, looking upward. As with HADCPs, they use multiple acoustic beams to measure one vertical profile of velocity associated to the instrument axis. An additional vertical acoustic beam and a pressure sensor are also used to measure the flow depth. Software is provided by the instrument manufacturers to convert these measurements to a discharge estimate. The conversion uses as input cross-section geometry and the index velocity method (as described above). Velocity accuracy can be as good as 0.5% or 2 mm/s, but discharge accuracy depends on reliability of velocity indexing.

7.6.2 Hydraulics-based velocity models

This method estimates the spatial distribution of flow velocity over the cross-section using vertical and horizontal velocity distributions obtained from hydraulic theory. Model parameters can be estimated from circumstantial information or inferred from direct discharge estimates acquired with hydrographic surveys. Continuously measured velocities are used to convert the theoretical non-dimensional profiles to actual velocities (log-law or different power laws are used for this purpose), yielding in a first phase a depth-averaged velocity, denoted as $\bar{u}$. Analytical approaches are then used to obtain the cross-section averaged velocity, U.

For illustration purposes, the velocity-model proposed by Sassi *et al.* (2011) is discussed below. This model assumes that the bed-attached boundary layer can be extended over the entire flow depth, and therefore velocities in a vertical, $u(\sigma) = f(z)$, are strongly related to the depth-averaged flow velocity at that vertical, $\bar{u}$. Close to the water surface, flow velocity may be affected by wind shear or by side wall effects, resulting in a velocity dip wherein the peak velocity occurs below the water surface. Systematic measurements can be used to capture the effect on the law of the wall. Such measurements led Sassi *et al.* (2011) to a generic relation for the vertical profile of streamwise velocity of the form:

$$\frac{u(\sigma)}{\bar{u}} = \frac{\ln(\sigma\,h) + \gamma\,\ln(1-\sigma) - \ln(z_0)}{\ln\left(\dfrac{h}{\exp(1+\gamma)}\right) - \ln(z_0)} \tag{7.6.3}$$

where h is the flow depth at a given spanwise location, $\sigma = (h-h(z))/h$ is the height above the bed where the velocity is modeled normalized with the flow depth, z_0 is the roughness height (taking into account bedform height), and γ is a velocity dip correction coefficient, which decreases with distance from the channel bank. This coefficient relates to the relative depth where the maximum velocity occurs (σ_{max});

$$\gamma = \frac{1}{\sigma_{max}} - 1 \tag{7.6.4}$$

The model parameters z_0 and γ can be obtained by fitting Equation (7.6.3) to vertical velocity profiles obtained from calibration surveys. When the maximum velocity occurs at the surface, $\gamma = 0$, and Equation (7.6.3) reduces to the law of the wall. The values of z_0 depend on bedform geometry and on the orientation of the bedforms relative to flow streamline direction. For tide-affected channels, the roughness height, z_0, reveals different values for the two tide phases (*i.e.*, ebb and flood) across the same location of the cross section (Figure 7.6.2). In inland alluvial channels, the bedforms tend to scale with depth, and $\ln(z_0)$ may show a linear dependence on depth (Sassi *et al.*, 2011).

As with vertical profiles of velocity, analytical or semi-empirical relationships are used to model the spanwise velocity distribution in the channel. These relationships are needed for situations where the instruments measure across the river width, but they do not cover the entire channel cross-section. This is the case for the HADCP approach, which does not measure near the probe and near the opposite bank. Among the methods proposed for extrapolating the HADCP measured velocity profile near the probe and the opposite bank are: a) a constant for the $\bar{u}/U$ ratio (where $\bar{u}$ is the depth-averaged velocity and U the mean cross-section velocity); b) constant Froude number across the channel; c) constant depth-averaged velocity across the channel; and, d) a constant value of $\bar{u}$ in the range beyond the measurement domain. These four methods were tested with HADCP data by Le Coz *et al.* (2010), leading to the conclusion that none of the methods increased accuracy compared to the use of the index-velocity method. However, this conclusion has to be further tested, as the ratio between measured and unmeasured fractions of the cross section is also a factor in the inference.

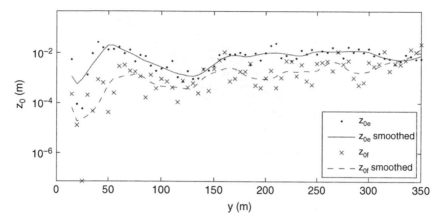

Figure 7.6.2 Cross-river profiles of the roughness length (z_0) inferred from ADCP data taken in a tidal river with dunes of about 1 m, during ebb (z_{0e}) and flood (z_{0f}). Adopted from Hoitink *et al.* (2009).

The assumption of a constant mean, spanwise velocity distribution can be employed for straight channels and relatively uniform bed morphology. For sinuous channels, however, the assumption does not hold, as systematic redistribution of flow momentum occurs resulting in a shift toward the outer bend of the peak of the spanwise velocity distribution with the increase in discharge. For such situations, more complex theoretical models are needed for replicating the spanwise velocity distribution (*e.g.*, Blanckaert & De Vriend, 2003). In a compound channel, the spanwise velocity profile of $\bar{u}/U$ is typically stage dependent, with the floodplain carrying an increasingly larger portion of the discharge with an increasing ratio of floodplain depth to main channel depth (*e.g.*, Van Prooijen *et al.*, 2005).

7.6.3 Hybrid approaches for velocity models

The regression- and analysis-based methods discussed above have advantages and limitations that make them appropriate for specific field monitoring situations. Hybrid approaches attempt to create synergy between the two methods, aiming for improved accuracy of the velocity models. When a regression model is being developed, knowledge about the physical system helps to appraise its predictive value. For example, horizontal extrapolation of the depth-mean flow velocity in shallow flows involves more uncertainty than using depth-averaged velocity models based on theoretical profiles. These flow situations require a trade-off accomplished through the adoption of a hybrid modeling approach in which distribution of the vertical velocity profile is related to $\bar{u}$ based on physical principles, and $\bar{u}$ is related to the cross-section average velocity, U, using a regression model (Hoitink *et al.*, 2009; Sassi *et al.*, 2011; Hidayat *et al.*, 2011). To ensure the applicability of the model in highly dynamic environments (*e.g.*, flash floods or tides), a regression model can be devised of the form:

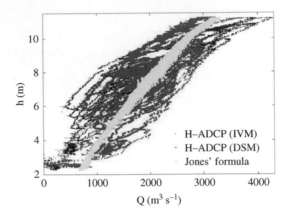

Figure 7.6.3 Comparison of the stage-discharge relations in a meandering river affected by backwater for Jones' formula, an index-velocity using HADCP data (IVM), and a semi-deterministic, semi-stochastic model based on HADCP data (DSM); from Hidayat *et al.* (2011).

$$U = \alpha(h)\overline{u}(\tau) + \beta \tag{7.6.5}$$

where τ is the time lag (or lead) between the variation of U and $\overline{u}$ and α, β are regression coefficients. The stage dependence of the regression coefficient α takes into account the systematic mean flow redistribution apparent in river bends and compound channels. Hoitink *et al.* (2009) and Sassi *et al.* (2011) show that such a semi-deterministic, semi-stochastic approach can outperform an index velocity model. Figure 7.6.3 exemplifies the differences between three alternative stage-discharge relations in a complex flow situation. Once developed, regression techniques are used for the estimation of the physical model parameters and of the flow velocity in the unmeasured areas across a channel.

Hydrodynamic models can capture the key physical processes governing the flow distribution across a channel, and therefore have a high potential for extrapolating flow measurements to yield discharge. Data assimilation in hydrodynamic models is emerging as a good approach for this area of hydrology, by exploiting effectively the increased computational power as well as the proficiency of the new measurement techniques. The model domain can be limited to the stretch of the geometry that impacts the flow distribution. For example, Nihei and Kimizu (2008) made a first attempt in adopting a data assimilation approach for discharge estimation from flow measurements, using HADCP data and a reduced complexity (2D-vertical) numerical model. Their simulations were based on a longitudinal momentum balance equation in which the acceleration terms and streamwise turbulent exchange of momentum were lumped in a single term, which was further optimized using the HADCP data. Sophisticated methods of data assimilation are increasingly being applied in many fields of earth sciences (Reichle, 2008). Weerts *et al.* (2010) introduced a generic data-assimilation toolbox (DATools), developed with the initial aim of flood forecasting. Estimating discharge from flow velocity measurements can be seen as now-casting, and a toolbox such as DATools can be readily employed to assimilate flow velocity data with a full-complexity 3D hydrodynamic model.

7.7 PRACTICAL ISSUES

Due to the importance of discharge measurement for water resources management, as well as for laboratory and field experiments in the area of hydraulics, a large variety of instruments and methods are available for discharge measurement. The accuracy of the laboratory instruments is well proven by comparison to fundamental methods (*e.g.*, volumetric measurement) or reference instruments housed in national standardization facilities. The protocols for instrument use and accuracy are covered by more than 160 standards developed under the auspices of the International Organization for Standardization (Reader-Harris, 2007). Many of the instruments for laboratory are highly accurate, with relative discharge uncertainty of less than 1%. The only practical concern related to the employment of laboratory instrumentation is to condition the flow prior to the discharge measurement location, as the instruments have to measure under similar conditions and flow ranges for which they have been calibrated. Section 4.3.3.2 in Volume I presents some of these flow conditioning devices.

Similarly to laboratory situations, the instruments and methods for measuring *in situ* discharges of water are only valid within their design assumptions. For instruments aimed at measuring discrete discharges, considerations have to be given first to verify if the flow at the site fulfills the assumptions underlying the instrument or method. For example, ADCPs cannot accurately measure flow in areas where the horizontal homogeneity assumption does not hold (Section 3.3.2.3.3 in Volume II). For instruments and methods leading to rating curves, the chief criterion for attaining accurate discharge estimations is to select a location for the gauging site without large deviations from steady and uniform flow. For the index-velocity method it is also important that the distribution of velocity across the section is similar for all flow ranges. Given that field measurements are much more challenging with respect to attaining a prescribed accuracy, four aspects relevant to the choice of a monitoring approach are discussed below: ease of deployment; accuracy and reliability; costs; and, ease of standardization.

For small streams, a high degree of accuracy can be achieved with discharge measurement structures such as broad-crested and sharp-crested weirs (Section 7.2.1). These structures create critical flow conditions that can be modeled in laboratory conditions with physical models. Sediment transport and floating debris can limit their accuracy, therefore they require regular maintenance. For large rivers, the most widely used methods are those based on rating curves. Rating curves have been and will remain widely used because of the ease of deployment of water level sensors. The reliability of discharge estimation with rating curves is only as good as the calibration data acquired to build the curves. Consequently, *a priori* knowledge about the status of the flow and its structure is an important factor in determining the reliability of calibration data. Moreover, changes in channel geometry due to natural or man-made interventions require recalibration of the ratings. The development of a sophisticated high-accuracy model to translate flow velocity and water level information to discharge can be justified only if it is likely that the geometry remains unchanged, especially when the model includes a regression component. When the latter condition is not met, the achieved accuracy during calibration may be offset by a loss of reliability.

In general, when the geometry of a channel section is regular (symmetric trapezoidal) and the bed slope is of the order 10^{-3} or steeper, the development of simple stage-discharge rating curves is an appropriate choice if the flow is in the subcritical regime.

When supercritical flow conditions occur, it becomes difficult to obtain calibration data as the water surface becomes irregular and cannot be accurately captured to obtain depths representative for the whole cross-section, using just one water level gauge. Moreover, the effect of wind on the water surface velocity is less disruptive on supercritical flows, favoring the use of surface-velocity monitoring techniques such as large-scale PIV (Section 3.7.4, Volume II). For mild channel slopes, such as in lowland areas, stage-discharge relations may become ambiguous, and additional velocity measurements are often needed to complement water level measurements, as these areas are prone to developing backwater. Acoustic travel-time meters remain popular, but require at least two measurement locations, set on opposite sides of the channel. Travel-time meters are currently increasingly being replaced by HADCPs and VADCPs that require just one deployment location.

The costs associated with the installation of a gauging station comprise the construction and installation, station maintenance, the expenses associated with the development of an appropriate methodology, and data post-processing and communication. Discharge measurement structures require a high investment and regular servicing, and are best applied to relatively small, non-alluvial channels. They can be also used for controlling the flow. For larger channels, the least expensive measurement methods are based on non-intrusive means (e.g., acoustic, radar, and optical probes). The use of an index-velocity method is increasingly popular as it is cost-effective and can be used in a variety of complex flow situations (e.g., backwater, unsteady flows) where the other methods cannot perform accurately. Use of hydrodynamic models, currently commercially-available (e.g., SIMK® package, Kölling, 2004), can further improve the accuracy of index-velocity methods while keeping the costs relatively modest. The use of hydrodynamic models is also attractive as they enable data assimilation, which is beneficial for adjusting discharge estimation in real time for effects of rapid spatio-temporal changes in the flow as soon as new data are produced.

Finally, the degree of standardization of a method is also an important factor when choosing a discharge monitoring strategy. Continuous monitoring in steady flows with measurement structures, stage-discharge and index-velocity ratings are well-covered by guidelines. Less practical guidance is available on the methods to diagnose the significance of flow unsteadiness for particular gauging sites and storm events, as well as which methods to apply for corrections (Muste et al., 2015). The new approaches based on hybrid methods (Section 7.6.3) are not yet widely applied, but it is expected that as they mature they will become an essential component of the new generation of "smart" methods for monitoring stream discharge in any flow situation.

Discharge monitoring in estuaries and tidal rivers poses additional challenges, under both normal and extreme hydrologic conditions. Salinity intrusion in coastal rivers complicates the usage of acoustic-based instruments, because density differences lead to refraction of acoustic rays. It is yet to be confirmed if acoustic tomography will prove to become a sufficiently robust tool for discharge monitoring under stratified conditions (Section 3.4, Volume II). Near the coast, tides and wind at the seaward river boundary (alone or combined) may affect water level and discharge variations. In these regions, artificial neural networks may be employed to predict discharge (Hidayat et al., 2014). Regarding hydrological extremes, radar techniques including UHF radar (Section 3.8 in Volume II) have the potential to monitor surface level and velocity in floodplains periodically inundated, but this approach is only at an initial research stage.

REFERENCES

Baker, R.C. (2000) *Flow Measurement Handbook: Industrial Designs, Operating Principles, Performance, and Applications*. New York, Cambridge University Press.

Blanckaert, K. & de Vriend, H.J. (2003) Nonlinear modeling of mean flow redistribution in curved open channels. *Water Resour Res*, 39, 1375, doi:10.1029/2003WR002068, 12.

Boiten, W. (2000) *Hydrometry*. Rotterdam, Balkema, 246.

Christensen, J.L. & Herrick, L.E. (1982) *Mississippi River test, Vol. 1. Final report DCP4400/300*. Prepared for the U.S. Geological Survey by AMETEK/Straza Division, El Cajon, California, under contract No. 14–08–001–19003, A5–A10.

Dalrymple, T. & Benson, M.A. (1984) Measurement of peak discharge by the slope-area method, published in Chapter A2 in *Techniques of Water-Resources Investigations* of the United States Geological Survey, US Geological Survey Books and Open-File Reports Section, Federal Center, Denver, CO.

Daugherty, R.L., Franzini, J.B., & Finnemore, E.J. (1985) *Fluid Mechanics with Engineering Applications*. New York, McGraw-Hill.

Dottori, F., Martina, M.L.V., & Todini, E. (2009) A dynamic rating curve approach to indirect discharge measurement. *Hydrol Earth Syst Sci*, 13(6), 847–863.

Eekhout, J.P.C., & Hoitink, A.J.F. (2015) Chute cutoff as a morphological response to stream reconstruction: The possible role of backwater. *Water Resour Res*, 51(5), 3339–3352

Frazier, A.H. (1974) *Water Current Meters*. Washington, D.C., Smithsonian Institution Press.

Fread, D.L. (1975) Computation of stage-discharge relationships affected by unsteady flow. *Water Resour Bull*, 11(2), 213–228.

Fulford, J.M. & Sauer, V.B. (1986) Comparison of velocity interpolation methods for computing open-channel discharge. In Subitsky, S.Y. (ed.) U.S. Geological Survey Water-Supply Paper 2290, p. 154.

Goldstein, R.J. (1996) *Fluid Mechanics Measurement*. Washington DC. Taylor & Francis.

Gonzalez-Castro, J.A. & Muste, M. (2007) Framework for estimating uncertainty of aDcp measurements from a moving boat by standardized uncertainty analysis. *J Hydraul Eng*, 133(12), 1390–1410.

Gordon, R.L. (1989) Acoustic measurement of river discharge. *J Hydraul Eng*, 115(7), 925–936.

Härdle, W.K., Müller, M., Sperlich, S., & Werwatz, A. (2012) *Nonparametric and Semiparametric Models*. Springer Science & Business Media.

Henderson, F.M. (1966) *Open Channel Flow*. New York, Macmillan Publishing Co.

Herschy, R. (2009) *Streamflow Measurement*. 3rd edition. Oxford, Taylor & Francis.

Hicks, D.M. & Mason, P.D. (1998) *Roughness Characteristics of New Zealand Rivers*, Water Resources Publications, Second edition, Murray Media, Miami Beach, FL; ISBN-10: 0477026087

Hidayat, H., Vermeulen, B., Sassi, M.G., Torfs, P.J.J.F., & Hoitink, A.J.F. (2011) Discharge estimation in a backwater affected meandering river. *Hydrol Earth Syst Sci*, 15(8), 2717–2728.

Hidayat, H., Hoitink, A.J.F., Sassi, M.G. and Torfs, P.J.J.F., 2014. Prediction of discharge in a tidal river using artificial neural networks. *Journal of Hydrologic Engineering*, 19(8), p.04014006.

Hoitink, A.J.F., Buschman, F.A., & Vermeulen, B. (2009) Continuous measurements of discharge from a horizontal acoustic Doppler current profiler in a tidal river. *Water Resour Res*, 45, W11406, doi:10.1029/2009WR007791.

Holmes, R. (2017) Streamflow Ratings. In: Singh, V.P. (ed.) *Handbook of Applied Hydrology*, Chapter 6. New York, McGraw Hill Education.

ISO 748 (2007) Hydrometry—*Measurement of Liquid Flow in Open Channels Using Current-Meters or Floats*. International standardization Organization, Geneva, Switzerland, 46.

Johnson, R.A. & Wichern, D.W. (1992) *Applied Multivariate Statistical Analysis* (Vol. 4). Englewood Cliffs, NJ: Prentice hall.

Jones, B.E. (1916) *A Method of Correcting River Discharge for a Changing Stage*. US Government Printing Office.

Le Coz, J., Hauet, A., Pierrefeu, G., Dramais, G., & Camenen, B. (2010) Performance of image-based velocimetry (LSPIV) applied to flash-flood discharge measurements in Mediterranean rivers. *J Hydrol, 394*(1–2), 42–52.

Le Coz, J., Blanquart, B., Pobanz, K., Dramais, G., Pierrefeu, G., Hauet, A., & Despax, A. (2016) Estimating the uncertainty of streamgauging techniques using field interlaboratory experiments. *J Hydraul Eng, 142*(7), 04016011.

Levesque, V.A. & Oberg, K.A. (2012) Computing discharge using the index velocity method. *Report on Techniques and Methods 3-A23*. US Geological Survey, 148. [Online] Available from http://pubs.usgs.gov/tm/3a23 [Accessed 5th November 2016].

Kennedy E.J. (1984) Discharge ratings at gaging stations. U.S. Geological Survey Techniques of Water-Resources Investigations, book 3, chap. A10, 59. [Online] Available from http://pubs.usgs.gov/twri/twri3-a10 [Accessed 17th January 2017].

Kean, J.W. & Smith J.D. (2005) Generation and verification of theoretical rating curves in the Whitewater River basin. Kansas. *J Geophys Res, 110*, F04012, doi:10.1029/2004JF000250.

Kölling, C. (2004) SIMK® – Calibration of Streamflow – Gauging Stations in Rivers and Canals. Fifth International Conference IGHEM 2004, Innovation in Hydraulic Efficiency Measurement, University of Applied Sciences, Lucerne, Switzerland.

Lee, K. & Muste, M. (2017) Refinement of Fread's Method for Improved Tracking of Stream Discharges during Unsteady Flows. *J Hydraul Eng*, doi: 10.1061/(ASCE)HY.1943-7900.0001280.

Moore, S.A., Jamieson, E.C., Rainville, F., Rennie, C.D., & Mueller, D.S. (2016) A Monte Carlo approach for uncertainty analysis of Acoustic Doppler Current Profiler discharge measurement by moving boat. *J Hydraul Eng*, paper 04016088, doi:10.1061/(ASCE)HY.1943-7900.0001249

Mueller, D.S., Wagner, C.R., Rehmel, M.S., Oberg, K.A., & Rainville, F. (2013) Measuring discharge with acoustic Doppler current profilers from a moving boat. (ver. 2.0, December 2013) *Report on Techniques and Methods* 3-A22, U.S. Geological Survey, 95. [Online] Available from: http://pubs.water.usgs.gov/tm3a22 [Accessed 28th March 2016].

Mueller, D.S. (2016) QRev—Software for computation and quality assurance of acoustic Doppler current profiler moving-boat streamflow measurements—Technical manual for version 2.8: *U.S. Geological Survey Open-File Report*, 2016–1068, 79., http://dx.doi.org/10.3133/ofr20161068

Muste, M., Vermeyen, T., Hotchkiss, R., & Oberg, K. (2007) Acoustic velocimetry for riverine environments. *J Hydraul Eng, 133*(12), 1297–1298.

Muste, M., Cheng, Z., Vidmar, P., & Hulme, J. (2015) Considerations on discharge estimation using index-velocity rating curves, In: *Proceedings of the 36th IAHR World Congress*, 28 June – 3 July, The Hague, The Netherlands.

Muste, M., Cheng, Z., Firoozfar, A.R., Tsai, H-W., Loeser, T., & Xu, H. (2016) Impacts of Unsteady Flows on Monitoring Stream Flows, *River Flow Conference*, July 12–15, 2016, St Louis, MO, USA

Muste M. & Lee K. (2013) Quantification of Hysteretic Behavior in Streamflow RCs. *Proceedings of the 35 IAHR World Congress*, September 8–13, Chengdu, China.

Nihei, Y. & Kimizu, A. (2008) A new monitoring system for river discharge with horizontal acoustic Doppler current profiler measurements and river flow simulation. *Water Resour Res, 44*, W00D20, doi:10.1029/2008WR006970.

NIST (2006) *NIST Calibration Services for Water Flowmeters*, National Institute of Standards and Technology, Gaithersburg, MD 20899

Petersen-Øverleir, A. (2006) Modelling stag – discharge relationships affected by hysteresis using the Jones formula and nonlinear regression. *Hydrol Sci J*, *51*(3), 365–388.

Ponce, V.M. & Simons, D.B. (1977) Shallow water propagation in open channel flow. *J Hydr Div*, *103*, 1461–1476.

Rantz, S.E. & others (1982a) *Measurement and Computation of Streamflow. Volume 1. Measurement of Stage and Discharge*, U.S. Geological Survey Water-Supply Paper No. 2175, Reston, VA., 313.

Rantz, S.E. & others (1982b) *Measurement and Computation of Streamflow: Volume 2. Computation of Discharge*, US Geological Survey Water Supply Paper 2175, Reston, VA, 373.

Reader-Harris, M. (2007) ISO flow measurement standards – Report on the ISO/TC 30 meeting in November 2006. *Flow Meas Instrum*, *18*, 114–120.

Reichle, R.H. (2008) Data assimilation methods in the Earth sciences. *Adv Water Resour*, *31* (11), 1411–1418.

Rennie, C.D., Millar, R.G., & Church, M.A. (2002) Measurement of bedload velocity using an acoustic Doppler current profiler. *J Hydraul Eng*, *128*(5), 473–483.

Ruhl, C.A. & Simpson, M.R. (2005) *Computation of Discharge Using the Index-Velocity Method in Tidally Affected Areas*: U.S. Geological Survey Scientific Investigations Report 2005–5004, 31.

Sassi, M.G., Hoitink, A.J.F., & Vermeulen, B. (2011) Discharge estimation from H-ADCP measurements in a tidal river subject to sidewall effects and a mobile bed. *Water Resour Res*, *47*, W06504, doi:10.1029/2010WR009972

Schmidt A.R. (2002) Analysis of stage-discharge relations for open channel flows and their associated uncertainties. *Ph.D. Thesis*, University of Illinois at Urbana-Champaign, Champaign, IL

Simpson, M.R. (2001) *Discharge Measurements using a Broadband Acoustic Doppler Current Profiler*. U.S. Geological Survey Open-File Report 01–01, 123. [Online] Available from: http://pubs.usgs.gov/of/2001/ofr0101 -[Accessed 25th March 2016].

Smith, C.F., Cordova, J.T., & Wiele S.M. (2010) *The Continuous Slope-Area Method for Computing Event Hydrographs*. USGS Science Investigation, Report 2010–5241.

Uchida, T. & Shoji, F. (2014) Numerical calculation for bed variation in compound-meandering channel using depth integrated model without assumption of shallow water flow. *Adv Water Res*, *72*, 45–56.

Tavoularis S. (2005) *Measurement in Fluid Mechanics*. Cambridge, Cambridge University Press.

Turnipseed, D.P. & Sauer, V.B. (2010) Discharge measurements at gaging stations, Report on Techniques and Methods 3-A8, 87. U.S. Geological Survey [Online] Available from: http://pubs.usgs.gov/tm/tm3-a8/ [Accessed 30th July 2015].

van Prooijen, B.C., Battjes, J.A., & Uijttewaal, W.S. (2005) Momentum exchange in straight uniform compound channel flow. *J Hydraul Eng*, *131*(3), 175–183.

Weerts, A.H., El Serafy, G.Y., Hummel, S., Dhondia, J., & Gerritsen, H. (2010) Application of generic data assimilation tools (DATools) for flood forecasting purposes. *Comp Geosci*, *36* (4), 453–463.

World Meteorological Association (2010a) *Manual on Stream Gauging, Volume I, Field Work*. WMO No. 1044 [Online] Available from: http://www.wmo.int/pages/prog/hwrp/publica tions/stream_gauging/1044_Vol_I_en.pdf [Accessed 5th November 2016].

World Meteorological Association (2010b) *Manual on Stream Gauging, Volume II, Computation of Discharge*. WMO No. 1044. [Online] Available from: http://www.wmo.int/pages/prog/hwrp/publications/stream_gauging/1044._Vol_II_en.pdf [Accessed 5th November 2016].

Chapter 8

Autonomous Underwater Vehicles as Platforms for Experimental Hydraulics

8.1 INTRODUCTION

Underwater robots are increasingly being used as platforms for experimental hydraulics. Compared to methods of data collection in surface or airborne environments, subaqueous settings present several unique challenges that drive the need for autonomous sampling platforms. Opportunities for human exploration are limited by the lethal nature of the environment and the need for life support systems. Remote sensing capabilities are restricted due to rapid electromagnetic wave attenuation in water; visible light transmission is at most tens of meters even in very clear waters, while wavelengths outside of the visible band are limited to 1 m or less. As a result, high-bandwidth radio frequency (RF) data transmission commonly used for remote communication and positioning in subaerial and outer space environments is typically replaced with relatively low-bandwidth acoustic methods in subaqueous environments. Therefore, unless a subaqueous platform has life support for human operators (*e.g.*, Lim *et al.*, 2010) or carries an umbilical leading back to the surface in the case of a Remotely Operated Vehicle (ROV; Forrest *et al.*, 2009) it is currently not possible to have real-time control or RF-based positioning (*e.g.*, GPS). This leads to the growing use of Autonomous Underwater Vehicles (AUVs) across multiple domains from experimental ecology (Williams *et al.*, 2012) to seafloor mapping (Trembanis *et al.*, 2012) to hydraulics research in both low (Forrest *et al.*, 2013) and high (Randeni *et al.*, 2016) energy environments. This section provides an overview of the considerations involved with the deployment, operation and recovery of AUVs for hydraulic measurements in offshore and ice-covered environments. Key considerations of vehicle performance for subsequent data processing involved with generating the final data products are also reviewed.

8.2 AUTONOMOUS UNDERWATER VEHICLES

Autonomous Underwater Vehicles (AUVs) are tetherless unmanned submersibles, preprogramed to execute a series of commands with minimal to no surface communications (*i.e.*, limited operator input). AUVs come in multiple shapes and sizes but are generally torpedo-shaped in order to minimize drag and maximize endurance (Figure 8.2.1). As operations are untethered, AUVs are increasingly being used in

Figure 8.2.1 Various types of AUVs: (a) Iver in Delaware Bay, DE, USA; (b) ISE Explorer in Indian Arm, BC, Canada; and (c) Gavia Scientific off of Baffin Island, NU, Canada.

environments that would be challenging to reach with traditional research vessels (*e.g.*, under-ice; Hayes & Morrison, 2002). Another advantage of operating untethered is that, once below the surface, AUVs are decoupled from both surface and ship motion. Decoupling the data collection platform from surface influence can significantly improve the collected data quality.

The use of AUVs, as both scientific and survey platforms, is becoming more widespread as the technology becomes increasingly robust. Applications can be roughly divided between commercial, military and academic sectors. The first of these, primarily the offshore oil industry, is concerned with seabed surveys and pipeline tracking (Evans *et al.*, 2003) whereas military involvement with AUVs has been almost entirely concentrated on port protection and mine countermeasures (Bovio *et al.*, 2006). Academic applications tackle a diversity of issues from tracking harmful algal blooms (Robbins *et al.*, 2006) to mapping deep-sea vents (Yoerger *et al.*, 2007) to space analog research (Forrest *et al.*, 2009).

From the perspective of limnology and oceanography (either biological, chemical or physical), one of the strengths of AUVs as platforms is the ability to travel horizontally to a vertical position tolerance not easily obtained by any other means. This allows characterization of horizontal variability at a resolution of centimeters over paths kilometers in length, which is often desirable. A combination of horizontal (AUV), vertical (traditional profilers), and temporal (traditional moorings) sampling techniques allows the three-dimensional, time-evolving nature of complex scalar fields in the water column to be characterized in a fashion that has previously not been possible. This is particularly relevant in ice-covered systems, as these are generally under-studied as a result of logistical and sampling challenges (*e.g.*, Forrest *et al.*, 2013).

AUV-based studies have focused on ocean settings (*e.g.*, Ramos *et al.*, 2007; Statham *et al.*, 2005), with a number of scientific deployments under sea-ice (*e.g.*, Ferguson *et al.*, 1999; Wadhams *et al.*, 2002; Hayes & Morrison, 2002; Brierley *et al.*, 2003; McEwan *et al.*, 2005; Nicholls *et al.*, 2006). Although there have been comparatively few applications of AUVs in lakes (*e.g.*, Laval *et al.*, 2000; Fong & Jones, 2006) there is an increasing effort to utilize AUVs under lake ice (*e.g.*, Forrest *et al.*, 2008; Forrest et al., 2012c; Forrest *et al.*, 2013).

AUVs have the capacity to carry a large suite of instrumentation, which can generally be grouped into four main categories: (1) optical imagery (*e.g.*, digital cameras, stereo-imagery, etc.); (2) acoustic imaging (*e.g.*, multibeam, sidescan sonar, etc.); (3) water sampling; and, (4) *in situ* measurements of water properties (*e.g.*, temperature, velocity, backscatter, fluorescence, etc.). Although this chapter is focused on these last two categories, as they are the most immediately applicable to hydraulics research, the application of AUVs for mapping surveys is of interest and is introduced briefly here.

Knowledge of the solid physical environment is critical to understanding the flow and properties of the surrounding waters. Production of an accurate bathymetric map of a water body is often the first step in environmental hydraulic studies, and AUVs: (1) reduce surface noise in the measurements; (2) allow data collection in shallow coastal waters (from 2–20 m water depth); (3) offer a potential cost-reduction over shipborne surveys by being deployable from shore or vessels of opportunity; and, (4) allow surveys to be conducted where otherwise unfeasible (*e.g.*, under-ice). Beyond bathymetric maps, AUVs can be equipped with optical and acoustic instrumentation that can provide insight into, for example, sediment and pollutant transport in near-shore environments, aquaculture and fisheries habitat, benthic impacts of coastal infrastructure, and flood mitigation strategies. AUVs have been used to provide high-resolution maps of seabed scour after the passage of a hurricane (Trembanis *et al.*, 2013), monitor the spread of benthic invasive species in a large lake (Forrest *et al.*, 2012a), produce digital terrain maps of the underside of sea-ice (Wadhams & Doble, 2008) and map the basal topography of an Antarctic ice shelf (Nicholls *et al.*, 2006). As sonar instrumentation continues to develop, there is also renewed interest in using AUV sonar systems for visualizing fluid mechanic processes in the water column (*e.g.*, Thorpe & Hall, 1983). A great advantage of AUVs is that mapping can often occur simultaneously with water quality sampling during a single deployment, providing a synoptic view of the physical environment and water column properties.

Water quality studies often require sample collection for further laboratory analysis, and the development of automated water sampling units allows the return of samples from specific points in space. Traditionally, water-sampling units are located inside the main body of the hull of an AUV. Water samplers that use this arrangement include the Gulper water sampler (Bird *et al.*, 2007), the automatic water sampler (Tamura *et al.*, 2000), and the mechanical syringe water sampler (Dowdeswell *et al.*, 2008). Vehicles with low internal volume space are generally unable to carry internal water samplers and there have been recent efforts to mount them externally, although external mounting of instrumentation significantly increases vehicle drag (Jagadeesh *et al.*, 2009).

From an environmental hydraulics perspective, the ability for real-time measurement of *in situ* water properties along specified transects, often at repeated intervals in time, is one of the greatest strengths of AUVs. These platforms can carry sensors for measurements of scalar properties of the water column including Conductivity-Temperature-Depth (CTD) profilers, optical backscatter units (*i.e.*, to measure turbidity, chlorophyll-*a* and dissolved organic matter), and fluorometers, as well as devices to measure velocities (*e.g.*, Acoustic Doppler Current Profilers; ADCPs; see Chapter 3.3, Volume II). Sensors aboard an AUV are usually no different than those deployed from surface vessels, and in some cases, such as for ADCP-derived velocity measurements (Fong & Jones, 2006), show the same along track biases as ship-based surveys. As with spatial surveys conducted from surface vessels, accurate positioning of an AUV is essential for geo-locating measurements, however, without the benefit of the real-time positioning available to surface vessels, AUV navigation is a much more complex topic. In the following sections an introduction to how an AUV navigates and determines its position underwater is presented, various aides used to improve position accuracy are discussed, factors that need to be considered that affect vehicle performance and sensor positioning are related, and different operational modes that can be utilized during mission planning to maximize collected data quality are introduced.

8.3 HORIZONTAL AND VERTICAL POSITIONING

Key in planning an AUV deployment is knowledge of how an AUV operates underwater. As real-time positioning is not possible underwater, a vehicle will pilot autonomously using different operational modes, aided through horizontal and vertical position information provided by onboard navigation sensors.

8.3.1 Operational modes

While there are many different strategies used for programming AUVs, mission planning typically involves uploading a series of spatial positions, given as depths (vertical locations) and waypoints (horizontal locations), into the AUV control software. Once the mission is initialized, the AUV then attempts to follow the path linking those points through real-time control of fins and rudders.

While there are several vehicle navigation modes for horizontal positioning, all rely on some form of dead-reckoning as GPS reception is not possible underwater. Dead-reckoning involves the integration of speed and direction along a path from the last known position. When the AUV is at the water surface, the position fix is typically obtained by a GPS unit located in a communications tower that extends above the water surface. When the AUV is submerged, known position can be determined using submerged acoustic transponders whose positions are known (*e.g.*, Long-BaseLine (LBL) or Ultra-Short BaseLine (USBL) acoustic positioning systems; see next section for more details). Between known position updates, the AUV position is ideally estimated by integrating AUV velocity over the earth (*i.e.*, speed and heading relative to the seabed) from the last known position. Estimates of vehicle speed through water are derived from propeller revolutions per minute (RPM) or the Doppler Velocity Log (DVL) is used to derive speed over the seabed when bottom track is available (*i.e.*, when the AUV is within acoustic range of the seabed). AUV heading information is typically provided by a flux-gate compass, which may be augmented by a DVL and possibly an Inertial Navigation System (INS), as detailed in the next section. As further discussed below, dead-reckoning without DVL bottom-lock can lead to extremely large position errors if there are significant water velocities. Periodic surfacing can be designed into the AUV mission planning in order to reset this position error through reacquiring new GPS positions, or LBL/USBL acoustic positioning systems can be used, although this method is of limited use in overhead environments (*e.g.*, under-ice) or where surface hazards are high (*e.g.*, near active shipping lanes).

In the vertical, two vehicle operation modes are often used: constant depth and constant altitude (height above bottom; see Figure 8.3.1). As the name implies, constant depth mode involves the AUV actively attempting to maintain a pre-programed fixed depth using an onboard hydrostatic pressure sensor. AUVs typically have

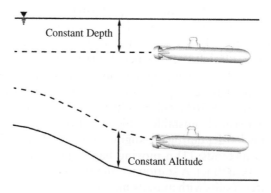

Figure 8.3.1 Operational modes of generic AUVs: constant depth (fixed depth below free surface) and constant altitude (fixed height from the bottom).

pressure sensors dedicated to navigation in addition to pressure sensors in scientific instrument packages such as CTDs. In constant-altitude mode the AUV attempts to maintain a set height above the seabed. AUV altitude can be determined using a dedicated acoustic altimeter or can be inferred from an onboard DVL. There are several post-processing approaches to determining the error in vertical position *a posteriori*. One approach is to use the mean and standard deviation calculated from the pressure data, for a constant depth mode, or altimetry data, for a constant altitude mode, to estimate a bound for the 95% confidence interval (*e.g.*, Forrest *et al.*, 2013). The vehicle's ability to maintain its vertical set point, as quantified by these 95% confidence interval bounds, is typically significantly better for constant depth than for constant altitude missions. For example, Forrest *et al.* (2008) reported a standard deviation of ± 0.05 m and ± 0.20 m for constant depth and constant altitude missions, respectively.

8.3.2 Horizontal positioning aiding

For many deployments, positioning of an AUV can be aided using an INS coupled with a DVL. An INS is a navigation aid that uses a combination of accelerometers and gyroscopes in each of the axes of rotation to continuously estimate the position, orientation and velocity of a moving object. These systems generally use the speed over ground, as estimated from the DVL when in range of the bottom, as additional input to a Kalman filter (defined further in the next section), to generate a significantly better position estimate (typically 0.1 % drift by distance traveled as compared to 1 – 10% using dead-reckoning using propeller RPM). The maximum range to bottom of a DVL is dependent on its frequency. For example, a 1200 kHz DVL has a typical range of 25–30 m whereas the range of a 150 kHz unit is typically 180–200 m. If the bottom is beyond DVL range (*i.e.*, in deep water applications), no DVL lock can be obtained and navigation error accumulates faster with an INS/DVL system compared to simple dead reckoning. In deep water (*i.e.*, >30 m for the 1200 kHz case) under-ice, a DVL can also be used to generate a speed over 'ground' by inverting the vehicle to track the bottom surface of the ice. Inverted flight is achieved by re-ballasting the AUV so that it floats upside-down. While ice-tracking has worked well under sea-ice (Doble *et al.*, 2009), poorer performance has been observed under lake-ice, perhaps due to the DVL failing to track the smoother ice surface and weaker water-ice density interface. Inverting the vehicle introduces additional complications to AUV performance and deployment, including submersion of the communication tower when the vehicle is at the surface. Future applications are aiming to enhance the functionality of upwards pointing Acoustic Doppler Current Profilers (ADCPs) to also include DVL functionality, and thus enabling both bottom-tracking and ice-tracking in the proper AUV flight orientation.

While robust at lower latitudes, problems with initial INS alignment should be expected at higher latitudes (> 70°). This results from the fact that the heading prediction the INS uses is increasingly unstable at higher latitudes and so heading alignment becomes proportionally harder during the alignment phase. Initial alignment of the INS is achieved using two methods; (1) a moving base alignment (fixed error) or, (2) a stationary alignment (fixed time). A stationary alignment, with an increased alignment time, tends to give better results with initially minimal drift (*e.g.*, Kaminski *et al.*, 2010).

In addition to using an INS to improve the vehicle horizontal position *prediction*, there are a number of techniques that use acoustic transponders to determine the *absolute*

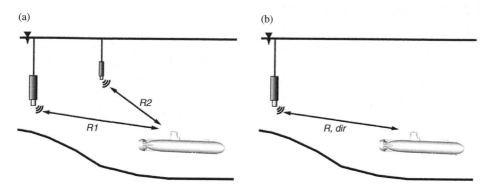

Figure 8.3.2 Examples of commonly used acoustic positioning systems: (a) long baseline (LBL) based on range only to the beacons (*R1* and *R2*) and (b) ultra short baseline (USBL) based on range and direction to the vehicle (*R* and *dir*).

vehicle position. While there are numerous methods utilizing acoustics for positioning, the two most common in AUV applications are acoustic Long BaseLine (LBL) and Ultra-Short BaseLine (USBL) positioning (see Figure 8.3.2). A LBL system works by the vehicle sending a signal to two or more acoustic transponders and then estimating the range to each. Knowing the average speed of sound in the water column, a range to each beacon can be calculated, and the position of the vehicle relative to those beacons can be determined by triangulation or trilateration. One limitation of LBL navigational systems that are currently used is the fact that multiple beacons are needed, increasing operational infrastructure requirements and complicating logistics (Figure 8.3.2a). In addition, vehicle motion between alternating range estimates introduces a continuous position offset on the order of 5–10 m but can have outlier positions exceeding 20–30 m. While the control software of an AUV will generally be able to correct for this, it can result in erratic estimates in horizontal positioning of the vehicle. A USBL system operates on similar principles but instead of the vehicle interrogating the topside transponder, the surface transponder has a tightly spaced transducer array (*e.g.*, < 0.30 m) that is used to generate ranges and directions to a transponder on the vehicle (Figure 8.3.2b). Knowing the attitude (*i.e.*, pitch, roll and yaw) of the surface transponder, it is then possible to derive the position of the vehicle. While these systems are generally less operationally difficult to establish, as only a single surface transponder is required, the horizontal positioning error can be greater than a LBL network (10–20 m). This estimate in position can be refined through simple least-squares regression of raw navigational data or through more advanced Kalman filters summarized in the next section.

8.3.3 Kalman filters

As a supplement or replacement of navigational aids, statistical methods, namely the linear Kalman Filter (KF) or the non-linear Extended Kalman Filter (EKF), are often used with AUVs to improve positional estimates based on raw dead reckoning data (*e.g.*, McEwan *et al.*, 2005; Morgado *et al.*, 2011). By minimizing the error variance associated with each of the vehicle attitude sensors, a KF uses sensor inputs to estimate the vehicle position.

8.3.3.1 Defining the Kalman filter

In a linear KF, the signal noise (*i.e.*, the statistical deviation from the mean) associated with each of the inputs is used to produce more precise estimates of the instantaneous state of the linear dynamic system (Maybeck, 1979). Following the seminal development of this type of filter (Kalman, 1960), it was almost immediately adapted for use in real-time, on-board navigation of the Apollo spacecraft (Grewal & Andrews, 2010). Following this successful application, the KF became an integral component for INS being incorporated into aerial and marine applications. In terms of application in the domain of AUV technologies, a KF is more appropriately thought of as an estimation algorithm rather than a true navigational filter.

The widespread adoption of the KF is mainly a result of it being a recursive process (*i.e.*, the calculation algorithm is only based on two time steps; t and $t - 1$) that has relatively low computational and storage requirements. The KF is also optimal as it utilizes available prior knowledge of the system and sensor performance (*i.e.*, characteristic noise levels) to statistically minimize the noise levels as it propagates the probability distribution of its state variables forward at each time step (Grewal & Andrews, 2010). While many variants exist, the KF is an extremely proven technique to estimate the state variables of a linear system (Maybeck, 1979).

To generate an optimal solution, a number of conditions must be met including: (1) a linear model must represent the system; (2) the system and measurement covariances (which are both descriptions of the tracked noise of the system) must both be white (uncorrelated in time); and (3) the distribution of this noise must be Gaussian (Maybeck, 1979). This last criterion is important so that, when the covariances are combined in the prediction step of the KF, their product displays a similar Gaussian distribution.

At each time step, a KF generates both prediction, based on the previous time step, and correction estimates, based on the current time step, using the error covariances. In the prediction step, *a priori* estimates, namely the current state and error covariance predictions, are projected forward to the next time step. In the correction step, *a posteriori* state and error covariance estimates are generated using the sensor input at that time step. This cycle is iterated with the objective of statistically minimizing the associated error at each time step. The transition of a linear system from the *a priori* state to the *a posteriori* state is referred to as the transition matrix. Typical transition matrices used to describe AUV performance usually include the hydrodynamic response to external forcings acting on the vehicle (*e.g.*, Fossen, 1994). While these matrices can sometimes be calculated directly (*e.g.*, Feldman, 1979), they are more often obtained from experimental testing.

8.3.3.2 Kalman filter equations

At each time update, the system state, or the state vector $\hat{x}_t^-$ is assumed to have evolved from the previous state vector $(\hat{x}_{t-1})$ at time $t - 1$, the *a priori* estimate. The error covariance matrix (P_t^-) is assumed to evolve similarly if the process noise covariance matrix, Q, is known (Welch & Bishop, 2006). The equations for the *a priori* estimates are given as:

$$\hat{x}_t^- = A\hat{x}_{t-1} + Bu_t \tag{8.3.1}$$

$$P_t^- = AP_{t-1}A^T + Q \tag{8.3.2}$$

where A is the state transition matrix previously mentioned and weights the effects of each of the system states (*e.g.*, accelerations, velocities, etc.) on $\hat{x}_t^-$, B is the input control matrix which weights the influence of the input vector on $\hat{x}_t^-$, and u_t is the input control vector.

These *a priori* estimates are then projected onto the *a posteriori* estimates through the use of the Kalman gain, K_t, which is a function of the transformation matrix, H, relating the state vector the measurement data at the present time step and the measurement noise covariance matrix, R:

$$K_t = P_t^- H'\left(HP_t^- H' + R\right)^{-1} \tag{8.3.3}$$

The *a posteriori* estimate of the state vector, $\hat{x}_t$, is then estimated knowing the true measurement vector, z_t, at a given time step:

$$\hat{x}_t = \hat{x}_t^- + K_t\left(z_t - H\hat{x}_t^-\right) \tag{8.3.4}$$

The influence of the K_t can be seen in Equation (8.3.4) in the residual, sometimes referred to as the *innovation* term, $(z_t - H\hat{x}_t^-)$. It is apparent that zero residual is a result of true agreement and that the response of the KF to a non-zero residual will be dependent on K_t. As the purpose of K_t is to blend the estimate and measurement at the present time step, the innovation term corresponds with what would be intuitively expected: as the estimate uncertainty decreases, K_t becomes smaller and, conversely, as the measurement noise decreases, K_t approaches 1. At the final step of the *a posteriori* estimate, P_t gets updated from the previous estimate using the identity matrix, I:

$$P_t = (I - K_t H)P_t^- \tag{8.3.5}$$

In practice (non-linear cases), the transition and transformation matrices might change with each time step, but A and H can be assumed to remain constant. The initial state estimate can be arbitrary as the KF will quickly optimize the error, but is typically taken as equal to the first available measurement. A can also contain a dynamic model of the physical system in state-space form, which allows state estimates to be made that are not directly related to the variable being measured. This approach also provides the KF with a means to predict system time, allowing state estimates to continue to be generated even if measurement data are lost. Similar to A and H, Q and R can also be set to a constant value. This system of equations can then be iterated for entire datasets where measurements have been made.

8.4 VEHICLE AND MISSION DESIGN CONSIDERATIONS FOR DATA COLLECTION

8.4.1 Vehicle induced flow field

An important aspect when sampling with an AUV is the influence of its hydrodynamic flow field on the accuracy of the sensor measurements. Two phenomena of particular interest are the boundary layer and pressure field generated by the AUV when it is in motion. The boundary layer will distort any measurement within it (*e.g.*, temperature, water velocity, etc.). Furthermore, the thermal discharge from electrical and mechanical components of the vehicle and the skin friction of the hull form are assumed to be contained within the boundary layer and the wake of the AUV. For these reasons, it is important for sampling to be conducted outside the boundary layer.

8.4.1.1 Boundary layer

Due to the viscosity of water and the adhesive force between the liquid and the AUV surface, the velocity profile of the flow field around the vehicle changes from zero at the hull to the free-stream velocity over a very short distance normal to the surface. The layer of liquid around the vehicle at which this change in velocity profile occurs is called the boundary layer.

Figure 8.4.1 illustrates the boundary layer around a torpedo-shaped AUV at zero angle of attack. As the flow passes the bow of the AUV, boundary shear stress is high, leading to the formation of a boundary layer with a high velocity gradient. As the fluid travels farther downstream along the surface, the retardation of fluid flow increases due to the shearing force and the boundary layer thickens. As the boundary layer increases in thickness, the velocity gradient and boundary shear distribution gradually decreases within the layer. If the region of rising pressure downstream becomes too severe, the flow in the boundary layer can be gradually slowed down to rest and reversed. This is called separation, and the detached flow leads to slow moving vortices forming in the wake.

The thickness of the boundary layer is defined as the distance from the boundary to the point where the velocity reaches the freestream velocity. While the thickness of the boundary layer can be estimated by analytical methods such as flat plate theory, it is also important to account for the curvature effects of the vehicle shape resulting from:

- Compression of the boundary layer at the AUV bow;
- Escalated thickening of the boundary layer at the AUV stern;
- Enhanced turbulence effects on concave surfaces; and
- Decreased turbulence effects on convex surfaces.

Therefore, it is recommended that estimates of boundary layer thickness based on flat plate theory be doubled to ensure that the sampling location along the length of the AUV is outside the boundary layer. For example, adapting Prandtl's turbulent boundary layer thickness estimate for an AUV gives:

Figure 8.4.1 Velocity flow field around an underwater vehicle.

$$\delta_{Ls} = 2\left(\frac{0.16L_s}{Re_{Ls}^{1/7}}\right) \tag{8.4.1}$$

where δ_{Ls} is the boundary layer thickness along the vehicle, L_S is the surface length and Re_{Ls} is the Reynolds number as a function of surface length.

8.4.1.2 Pressure field

An AUV generates hydrodynamic pressure fields around itself when it is in motion. For a torpedo-shaped AUV, these pressure fields tend to propagate around the bow and stern of the vehicle while the pressure field around the parallel mid-body remains fairly neutral. Figure 8.4.2 shows the pressure fields generated by a moving torpedo-shaped AUV at zero angle of attack. The pressure fields are generally non-dimensionalized:

$$C_p = \frac{p - p_\infty}{0.5\rho U_\infty} \tag{8.4.2}$$

where C_p is the pressure coefficient, p is the pressure at the location, p_∞ is the freestream pressure, ρ is the fluid density and U_∞ is the freestream velocity.

The red and blue contours in Figure 8.4.2 give the positive and negative pressure fields around the vehicle respectively, with the pressure coefficient contour boundaries clipped to values of ±0.05 to indicate the extent of the pressure fields generated by the vehicle that can impede pressure measurements. For example, the AUV in Figure 8.4.2 travelling at a cruise speed of 2.0 m/s in freshwater ($\rho = 1000$ kg/m^3) produces ±102 Pa in hydrodynamic pressure at the boundaries of contours. The hydrodynamic pressure increases to a maximum value at the center of the contours, ranging from −1200 Pa to 2000 Pa. For a depth sensor located where the peak values occur, this equates to approximately 0.12 to 0.20 m of hydrostatic pressure. Considering the *Gavia*-class AUV that is 0.20 m in diameter, the influence of the hydrodynamic pressure field can cause undesirable oscillations in depth control as the vehicle changes its velocity. Therefore, it is desirable to place pressure sensitive sensors around amidships of a torpedo-shaped AUV. If pressure sensors are to be placed elsewhere (*e.g.*, at the bow of the AUV in the case of the CTD) pressure corrections inherent to the sensor design and operation need to be considered.

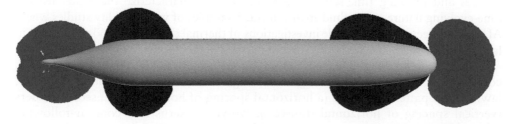

Figure 8.4.2 Pressure flow field around an underwater vehicle with red and blue contours representing the positive and negative pressure fields around the vehicle, respectively.

8.4.2 Mission design

AUVs usually carry an array of scientific sensors, and payloads can often be modified depending on the focus of the scientific investigation. Although the scientific payload may vary with each deployment, AUV missions can be generalized into: 1) mapping missions and 2) water quality missions. Mapping missions aim to create a georeferenced optical or acoustic image of the benthic terrain or substrate, whereas water quality missions are focused on the properties of the water column, the spatial and temporal variation of measurable scalars (*e.g.*, temperature, salinity, biomass, etc.) and vectors (*i.e.*, water velocities). While the two mission types are often completed simultaneously, mission design must take into consideration the specific requirements of the variables to be measured, how AUV mission design will affect data quality and what navigational instrumentation can be used.

In general, mapping missions require a high degree of precision in vehicle position and orientation. For this reason, mapping missions are almost exclusively INS aided and are designed as constant altitude missions to maintain consistent data spatial resolution and image or acoustic swath width. However, constant depth missions are sometimes preferred over (or under) extremely rough or abrupt topography (*e.g.*, coral reefs or sea-ice ridges) or in scenarios where uncharted obstacles or overhead environments pose a risk of vehicle loss or entrapment (*e.g.*, under ice shelves). Operator-controlled variables, such as sonar or camera settings, line spacing, ping or shutter rates, and vehicle altitude and speed must be carefully controlled. This is of particular relevance for conducting repeat surveys to investigate the temporal evolution of substrate features to ensure changes are real and not by-products of survey design or post-processing techniques (Trembanis *et al.*, 2013). In addition, external factors such as ambient light conditions in shallow water optical surveys (Forrest *et al.*, 2012a) or the hydrodynamics of the surrounding water body must be accounted for to eliminate vehicle performance artifacts in the generated product. For example, recent work has shown that low frequency oscillations due to pressure variations induced by ocean swells in shallow coastal waters can introduce vertical offsets into mapping data that must be accounted for in post-processing (Schmidt *et al.*, 2010). Analysis of vehicle engineering flight records and vehicle performance optimization are critical for production of high quality, high-resolution acoustic mapping products.

Water quality studies are generally focused on the investigation of fluid mechanic processes in the water column. These missions are often designed for the AUV to follow repeat transects at constant depth to look at temporal and spatial variations of scalars and vector fields, and are usually executed in concert with the collection of vertical profiles and mooring time-series data to completely characterize the 3-dimensional time-evolving nature of the fluid environment. Examples of water quality studies where AUVs have been utilized include investigations of thermal convection under ice-cover in lakes (Forrest *et al.*, 2008) and tracking of harmful algal blooms (Robbins *et al.*, 2006).

Missions can follow any sequence of waypoints programed into the AUV; however, commonly used geometrical designs are single straight line horizontal transects, lawnmower patterns or grids (a horizontal spacing of horizontal transects), ladders (vertical spacing of horizontal transects), 'yo-yo' patterns and contour-following (although this feature is not yet available in many AUVs; Bennett & Leonard, 2002) depending on the mission design criteria.

To meet the desired spatial resolution and coverage requirements of a dataset various factors must be considered, including sensor sampling frequency, cruising speed of the AUV, water velocity, vehicle range and power limitations. The spatial resolution of, for example, temperature and salinity properties along a single transect can be addressed by the relative speed of the vehicle through the water and the sampling frequency of the CTD. Typical cruising speeds of torpedo-type AUVs are between 0.75–2.5 ms^{-1} and so a CTD sampling at 16 Hz would yield an along-track resolution of 0.05–0.15 m between data points. In addition to along-track resolution, another consideration is the overall size of the area of interest and the resolution of coverage within that domain. The completeness of coverage, as determined by the number and spacing of AUV transects within the sample region, will be determined by the range of the vehicle.

8.5 MAPPING UNDER ICE

Despite the risk of vehicle loss (*e.g.*, Au-tosub2; Strutt, 2006), there have been increasing efforts in recent years to deploy AUVs under ice in order to collect otherwise unobtainable datasets (Forrest *et al.*, 2008; Doble *et al.*, 2009). The ability to measure horizontal variability of environmental parameters beneath the ice surface increases both the quality of data being collected and the range of observation. To date there have been relatively few scientifically driven under-ice deployments (*e.g.*, Ferguson *et al.*, 1999; Wadhams *et al.*, 2002; Brierley *et al.*, 2003; McEwan *et al.*, 2005; Nicholls *et al.*, 2006; Forrest *et al.*, 2013). The majority of these deployments are in marine and coastal settings, with fewer in lacustrine settings. This is, in part, due to the elevated infrastructure costs of under-ice deployment. From the perspective of hydraulics research, fine-scale horizontal measurements under-ice are really only possible with AUVs. The reader is invited to review the growing body of literature, but specific considerations for experimentation are presented below.

8.5.1 Ice environments

There are many different types of ice environments that are of interest in terms of hydraulics research including river ice, lake ice (Forrest *et al.*, 2013), sea-ice (both fast and drifting; Hayes & Morrison, 2002; Doble *et al.*, 2009); icebergs and ice islands (Forrest *et al.*, 2012a); ice shelves (Dutrieux *et al.*, 2014); and, ice tongues (Stevens *et al.*, 2014). One of the major challenges with these environments is safe navigation for the successful deployment and retrieval of the vehicle. A critical element to this is communications during under-ice operations.

Through-ice data communication is of great interest to AUV ice operations because it could allow relatively high bandwidth communication with the AUV without drilling holes and deploying acoustic transducers. In emergency situations acoustic telemetry has typically been used for data communications between a control console and an AUV. Acoustic communication data bandwidth is quite low which makes control of the AUV difficult. A through-ice, high bandwidth communications link would be a significant system improvement. While it is sometimes possible to have wireless communications through fresh lake ice under the right conditions (no snow and black ice while the vehicle was in contact with the ice), it is difficult to obtain these conditions for most applications (*e.g.*, Forrest *et al.*, 2012b). The same luxury is not available when operating under sea ice, of course, due to the elevated salt content in sea ice. Thus, lake ice is an ideal test-bed for under-ice AUV operations.

Wireless Fiber Systems (WFS) manufactures electromagnetic modems that can operate through seawater and ice at a data rate of up to 100 Kbit/sec. Trials in 2009 (Kaminski *et al.*, 2009) demonstrated their operating range through ice to be extremely limited (<10 m) and only performed under certain conditions. For example, the WFS modem could provide a high-speed data link to the AUV if the vehicle surfaced under smooth, relatively thin, annual ice. 100 Kbits/sec data rates were routinely observed during the tests. The performance of modems communicating through thick ice has not been proven but it is not likely they would work through more than a few meters. Even though the performance results of the WFS modems are promising, they cannot be relied upon as the only means to communicate with the AUV in an emergency situation. This is because the AUV may surface in an area with rough, thick ice that would make using the WFS modems difficult or impossible.

In an emergency scenario, it may be necessary to locate the AUV under the ice with enough accuracy that it would be possible to stand on the ice directly above the vehicle. This is necessary to establish through-ice communications. The two scenarios where this requirement is envisioned are: when it is necessary to use the WFS through-ice modems (recall that the surface modem must be placed on the ice within a range of 4–5 m of the subsurface modem located in the AUV); and, for emergency recovery or charging with a cable that is terminated with a wet-mateable connector.

Doble *et al.* (2009) were able to locate their AUV under both lake- and sea-ice by placing a commercial avalanche rescue beacon inside the hull. These beacons are used to locate skiers and hikers buried by avalanches at search ranges of 50–70 m depending on the brand of beacon being used. For example, Forrest *et al.* (2007) routinely located their AUV at ranges of 30–40 m during a 2007 deployment under lake ice (~0.50 m thick). The beacons used for this work were an Ortovox F1 Beacon (230 g dry weight and powered by 2 AA cells) and a Mammut Pulse Barryvox Beacon (210 g dry weight powered by 3 AAA cells). The operating frequency of both of these units was 457 kHz, and both of the units can function in either transmit or search modes. Under first year sea-ice (< 2 m), results were much less impressive when the transponders were placed in an AUV hull (< 15 m; Kaminski *et al.*, 2009), although this technique could still present a useful tool for determining a drill site in a vehicle rescue based scenario.

8.5.2 Under-ice AUV operations

The deployment method of an AUV for under-ice missions is dictated by the project infrastructure and the vehicle size; the two most common approaches are deployment from a surface vessel through a natural open water lead in ice or beyond the ice margin (*e.g.*, Jakuba *et al.*, 2008; Williams *et al.*, 2013) or through holes cut through the ice surface (*e.g.*, Stone *et al.*, 2010; Forrest *et al.*, 2013). In either case, a number of different strategies have been implemented to get the vehicle clear of the ice surface, including triggering mission deployments with acoustic commands (Kaminski *et al.*, 2010), self-localizing sonar mapping (SLAM; Stone *et al.*, 2010), or direct launches from the surface (Forrest *et al.*, 2007). Whichever strategy is used, the vehicle will need a method to then localize itself under-ice in the absence of a GPS signal. This is generally done with either LBL or USBL techniques.

Work has been done to combine the LBL location with INS positioning in order to maintain a relative position for under sea-ice deployments where the surface can be concurrently drifting (up to 10 km day^{-1}) and rotating (2 – 3° day^{-1}; Williams *et al.*, 2013). An alternative method would be acoustic homing through minimization of range to a single LBL beacon lowered in a receiving net at the end of a given mission (Forrest *et al.*, 2007; Kaminski *et al.*, 2010). This would allow the vehicle to be returned to the deployment hole while continuously knowing its absolute position throughout the entirety of the mission (provided the INS drift was still low). For vehicles with only a downward looking DVL aid for the INS it is possible to run in inverted flight in order to minimize navigational error by measuring the 'bottom' velocity along the ice surface (Doble *et al.*, 2009).

8.5.3 Case study: Erebus Glacial Tongue

To summarize the various elements discussed in this chapter, a case study is presented. From October to November 2010, field observations were made with *UBC-Gavia*, a *Gavia*-class AUV, off the northern side of the Erebus Glacier Tongue (EGT). The EGT is a floating extension of the Erebus Glacier that extended 12 km out into southern McMurdo Sound (Stevens *et al.*, 2014) from the western coast of the Ross Islands, Antarctica, at the time of the study. The aim of the AUV deployment was to characterize the near-field turbulent conditions around the EGT. The AUV, *UBC-Gavia*, was deployed through a 1x3m hole (Figure 8.5.1) in the sea-ice within 100 m of the sidewall of the EGT and programed to follow a straight line between two waypoints at specified constant depths ranging between 10–25 m to a distance ~400 m from the deployment hole.

The vehicle payload included a Seabird Conductivity-Temperature-Depth (CTD) profiler, an optical backscatter meter, up and downwards facing ADCP, side-scan sonar and a high-resolution camera. An LBL navigation system consisting of two acoustic beacons moored below the stationary pack ice was used to calculate ranges and update the position of the AUV in real-time. However, the range estimates were found to be inaccurate because the sound speed velocity used by the navigational system onboard the *UBC-Gavia* to calculate the ranges was incorrect, leading to incorrect LBL positions. The following section describes the method used to improve the post-processed position estimates.

8.5.3.1 Correction for sound velocity

The true value of the speed of sound, c, at the time of sampling was estimated from water column properties as

$$c = a_1 + a_2 T + a_3 T^2 + a_4 T^3 + a_5 (S - 35) + a_6 z + a_7 z^2 + a_8 T(S - 35) + a_9 T z^3$$

$$(8.5.1)$$

where T is the temperature of the sea water, S is the salinity, z is the water depth and a_1, a_2, a_3, a_4, a_5, a_6, a_7, a_8 and a_9 are constants (Fofonoff & Millard, 1983). Equation (8.5.1) assumes uniform water properties between the AUV and the acoustic beacons, which

Figure 8.5.1 *UBC-Gavia* being readied for deployment through a 1x3 m hole in the sea-ice in McMurdo Sound, Antarctica, 2010.

was considered valid as the surface layer (down to ~50 m) was relatively well-mixed (Stevens *et al.*, 2014) and the vehicle was running at depths shallower than this. After the correct value of c was estimated corrected ranges, r_c, between the beacons and the vehicle were calculated using the following correction factor (where 1424 m/s was the initial uncorrected estimate of c and r was the initial uncorrected range).

$$r_c = r \frac{c}{1424} \tag{8.5.2}$$

Once the new values of r_c were calculated, new LBL positions could be estimated for each time step the vehicle was transiting beneath the ice. As will be shown below, these positions, assuming a zero variance, could then be fused with the state vector to generate new positions using a Kalman filter.

8.5.3.2 Applied filters

An example of the application of these filters is presented on a single line transect from 27 Oct 2010. Initially, the corrected ranges to each of the LBL transponders (Equation (8.5.2)) were used to estimate the position of the vehicle along the transect. The position estimates made by the vehicle (in latitude / longitude) were converted to UTM and positioned relative to the deployment hole, which was centered at (0,0). This initial coordinates, in conjunction with the initial depth coordinates, which were non-zero, became the prior state vector, $\hat{x}_{t-1}$, for the Kalman filter defined in Equation (8.3.1). The prior error covariance P_{t-1}, transition A, and transformation H matrices were all assumed to be unity whereas the noise covariance Q and measurement covariance R matrices were determined for the run. This KF application was controlled by the process and measurement noise of the system. As such, the KF could be 'tuned' until the generated navigational solution matched the known LBL path.

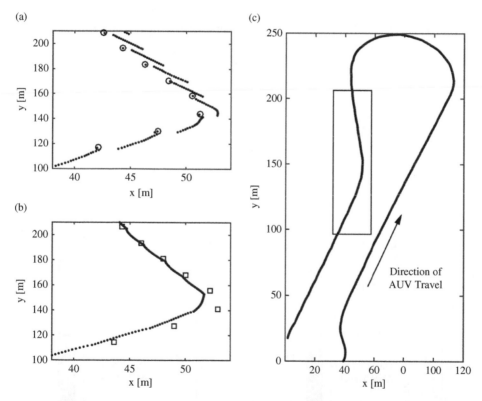

Figure 8.5.2 (a) Detailed view of raw navigation data (blue dots) with uncorrected LBL data (non-filled circles) on a short radius turn on an AUV transect collected on 27 Oct. 2010; (b) view of processed navigation data using the discrete Kalman Filter (blue dots) after corrected LBL positioning with modified sound speed (open squares); and, (c) complete return transect from the deployment hole through the ice located at the point of origin with direction of travel of AUV and the site of detailed view as shown. Note that the scales have been distorted to better illustrate smoothing of the navigation data.

Figure 8.5.2a represents a small radius turn in the AUV path selected from the run conducted on 27 Oct 2010. This figure shows the unfiltered navigation path. The jumps in position data at each of the LBL solutions (circles) are where the vehicle control system corrects the position from where it estimates the vehicle to be (dotted line) to the absolute position being generated from the LBL solutions (*i.e.*, a position of zero covariance). Once the corrected LBL positions were calculated (squares), the final solution was generated by the KF (dashed line) using Equations (8.3.1) and (8.3.2). As shown in Figure 8.5.2b, the resulting solution still has a slight horizontal oscillation to it, but is significantly better than the original estimations. Figure 8.5.2c then shows the overall result for the entire run.

Having the corrected position estimates from the KF, it is then possible to geolocate the scientific payload data. As an example, temperature and salinity data collected with a SBE49 CTD sensor are presented below (Figure 8.5.3). While at the limit of the temperature and conductivity accuracy of the instrument (temperature resolution and accuracy of < 0.0001°C and ± 0.005 °C; conductivity resolution and accuracy of 0.05 mS/m and ±0.5 mS/m), coherent structure is observed in both parameters (Figure 8.5.3a and b) as is immediately obvious in the transition from hot to cool colors in both panels. The simultaneous change between both temperature and salinity also indicates a transition between different water masses. Observing such lateral changes (*e.g.*, decimeter scale) is really only possible with sampling profilers such as this.

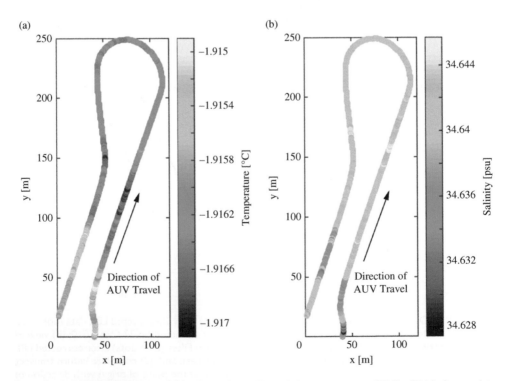

Figure 8.5.3 (a) Temperature and (b) salinity data collected along transect on 27 Oct 2010 after applying sound speed corrections and KF.

REFERENCES

Bennett, A.A. & Leonard, J.J. (2002) A behavior-based approach to adaptive feature detection and following with autonomous underwater vehicles. *IEEE J Ocean Eng*, 30(2), 213–226.

Bird, L.E., Sherman, A.D., & Ryan, J.P. (2007) Development of an active, large volume, discrete seawater sampler for Autonomous Underwater Vehicles. *Proceedings of the Oceans MTS/ IEEE Conference 2007, 29 September 4. October 2007, Vancouver, Canada*. pp. 1–5.

Bovio, E., Cecchi, D., & Baralli F. (2006) Autonomous Underwater Vehicles for scientific and naval operations. *Annu Rev Control*, 30(2), 117–130.

Brierley, A.S., Fernandes, P.G., Brandon, M.A., Armstrong, F., Millard, N.W., McPhail, S.D., Stevenson, P., Pebody, M., Perrett, J., Squires, M., Bone, D.G., & Griffiths, G. (2003) An investigation of avoidance by Antarctic krill of RRS James Clark Ross using the Autosub-2 autonomous underwater vehicle. *Fish Res*, 60(2–3), 569–576.

Doble, M.J., Forrest, A.L., Wadhams, P., & Laval, B.E. (2009) Through-ice AUV deployment: Operational and technical experience from two seasons of Arctic fieldwork. *Cold Reg Sci Technol*, 56(2–3), 90–97.

Dowdeswell, J.A., Evans, J., Mugford, R., Griffiths, G., McPhail, S., Millard, N., Stevenson, P., Brandon, M.A., Banks, C., Heywood, K.J., Price, M.R., Dodd, P.A., Jenkins, A., Nicholls, K. W., Hayes, D., Abrahamsen, E.P., Tyler, P., Bett, B., Jones, D., Wadhams, P., Wilkinson, J.P., Stansfield, K., & Ackley, S. (2008) Autonomous underwater vehicles (AUVs) and investigations of the ice-ocean interface in Antarctic and Arctic waters. *J Glaciol*, 54(187), 661–672.

Dutrieux, P., De Rydt, J., Jenkins, A., Holland, P.R., Ha, H.K., Lee, S.H., Steig, E.J., Ding, Q., Abrahamsen, E.P., & Schröder, M. (2014) Strong sensitivity of Pine Island ice-shelf melting to climatic variability. *Science*, 343(6167), 174–178.

Evans J., Petillot, Y., Redmond, P., Wilson, M., & Lane, D. (2003) AUTOTRACKER: AUV embedded control architecture for autonomous pipeline and cable tracking. In: *Proceedings of the Oceans MTS/IEEE Conference 2003, 22–26 September 2003*, San Diego, California, 5, pp. 2651–2658.

Feldman, J. (1979) *DTNSRC Revised Standard Submarine Equations of Motion*. Tech. Rep. DTNSRDC/SPD-0393-09, David W. Taylor Naval Ship Research and Development Center, Bethesda, Md.

Ferguson, J., Pope, A., Butler, B., & Verrall, R.I. (1999) Theseus AUV – Two record breaking missions. *Sea Technol*, 40(2), 65–70.

Fofonoff, N.P. & Millard, R.C. (1983) *Algorithms for computation of fundamental properties of seawater*. Unesco technical papers in marine science 44, 53.

Fong, D.A. & Jones, N.L. (2006) Evaluation of AUV-based ADCP measurements. *Limnol Oceanogr Methods*, 4, 58–67.

Forrest, A.L., Bohm, H., Laval., B.E., Magnusson, E., Yeo, R., & Doble, M.J. (2007) Investigation of under-ice thermal structure: Small AUV deployment in Pavilion Lake, BC, Canada. *Proceedings of the Oceans MTS/IEEE Conference 2007, 29 September 4. October 2007*, Vancouver, Canada. pp. 1–9.

Forrest, A.L., Laval, B.E., Lim, D.S.S., Williams, D.R., Trembanis, A.C., Marinova, M.M., Shepard, R. Brady, A.L., Slater, G.F., Gernhardt, M.L., & McKay, C.P. (2009) Performance evaluation of underwater platforms in the context of space exploration. *Planet Space Sci*, 58(4), 706–716.

Forrest, A.L., Laval, B.E., Pieters, R., & Lim, D.S.S. (2008) Convectively driven transport in temperate lakes. *Limnol Oceanogr*, 53(5, part 2), 2321–2332.

Forrest, A.L., Laval, B.E., Pieters, R., & Lim, D.S.S. (2013) A cyclonic gyre in an ice-covered lake. *Limnol Oceanogr*, 58(1), 363–375.

Forrest, A.L., Wittmann, M.E., Schmidt, V., Raineault, N.A., Hamilton, A., Pike, W., Schladow, S.G., Reuter, J.E., Laval, B.E., & Trembanis, A.C. (2012a) Quantitative assessment of invasive species in lacustrine environments through benthic imagery analysis. *Limnol Oceanogr Methods*, 10, 65–74.

Forrest, A.L., Hamilton, A.K., Schmidt, V.E., Laval, B.E, Mueller, D., Crawford, A.J., Brucker, S., & Hamilton, T. (2012b) Digital terrain mapping of Petermann Ice Island fragments in the Canadian High Arctic. *Proceedings of the 21st IAHR International Symposium on Ice 2012, Dalian, China*. Dalian University of Technology Press. pp. 1–12.

Forrest, A.L., Andradóttir, H.Ó., & Laval, B.E. (2012c) Preconditioning of an underflow during ice-breakup in a subarctic lake. *Aquat Sci*, 74(2), 361–374.

Fossen, T.I. (1994) *Guidance and Control of Ocean Vehicles*. Chichester, John Wiley & Sons.

Grewal, M.S. & Andrews, A.P. (2010) Applications of Kalman Filtering in Aerospace 1960 to the Present [Historical Perspectives]. *IEEE Control Systems*, 30(3), 69–78.

Hayes, D.R. & Morrison, J.H. (2002) Determining turbulent vertical velocity, and fluxes of heat and salt with an Autonomous Underwater Vehicle. *J Atmos Oceanic Tech*, 19(5), 759–779.

Jagadeesh, P., Murali, K., & Idichandy, V.G. (2009) Experimental investigation of hydrodynamic force coefficients over AUV hull form. *Ocean Eng*, 36(1), 113–118.

Jakuba, M.V., Roman, C.N., Singh, H., Murphy, C., Kunz, C., Willis, C., Sato, T., & Sohn, R.A. (2008) Long-baseline acoustic navigation for under-ice autonomous underwater vehicle operations. *J Field Robot*, 25(11–12), 861–879.

Kalman, R.E. (1960) A new approach to linear filtering and prediction problems. *Trans. ASME – J Basic Eng*, 82, 35–45.

Kaminski, C., Crees, T., Ferguson, J., Forrest, A., Williams, J., Hopkin, D., & Heard, G. (2010) 12 days under ice–an historic AUV deployment in the Canadian High Arctic. *Proceedings of the 2010 IEEE/OES Autonomous Underwater Vehicles, Monterey, CA*. pp. 1–11.

Kaminski, C., Laframbroise, J.M. & Forrest, A.L. (2009) *Arctic 2009*. Quicklook Report (DRDC Research Report).

Laval, B., Bird, J.S., & Helland, P.D. (2000) An autonomous underwater vehicle for the study of small lakes. *J Atmosph Ocean Technol*, 17, 69–76.

Lim, D.S.S., Warman, G.L., Gernhardt, M.L., McKay, C.P., Fong, T., Marinova, M.M., Davila, A.F., Andersen, D., Brady, A.L., Cardman, Z., Cowie, B., Delaney, M.D., Fairén, A.G., Forrest, A.L., Heaton, J., Laval, B.E., Arnold, R., Nuytten, P., Osinski, G., Reay, M., Reid, D., Schulze-Makuch, D., Shepard, R., Slater, G.F., & Williams, D. (2010) Scientific field training for human planetary exploration. *Planet Space Sci*, 58(6), 920–930.

Maybeck, P.S. (1979) *Stochastic Models, Estimation, and Control*. New York, Academic Press.

McEwan, R., Thomas, H., Weber, D., & Psots, F. (2005) Performance of an AUV navigation system in Arctic latitudes. *IEEE J Ocean Eng*, 30(2), 443–454.

Morgado, M., Batista, P., Oliveira, P., & Silvestre, C. (2011) Position USBL/DVL sensor-based navigation filter in the presence of unknown ocean currents. *Automatica*, 47, 2604–2614.

Nicholls, K.W., Abrahamsen, E.P., Buck, J.J.H., Dodd, P.A., Goldblatt, C., Griffiths, G., Heywood, K.J., Hughes, N.E., Kaletzky, A., Lane-Serff, G.F., McPhail, S.D., Millard, N.W., Oliver, K.I.C., Perrett, J., Price, M.R., Pudsey, C.J., Saw, K., Stansfield, K., Stott, M.J., Wadhams, P., Webb, A.T., & Wilkinson, J.P. (2006) Measurements beneath an Antarctic ice shelf using an autonomous underwater vehicle. *Geophys Res Letters*, 33, L08612, doi:10.1029/2006GL025998.

Randeni, P., Supun, A.T., Forrest, A.L., Cossu, R., Leong, Z.Q., King, P.D., & Ranmuthugala, D. (2016) Autonomous Underwater Vehicle Motion Response: A nonacoustic tool for blue water navigation. *Mar Technol Soc J*, 50(2), 17–26.

Ramos, P.A., Neves, M.V., & Pereira, F.L. (2007) Mapping and initial dilution estimation of an ocean outfall plume using an autonomous underwater vehicle. *Cont Shelf Res*, 27(5), 583–593.

Robbins, I.C., Kirkpatrick, G.J., Blackwell, S.M., Hillier, J., Knight, C.A., & Moline, M.A. (2006) Improved monitoring of HABs using autonomous underwater vehicles (AUV). *Harmful Algae*, 5(6), 749–761.

Schmidt, V.E., Raineault, N.A., Skarke, A., Trembanis, A.C., & Mayer, L.A. (2010) *Correction of bathymetric survey artifacts resulting from apparent wave-induced vertical position of an AUV*. Report for University of New Hampshire Center for Coastal and Ocean Mapping/Joint Hydrographic Center, Durham, NH, 13.

Statham, P.J., Connely, D.P., German, C.R., Brand, T., Overnell, J.O., Bulukin, E., Millard, N., McPhail, S., Pebody, M., Perret, J. Squire, M., & Webb, A. (2005) Spatially complex distribution of dissolved manganese in a fjord as revealed by high-resolution in situ sensing using the Autonomous Underwater Vehicle Autosub. *Environ Sci Technol, 39*(24), 9440–9445.

Stevens, C.L., McPhee, M.G., Forrest, A.L., Leonard, G.H., Stanton, T. & Haskell, T.G. (2014) The influence of an Antarctic glacier tongue on near-field ocean circulation and mixing. *J Geophys Res Oceans, 119*(4), 2344–2362.

Stone, W., Hogan, B., Flesher, C., Gulati, S., Richmond, K., Murarka, A., Kuhlmann, G., Sridharan, M., Siegel, V., Price, R.M., Doran, P.T., & Priscu, J. (2010) Design and deployment of a four-degrees-of-freedom hovering autonomous underwater vehicle for sub-ice exploration and mapping. *Proc Inst Mech Eng, Part M: J Eng Maritime Environ, 224*(4), 341–361.

Strutt, J.E. (2006) *Report of the inquiry into the loss of Autosub2 under the Fimbulisen.* Southampton, UK, National Oceanography Centre Southampton, 39. (National Oceanography Centre Southampton Research and Consultancy Report, 12).

Tamura, K., Aoki, T., Nakamura, T., Tsukioka, S., Murashima, T., Ochi, H., Nakajoh, H., Ida, T., & Hyakudome, T. (2000) The development of the AUV-Urashima. *Proceedings of the OCEANS 2000 MTS/IEEE Conference and Exhibition (Cat. No.00CH37158)*, Providence, RI. Vol *1*, pp. 139–146, doi:10.1109/OCEANS.2000.881249.

Thorpe, S.A. & Hall, A.J. (1983) The characteristics of breaking waves, bubble clouds, and near-surface currents observed using side-scan sonar. *Cont Shelf Res, 1*(4), 353–384.

Trembanis, A.C., Forrest, A.L., Miller, D.C., Lim, D.S., Gernhardt, M.L., & Todd, W.L. (2012) Multiplatform ocean exploration: Insights from the NEEMO space analog mission. *Mar Technol Soc J, 46*(4), 7–19.

Trembanis, A., DuVal, C. Beaudoin, J., Schmidt, V., Miller, D., & Mayer, L. (2013). A detailed seabed signature from Hurricane Sandy revealed in bed forms and scour. *Geochem Geophys, 14*, 4334–4340, doi:10.1002/ggge.20260.

Wadhams, P. & Doble, M. J. (2008) Digital terrain mapping of the underside of sea ice from a small AUV. *Geophys Res. Letters, 35*, L01501, doi:10.1029/2007GL031921.

Wadhams, P., Holfort, J. Hansen, E., & Wilkinson, J.P. (2002) A deep, convective chimney in the winter Greenland Sea. *Geophys Res Letters, 29*(10), 76-1-76-4.

Welch, G. & Bishop, G. (2006) *An Introduction to the Kalman Filter*. Department of Computer Science, University of North Carolina.

Williams, G.D., Maksym, T., Kunz, C., Kimball, P., Singh, H., Wilkinson, J., Lachlan-Cope, T., Trujillo, E., Steer, A., Massom, R., Meiners, K., Heil, P., Lieser, J., Leonard, K., & Murphy, C. (2013) Beyond Point Measurements: Sea Ice Floes Characterized in 3-D. *Eos, Trans Am Geophys Union, 94*(7), 69–70.

Williams, S.B., Pizarro, O.R., Jakuba, M.V., Johnson, C.R., Barrett, N.S., Babcock, R.C., Kendrick, G.A., Steinberg, P.D., Heyward, A.J., Doherty, P.J., Mahon, I., Johnson-Roberson, M., Steinberg, D., & Friedman, A. (2012) Monitoring of benthic reference sites: Using an AUV. *IEEE Robot Autom Mag, 19*(1), 73–84.

Yoerger, D.R., Jakuba, M., Bradley, A.M., & Bingham, B. (2007) Techniques for deep sea near bottom survey using an Autonomous Underwater Vehicle. *Robotics Res, 28*, 416–429.

Subject Index

ABI *see* acoustic backscattering instrument
acceleration 333, 361
accelerometer **190**, 267, 333, 382
acoustic attenuation 78–79, 81–82, 225, 284, 287–295, 297–298
acoustic backscatter **36**, 38–39, 82, 283–284, 286–294
  intensity 82, 284
  suspended sediment concentration 283–298
acoustic backscattering instrument (ABI) 5, **36**, 37–59, 81–86, 283, 286–290
  1D profiles 42
  2D–3D profiles 50
  backscatter *see* acoustic backscatter
  bistatic *see* also multistatic 45 **46**, 85
  concentration measurement 283, 286–290
  emitter 38, **39**
  error sources 52–59, 286
  Doppler
    shift frequency 39, 83
    noise 52–56
    noise correction 54–56
  gate **40**, 42, **47**, 84
  gate volume 84
  monostatic 38, 41–43, 85
  multistatic 38, 43–46, 85
  multibeam 43
  receiver 38, **39**
  transducer
    characteristics 84
    configurations 42, 44, 46–50
  velocity profiler 43, 50
acoustic beam *see* acoustic backscattering instrument:transducer characteristics
acoustic Doppler current profiler *see* ADCP
acoustic Doppler flux profiler (ADFP) 285
acoustic Doppler velocimeter (ADV) 5, **36**, 37, 46–49, 89–**90**
  three-receiver configuration 46
  FlowTracker 89–90
  four-receiver configuration 48

transformation matrix 47–49
turbulence characterization 59
Vectrino 49
Vector 89
acoustic Doppler velocity profiler (ADVP) **36**, 37, 50–51, 285
  spatial averaging 58
acoustic footprint 78, 215–216, 225, 227, 229, 273–274
acoustic inversion methods 288–298
  iterative inversion method 288
  explicit inversion method 288
acoustic mapping velocimetry 271–274
  spatial and temporal resolution 273
  error sources 273–274
acoustic maps 271–272
acoustic signal 61, **262**, 268, 313
acoustic tomography 79–81, **346**, 350–351, 372
acoustic travel-time discharge meters 344, **346**, 350, 372
acoustic travel-time tomography 5, 37, 79–81, **346**, 350–351, 372
  tomographic pair 81
active transponder 274–275
adaptive Gaussian window 163–164, **165**
ADCP 5–6, **36**, 37, 59–79, 284, 353–355, 380
  apparent bedload velocity 269–271
  beam configurations 68
  bedload
    calibration 270
    error sources 270–271
    measurement 269–271
    resolution 270
    spatial distribution 270
  blanking distance 64, **66**, **67**, 66–68, 189
  bottom tracking 62, 74, 269–270
  broadband 64–66
  common configurations **60**, 61–62, 68–70
  concentration measurement 284, 290–298
  discharge measurement 353–355

ADCP (*cont.*)
  error sources 75–77, **76**
  horizontal ADCP (HADCP) 75, 77–79,
    284, 297–298, 364, 366–368, **370**,
    370, 372
  narrowband 63–64
  position and orientation 62, 74–75, 92,
    269, 273,353, 382
  processing techniques 63–66
  range gates and depth cells **66**, 66–67
  sampling frequency 63
  sediment concentration 284–285, 290–298
  side lobe interference **67**, 68, **76**, 284
  unmeasured zones **67**, 68
  velocity profile **69**, 70
ADCP deployment methods 70–74
  bottom tracking 62, 74, 269–270
  compass considerations 75
  GPS tracking 74, 269
  manned boat 71
  remote controlled boat 73–74
  tethered boat 71–73
ADCP transects **346**, 353–355, **354**
aerial LSPIV 168, 172, 173
aerial photogrammetry *see* photogrammetry:
    aerial
AGW *see* adaptive Gaussian window
air bubbles 57, **109**, 109, **130**, 131
airborne lidar *see* lidar
aliasing 56–57, 85, 138, 142, 145–146
anemometry *see* LDV
armor layer 252, 315
atmospheric pressure *see* pressure:
    atmospheric
attenuation compensation 289–290
  backward and forward scattering 289
  two-frequency method 289–290
automated transverse system **106**, 237–238
autonomous underwater vehicle (AUV) 6,
    377–380, **378**
  advantages of 379–380
  boundary layer 386–387, **386**
  mission design 385, 388–389
  operational modes 381–382, **381**
  positioning 380–383
  pressure field 386–387, **387**
  types of **378**
  underwater robots 377
AUV *see* autonomous underwater vehicle

backscatter, acoustic *see* acoustic backscatter
backscatter, optical *see* optical backscatter
backscattering, laser 216
backscattering, acoustic *see* acoustic
    backscattering
BASEGRAIN 244–247

bathymetry 5, 173–173, 211–259, 272, 279,
    309, **310**, 313, 352, 365
bathymetry measurement 212, 216, **227**, **228**
beam *see* laser beam
beat signal 97
bed
  material transport *see* sediment transport
  reference level *see* reference level
  roughness 174, 211, 230, 269
  shear stress 6, 323–332
  slope 315, 358, 371
bedforms 221, **222**, 271, **272**, 315, 368
  movement 271–274
  scanning **222**, 271
bedload *see* sediment transport:bedload
  flux 268–269
  measurements, ADCP *see* ADCP:bedload
    measurement
  particle tracer 274
  particle velocity 261, 269
  samplers **262**, 262–266
  samples 261–262
  transport 262, 265–272, 274, 285
  transport rate 271–274
Bernoulli 2
Biot-Savart Law 25
bistatic *see* multistatic
bottom tracking *see* ADCP:deployment
    methods
bottom-mounted, ADCP 367
boundary layer 22, **23**, 26–27, 79, 120, **323**,
    326, 328, 332, 368, 386–387
Bourdon pressure gauge **317**, 318
Bragg cell **101**–102, 104
broad-crested weirs *see* weirs: broad-crested
BTMA-2 sampler 265
bulk average velocity *see* cross-sectional aver-
    age velocity
bulk velocity *see* cross-sectional average
    velocity
Bunte-type sampler 265–266
buoyancy 10, 17, 21–22, 108, 129–130, 336
burst 110–114, 116–117, 122, 125, 127, 213
  detection 111–112

cameras 13, 15, 134–135, 141, 151–152,
    154–155, 165–166, 184, 231, 234–235
  exterior camera orientation 231
  interior camera geometry 231
cantilever in bending **334**, 336
capacitive sensor 316, **317**, 318
CCD 13–14, 152, 185, 231
circulation 25
CMOS 13–14, 184–185, 231
coherent light 94, 102–103, 124, 127, 216
coherent turbulent shear flow 9

coincidence
  data rates 121–122
  window 113–114, 122
  mode 114, 122
collimating optics 103
collinearity equation 169, 231, 234
complex experiments 3
concentration *see* sediment concentration,
    suspended
concentration measurements 276–298
concentration measurements, ADCP *see*
    ADCP:concentration measurement
conductivity 89, 191, **310**, 336–338, 344,
    380, 391, 394
conductivity-temperature-depth (CTD)
    336–**338**, 380, 382, 386, 389,
    391, 394
constant temperature anemometry (CTA) 35,
    329
continuous bedload monitoring 268,
    274–275
continuous discharge monitoring 187, 344,
    347, **345–347**, 372
continuous slope-area method (CSAM) **347**,
    356, 362, **363**
controls, channel 357–358
conveyor belt sampler 263, 265
Coulomb 2
critical points 23, 25, 27–31
cross-correlation function 137, 142, 146, 156,
    161, 170–171, 176, 179, 271
cross-sectional average velocity 343, 350,
    359, 364, 365–369
CTD *see* conductivity-temperature-depth

Da Vinci 9, **10**, 167
data collection by AUV 377, 379–380, 385,
    388–389
  acoustic imaging 379–380
  conductivity-temperature-depth (CTD)
    profilers 380, 382, 386, 389, 391, 394
  horizontal AUV transects **394**
  optical backscatter 380, 391
  optical imaging 379–380
  water quality 380, 388
  water sampling 379–380
data modeling 223–224
dealiasing 56–57
DEM *see* digital elevation model
density measurement 336–337
deployment methods, point velocity meters 92
  wading measurements 92
  cableway 92
  manned boat 92
  ice 92
depth 6, 309–315, 338

depth-integrating sediment samplers **277**,
    278–279
depth-slope product **323**, 330
despiking *see* spike removal
detection volume 114
differential pressure flow meters 345, 349
  principles **345**, **346**, **347**
digital elevation model 6, 211–213, **223**, 230,
    237, 271–274
dilution methods 5, 37, 191–193, 310, **346**,
    351
dilution velocity measurement
  slug release 191, **192**
  continuous release 192–193
  sodium chloride 191, **192**
  rhodamine 191
direct linear transformation (DLT) 168
discharge *see* discharge measurements
discharge components, ADCP
  edges **354**
  in-bin 353
  measured **354**
  unmeasured top **354**
  unmeasured bottom **354**
  total 355
discharge measurements 6, 37, 343–376
  accuracy 344, 356, 364, 367–369, 371–372
  discrete 343–347, 351–355, 362, 371
  calibration 343–344, 348, 350, 356–357,
    359–360, 364, **365**, 371
  continuous 343–347, 356–370, 372
  laboratory methods **345–346**
  field methods **346–347**
displacement meter **212**, 236–238
distribution, LDV signal
  negative exponential 116–117
  Poisson 116
Doppler, LDV
  burst **112**
  frequency 95, 98–99, 104, 110–113, 122,
    126,
  shift 94–97, 110, 125
Doppler noise *see* acoustic backscattering
    instrument
drag force 6, 190, 310, 332–336
  drag coefficient 332
  drag plate **334**
  dynamometer **334**
  force-balance system **334**
  force-gauges 333, **334**
  load cell 333, **334**, 335
  piezoelectric force transducers 333, 335
  strain gauges 335
  torque wrench **334**
  transducer alignment 335
drifters 5, 37, 189–190

drogues 5, 37, 189–190
  holey sock 190
  kite 190
drone 5
dune tracking 271–274
dunes 271–274
dynamic range, LDV 126
dynamometer *see* drag force

electric analog 4
electromagnetic current meter 19, 37, 86, **87**,
      89, 91–92, 94, 350
electromagnetic discharge meters
      346, 350
Elwha sampler **264**, 265
emitting probe 102–106
energy spectral density 112, **113**
equal-discharge increment 278–279
equal-width increment 279
Erebus Glacier Tongue 379–380, 391
Exner equation 271

FFT 112, 126, 131, **138**, 142–143, 178
floats 179, **346**
flow lines 12–13
flow obstacle techniques 326–328, 330
  FST-hemispheres **324**, 327–328, **328**, 330,
      332
  Preston tube **324**, 326, **327**, 332
  Stanton gauge **324**, 326, 332
  sublayer fence **324**, 326–327
flow rate measurements *see* discharge
      measurements
flow visualization 5, 9–34, 131
flow visualization techniques 15–23
  dyes and aqueous tracers 15–21
  hydrogen bubble 21–23
  near solid surfaces 23–25
  oil film 23–24
  particle trajectories (PIV) 21
  summary of methods **16**
  tell-tales 24
  tufts 24
fluctuating pressure measurements 318–321
fluorescein 16–17
fluorescent dyes 16–17, 191
focal length 14, 103, 106, 108, 114, 231, **232**
focused beam reflectance **262**, 283
forward scatter 39, 105–107, 127, 285, 289
frame straddling 135
free-surface, LDV 104–105, 108
frequency, LDV 98, 99, 101–102, 104,
      113–116, 118, 122, 125–126, 128, 268
  analysis 112
  broadening 105
  cut-off 131

Doppler *see* Doppler: frequency
  Nyquist 122
  resolution 127
  shift/shifting 101–102, 104, 125–126
  tracking 112
fringes, LDV
  distortion 105, 110
  model 98, 111, 124
  pattern 94 103–105, 110–111, 114, 122,
      124–126
functions
  autocorrelation 64, 118, 120–121, 123,
      136
  kernel 121, 140, 142
  power spectral density 30, 54, 114,
      117–118, 120–123
  structure 118
fundamentals and methods 1

gas lasers 102
Gaussian beam 30, 102, 104, 111
geared and lobed impellers **345**
geophone **262**, 265, **267**, 267–269
geo-referencing 219–221, 226, **228**, 244
  GNSS 214, 217, **218**, 220–221, 223
  GPS 221, **222**, 226, 227, 229
global positioning system *see* GPS
global shutter 14, 184
GPS 74–76, 81, 92, 189–190, 221–222, 226,
      227, 229, 239, 269–270, **353**, 358,
      377, 381, 391
  real time kinematic (RTK) **222**, 358
grain size class **250**, 261, 265, 269
grain size distribution 5, 128, 131, 211, 230,
      240–247, 252, 276, 288
  frequency distribution 241, 248
  sediment size classification 248
  Wentworth scale 249
  ϕ-scale 241, 249, **250**, 251

HADCP *see* ADCP:horizontal ADCP
handles 28
heat transfer techniques 328–332
  hot-film probes 329
  thermal probes 329
Helley-Smith sampler 264
heterodyne principle 97, 114, 124
high-frequency radar 5, 37, 179, 187–189,
      313, 356, 364, 372
  Bragg-scattering 187–188
  Bragg resonance 187–189
  comparison with ADCP measurements
      188–189, **189**
h-level crossing 271
holes 28
Hooke's law 333

horizontal ADCP *see* ADCP:horizontal ADCP
hot film anemometer 4, 35, **36**, 186, **187**, 329
hydraulics applications 2
hydrogen bubble visualization 12, **13**, **16**,
     21–23, 59, 285–286
hydrophone 225, **262**, 267–268

ice 6, 92, 132, 379–382, 388–392, **392**
illumination 14–15, 132–133, 141, 152, 155,
     174, 236, 321
image analysis 223, 229–236
image velocimetry, acoustic mapping 272
imaging sensors *see also* cameras
     13–14
  dynamic range 14
  sensitivity 14
impact pipe **262**, 267–268
impulse, impact sensors 268
in situ field experiments 3
incipient motion **310**, **323**, 326
index-velocity method 75, 364–368, **370**,
     371–372
instrumentation and measurement
     techniques 1
  role of 2
inter-arrival time 114, 116–117,
     122, 131
interference filters 103–105
intrusiveness 94, 123
isokinetic sampling 276–279

Japanese pipe microphone 267–268

Kalman filters 382, 383–385, 392–393
  application of 392–394, **393**
  defining 384
  equations 384–385

Lagrangian,
  bedload 274
  drifters 190
  LSPIV 171
large scale particle image velocimetry (LSPIV)
     5, **36**, 167–185, 181–183, **346**, 366,
     372
  aerial 168, 172, **173**
  discharge estimation 172–174, **173**
  error sources 174–177
  global error sources 174
  ground-control points (GCP) 169
  image processing 170–171
  index velocity 173–174, 183
  interrogation area (IA) 171
  local error sources 174
  mapping relation 168
  ortho-rectified image 170

search area 176
two–dimensional collinearity 168
laser
  Ar-Ion (Argon) 15–16, 20, 102, **106**
  beams 94, 97–98, 100, 102, 104–106, 108,
     110–111, 116, 124, 127
  diode **31**, 102
  He-Ne (Helium-Neon) **31**, 97, 102, 106,
     107
  Neodymium-doped Yttrium Aluminum
     garnet (Nd:YAG) 15, 20, **31**, 102,
     133–134, 141, 152–153, 155, 158, 216
  solid-state 15, 102
laser beam
  alignment 102–107, 122–123, 135, 155,
     283
  divergence 30, 102–104, 136, 215, 217,
     218, **222**
  focus 104–105
  Gaussian 30, 102, 104, 111
  parameter products (BPP) 104
  positioning 99–100
  waist 30, 103–105, 115–116, 122,
     135, 136
laser diffraction, particle size measurement
     282
laser Doppler anemometry *see* LDV
laser Doppler velocimetry *see* LDV
laser
  beam divergence 215–216
  laser scanning 216–217
  laser wavelength 216
  point quality 220–222, **222**
  ranging 213–215
  sampling rate 220–222, **222**
  spatial density 220–222, **222**
laser-induced fluorescence 13, 16,
     20–21, 37
lasers 15, 30, **31**
LDA *see* LDV
LDV 5, **36**, 37, 94–131
  bias 117
  dual-beam 94, 97–99, 124–125
  measuring volume 94, 97, 99–100,
     103–104, 106–107, 110–116, 122,
     124–125, 127, 131
  noise 111–112, 114–115, 118, 121–122,
     128
  signal distribution 116–117
  single beam 97
  spatial resolution 105, 110, 114–116
  step noise 118
  time-series 114, 117, 122
  time-series, equal spacing 118, 122
  tracer velocity 99, 125, 129
  velocity vector 99, 106

lenses 14–15, **101**, 103, 113, 135–136, **135**
  f-number 14, 136, **149**, 185
  focal length 14, 103, **106**, 108, 114, 231
  image distance 14
  magnification 14
  object distance 14
lidar 35, **213**, 220, 313
LIF *see* laser-induced fluorescence
light
  backscatter 103–105, 127–128, 216–217, 280–283
  coherent *see* coherent light
  intensity 104, 108, 111, 128
  laser 30–31, 94–97, 110, 127
  scattering 108, 114, 127, 128
  speed of 95–96, 99, 101
  wavelength 98, 128
light detection and ranging *see* lidar
line measurements 239
LISST *see* laser diffraction
local control 357–358

M2 parameter 104
Manning's equation 358–359
mass-transfer techniques 328–332
  mass-transfer probes 329
mean-section discharge estimation method 351–353
mechanical current meter 37, 86–88, **87**, 91–94
mechanical profilers 238–239
membrane 316–318
metal basket 265
microelectromechanical systems (MEMS) 324, 330, 335
micro-optical sensors **324**, 329–330
microphone 267–268, 318
mid-section discharge estimation method 351–353
Mie scattering regime 104, 110, 114, 127–128, 282
mineral powders 109
mixing, CTD measurements 336
moving boat, ADCP 353

navigation of AUVs 380–383
  dead reckoning 381
  Doppler velocity log (DVL) 381, 391
  inertial navigation system (INS) 381–382, 388
  long-baseline (LBL) acoustic positioning systems 381, **383**, 391–394
  ultra-short baseline (USBL) acoustic positioning systems 381, **383**, 391
nodes 27–30

noise removal, ABI *see* acoustic backscattering instrument
non-intrusive measurement (terrestrial laser scanning) **222**
nozzles **345**
nutating disks **345**

OBS *see* optical backscatter sensor
optical
  fibers 102–104
  filtering 16, 21, 99
  heterodyne 97, 114, 124
optical backscatter 6, **262**, 280
optical backscatter sensor (OBS) **262**, 280–281, 285, 296, **296**, 380, 391
optical measurements of sediment concentration 280–283
  optical backscatter **280**, 281
  optical transmission 280–281
  particle-size dependence 281
  turbidity 280
  transmissometer 280
optics 94, 98, 101, 103–106, 114, 124, 127
  collimating 103
orifice plates **345**, 349

particle
  density 108, 110
  diameter 11, 110, 128, 131, 134, 283
particle displacement 275
particle image velocimetry 5, 21, **31**, 35, **36**, 37, 132–156, 185–186, **187**, 230, 233, 321
  configurations 148–156
  correlation methods 136–140
  error sources 144–148
  field applications 153–156
  image capture 134–136, 185–186
  iterative discrete shift 140
  iterative deformation method 140
  light sheet formation 134–136
  noise *see* particle image velocimetry:error sources
  optics *see* particle image velocimetry:light sheet formation
  resolution 140–144
  stereoscopic 134, 149–152, **150**
  sub-pixel accuracy *see* particle image velocimetry:error sources
  three-component, two-dimensions 151–152
  three-component, three dimensions 152–153
  tomographic 134, 149, **150**, 152–153
  tracer particles 108–111, 132–134
  two-component, two-dimensions 149–151
  velocity calculation *see* particle image velocimetry:correlation methods

particle tracer, bedload *see* bedload particle tracer
particle tracking velocimetry 5, 21, **36**, 37, 132, 156–167, 183–185, 233
    centroid estimators 159–161, **160**
    error sources 162–163
    large-scale 183–185
    methodology 157–161
    multi-frame tracking 140–141, 157, 164–167
    three-dimensional 166–167
    velocity averaging 163–164
particle velocity *see* bedload particle velocity
passive acoustic measurement of bedload 266–269, 274
    calibration 267–269
    principles 266–267
    error sources 268–269
passive integrated transponder 274
pathline 12, **13**
phase Doppler anenometry (PDA) 37
photo-detector 101, 104–105, 108, 111
photogrammetric methods **212**, 229–236, 272–273, **311**
    aerial 230, 235
    close range 230, 235
    errors 235–236
    multiview stereo 230, 247–248
    rectification 232
    reference points 231–232, 234
    structure from motion 230, 247–248
photo-sieving 244–247
physical sediment samples 276–280
    error 277, 279–280
physical traps 261–264, **262**, 274
    error sources 265
piezoelectric **310**, **317**, 317–319, 333, 335, 338
piezoelectric sensor (bedload) 267, 269
piezometer **311**, 312, **317–318**, 317–318
piezo-resistive **317**, 319–320, **319–320**
pinhole 100–101, 103–104, 107–108, 114, 319
pit sampler **262**, 263–265
Pitot 2
Pitot tube **4**, **36**, **317–318**, 318, 326
PIV *see* particle image velocimetry
planar laser-induced fluorescence *see* laser-induced fluorescence
plate geophone **262**, 265, **267**, 267–269
PLIF *see* laser-induced fluorescence
point cloud 216–217, 219, 223–224
point measurement instruments *see* velocimeters
point velocity meters *see* velocimeters

point-integrating (suspended sediment) **277**, 276–278
position measurements *see* surveying
power meter 100–101, 104, 107–108
pressure 6, **36**, **311**, 315–322, 338
    absolute pressure 316–317
    atmospheric pressure 316–317, 319–320
    differential pressure 316–317, 319
    fluctuating measurements 318–321
    gauge pressure 316–318
    manometer 316–318
    pressure sensing instruments 316–318
    relative pressure 316, **317**
    steady measurements 316–318
    transducers 316, 318–322, 338
    units 315, **316**
pressure drag 332
pressure sensitive paint **317**, 321
Preston tube 326, **327**
profiler *see* ultrasonic velocity profiler (UVP) and acoustic Doppler velocity profiler (ADVP)
profilers, mechanical **212**, 238–239
    contour gauges 238–239
    point gauges 238–239
    thousand-footer **238**, 238–239
PTV *see* particle tracking velocimetry
pulse repetition frequency, ABI 40
pulsed wall probes **324**, 329
pulse-to-pulse coherent **40**, **44**, **50**, 84–85

radar *see* high-frequency radar
radial velocity 41, 56, 68, **68**, 83–85, 187
radio frequency identification (RFID) 274–276, **275**
random spikes *see* spikes
range velocity ambiguity 43, 51, 56, 85–86
    monostatic 85
    multistatic 85
rating curves 347, 356–364
real time kinematic *see* GPS:real time kinematic
recovery rate, bedload 275
reference frame, LDV 96, 104, 108
reference level 310, 313–314
reflectivity, terrestrial laser scanning 215–216
refractive index 101, 108–109, 127, 128, 148
Reid-type (Birkbeck) sampler **263**
remote sensing 211, **213**, 239, **311**, 313–314
remotely operated vehicle (ROV) 377
reservoir 107, 261–262
retention basin (sediment) **262**, 262–263, 265, 269, 275
Reynolds 9, **10**

Reynolds
  decomposition 38
  number 26, 91, 120, 129, 174, 332–333,
    349, 387
  stress tensor 99
  stresses 47, 52, 54–55, 58, 99, 118, **323**
rhodamine 16–17, 20, **26**, 191
rolling shutter 14, 184
rough bed flows 238–239, 310, 323, 325,
    328, 330–332
roughness 6, 59, 91, 174, 192, 211–212,
    212, 217, **222**, 224, 230, 240, 248,
    268–269, 309–310, **310**, 356–360,
    368, **369**
RTK *see* GPS: real time kinematic

saddles **24**, 27–30
salinity **76**, 79, 81, 229, 309, 336, 338, 372,
    388–389, 391, 394
Sample-and-Hold 118, **119**
samplers, bedload *see* bedload samplers
samplers, suspended load *see* suspended
    sediment samplers
sampling (grain size distribution) 240–252
  areal samples 240
  freeze cores 241
  grid counts 240
  heel-to-toe walks 240
  line-by-number sampling 240, 242–244
  pebble count 240, 242–243
  photogrammetric sampling 247
  surface sampling 240, 242–247
  volumetric sampling 240–242
sampling duration 90–92, 191, 219, 265, 270
sampling volume **40**, 43, 52–53, 57–59, 70,
    84, 89–90, **90**, 240–241, 248, 281,
    284–285
scattering volume 40, **40**, 45, 84
Schmidt Number 11
section-by-section, ADCP transects 353
sediment concentration, suspended
    276–298
  acoustic methods 283–298
  acoustic method error sources 286
  ADCPs 290–298
  concentration error 277
  depth-integrating samplers **277**, 278–279
  manual sampling 276–280
  optical methods 280–283
  profiling 286
sediment load *see* also sediment transport
    261, 269, 276–278
sediment samples *see* physical sediment
    samples
sediment size classification *see also* grain size
    distribution 248

sediment transport 6, 27, 38, 59, 61, 74, 129,
    212, 224, 240, 261–298, 331, 353,
    358, 371
  bedload 6, 261–275, **262**, 285
  suspended load 6, 261, **262**, 276–286
  washload 261, 277
sediment transport measurements 261–307
  acoustic measurements 283–298
  bedload 261–276
  optical measurements 280–283
  physical sampling, bedload 261–266
  physical sampling, suspended load 276–
    280
  suspended load 276–298
seeding particles *see* tracer particles
settling basin 262–263
shear stress 322–332
  flow obstacle techniques *see* flow obstacle
    techniques
  shear plates 324–325, 330–332
shear velocity 322–323, **323**, 330
side lobe interference *see* ADCP:side lobe
    interference
side-looking *see* ADCP:horizontal ADCP
side-wall correction 330
sieving 240–244, 251
signal, LDV
  pedestal 110–112, 114
  processing 101, 111–112, 114–116, 131
signal drift, load cells 335
signal-to-noise ratio (SNR) 57, 89–90, 104,
    106, 112, 225, 290–291
slope–area **347**, 356, 359, 362, **363**, 364
slot sampler 263
slotting 120–121
sluice gate **345**, 349
SODAR *see* sonic detection and ranging
sonar 172, **212**, 223–229, 380
  active sonar 224
  error sources 229
  multibeam sonar **212**, 226
  noise 229
  passive sonar 224
  ping 225
  side-scan sonar **212**, 226
  single beam sonar **212**, 226
sonic detection and ranging (SODAR) 35
sound velocity 391
  correction of 391–392, **393**
  errors in 229
space-time image (STI) 177–179
space-time image velocimetry (STIV) **36**,
    177–183
  discharge estimates 181–183
  methodology 177–180
  performance 181–183

spatial variation 3, 310, 346, 388
specific gravity **109**, 174
spike removal 58, 122–123
spikes 52, 57, 122
spiking 122
stage-fall **347**, 347, 356, 360
stage measurements 356–364
stage-area rating 365
stage-discharge methods (SDMs) **173**, 356–365, 367, 370, **370**, 372
stage-discharge rating curves
    calibration 356–357, 359–360, 364, **365**
    complex 360–364
    establishing 347
    hysteresis 361–362, **363**, 367
    selection of location 357, 359–360, 371
    simple 356–360
Stanton gauge 326
steady pressure measurements *see* pressure: steady measurements
stereoscopic particle image velocimetry *see* particle image velocimetry: stereoscopic
Stokes Number 10–11, 129
strain gauge 190, **310**, 318, 333–336, 338
strain gauge transducers 318, 333–336
streakline 12, **13**, **18**, 22
stream gauging 90–91, 356–364
streamline 12, 27–28
subsurface sediment layer 240, 242–245, 251, **252**
surface coating methods 325–26, 330–332
    ferrofluid **324**, 325, 330–331
    liquid crystal coating **324**, 326, 330–331
    oil-film interferometry **324**, 325, 332
surface elevation, water 6, 226, 238, **311**, 312–315
surface sediment layer 240, 242–245, 248, 251, **252**
surrogate techniques, sediment 261, **262**, 266–270, 281, 285, 287–288, 296–297
surveying **212**, 239, 313, 343, 358–359
suspended load *see* sediment transport: suspended load
suspended sediment *see* sediment transport: suspended load
    concentration *see* sediment concentration, suspended
    load *see* sediment transport:suspended load
    particles 288
    sampler 276–280
Swiss plate geophone *see* geophone

temperature 190, **310**, 336–338
terrestrial laser scanning (TLS) 6, **212**, 211–224
    post-processing 223–224
thermistor 338
timeline 12, **13**, 22
time-of-flight 213–215
tomographic particle image velocimetry *see* particle image velocimetry: tomographic
tomography *see* acoustic travel-time tomography
topographic measurement 211–213, 224
topography 5, 211–259, **310**, 380
total sediment load *see also* sediment transport 261, 277
Toutle River II sampler **264**, 264–265
towing tank 333, **334**
tracer particles 10–12, 94–95, 108–111, 127–131, 133–134, 157–159, 161, 275–276
    density **109**, 110, 112
    velocity 99, 125, 129
    tracking ability 108, 110, 128–131, 157
tracers 10–12
    dye 11–12, 15–20
    hydrogen bubbles **13**, **16**, 21–23, 59, 285–286
    particles *see* also tracer particles 10–12, **21**
    passive 10–12
    response 10–12
    Stokes law 10–11
travel-time tomography *see* acoustic travel-time tomography
traversing system **106**, 237–238
triangular weirs *see* weirs: triangular
triangulation 213, 224, 231, **232**, 233, 237, 383
trough sampler **262**
turbidity meter 6, **280**, 296–298, **297**
turbulence 6, 9, 23, 38, 52–55, 58–59, 61, 76, 91, 95, 105, 107, 110–111, 115, 120, 123, 129, 131, 322–323, **323**, 330
    Taylor's frozen 115, 131
    turbulent kinetic energy 52, 142, **323**
Tyrolean weir 263

UHF radar *see* high-frequency radar
ultrasonic transducers 224–229
ultrasonic velocity profiler (UVP) **36**, 37–38, 43, 54, 85–86, 289
under ice operations 389–391, **392**
unsampled zone 276, 278
UVP *see* ultrasonic velocity profiler

variable-area meter 346
velocimeters, point 36, 87
  acoustic Doppler velocimeters (ADV) 36,
    43, 46–49, 58–59, 89–94, 90, 93
  calibration 88, 92–94
  comparison 88, 91
  electromagnetic current meters (ECM) 19,
    37, 86–87, 89, 91–92, 94
  laser-Doppler velocimeters (LDV/LDA)
    94–131
  mechanical current meters 87–88,
    87–88
  Pitot tubes 317–318, 318, 326
velocimeters, multi-point 36
  acoustic Doppler current profiler (ADCP)
    5–6, 36, 37, 59–79, 284, 353–355, 380
  acoustic Doppler velocity profiler (ADVP)
    36, 37, 50–51, 285
  ultrasonic velocity profiler (UVP) 36,
    37–38, 43, 54, 85–86, 289
  particle image velocimetry (PIV) 131–156,
    185–186
  particle tracking velocimetry (PTV)
    156–167
  large scale particle image velocimetry
    (LSPIV) 167–177
  space-time image velocimetry (STIV)
    177–183
  large-scale particle tracking velocimetry
    (LPTV) 183–185
velocity measurements 5, 36, 87
velocity, reach averaged 191, 351
velocity dilution method see dilution velocity
    measurement
velocity integration methods 351–355
velocity models
  hydraulics-based 367–369
  hybrid approaches 369–370
  hydrodynamic 370
velocity profile, bed shear stress 323

velocity-index method see index-velocity
    method
Venturi meters 345, 349
vessel based measurements 226, 229, 273,
    313, 390
viscous sublayer 323–324, 326–332
v-notch weirs see weirs: triangular
volume flow rate measurements see discharge
    measurements
vortex
  line 25–27
  skeleton model 25–27
  strength 25
  topology 25–27
  tube 25–27

wall shear stress see bed shear stress
washload see sediment transport:
    washload
water depth 309–315, 311
water depth gauges 311, 311–315
  capacitance gauge 311, 311–315
  float gauges 311, 311–315
  lead lines 311, 311–315
  piezometer 311, 311–315
  point gauges 311, 311–315
  resistance gauge 311, 311–315
  rulers 311, 311–315
  sounding poles 311, 311–315
  ultrasonic sensors 311, 311–315
water surface slope 315
weighers 345
weighting 55, 114, 117–118, 163–164, 180
  averaging 117, 163–164
  factor 117–118
weighting window 138, 140, 142–143, 146
weirs 345, 347, 348–349, 371
  triangular (v-notch) 348
  broad-crested 347, 349, 371
Wheatstone bridge 318, 335

Printed and bound by CPI Group (UK) Ltd, Croydon, CR0 4YY

24/10/2024

01778290-0003